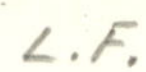

DIGITAL AND ANALOG COMMUNICATION SYSTEMS

DIGITAL AND ANALOG COMMUNICATION SYSTEMS

K. SAM SHANMUGAM

Wichita State University

John Wiley & Sons

New York / Chichester / Brisbane / Toronto

Library of Congress Cataloging in Publication Data:

Shanmugam, K. Sam.
Digital and analog communication systems.

Includes bibliographical references and index.
1. Telecommunication.
2. Digital communications.
3. Information theory.
4. Signal theory (Telecommunication) I. Title.

TK5101.S445 621.38 78-26191
ISBN 0-471-03090-2

Printed in the United States of America

10 9 8 7 6 5 4 3 2 1

To

Radha,
Kannon, and
Ravi

PREFACE

With the increasing importance of digital communication, there is a need for a communication theory text that emphasizes the basic aspects of information theory, discrete modulation techniques, and coding theory at a level appropriate for advanced undergraduates and first-year graduate students. The purpose of this book is to present an introductory level treatment of digital and analog communication systems with an emphasis on digital communication systems.

A study of communication systems covers many varied aspects—from the mathematical and statistical abstraction of information theory, modulation theory, and coding theory, to the electronic considerations associated with the building of functional blocks to perform various signal processing tasks. I have attempted to present in this book a unified treatment of these many diverse aspects of communication systems.

Throughout the book we consider the process of electrical communication as consisting of a sequence of signal processing operations. Each functional block in the system performs a specific signal processing operation. For each functional block, we define the input/output requirements and the parameters we have at our disposal. We derive expressions that relate the performance of each functional block to its parameters and use these relationships to optimize the design of the functional block. While we occasionally present examples of physical blocks (or circuits) that correspond to specific functional blocks, less emphasis is placed on this aspect of the problem—namely, the physical realization of functional blocks. The major reasons for neglecting this problem are: (1) specific circuits and devices become obsolete in a short time due to rapid technological developments, and (2) the design of circuits could be taught better in a circuit design course rather than in a communication systems course.

The material presented in this book is arranged into three study areas: a review section, the study of digital communication systems, and the study of analog communication systems. The review section (Chapters 2 and 3) covers signal models, systems analysis, random variables, and random processes. A cursory treatment of modulation and demodulation techniques and signal-to-noise ratios are included in the review. These topics are covered in detail in later chapters.

Chapters 4, 5, 8, and 9 deal with the analysis and design of digital communication systems. Chapter 4 presents information theory and its implications for digital communication systems. Discrete pulse and carrier wave modulation schemes are discussed in Chapters 5 and 8. Finally, basic ideas in coding theory are dealt with in Chapter 9.

Analog communication systems are discussed in Chapters 6 and 7. Baseband analog signal transmission and continuous wave (CW) modulation techniques for analog signal transmission are covered in Chapter 6. The effect of noise in CW modulation schemes is discussed in Chapter 7. This chapter also contains a comparison of various CW modulation schemes.

Finally, Chapter 10 deals with digital transmission methods for analog signals. Sampling, quantizing, and encoding of analog signals for transmission over digital systems are described here.

It is assumed that the student has some knowledge of circuit analysis and linear systems analysis. Prior exposure to Fourier transforms, random variables, and random processes would be helpful, but not essential. With this background, it should be possible to organize and teach several possible courses out of this text. First, this book can be used, as a whole, for a two-semester course on digital and analog communication systems. It is also possible to teach a one-semester course on digital and analog communication systems by covering Chapters 2 and 3 rapidly and selecting materials from Chapters 5, 6, 7, and 8. Finally, an elective course in digital communication systems may be taught using the material in Chapters 4, 5, 8, 9, and 10. Selective double coverage of topics in the book will allow the instructor considerable flexibility in structuring these courses.

Considerable effort has been made to present material at such a level that there is consistent progression from concepts to design considerations without getting mired in too much theoretical details. Proofs of theorems and statements are included only when it is felt that they contribute sufficient insight into the problem being addressed. Proofs are omitted when they involve lengthy theoretical discourse of material at a level beyond the scope of this text. In such cases, outlines of proofs with adequate references to outside material is presented so that the ambitious reader can labor through the details.

Each chapter contains a number of examples and problems. The examples and problems cover practical and theoretical aspects of analysis and design of communication systems. An instructor's manual giving complete solutions to the problems is available from the publisher on request.

ACKNOWLEDGMENTS

I am indebted to many people for their advice, assistance, and contributions to the development of this text. First, I wish to thank the Air Force Communication Service and the U.S. Air Force Rome Air Development Center for their support of my research activities in the field of communication systems. Several problems and concepts presented in this text came out of these research activities and I am deeply indebted to these and other agencies, including the National Science Foundation, NASA, and the Department of Energy, for their research support.

Several individuals have read the manuscript and their comments have helped me considerably to improve the accuracy and clarity of the text. I am pleased to acknowledge the assistance of Professor Arthur M. Breipohl, University of Kansas, and Professor M. Lal, Wichita State University, who supplied many helpful suggestions while teaching from the earlier drafts of the manuscript. The comments and suggestions from the reviewers, Professors Arthur M. Breipohl (University of Kansas), R. Frank Quick, Jr. (Carnegie Mellon University), Robert A. Gabel (University of Colorado), and Gene A. Davenport, Editor, John Wiley and Sons, have also been very helpful. I am also grateful to my graduate students, J. M. Naik, Art Frankowski, and Mark A. Miller for proofreading the final manuscript.

I also wish to thank the staff and administration of the College of Engineering at Wichita State University for providing me with typing and other help in the earlier stages of the development of the manuscript. In particular, I wish to thank Marta Manny who typed the entire manuscript in expert fashion, not once, but several times.

Finally, I am grateful to my wife and children for their patience and understanding during the time I devoted to writing this book.

K. Sam Shanmugam

CONTENTS

DIGITAL AND ANALOG COMMUNICATION SYSTEMS

1

INTRODUCTION

Down through the ages, people have devised numerous methods for communicating their thoughts and needs to others. In primitive days, when human beings lived in small groups distributed over a relatively small geographical area, communication within the group took place through speech, gestures, and graphical symbols. As these groups became larger and civilizations spread over large geographical areas, it was necessary to develop methods of long distance communication. Early attempts at long distance communication included such things as smoke signals, light beams, carrier pigeons, and letters transported by a variety of means. With the beginning of the industrial revolution, the need for fast and accurate methods of long distance communication became more compelling. Communication systems employing electrical signals to convey information from one place to another over a pair of wires provided an early solution to the problem of fast and accurate means of long distance communication. The field of electrical communication engineering received tremendous attention during and after World War II. Significant developments during this era includes radar and microwave systems, transistor and miniaturized integrated circuits, communication satellites, and lasers. Today, electrical communication systems span the entire world carrying voice, text, pictures, and a variety of other information.

During the post-war era there was also a vast growth in the automation and computer industries. This growth made it necessary for computers and other machines to communicate not only with people but also with other machines. In many cases the information to be exchanged between people and machines, and between machines is digital or numerical in contrast to the predominantly analog information exchanged in personal communications. Irrespective of the nature of information transmitted, and the actual method of transmission, we can use the following model to describe a communication system.

1.1 MODEL OF A COMMUNICATION SYSTEM

Figure 1.1 shows the basic functional blocks of a communication system. The overall purpose of this system is to transfer information from one point in space and time, called the source, to another point, the user destination. As a rule, the message produced by a source is not electrical. Hence an input transducer is required for converting the message to a time-varying electrical quantity called a message signal. At the destination point, another transducer converts the electrical waveform to the appropriate message.

The information source and the destination point are usually separated in space. The channel provides the electrical connection between the information source and the user. The channel can have many different forms such as a microwave radio link over free space, a pair of wires, or an optical fiber. Regardless of its type, the channel degrades the transmitted signal in a number of ways. The degradation is a result of signal distortion due to imperfect response of the channel and due to undesirable electrical signals (noise) and interference. Noise and signal distortion are two basic problems of electrical communication. The transmitter and the receiver in a communication system are carefully designed to avoid signal distortion and minimize the effects of noise at the receiver so that a faithful reproduction of the message emitted by the source is possible.

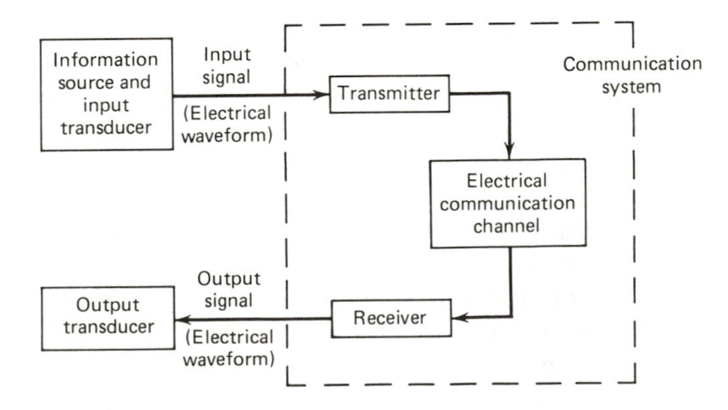

Figure 1.1 Model of an electrical communication system.

The transmitter couples the input message signal to the channel. While it may sometimes be possible to couple the input transducer directly to the channel, it is often necessary to process and modify the input signal for efficient transmission over the channel. Signal processing operations performed by the transmitter include amplification, filtering, and modulation. The most important of these operations is modulation—a process designed to match the properties of the transmitted signal to the channel through the use of a carrier wave.

Modulation is the systematic variation of some attribute of a carrier waveform such as the amplitude, phase, or frequency in accordance with a function of the message signal. Despite the multitude of modulation techniques, it is possible to identify two basic types of modulation: the continuous carrier wave (CW) modulation and the pulse modulation. In continuous wave (CW) carrier modulation the carrier waveform is continuous (usually a sinusoidal waveform), and a parameter of the waveform is changed in proportion to the message signal. In pulse modulation the carrier waveform is a pulse waveform (often a rectangular pulse waveform), and a parameter of the pulse waveform is changed in proportion to the message signal. In both cases the carrier attribute can be changed in continuous or discrete fashion. Discrete pulse (digital) modulation is a discrete process and is best suited for messages that are discrete in nature such as the output of a teletypewriter. However, with the aid of sampling and quantization, continuously varying (analog) message signals can be transmitted using digital modulation techniques.

Modulation is used in communication systems for matching signal characteristics to channel characteristics, for reducing noise and interference, for simultaneously transmitting several signals over a single channel, and for overcoming some equipment limitations. A considerable portion of this book is devoted to the study of how modulation schemes are designed to achieve the above tasks. The success of a communication system depends to a large extent on the modulation.

The main function of the receiver is to extract the input message signal from the degraded version of the transmitted signal coming from the channel. The receiver performs this function through the process of demodulation, the reverse of the transmitter's modulation process. Because of the presence of noise and other signal degradations, the receiver cannot recover the message signal perfectly. Ways of approaching ideal recovery will be discussed later. In addition to demodulation, the receiver usually provides amplification and filtering.

Based on the type of modulation scheme used and the nature of the output of the information source, we can divide communication systems into three categories:

1. analog communication systems designed to transmit analog information using analog modulation methods
2. digital communication systems designed for transmitting digital information using digital modulation schemes and
3. hybrid systems that use digital modulation schemes for transmitting sampled and quantized values of an analog message signal.

Other ways of categorizing communication systems include the classification based on the frequency of the carrier and the nature of the communication channel.

With this brief description of a general model of a communication system, we will now take a detailed look at various components that make up a typical communication system using the digital communication system as an example. We will enumerate the important parameters of each functional block in a digital communication system and point out some of the limitations of the capabilities of various blocks.

1.2 ELEMENTS OF A DIGITAL COMMUNICATION SYSTEM

Figure 1.2 shows the functional elements of a digital communication system. The overall purpose of the system is to transmit the messages (or sequences of symbols) coming out of a source to a destination point at as high a rate and accuracy as possible. The source and the destination point are physically

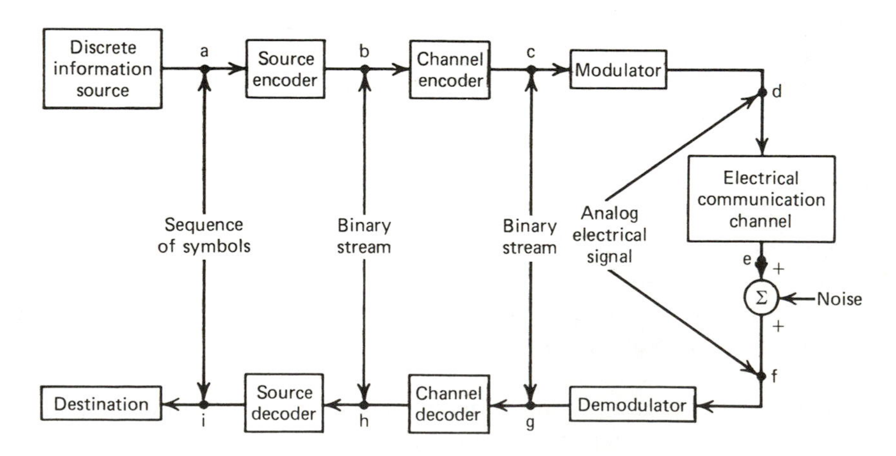

Figure 1.2 Functional blocks of a digital communication system.

separated in space and a communication channel of some sort connects the source to the destination point. The channel accepts electrical/electromagnetic signals, and the output of the channel is usually a smeared or distorted version of the input due to the nonideal nature of the communication channel. In addition to the smearing, the information bearing signal is also corrupted by unpredictable electrical signals (noise) from both man-made and natural causes. The smearing and noise introduce errors in the information being transmitted and limits the rate at which information can be communicated from the source to the destination. The probability of incorrectly decoding a message symbol at the receiver is often used as a measure of performance of digital communication systems. The main function of the coder, the modulator, the demodulator, and the decoder is to combat the degrading effects of the channel on the signal and maximize the information rate and accuracy.

With these preliminaries in mind, let us now take a detailed look at each of the functional blocks in the system.

1.2.1 Information Source

Information sources can be classified into two categories based on the nature of their outputs: analog information sources and discrete information sources. Analog information sources, such as a microphone actuated by speech, or a TV camera scanning a scene, emit one or more continuous amplitude signals (or functions of time). The output of discrete information sources such as a teletype or the numerical output of a computer consists of a sequence of discrete symbols or letters. An analog information source can be transformed into a discrete information source through the process of sampling and quantizing. Discrete information sources are characterized by the following parameters:

1. Source alphabet (symbols or letters)
2. Symbol rate
3. Source alphabet probabilities
4. Probabilistic dependance of symbols in a sequence

From these parameters, we can construct a probabilistic model of the information source and define the source entropy (H) and source information rate (R) in bits per symbol and bits per second, respectively. (The term *bit* is used to denote a binary digit.)

To develop a feel for what these quantities represent, let us consider a discrete information source—a teletype having 26 letters of the English alphabet plus six special characters. The source alphabet for this example consists of 32 symbols. The symbol rate refers to the rate at which the teletype produces characters; for purposes of discussion, let us assume that the teletype operates at a speed of 10 characters or 10 symbols/sec. If the teletype is producing messages consisting of symbol sequences in the English language, then we know that some letters will appear more often than others. We also know that the occurrence of a particular letter in a sequence is somewhat dependent on the letters preceding it. For example, the letter E will occur more often than the letter Q and the occurrence of Q implies that the next letter in the sequence will most probably be the letter U, and so forth. These structural properties of symbol sequences can be characterized by probabilities of occurrence of individual symbols and by the conditional probabilities of occurrence of symbols.

An important parameter of a discrete source is its entropy. The entropy of a source, denoted by H, refers to the average information content per symbol in a long message and is given the units of bits per symbol where bit is used as an abbreviation for a binary digit. In our example, if we assume that all symbols occur with equal probabilities in a statistically independent sequence, then the source entropy is five bits per symbols (i.e., $2^5 = 32$; we will give a mathematical definition for source entropy in Chapter 4). However, the probabilistic dependence of symbols in a sequence, and the unequal probabilities of occurrence of symbols considerably reduce the average information content of the symbols. Intuitively we can justify the previous statement by convincing ourselves that in a symbol sequence QUE, the letter U carries little or no information because the occurrence of Q implies that the next letter in the sequence has to be a U. We will establish these concepts rigorously in Chapter 4.

The source information rate is defined as the product of the source entropy and the symbol rate and has the units of bits per second. The information rate, denoted by R, represents the minimum number of bits per second that will be needed, on the average, to represent the information coming out of the discrete source. Alternately, R represents the minimum average data rate needed to convey the information from the source to the destination.

1.2.2 Source Encoder/Decoder

The input to the source encoder (also referred to as the source coder) is a string of symbols occurring at a rate of r_s symbols/sec. The source coder converts the symbol sequence into a binary sequence of 0's and 1's by assigning code words to the symbols in the input sequence. The simplest way in which a source coder can perform this operation is to assign a fixed-length binary code word to each symbol in the input sequence. For the teletype example we have been discussing, this can be done by assigning 5-bit code words 00000 through 11111 for the 32 symbols in the source alphabet and replacing each symbol in the input sequence by its preassigned code word. With a symbol rate of 10 symbols/sec, the source coder output data rate will be 50 bits/sec.

Fixed-length coding of individual symbols in a source output is efficient only if the symbols occur with equal probabilities in a statistically independent sequence. In most practical situations symbols in a sequence are statistically dependent, and they occur with unequal probabilities. In these situations the source coder takes a string of two or more symbols as a block and assigns variable-length code words to these blocks. The optimum source coder is designed to produce an output data rate approaching R, the source information rate. Due to practical constraints, the actual output rate of source encoders will be greater than the source information rate R. The important parameters of a source coder are block size, code word lengths, average data rate, and the efficiency of the coder (i.e., actual output data rate compared to the minimum achievable rate R).

At the receiver, the source decoder converts the binary output of the channel decoder into a symbol sequence. The decoder for a system using fixed-length code words is quite simple, but the decoder for a system using variable-length code words will be very complex. Decoders for such systems must be able to cope with a number of problems such as growing memory requirements and loss of synchronization due to bit errors.

1.2.3 Communication Channel

The communication channel provides the electrical connection between the source and the destination. The channel may be a pair of wires or a telephone link or free space over which the information bearing signal is radiated. Due to physical limitations, communication channels have only finite bandwidth (B Hz), and the information bearing signal often suffers amplitude and phase distortion as it travels over the channel. In addition to the distortion, the signal power also decreases due to the attenuation of the channel. Furthermore, the signal is corrupted by unwanted, unpredictable electrical signals referred to as noise. While some of the degrading effects of the channel can be removed or compensated for, the effects of noise cannot be completely removed. From this point of view, the primary objective of a communication system design should be to suppress the bad effects of the noise as much as possible.

One of the ways in which the effects of noise can be minimized is to increase the signal power. However, signal power cannot be increased beyond certain levels because of nonlinear effects that become dominant as the signal amplitude is increased. For this reason the signal-to-noise power ratio (S/N), which can be maintained at the output of a communication channel, is an important parameter of the system. Other important parameters of the channel are the usable bandwidth (B), amplitude and phase response, and the statistical properties of the noise.

If the parameters of a communication channel are known, then we can compute the channel capacity C, which represents the maximum rate at which nearly errorless data transmission is theoretically possible. For certain types of communication channels it has been shown that C is equal to $B \log_2(1 + S/N)$ bits/sec. The channel capacity C has to be greater than the average information rate R of the source for errorless transmission. The capacity C represents a theoretical limit, and the practical usable data rate will be much smaller than C. As an example, for a typical telephone link with a usable bandwidth of 3 kHz and $S/N = 10^3$, the channel capacity is approximately 30,000 bits/sec. At the present time, the actual data rate on such channels ranges from 150 to 9600 bits/sec.

1.2.4 Modulator

The modulator accepts a bit stream as its input and converts it to an electrical waveform suitable for transmission over the communication channel. Modulation is one of the most powerful tools in the hands of a communication systems designer. It can be effectively used to minimize the effects of channel noise, to match the frequency spectrum of the transmitted signal with channel characteristics, to provide the capability to multiplex many signals, and to overcome some equipment limitations.

The important parameters of the modulator are the types of waveforms used, the duration of the waveforms, the power level, and the bandwidth used. The modulator accomplishes the task of minimizing the effects of channel noise by the use of large signal power and bandwidth, and by the use of waveforms that last for longer durations. While the use of increasingly large amounts of signal power and bandwidth to combat the effects of noise is an obvious method, these parameters cannot be increased indefinitely because of equipment and channel limitations. The use of waveforms of longer time duration to minimize the effects of channel noise is based on the well-known statistical *law of large numbers*. The law of large numbers states that while the outcome of a single random experiment may fluctuate wildly, the overall result of many repetitions of a random experiment can be predicted accurately. In data communications, this principle can be used to advantage by making the duration of signalling waveforms long. By averaging over longer durations of time, the effects of noise can be minimized.

To illustrate the above principle, assume that the input to the modulator consists of 0's and 1's occurring at a rate of 1 bit/sec. The modulator can assign waveforms once every second. For example, the waveforms $A \cos \omega_1 t$ $(0 \leq t < 1)$ and $A \cos \omega_2 t$ $(0 \leq t < 1)$ can be assigned to bits 0 and 1, respectively. Notice that the information contained in the input bit is now contained in the frequency of the output waveform. To employ waveforms of longer duration, the modulator can assign waveforms once every four seconds. This would require 16 waveforms, $A \cos \omega_1 t$, $A \cos \omega_2 t, \ldots, A \cos \omega_{16} t$, each lasting four seconds; these 16 waveforms can represent the sixteen 4-bit combinations 0000, 0001, . . . , 1111. The limitation of this technique is now obvious. The number of distinct waveforms the modulator has to generate (hence the number of waveforms the demodulator has to detect) increases exponentially as the duration of the waveforms increases. This leads to an increase in equipment complexity and hence the duration cannot be increased indefinitely. The number of waveforms used in commercial digital modulators available at the present time ranges from 2 to 16.

1.2.5 Demodulator

Modulation is a reversible process, and the extraction of the message from the information bearing waveform produced by the modulator is accomplished by the demodulator. For a given type of modulation, the most important parameter of the demodulator is the method of demodulation. There are a variety of techniques available for demodulating a given modulated waveform; the actual procedure used determines the equipment complexity needed and the accuracy of demodulation. Given the type and duration of waveforms used by the modulator, the power level at the modulator, the physical and noise characteristics of the channel, and the type of demodulation, we can derive unique relationships between data rate, power bandwidth requirements, and the probability of incorrectly decoding a message bit. A considerable portion of this text is devoted to the derivation of these important relationships and their use in system design.

The characteristics of the modulator, the demodulator, and the channel establish an average bit error rate between points c and g in Figure 1.2. More often than not, this bit error rate and the corresponding symbol error rate will be higher than desired. A lower bit error rate can be accomplished by redesigning the modulator and the demodulator (often referred to as *modems*) or by using error control coding.

1.2.6 Channel Encoder/Decoder

Digital channel coding is a practical method of realizing high transmission reliability and efficiency that otherwise may be achieved only by the use of signals of longer duration in the modulation/demodulation process. With digital coding, a relatively small set of analog signals, often two, is selected for transmission over the channel and the demodulator has the conceptually simple task of distinguishing between two different waveforms of known shapes. Error control is accomplished by the channel coding operation that consists of systematically adding extra bits to the output of the source coder. While these extra bits themselves convey no information, they make it possible for the receiver to detect and/or correct some of the errors in the information bearing bits.

There are two methods of performing the channel coding operation. In the first method, called the block coding method, the encoder takes a block of k information bits from the source encoder and adds r error control bits. The number of error control bits added will depend on the value of k and the error control capabilities desired. In the second method, called the convolutional coding method, the information bearing message stream is encoded in a continuous fashion by continuously interleaving information bits and error control bits. Both methods require storage and processing of binary data at the encoder and decoder. While this requirement was a limiting factor in the early days of data communication, it is no longer such a problem because of the availability of solid state memory and microprocessor devices at reasonable prices.

The important parameters of a channel encoder are the method of coding, rate or efficiency of the coder (as measured by the ratio of data rate at input to the data rate at the output), error control capabilities, and complexity of the encoder.

The channel decoder recovers the information bearing bits from the coded binary stream. Error detection and possible correction is also performed by the channel decoder. The decoder operates either in a block mode or in a continuous sequential mode depending on the type of coding used in the system. The complexity of the decoder and the time delay involved in the decoder are important design parameters.

1.2.7 Other Functional Blocks

A number of other functional blocks, not shown in Figure 1.2, exist in practical data communication systems. Examples of such blocks are equalizers, clock recovery networks, and scramblers/unscramblers, to name a few.

Equalizers compensate for changes in channel characteristics, and the clock recovery networks extract timing information at the receiver to maintain synchronization between the transmitter and receiver. Scramblers are used in systems to prevent unwanted strings of symbols (such as periodic bit strings in data, or long strings of ones or zeros) from appearing at the input of the channel. Unscramblers are used at the receiver to recover the original bit stream or symbol sequence from its scrambled version.

It must also be pointed out that the functional blocks might have a variety of different structures depending on the signaling schemes used. For example, in baseband PAM systems the modulator actually consists of a pulse generator and a pulse shaping filter called the transmitting filter. The demodulator in such systems will have a shaping filter called a receiving filter followed by an analog-to-digital (A/D) converter.

From a user's point of view, a data communication system consists of the physical blocks (or pieces of equipment) shown in Figure 1.3. The data terminal equipment, usually a terminal or a computer, sends and receives data from other data terminal equipment (DTE). The DTE also generates and receives control signals that regulate the information flow between the terminals. These control signals might include requests to send data, an indication of readiness to receive data, and a busy data terminal indication. The flow of data and control signals to and from the DTE takes place through standard interfaces (such as the RS-232C specified by the Electronics Industries Association). The DTE is connected to the modem, which consists of a modulator for sending data and a demodulator for receiving data.

The modem is connected to the channel and controls the interchange of information through the channel. The overall control of information flow, error detection, error correction, and various coding tasks are performed by the DTE.

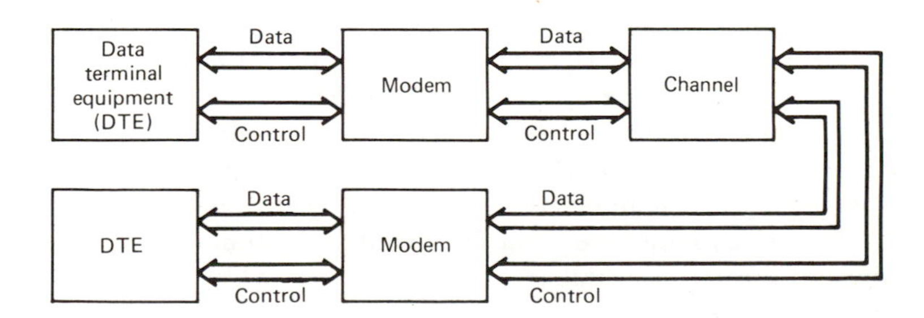

Figure 1.3 Important physical blocks in a point to point data communication system.

1.3 ANALYSIS AND DESIGN OF COMMUNICATION SYSTEMS

A communication systems engineer has to deal with two major technical tasks: analysis of communication systems and design of communication systems. Analysis consists of evaluating the performance of a given communication system, whereas design consists of arriving at the details of a system to perform a given task satisfactorily. While it is often not possible to clearly define where analysis ends and design begins, we can safely say that one has to learn how to analyze communication systems before learning to design them.

1.3.1 Analysis of Communication Systems

For the purpose of analysis, we will consider the communication system as an interconnection of various subsystems wherein each subsystem performs a specific signal processing task. Information transmission is accomplished through a series of signal transformations. Thus, to analyze the system, we need mathematical descriptions of signals and signal transformations.

In this systems approach, we will focus our attention on the analysis (and design) of subsystems (or functional blocks). For each functional block, we will define the input/output requirements and list the parameters we have at our disposal. We will then derive relationships that relate the functional performance of the block to its parameters and use these relationships to optimize the performance of the functional blocks.

1.3.2 Design of Communication Systems

When planning a communication system, the designer's task is one of deciding on a particular type of communication system for a given application. The system he proposes should meet a specified set of *performance requirements*. For analog message signals, the system performance is specified in terms of the ratio of the average message signal power to noise power at the destination point. For discrete message signals the probability of incorrectly decoding a message symbol at the receiver is used as a performance measure.

While designing a communication system the engineer faces several *constraints*. These constraints are: time-bandwidth constraints, noise limitation, and equipment (cost) constraints.

Time-Bandwidth Constraint. For efficient utilization of a communication system, it is necessary to send as much information as possible in as short a time span as possible. A convenient measure of signaling speed in a communication system is the bandwidth of the signal. When signal transmission takes place in real time over a communication channel, the design should ensure that the signal bandwidth is less than or equal to the channel bandwidth. If this constraint cannot be satisfied, then it may be necessary to decrease signaling speed and thereby increase transmission time.

The channel frequency and bandwidth allocations are regulated by the government for public channels and by regulations imposed by private companies if the channel is leased. In general, larger bandwidths will be available with higher carrier frequencies. As a rough guideline, available bandwidth may be taken as 10 percent of the carrier frequency. A listing of portions of the electromagnetic spectrum that are now in use is given in Appendix B.

Noise Limitation. Noise refers to unwanted, often unpredictable, electrical waveforms that corrupt the message signal. Noise can be classified into two broad categories depending on its source. Noise generated by components within a communication system is referred to as internal noise. The second category of noise, called external noise, results from sources external to the communication system. External noise sources include man-made sources and extraterrestrial natural sources.

Noise is inevitable in an electrical communication system since there is always thermal noise associated with conduction and radiation. Noise limits our ability to correctly identify the intended message signal and thereby limits information transmission. Typical noise variations are measured in microvolts. If the signal variations are quite large in comparison, then the noise may be ignored. Indeed, in many practical systems the effects of noise are unnoticeable. However, in long-range systems operating with a limited amount of signal power, the signal may be as small as or even smaller than the noise. In such situations, the presence of noise severely limits the capabilities of a communication system. We will see later on that it is possible to design signaling schemes that use a wider bandwidth to reduce the effects of noise (i.e., systems which exchange bandwidth for signal-to-noise ratio).

Equipment Limitation. Time-bandwidth and noise limitations dictate what can or cannot be achieved in terms of performance in a communication system. This theoretical limit, however, may not be reached in a practical system due to equipment limitations. For example, theory might call for a band pass filter with a quality factor of 100 at a center frequency of 1 kHz and a cut-off rate of 90 dB/decade. Such a filter cannot be realized in practice. Even if a filter with nearly identical characteristics can be built, the cost may exceed what the user of the communication system is willing to pay.

In summary, bandwidth and signal-to-noise ratio constraints limit the maximum rate at which information transfer can take place in a communication system. This upper limit is always finite. In addition to this theoretical constraint, the designer is given additional constraints in equipment cost and complexity. The designer has to come up with a signaling scheme that offers the best compromise between transmission time, transmitted power, transmission bandwidth, and equipment complexity while maintaining an acceptable level of performance.

1.4 ORGANIZATION OF THE BOOK

We begin our study of the analysis and design of communication systems with a review of signal theory and linear systems in Chapter 2. The emphasis here will be on the frequency domain approach leading to the concepts of power spectral density and bandwidth. In Chapter 3, we review the theory of probability, random variables, and random processes as applied to random signals of interest in communication systems. Random processes will be used throughout our study to model message signals and noise.

After developing the mathematical tools, we proceed to develop the principles of information theory and its implications for electrical communication systems. In Chapter 4 we look at statistical models for information sources and develop the concepts of information contents of messages, information rates of sources, and source coding. In Chapter 4, we also present statistical models for discrete and continuous communication channels and establish some very important theoretical limits within which a communication system has to perform. The system designer and analyst must know these performance limits which can possibly be achieved but never exceeded.

In Chapter 5 we look at methods of transmitting digital information in which the transmitted signal spectrum occupies a band of frequencies, including DC. These methods use pulse amplitude modulation (PAM) and we look at the problem of optimizing baseband PAM systems. Direct baseband transmission of analog message signals is treated in Chapter 6.

Since only a few real channels transmit energy near zero frequency, it is usually necessary to shift the spectrum of the message signal to a more appropriate location in the frequency domain using a carrier. A variety of schemes for modulating a carrier for transmitting analog message signals are discussed in Chapter 6. These schemes include linear modulation as well as exponential modulation. In Chapter 7 we deal with the problem of signal recovery in the presence of noise in analog carrier modulation schemes. A comparison of these schemes (in terms of power-bandwidth requirements, equipment complexity, and performance) is given in Chapters 6 and 7.

Digital carrier modulation schemes are covered in Chapter 8. In these methods—the amplitude shift-keying method (ASK), frequency shift-keying method (FSK), and phase shift-keying method (PSK)—the digital information is carried in the amplitude, frequency, and phase, respectively, of a carrier waveform. Optimal and suboptimal receiver structures for ASK, FSK, and PSK signaling methods are also discussed in Chapter 8, where we also compare these digital signaling schemes in terms of data rate, error rate, power-bandwidth requirements, and equipment complexity. With examples, we illustrate how to trade off parameter values while maintaining the same quality of performance.

Chapter 9 deals with error control and information flow control in data communication systems. Block coding and convolutional coding schemes for controlling random errors and burst errors are discussed in this chapter.

Finally, in Chapter 10 we look at the concepts involved in transmitting analog data using digital signaling methods. Principles of sampling, quantizing, and coding and time division multiplexing of analog data are presented in this chapter. The performances of several practical communication systems are compared with the performance of an ideal system.

Throughout the book we focus our attention on the design and analysis of *functional blocks* in the system. For each functional block, we define the input/output requirements and the parameters we have at our disposal. We derive relationships that relate the functional performance of the block to its parameters and use these relationships to optimize the design of the functional blocks. While we occasionally present examples of *physical blocks* (or circuits) that correspond to specific functional blocks, less emphasis is placed on this aspect of the problem—namely, the physical realization of functional blocks. The major reasons for neglecting this problem are: specific circuits and devices become obsolete in a short time due to rapid technological developments, and the design of circuits could be taught more effectively in a circuit design course rather than in a communication systems course.

Each chapter contains a number of examples and exercises for the student. Proofs of theorems and statements are included only when it is felt that they contribute sufficient insight into the problem being addressed. Proofs are omitted when they involve lengthy theoretical discourse of material at a level beyond the scope of this text. In such cases, an outline of proofs with adequate references to outside material is presented so that the ambitious reader can labor through the details. Supplementary material including tables of mathematical relationships and other numerical data are included in the apppendixes at the back of the book.

2

SYSTEMS AND SIGNAL ANALYSIS

Communication systems consist of a set of interconnected functional blocks that transfer information between two points by a sequence of signal processing operations. Communication theory deals with mathematical models and techniques that can be used in the study of communication systems. In this chapter we will develop mathematical models for functional blocks in communication systems as well as models for signals that we will encounter in our study of communication systems.

In electrical communication systems, signals are time-varying quantities such as voltages or currents. The functional elements of the system are electrical networks. Electrical networks and signals can be described in the *time domain* where the independent variable is time, or in the frequency domain where the independent variable is frequency. In the study of communication systems, it is often necessary and convenient to describe systems and signals in the *frequency domain.* Hence we will emphasize the frequency domain approach leading to the concepts of *spectrum* and *bandwidth.*

It is assumed that the reader has some familiarity with spectral analysis—the description of systems and signals in the frequency domain, and the correspondence between time domain and frequency domain descriptions. The presentation in this chapter is intended as a review.

We begin our review of systems and signal analysis techniques with several classifications of signals and systems, and frequency domain representation for signals using the complex exponential Fourier series and the Fourier transform. Concepts of spectral densities, particularly the power spectral density, are introduced next. Finally, frequency domain models for systems, and techniques for analyzing the effects of systems on signals are discussed. We also include a brief treatment of how the Fourier transform or the Fourier series coefficients of a signal can be obtained via numerical integration on a digital computer.

We conclude our brief review of systems and signal analysis techniques with a quick look at the typical signals and signal processing operations we will encounter in communication systems.

The reader should note that what we describe here are mathematical models for signals and systems—an idealized description of signals and systems that are pertinent to the problem in hand. It may be possible to develop several different models for a given problem. The choice of a model for a particular problem will have to be based on an understanding of the physical phenomena involved and the limitations of various models.

2.1 SYSTEMS AND SIGNALS

We use the word "system" to describe a set of elements (or functional blocks) that are connected together in such a manner as to achieve a desired objective. In communication systems the objective is the transfer of information. The system responds to a signal appearing at its input (see Figure 2.1). For our purposes, we define a signal to be any ordinary function of time. The value of the signal at a given time may be real or complex. Signals that we observe physically (using voltmeters, ammeters, oscilloscopes, etc.) are real valued. However, we will often use complex notation to treat real-valued signals since certain mathematical models and calculations are somewhat simplified if we use complex notation. With this brief introduction, let us look at a few methods of classifying signals and systems.

2.1.1 Classification of Signals

Deterministic and Random Signals. In communication systems, we encounter two broad classes of signals referred to as deterministic and

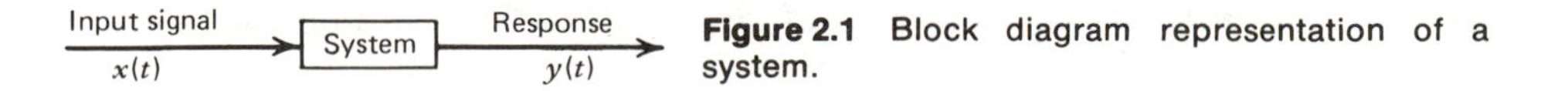

Figure 2.1 Block diagram representation of a system.

random signals. Deterministic signals can be modeled by explicit mathematical expressions. A signal of the form, say, $x(t) = 5 \sin 20t$ is a deterministic signal. A random signal is one about which there is some degree of uncertainty before it actually occurs. An example of a random signal is the output of a radio receiver that is tuned to a frequency at which there is no broadcasting—the receiver is responding to noise arising from disturbances in the atmosphere and its internal circuitry. We will deal with models for deterministic signals in this chapter, deferring the treatment of random signals until the next chapter.

Periodic and Nonperiodic Signals. A signal $x(t)$ is called periodic if there exists a constant $T > 0$ such that

$$x(t) = x(t + T), \quad -\infty < t < \infty \tag{2.1}$$

The smallest value of $T > 0$ that satisfies Equation (2.1) is called the period of the signal. A signal for which there is no value of T satisfying (2.1) is called nonperiodic or aperiodic. We will encounter both periodic and nonperiodic signals in our study of communication systems.

Energy and Power Signals. If a signal $x(t)$ is a voltage across a 1 ohm resistor, then the instantaneous value of power is $|x(t)|^2$. The magnitude square is used here to allow the possibility of $x(t)$ being a complex-valued signal. The energy dissipated by the signal during the time interval $(-T/2, T/2)$ is given by

$$E_x^T = \int_{-T/2}^{T/2} |x(t)|^2 \, dt$$

and the average power dissipated by the signal during the interval is

$$S_x^T = \frac{1}{T} \int_{-T/2}^{T/2} |x(t)|^2 \, dt$$

We define $x(t)$ to be an *energy signal* if and only if $0 < E_x < \infty$, where

$$E_x = \lim_{T \to \infty} \int_{-T/2}^{T/2} |x(t)|^2 \, dt \tag{2.2}$$

The signal is defined to be a power signal if and only if $0 < S_x < \infty$, where

$$S_x = \lim_{T \to \infty} \frac{1}{T} \int_{-T/2}^{T/2} |x(t)|^2 \, dt \tag{2.3}$$

Impulse Signals. There is a class of signals described by *singularity func-*

tions that play a very important role in signal analysis. Singularity functions (also known as *generalized functions* or *distributions*) are mathematical abstractions and, strictly speaking, do not occur in physical systems. However, they are useful in approximating certain limiting conditions in physical systems.

A singularity function that we will use quite frequently in analyzing signals in communication systems is the *unit impulse* or *Dirac delta* function $\delta(t)$. The unit impulse $\delta(t)$ is not a function in a strict mathematical sense and it is usually defined by an integral relationship. Specifically, if $x(t)$ is a function that is continuous at $t = t_0$, then $\delta(t - t_0)$ is defined by

$$\int_a^b x(t)\delta(t - t_0)\,dt = \begin{cases} x(t_0), & \text{if } a < t_0 < b \\ 0 & \text{otherwise} \end{cases} \tag{2.4}$$

The impulse function defined in Equation (2.4) has the following properties:

1. $\displaystyle\int_{-\infty}^{\infty} \delta(t)\,dt = 1$ (2.5)

2. $\delta(t) = 0$ for $t \neq 0$, and
   $\delta(t)$ at $t = 0$ is undefined (2.6)

3. $\delta(at) = \dfrac{1}{|a|}\delta(t), \quad a \neq 0$ (2.7)

4. $\delta(t) = \dfrac{d}{dt}[u(t)]$ (2.8)

   where $u(t)$ is the *unit step* function defined as

$$u(t) = \begin{cases} 1 & \text{for } t > 0 \\ 0 & \text{for } t < 0 \end{cases}$$

5. Numerous conventional functions approach $\delta(t)$ in the limit. For example,

$$\lim_{a\to 0} \frac{1}{a}\left[u\left(\frac{t}{a} + \frac{1}{2}\right) - u\left(\frac{t}{a} - \frac{1}{2}\right)\right] = \delta(t) \tag{2.9a}$$

$$\lim_{\epsilon\to 0} \frac{1}{\pi t} \sin\frac{\pi t}{\epsilon} = \delta(t) \tag{2.9b}$$

$$\lim_{B\to\infty} \int_{-B}^{B} \exp(\pm j2\pi f t)\,df = \delta(t) \tag{2.9c}$$

We can treat $\delta(t)$ as an ordinary function provided that all conclusions are based on the integration properties of $\delta(t)$ stated in Equation (2.4).

2.1.2 Classification of Systems

Mathematically, a "system" is a functional relationship between the input $x(t)$ and output $y(t)$. We can write the input–output relationship as

$$y(t_0) = f[x(t); -\infty < t < \infty]; -\infty < t_0 < \infty \tag{2.10}$$

Based on the properties of the functional relationship given in Equation (2.10) we can classify systems as follows:

Linear and Nonlinear Systems. A system is said to be linear if superposition applies. That is, if

$$\begin{aligned} y_1(t) &= f[x_1(t)] \\ y_2(t) &= f[x_2(t)] \end{aligned} \tag{2.11a}$$

then, for a linear system,

$$f[a_1x_1(t) + a_2x_2(t)] = a_1y_1(t) + a_2y_2(t) \tag{2.11b}$$

Any system in which superposition does not apply is called a nonlinear system.

Time-Invariant and Time-Varying Systems. A system is time invariant if a time shift in the input results in a corresponding time shift in the output so that, if

$$y(t) = f[x(t)] \tag{2.12a}$$

then

$$y(t - t_0) = f[x(t - t_0)]; -\infty < t, t_0 < \infty \tag{2.12b}$$

Any system not meeting the requirement stated above is called a time-varying system.

Causal and Noncausal Systems. A causal (or physical) system is one whose response does not begin before the input function is applied. Stated in another way, the value of the output at $t = t_0$ depends only on the values of the input $x(t)$ for $t \leq t_0$, that is,

$$y(t_0) = f[x(t); t \leq t_0]; -\infty < t, t_0 < \infty \tag{2.13}$$

Noncausal systems do not satisfy the condition given above. They do not exist in the real world but can be approximated by the use of time delay.

The classification of systems and signals given above will help us in finding a suitable mathematical model for a given system that is to be analyzed.

2.2 SIGNAL REPRESENTATION USING FOURIER SERIES

The communication systems engineer is often concerned with the signal location in the frequency domain and the signal bandwidth rather than transient analysis. Thus, we will be interested in steady-state analysis much of the time. Fourier series provides a frequency domain model for periodic signals that is useful for analyzing their frequency content and for calculating the steady-state response of networks for periodic input. Signals having finite energy over a finite interval (t_1, t_2), and signals that are periodic with a finite energy within each period can both be represented by the Fourier series.

2.2.1 Complex Exponential Fourier Series

A signal $x(t)$ having finite energy over the time interval (t_1, t_2) can be represented for values of t almost everywhere in the interval (t_1, t_2) by a sum of complex exponentials of the form

$$x(t) = \sum_{n=-\infty}^{\infty} C_x(nf_0) \exp(j2\pi nf_0 t), \quad t_1 < t < t_2; \quad j = \sqrt{-1}, \quad f_0 = \frac{1}{t_2 - t_1} \tag{2.14a}$$

where $\exp(y)$ denotes e^y.
The coefficients of the series expansion are given by

$$C_x(nf_0) = \frac{1}{t_2 - t_1} \int_{t_1}^{t_2} x(t) \exp(-j2\pi nf_0 t)\, dt \tag{2.14b}$$

Equation (2.14) states that a signal defined over a time interval of width $T_0 = t_2 - t_1$ can be expanded using only those frequency components that are integer multiples of the *fundamental frequency* $f_0 = 1/T_0$. The coefficients $C_x(nf_0)$ are complex numbers and are called the *spectral components* of $x(t)$. The frequencies nf_0 for $n \geq 2$ are called *harmonic* frequencies.

If $x(t)$ is piecewise continuous, then the series (summation) given in Equation (2.14) will converge to $x(t)$ everywhere it is finite and continuous. Since all physical signals are finite and continuous (even though their mathematical models sometimes are not!), we will view the series as being identical to $x(t)$ in the interval (t_1, t_2). Outside this interval (t_1, t_2) the function and its Fourier series need not be equal. Furthermore, if $x(t)$ has a finite discontinuity at $t = t_0$, $t_0 \in (t_1, t_2)$, then the Fourier series converges to the arithmetic average of the values of the function on either side of the discontinuity; that is, the series converges to

$$\lim_{\epsilon \to 0} \frac{x(t_0 - \epsilon) + x(t_0 + \epsilon)}{2}$$

While the Fourier series can be used to represent any given function with finite energy over a finite interval (t_1, t_2), it is a particularly useful model for periodic signals having finite energy within each period.

2.2.2 Fourier Series Representation of Periodic Signals

A periodic power signal $x(t)$ with a period T_0 can be represented by an exponential Fourier series of the form

$$x(t) = \sum_{n=-\infty}^{\infty} C_x(nf_0) \exp(j2\pi n f_0 t); -\infty < t < \infty \tag{2.15a}$$

where

$$C_x(nf_0) = \frac{1}{T_0} \int_{-T_0/2}^{T_0/2} x(t) \exp(-j2\pi n f_0 t)\, dt \tag{2.15b}$$

and

$$f_0 = 1/T_0 \tag{2.15c}$$

Notice that this series has the same form as the one given in Equation (2.14) with $t_1 = -T_0/2$, $t_2 = T_0/2$. The important difference is that the series in (2.14) is valid for $t \in (t_1, t_2)$ whereas the series in (2.15) is valid for all t. This extension is due to the periodicity of $x(t)$.

Properties of the Fourier Series. Some of the important properties of the Fourier series are listed below:

1. If $x(t)$ is real, then

$$C_x(nf_0) = C_x^*(-nf_0) \tag{2.16}$$

   where the asterisk denotes complex conjugate.

2. If a real-valued signal $x(t)$ has either even $(x(t) = x(-t))$ or odd $(x(t) = -x(-t))$ time symmetry, then

$$\text{Imaginary part of } C_x(nf_0) = 0 \text{ if } x(t) \text{ is even} \tag{2.17}$$

$$\text{Real part of } C_x(nf_0) = 0 \text{ if } x(t) \text{ is odd} \tag{2.18}$$

3. For $x(t)$ real or complex,

$$\frac{1}{T_0} \int_{-T_0/2}^{T_0/2} |x(t)|^2\, dt = \sum_{n=-\infty}^{\infty} |C_x(nf_0)|^2 \tag{2.19}$$

   The preceding equation is known as *Parseval's theorem*.

4. A real-valued signal $x(t)$ can be represented by a trigonometric Fourier series of the form

$$x(t) = A_0 + \sum_{n=1}^{\infty} A_n \cos 2\pi n f_0 t + \sum_{n=1}^{\infty} B_n \sin 2\pi n f_0 t \tag{2.20}$$

The coefficients of this series are real, and are related to the coefficients of the complex exponential series of $x(t)$ by

$$C_x(nf_0) = \frac{\sqrt{A_n^2 + B_n^2}}{2} \exp(-j\phi_n), \quad n = 1, 2, 3, \ldots \tag{2.21a}$$

where

$$\phi_n = \tan^{-1}(B_n/A_n), \quad n = 1, 2, 3, \ldots \tag{2.21b}$$

Proofs of Equations (2.16) through (2.21) are left as an exercise for the reader (Problems 2.5–2.8).

Spectra of Periodic Signals. The coefficients of the Fourier series of a signal are displayed in a set of two plots in the frequency domain, one showing the amplitude of the spectral components versus frequency and the other showing the phase angles versus frequency. These two plots are called *amplitude* and *phase spectra*, respectively. They are also called the *line spectra*.

Another plot that is often very useful in frequency domain analysis is the *power spectral density* (*psd*) $G_x(f)$ of the signal $x(t)$. Parseval's theorem for periodic power signals states that

$$\lim_{T\to\infty} \frac{1}{T} \int_{-T/2}^{T/2} |x(t)|^2 \, dt = \sum_{n=-\infty}^{\infty} |C_x(nf_0)|^2$$

The left-hand side of the preceding equation is the normalized average power of the signal and $|C_x(nf_0)|^2$ gives the spectral distribution of power. Now, if we define the power spectral density (psd) function $G_x(f)$ to be a real, even, and nonnegative function of frequency that gives the distribution of power in the frequency domain, then we have

$$G_x(f) = \sum_{n=-\infty}^{\infty} |C_x(nf_0)|^2 \delta(f - nf_0) \tag{2.22a}$$

The power spectral density of a periodic power signal is impulsive in nature indicating that nonzero units of power are concentrated at discrete frequencies. The reader can verify that

$$S_x = \lim_{T\to\infty} \frac{1}{T} \int_{-T/2}^{T/2} |x(t)|^2 \, dt = \int_{-\infty}^{\infty} G_x(f) \, df \tag{2.22b}$$

Spectral plots are called *two-sided* plots if components at both positive and negative frequencies are shown in the plots. Spectral plots where the sum of the magnitudes of the components at $f = +nf_0$ and $f = -nf_0$ are shown displayed at $f = nf_0$ are called *one-sided spectral plots.* We shall find it convenient to use the two-sided spectral plots.

Example 2.1. For the rectangular waveform shown in Figure 2.2, obtain the complex exponential Fourier series representation and plot the amplitude, phase and power spectral densities.

Solution. The coefficients of the complex exponential Fourier series for $x(t)$ are given by

$$\begin{aligned} C_x(nf_0) &= \frac{1}{T_0}\int_{-T_0/2}^{T_0/2} x(t)\exp(-j2\pi nf_0 t)\,dt \\ &= \frac{1}{T_0}\int_{-\tau/2}^{\tau/2} A\exp(-j2\pi nf_0 t)\,dt \\ &= \frac{A}{-j2\pi nf_0 T_0}\exp(-j2\pi nf_0 t)\Big|_{t=-\tau/2}^{t=\tau/2} \\ &= \frac{A}{\pi n}\sin(\pi nf_0\tau) \end{aligned}$$

since $f_0T_0 = 1$ and $e^{j\theta} - e^{-j\theta} = 2j\sin\theta$. We can rewrite $C_x(nf_0)$ as

$$C_x(nf_0) = Af_0\tau\left(\frac{\sin(\pi nf_0\tau)}{\pi nf_0\tau}\right) = Af_0\tau\,\mathrm{sinc}(nf_0\tau)$$

where the "*sinc*" *function* is defined as

$$\mathrm{sinc}(y) = \frac{\sin \pi y}{\pi y}$$

Plots of $|C_x(nf_0)|$, phase of $C_x(nf_0)$, and the psd of $x(t)$ are shown in Figure 2.3 for the value of $\tau = T_0/5$. (The ratio τ/T_0 is called the *duty cycle* of the pulse waveform.)

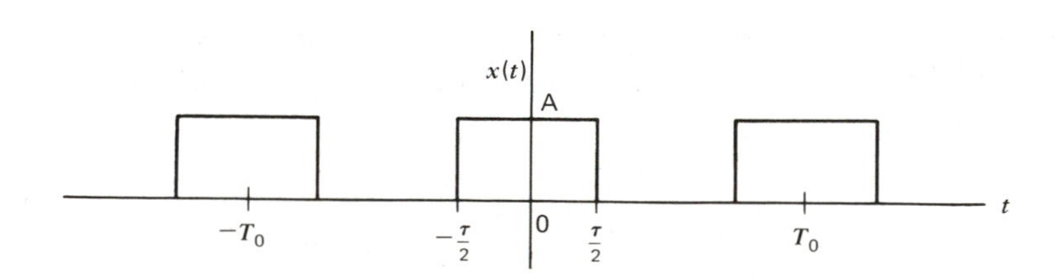

Figure 2.2 Rectangular pulse waveform.

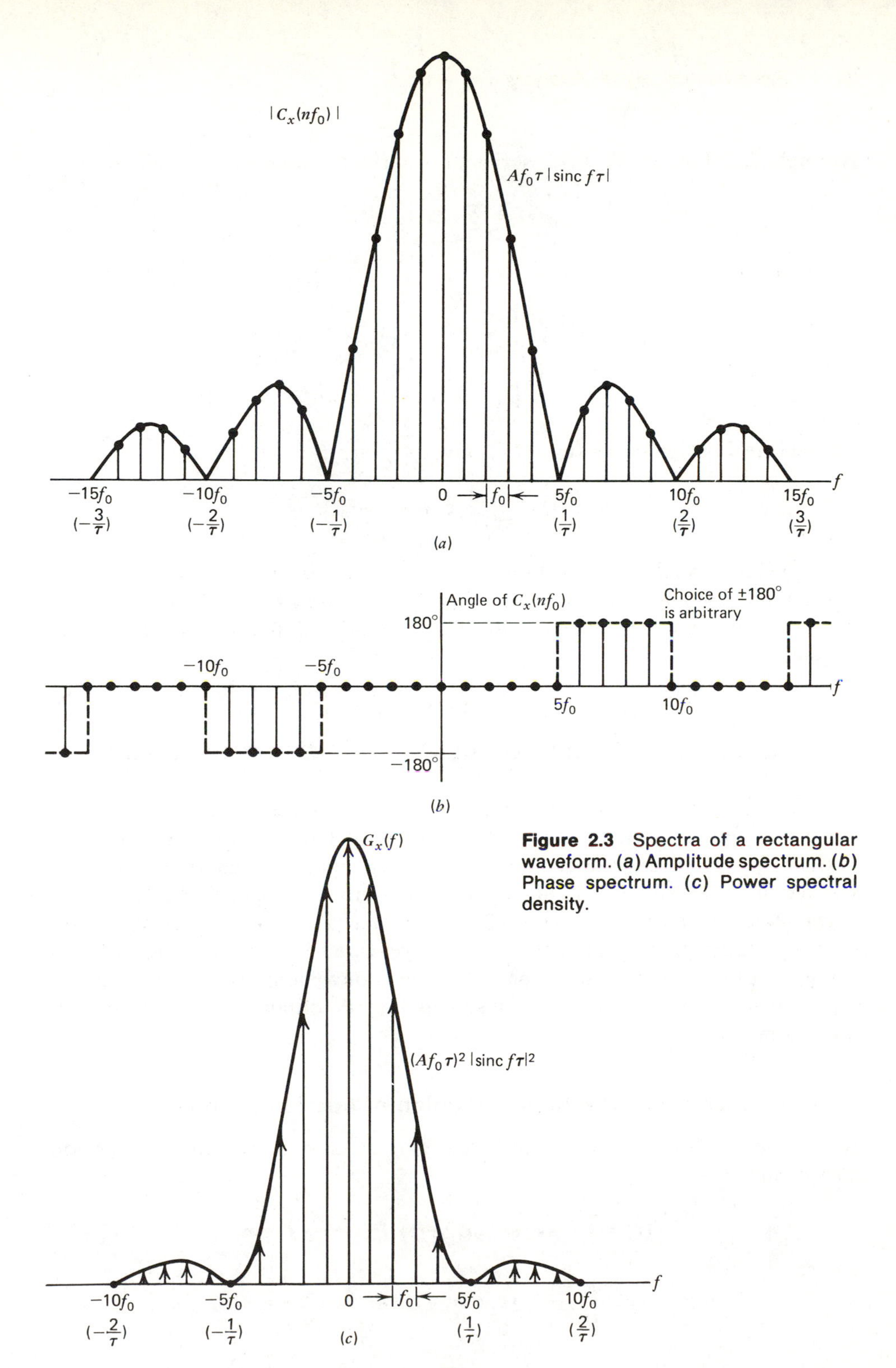

Figure 2.3 Spectra of a rectangular waveform. (*a*) Amplitude spectrum. (*b*) Phase spectrum. (*c*) Power spectral density.

Example 2.2. Find the Fourier series of a periodic impulse train defined by

$$x(t) = \sum_{k=-\infty}^{\infty} \delta(t - kT_0)$$

Solution.

$$\begin{aligned} C_x(nf_0) &= \frac{1}{T_0}\int_{-T_0/2}^{T_0/2} x(t)\exp(-j2\pi nf_0t)\,dt \\ &= \frac{1}{T_0}\int_{-T_0/2}^{T_0/2} \delta(t)\exp(-j2\pi nf_0t)\,dt = \frac{1}{T_0} \end{aligned}$$

Hence, the Fourier series for $x(t)$ is given by

$$x(t) = \sum_{-\infty}^{\infty} \frac{1}{T_0}\exp(j2\pi nf_0t)$$

In systems analysis, the Fourier series can be used to calculate the steady-state output of a linear time-invariant system responding to an arbitrary periodic waveform. We will discuss this in detail in a later section of this chapter.

2.3 SIGNAL REPRESENTATION USING FOURIER TRANSFORMS

The Fourier series discussed in the preceding section can be used to represent a periodic power signal over the time interval $(-\infty, \infty)$. In communication systems we often encounter aperiodic energy signals. Such signals may be *strictly time limited* ($x(t)$ is identically zero outside a specified interval), or *asymptotically time limited* ($x(t) \to 0$ as $|t| \to \infty$). If $x(t)$ is an energy signal that is strictly time limited, then it can be represented in the frequency domain using Fourier series. However, a more convenient frequency domain representation for aperiodic energy signals may be obtained using the Fourier transform.

2.3.1 Fourier Transform Representation of Aperiodic Signals

According to the Fourier integral theorem, we can represent an aperiodic energy signal by

$$x(t) = \int_{-\infty}^{\infty} X(f)\exp(j2\pi ft)\,df; -\infty < t < \infty \tag{2.23a}$$

where

$$X(f) = \int_{-\infty}^{\infty} x(t)\exp(-j2\pi ft)\,dt; -\infty < f < \infty \tag{2.23b}$$

$X(f)$ is called the *Fourier transform* of $x(t)$ and the inverse operation defined in Equation (2.23b), which states that $x(t)$ can be obtained from $X(f)$, is called the *inverse Fourier transform* of $X(f)$. The pair $x(t)$ and $X(f)$ is called a Fourier transform pair and we signify this by writing

$$x(t) \leftrightarrow X(f)$$

or

$$X(f) = F\{x(t)\}, \qquad x(t) = F^{-1}\{X(f)\}$$

(Functions denoted by upper case letters will be used to represent Fourier transforms.)

Comparison of Equations (2.14) and (2.23) reveals that the Fourier series and Fourier transform are somewhat similar. In the frequency domain, the Fourier series provides a *discrete spectral* representation for a periodic power signal. The Fourier transform provides a *continuous spectral* representation for an aperiodic signal. In the periodic case we return to the time domain from the frequency domain by summing spectral components. In the aperiodic case we integrate the continuous function $X(f)\exp(j2\pi ft)$.

The Fourier transform $X(f)$ plays the same role for aperiodic signals that $C_x(nf_0)$ plays for periodic signals. Several important properties of the Fourier transform* are given below.

1. If $x(t)$ is real, then

$$X(f) = X^*(-f) \tag{2.24}$$

2. If $x(t)$ is real and has even or odd time symmetry, then

$$X(f) = \begin{cases} 2\int_0^\infty x(t)\cos(2\pi ft)\,dt & \text{when } x(t) \text{ is even} \quad (2.25) \\ -2j\int_0^\infty x(t)\sin(2\pi ft)\,dt & \text{when } x(t) \text{ is odd} \quad (2.26) \end{cases}$$

3.

$$\int_{-\infty}^{\infty} |x(t)|^2\,dt = \int_{-\infty}^{\infty} |X(f)|^2\,df \tag{2.27}$$

This relationship is known as *Parseval's* (or sometimes called *Rayleigh's*) *energy theorem.*

Spectra of Aperiodic Energy Signals. The Fourier transform is a method of expressing an aperiodic energy signal in terms of a continuous set of complex exponential frequency components. The resulting spectral plots,

*See Appendix C for a table of Fourier transform pairs.

plots of the amplitude and phase of $X(f)$, will be continuous plots defined for all values of f. Therefore, aperiodic energy signals will have *continuous spectra* whereas periodic signals have *discrete line spectra.*

For periodic signals, we defined a power spectral density function based on Parseval's power theorem for periodic signals. For aperiodic energy signals, Parseval's energy theorem (Equation (2.27)) states that the total energy is obtained by integrating $|X(f)|^2$ over all frequencies; that is, $|X(f)|^2$ gives the distribution of signal energy in the frequency domain. Based on this relationship we define $|X(f)|^2$ as the *energy spectral density* of an aperiodic energy signal.

Example 2.3. Find the Fourier transform of the rectangular pulse shown in Figure 2.4. Sketch the amplitude and phase spectrum.

Solution. The Fourier transform of $x(t)$ shown in Figure 2.4 can be obtained from

$$\begin{aligned} X(f) &= \int_{-\infty}^{\infty} x(t) \exp(-j2\pi ft)\, dt \\ &= \int_{-\tau/2}^{\tau/2} \exp(-j2\pi ft)\, dt = 2\int_{0}^{\tau/2} \cos(2\pi ft)\, dt \\ &= \tau \operatorname{sinc}(f\tau) \end{aligned} \tag{2.28}$$

Amplitude and phase plots of $X(f)$ are shown in Figure 2.5. Comparison of these plots with the plots shown in Figure 2.3 shows that the Fourier transform leads to continuous spectra whereas the Fourier series yields discrete line spectra.

Figure 2.5 also shows that $|X(f)| \ll |X(0)|$ for $|f| > 1/\tau$. Hence we may take $1/\tau$ as the "spectral width" of $x(t)$, a result that we will refer to several times when we study "pulsed" signaling schemes used in digital communication systems. It must also be pointed out here that shorter pulses have broad spectra and longer pulses have narrow spectra (see Problem 2.12).

Example 2.4. Find the Fourier transform of the sign (or signum) function defined as

$$\operatorname{sgn}(t) = \frac{t}{|t|} = \begin{cases} 1 & \text{for } t > 0 \\ -1 & \text{for } t < 0 \end{cases}$$

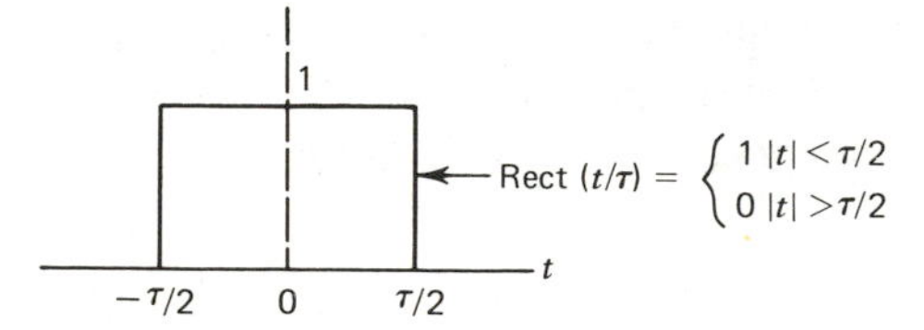

Figure 2.4 Rectangular pulse $x(t) = \text{Rect}(t/\tau)$

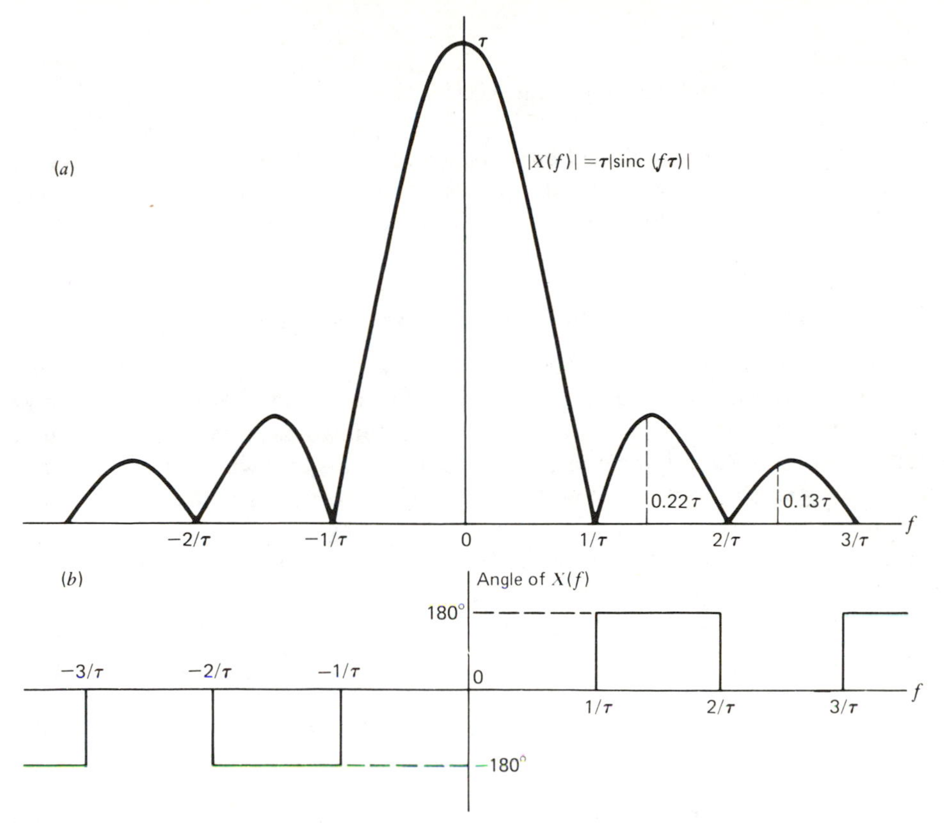

Figure 2.5 Amplitude and phase spectrum of a rectangular pulse. (*a*) Amplitude spectrum. (*b*) Phase spectrum.

Solution. The sign function is one of many power signals for which the Fourier transform exists. In many such cases, the Fourier transform is obtained by a limiting process. For example, define $x(t, \alpha)$ to be

$$x(t, \alpha) = \exp(-\alpha|t|)\,\mathrm{sgn}(t), \quad \alpha > 0$$

Then $\lim_{\alpha \to 0} x(t, \alpha) = \mathrm{sgn}(t)$, and we may attempt to obtain the Fourier transform of $\mathrm{sgn}(t)$ from the Fourier transform of $x(t, \alpha)$ by letting $\alpha \to 0$.

$$\begin{aligned}
F\{x(t, \alpha)\} &= \int_{-\infty}^{\infty} x(t, \alpha) \exp(-j2\pi ft)\, dt \\
&= \int_{-\infty}^{0} -\exp(\alpha t) \exp(-j2\pi ft)\, dt + \int_{0}^{\infty} \exp(-\alpha t) \exp(-j2\pi ft)\, dt \\
&= \frac{-4\pi jf}{\alpha^2 - (2\pi jf)^2},
\end{aligned}$$

and hence

$$F\{\text{sgn}(t)\} = \lim_{\alpha \to 0} F\{x(t, \alpha)\} = \frac{2}{2\pi jf} = \frac{1}{\pi jf}$$

In the preceding approach, we have assumed that the limiting and integration operations can be interchanged, an assumption which holds, for example, when $x(t, \alpha)$ is continuous in t and α for all values of t and α.

2.3.2 Fourier Transforms of Periodic Power Signals

In the preceding sections of this chapter we saw that periodic power signals can be represented by Fourier series and that aperiodic energy signals can be represented by the Fourier transform. Very often we have signals that contain both periodic and aperiodic components and it seems that we have something of a quandary since the periodic and nonperiodic parts require different representations. This confusion can be avoided by using impulse functions in the frequency domain to represent discrete spectral components of periodic power signals.

Important Fourier transform pairs involving (infinite energy) power signals are

$$A\delta(t) \leftrightarrow A \tag{2.29}$$

$$A\delta(t - t_0) \leftrightarrow A \exp(-j2\pi f t_0) \tag{2.30}$$

$$A \leftrightarrow A\delta(f) \tag{2.31}$$

$$A \exp(j2\pi f_0 t) \leftrightarrow A\delta(f - f_0) \tag{2.32}$$

We can prove (2.29) easily by using the integral property of the unit impulse:

$$F\{A\delta(t)\} = \int_{-\infty}^{\infty} A\delta(t) \exp(-j2\pi f t)\, dt = A$$

The transform pair given in (2.31) can be established similarly. Proofs of Equations (2.30) and (2.32) are left as exercises for the reader.

In using the transform pairs given in Equations (2.29) to (2.32) we must remember that we are dealing with functions having infinite or undefined energy content and the concept of energy spectral density no longer applies.

Using Equation (2.32), the reader can easily verify that an arbitrary periodic signal having a complex Fourier series of the form

$$x(t) = \sum_{n=-\infty}^{\infty} C_x(nf_0) \exp(j2\pi n f_0 t) \tag{2.33}$$

has a Fourier transform of the form

$$X(f) = \sum_{n=-\infty}^{\infty} C_x(nf_0)\delta(f - nf_0) \tag{2.34}$$

Thus, the impulse function (which does not really exist in the real world, and is not even defined explicitly) provides a unified method of describing aperiodic and periodic signals in the frequency domain using Fourier transforms.

Finally, to return to time domain using the Fourier transform given in Equation (2.34), we integrate the right-hand side of Equation (2.34) to get

$$x(t) = F^{-1}\{X(f)\} = \int_{-\infty}^{\infty} \left(\sum_{n=-\infty}^{\infty} C_x(nf_0)\delta(f - nf_0) \right) \exp(j2\pi ft)\, df$$
$$= \sum_{n=-\infty}^{\infty} C_x(nf_0) \int_{-\infty}^{\infty} \delta(f - nf_0) \exp(j2\pi ft)\, df$$
$$= \sum_{n=-\infty}^{\infty} C_x(nf_0) \exp(j2\pi nf_0 t)$$

which is the Fourier series representation of $x(t)$ given in Equation (2.33).

2.3.3 Transform Theorems

Several useful theorems relating time domain signal processing operations to appropriate frequency domain operations involving Fourier transforms can be proved. We look at three of these theorems now. The reader is referred to Appendix C for additional theorems.

Convolution. Convolution is a mathematical operation that is an important analytical tool used by communication engineers. The mathematical operation of convolving two real-valued functions of the same variable is defined as

$$p(t) * q(t) \triangleq \int_{-\infty}^{\infty} p(\lambda)q(t - \lambda)\, d\lambda$$
$$\triangleq \int_{-\infty}^{\infty} q(\lambda)p(t - \lambda)\, d\lambda \tag{2.35}$$

The notation $*$ denotes convolution (not complex conjugation) and $\triangleq$ stands for "equal by definition." Later in this chapter, we will discuss the use of convolution to describe input-output relationships in a linear time-invariant system. Right now, we are interested in seeing the frequency domain implication of time domain convolution.

If we take the Fourier transform of both sides of Equation (2.35), we have

$$F\{p(t) * q(t)\} = \int_{-\infty}^{\infty} \exp(-j2\pi ft) \left(\int_{-\infty}^{\infty} p(\lambda)q(t - \lambda)\, d\lambda \right) dt$$
$$= \int_{-\infty}^{\infty} p(\lambda) \exp(-j2\pi f\lambda) \left(\int_{-\infty}^{\infty} q(t - \lambda) \exp(-j2\pi f(t - \lambda))\, dt \right) d\lambda$$

If we let $t - \lambda = \alpha$ in the second integral, we have

$$F\{p(t) * q(t)\} = \int_{-\infty}^{\infty} p(\lambda) \exp(-j2\pi f\lambda) \left(\int_{-\infty}^{\infty} q(\alpha) \exp(-j2\pi f\alpha)\, d\alpha \right) d\lambda$$

which simplifies to

$$F\{p(t) * q(t)\} = P(f)Q(f)$$

That is, convolution of two signals in time domain corresponds to the multiplication of their Fourier transforms in the frequency domain. We can easily show that convolution in frequency domain corresponds to multiplication in time domain. We state these properties as

$$p(t) * q(t) \leftrightarrow P(f)Q(f) \tag{2.36a}$$

and

$$p(t)q(t) \leftrightarrow P(f) * Q(f) \tag{2.36b}$$

Frequency Translation. In communication systems we often encounter signals of the form

$$y(t) = x(t) \cos 2\pi f_0 t$$

Such signals are called modulated signals. The Fourier transform of $y(t)$ is related to the Fourier transform of $x(t)$ by

$$Y(f) = \tfrac{1}{2}[X(f - f_0) + X(f + f_0)]$$

To prove the preceding equation, let us begin with

$$\cos 2\pi f_0 t = \tfrac{1}{2}[\exp(j2\pi f_0 t) + \exp(-j2\pi f_0 t)]$$

Now,

$$\begin{aligned} F\{x(t) \exp(j2\pi f_0 t)\} &= \int_{-\infty}^{\infty} x(t) \exp(j2\pi f_0 t) \exp(-j2\pi f t)\, dt \\ &= \int_{-\infty}^{\infty} x(t) \exp(-j2\pi(f - f_0)t)\, dt \\ &= X(f - f_0) \end{aligned}$$

Similarly,

$$F\{x(t) \exp(-j2\pi f_0 t)\} = X(f + f_0)$$

and hence

$$F\{x(t) \cos(2\pi f_0 t)\} = \tfrac{1}{2}[X(f - f_0) + X(f + f_0)] \tag{2.37}$$

The preceding equation shows that given the Fourier transform $X(f)$ of $x(t)$,

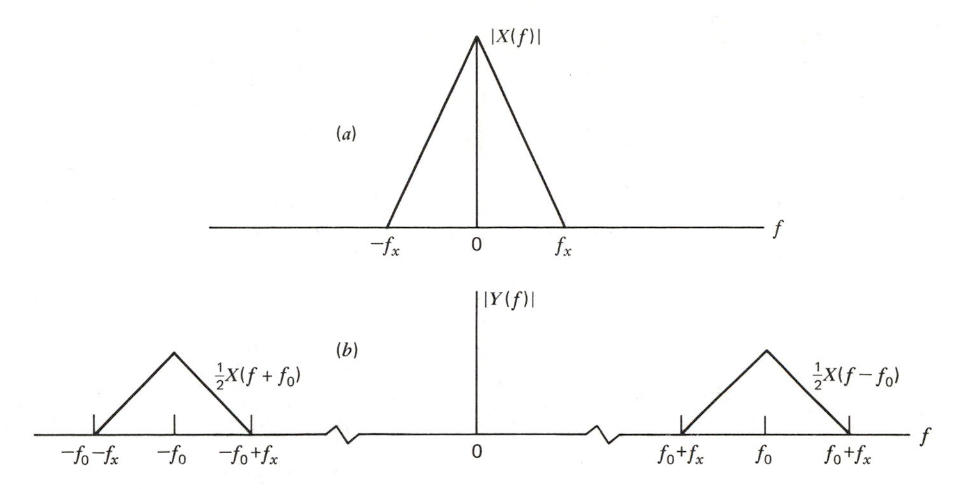

Figure 2.6 (*a*) Amplitude spectrum of $x(t)$. (*b*) Amplitude spectrum of $x(t) \cos 2\pi f_0 t$.

we can obtain the Fourier transform of $x(t) \cos 2\pi f_0 t$ as follows: Divide $X(f)$ by 2, shift the divided plot to the right by f_0, and to the left by f_0, and add the two shifted plots. Because of the shifting of the spectrum in the frequency domain, the result given in Equation (2.37) is known as the *frequency translation* (*or modulation theorem*). It is conveniently restated as

$$x(t) \cos 2\pi f_0 t \leftrightarrow \tfrac{1}{2}[X(f - f_0) + X(f + f_0)] \tag{2.38}$$

Figure 2.6 illustrates the result stated in Equation (2.38). The reader should become familiar with the frequency translation theorem since we will be using it quite often in our study of modulation schemes.

Sampling. Another signal processing operation that occurs often in communication systems is sampling. The sampling operation is mathematically described by

$$x_s(t) = x(t) \sum_{n=-\infty}^{\infty} \delta(t - nT_s) \tag{2.39}$$

where $x(t)$ is the signal being sampled, and $x_s(t)$ is the sampled signal* that consists of a sequence of impulses separated in time by T_s. The reciprocal of T_s, $f_s = 1/T_s$, is called the sampling frequency.

The Fourier transform of the sampled signal can be obtained as

$$F\{x_s(t)\} = F\{x(t) \sum_{n=-\infty}^{\infty} \delta(t - nT_s)\}$$

*$x_s(t)$ is often written as $x_s(t) = \sum_{n=-\infty}^{\infty} x(nT_s)\delta(t - nT_s)$ since $\delta(t) = 0$ everywhere except at $t = 0$.

$$= X(f) * F\left(\sum_{n=-\infty}^{\infty} \delta(t - nT_s)\right)$$

$$= X(f) * \left(f_s \sum_{n=-\infty}^{\infty} \delta(f - nf_s)\right)$$

In deriving the preceding step, we have used the result derived in Example 2.2. Completing the convolution operation in the last step, we obtain

$$F\{x_s(t)\} = f_s \sum_{n=-\infty}^{\infty} X(f - nf_s) \tag{2.40}$$

or

$$\begin{aligned} X_s(f) = f_s X(f) + f_s\{&X(f - f_s) + X(f + f_s) \\ &+ X(f - 2f_s) + X(f + 2f_s) \\ &+ X(f - 3f_s) + X(f + 3f_s) + \cdots\} \end{aligned}$$

Given the Fourier transform $X(f)$ of $x(t)$, we can obtain the Fourier transform $X_s(f)$ of $x_s(t)$ by translating $f_sX(f)$ by 0, $\pm f_s$, $\pm 2f_s, \ldots$ and adding the translated components together. Examples shown below in Figures 2.7 and 2.8 illustrate the sampling operation in the time domain and frequency domain, respectively.

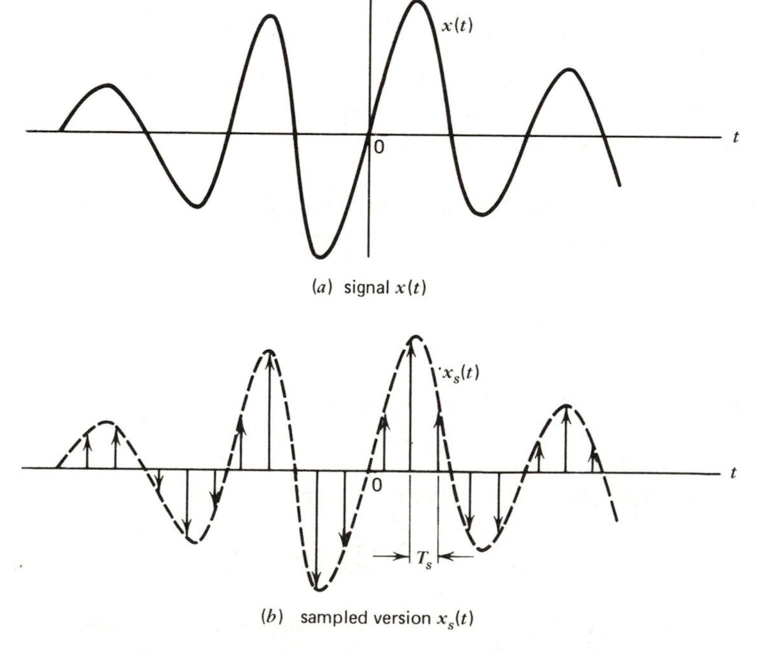

Figure 2.7 Sampling operation in time domain.

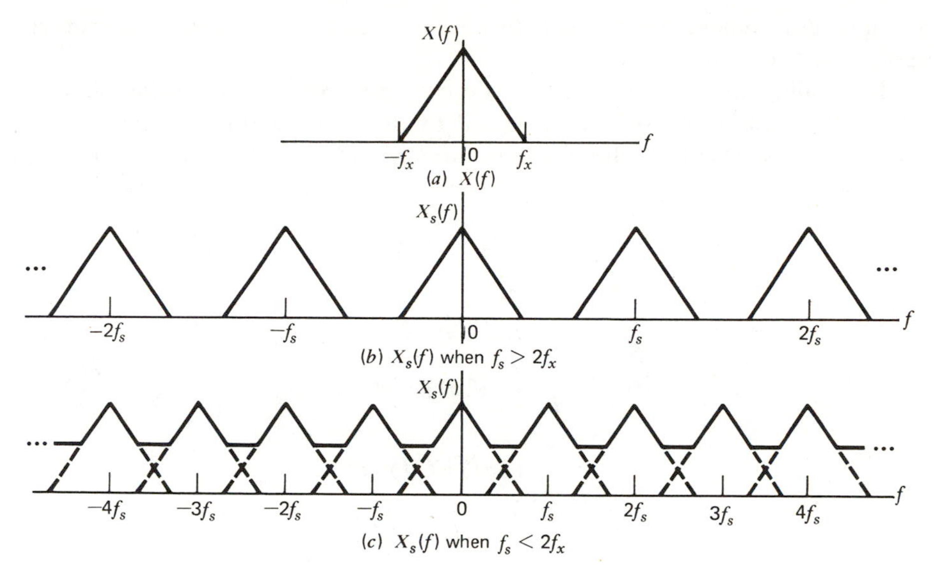

Figure 2.8 Sampling operation shown in frequency domain. (*a*) $X(f)$. (*b*) $X_s(f)$ when $f_s > 2f_x$. (*c*) $X_s(f)$ when $f_s < 2f_x$.

Figure 2.8 reveals that if a signal spectrum extends to f_x and is sampled at a rate higher than $2f_x$, then the spectrum of the sampled signal consists of periodic repetitions of $X(f)$. If the sampling rate is lower than $2f_x$, then the shifted components of $X(f)$ overlap and the spectrum of the sampled signal has no resemblance to the spectrum of the original signal. The spectral overlap effect is known as *aliasing*, and the sampling rate $2f_x$ below which aliasing occurs is called the *Nyquist rate.*

If the sampling rate is greater than the Nyquist rate, then $X_s(f)$ contains $X(f)$ intact. Hence, it is at least theoretically possible to recover $X(f)$ and $x(t)$ from $X_s(f)$ by passing the sampled signal through a device (known as a lowpass filter) which will pass only those spectral components that lie in the range $-f_x$ to f_x. We will say more about this in Chapter 10 when we discuss the use of sampling in Pulse Code Modulation (PCM) schemes.

Scaling-Uncertainty Principle. We can easily show that a signal $x(t)$ and its Fourier transform $X(f)$ are related to each other by

$$F\{x(at)\} = \frac{1}{a} X\left(\frac{f}{a}\right); \quad a > 0 \tag{2.41}$$

This relationship is known as the scaling theorem and it implies that the signal and its transform cannot both be of short duration. A signal that exists (i.e., has nonzero values) for a short duration yields a wider spectrum and vice versa (see Problem 2.12)

The scaling theorem is used to derive the "uncertainty principle" that relates the "duration" of a time signal to the "bandwidth" occupied by its spectrum. If we use the following measures α and β for the time duration and bandwidth

$$\alpha^2 = \frac{\int_{-\infty}^{\infty} (2\pi t)^2 x^2(t)\, dt}{\int_{-\infty}^{\infty} x^2(t)\, dt} = \frac{1}{E_x} \int_{-\infty}^{\infty} (2\pi t)^2 x^2(t)\, dt \tag{2.42a}$$

$$\beta^2 = \frac{\int_{-\infty}^{\infty} (2\pi f)^2 |X(f)|^2\, df}{\int_{-\infty}^{\infty} |X(f)|^2\, df} = \frac{1}{E_x} \int_{-\infty}^{\infty} (2\pi f)^2 |X(f)|^2\, df \tag{2.42b}$$

then the uncertainty principle is stated as

$$\alpha\beta \geqslant \pi \tag{2.42c}$$

(For a proof of this relationship the reader is referred to Papoulis, Chapter 8, Section 2.)

The relationship given in (2.42c) is called the "uncertainty principle" by analogy to the uncertainty principle of quantum mechanics. The latter states that the position and velocity of atomic particles cannot be simultaneously measured to any arbitrary degree of accuracy desired. Precise determination of one parameter can be had only at the expense of the other. Equation (2.42c) implies a similar concept, namely, that both the time duration and the bandwidth of a signal cannot be made arbitrarily small simultaneously.

2.4 POWER SPECTRAL DENSITY

There is an important class of signals, namely, the aperiodic power signals, for which we have not yet developed a frequency domain model. The Fourier series does not exist for aperiodic power signals that are not time limited. For such signals the Fourier transform may or may not exist, and the concept of energy spectral density does not apply. However, since the average power* is assumed to be finite, we may use power spectral density functions to describe aperiodic power signals in the frequency domain.

Suppose we are given a power signal $x(t)$. Let us form a truncated version $x_T(t)$ of $x(t)$ as shown in Figure 2.9. The normalized average power* of $x_T(t)$ is given by

$$S_x^T = \frac{1}{T}\int_{-T/2}^{T/2} |x_T(t)|^2\,dt \tag{2.43}$$

Now let us introduce the *autocorrelation* function $R_{xx}^T(\tau)$ of $x_T(t)$ defined as

$$R_{xx}^T(\tau) = \frac{1}{T}\int_{-T/2}^{T/2} x_T(t)x_T(t+\tau)\,dt \tag{2.44}$$

The Fourier transform of $R_{xx}^T(\tau)$ is

$$\begin{aligned}
\int_{-\infty}^{\infty} R_{xx}^T(\tau)\exp(-j2\pi f\tau)\,d\tau &= \frac{1}{T}\int_{-\infty}^{\infty}\exp(-j2\pi f\tau)\left(\int_{-\infty}^{\infty} x_T(t)x_T(t+\tau)\,dt\right)d\tau \\
&= \frac{1}{T}\int_{-\infty}^{\infty}\left(\int_{-\infty}^{\infty} x_T(t)x_T(t+\tau)\right. \\
&\qquad \left.\times \exp(-j2\pi f(t+\tau))\exp(j2\pi ft)\,dt\right)d\tau \\
&= \frac{1}{T}\int_{-\infty}^{\infty} x_T(t)\exp(j2\pi ft) \\
&\qquad \times\left(\int_{-\infty}^{\infty} x_T(t+\tau)\exp(-j2\pi f(t+\tau))\,d\tau\right)dt \\
&= \frac{1}{T}X_T(-f)X_T(f)
\end{aligned} \tag{2.45}$$

where $X_T(f)$ is the Fourier transform of $x_T(t)$. Since $x(t)$ is real, we have $X_T(-f) = X_T^*(f)$ and hence

$$\int_{-\infty}^{\infty} R_{xx}^T(\tau)\exp(-j2\pi f\tau)\,d\tau = \frac{1}{T}|X_T(f)|^2$$

Letting $R_{xx}(\tau) = \lim_{T\to\infty} R_{xx}^T(\tau)$, and taking the limit $T\to\infty$ on both sides of the preceding equation, we have

$$\int_{-\infty}^{\infty} R_{xx}(\tau)\exp(-j2\pi f\tau)\,d\tau = \lim_{T\to\infty}\frac{|X_T(f)|^2}{T} \tag{2.46}$$

*Average power is expressed in the units of dBw or dBm as $(S)_{\text{dBw}} = 10\log_{10}$ (S in watts) and $(S)_{\text{dBm}} = 10\log_{10}$ (S in milliwatts).

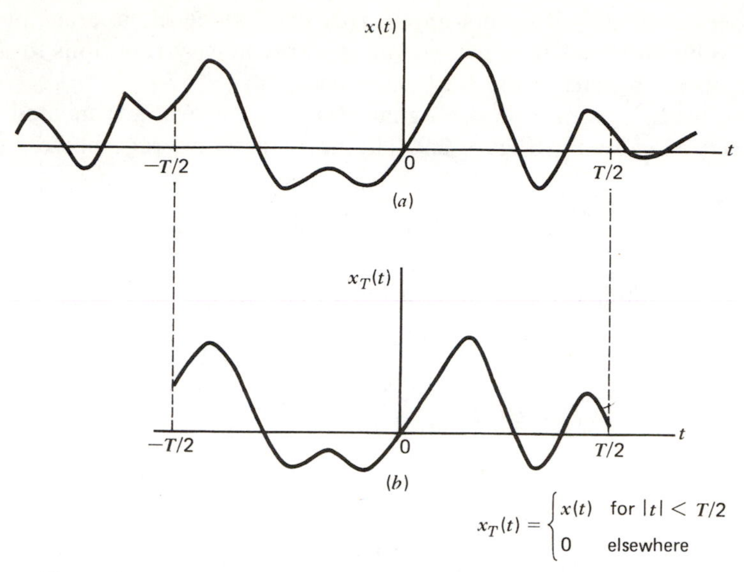

Figure 2.9 (*a*) A power signal. (*b*) Truncated version of the signal.

The reader should note that the limit on the right-hand side of Equation (2.46) may not always exist. The left-hand side of Equation (2.46) is the Fourier transform of the autocorrelation function of the power signal $x(t)$. On the right-hand side we have $|X_T(f)|^2$, which is the energy spectral density of $x_T(t)$, and hence $|X_T(f)|^2/T$ gives the distribution of power in the frequency domain. Hence we can use the relationship given in Equation (2.46) to define the power spectral density (psd) $G_x(f)$ of $x(t)$ as

$$G_x(f) = F\{R_{xx}(\tau)\} \tag{2.47a}$$

$$= \lim_{T\to\infty} \frac{|X_T(f)|^2}{T} \tag{2.47b}$$

The relationship given in Equation (2.47*a*) is called the *Wiener–Khintchine theorem* and has great significance in random signal theory.

We will use the psd defined in Equation (2.47*a*) as the frequency domain description of aperiodic power signals. Actually, we can also use the psd to describe periodic power signals. For periodic power signals the psd has been defined previously in Equation (2.22). The reader can verify that the definition given in Equation (2.47*a*) indeed reduces to Equation (2.22) for periodic power signals. In both cases, for a power signal, we have

$$
\begin{aligned}
S_x &= \lim_{T\to\infty} \frac{1}{T} \int_{-T/2}^{T/2} x^2(t)\, dt \\
&= R_{xx}(0) \\
&= \int_{-\infty}^{\infty} G_x(f)\, df
\end{aligned}
\tag{2.48}
$$

If $x(t)$ is a current or voltage waveform feeding a one ohm load resistance, then S_x has the units of watts and hence $G_x(f)$ is given the units of watts per Hertz. If the load resistance has a value other than one ohm, then $G_x(f)$ is usually specified in terms of volts^2 per Hertz.

It must be pointed out here that the power spectral density function (and the autocorrelation function) does not uniquely describe a signal. The psd retains only the magnitude information and all phase information is lost. Thus, for a given power signal there is a power spectral density, but there are many signals having the same power spectral density. In contrast, the Fourier series and Fourier transforms of signals, when they exist, uniquely describe a signal at all points of continuity.

Example 2.5 The autocorrelation function of an aperiodic power signal is

$$R_{xx}(\tau) = \exp(-\tau^2/2\sigma^2)$$

Find the psd and the normalized average power content of the signal.

Solution. By the Wiener–Khintchine theorem, the psd of the signal is given by

$$
\begin{aligned}
G_x(f) &\overset{\Delta}{=} \int_{-\infty}^{\infty} \exp(-\tau^2/2\sigma^2) \exp(-j2\pi f\tau)\, d\tau \\
&= \sqrt{2\pi\sigma^2} \exp(-(2\pi f\sigma)^2/2)
\end{aligned}
$$

The normalized average power is given by

$$
\begin{aligned}
S_x &= \lim_{T\to\infty} \frac{1}{T} \int_{-T/2}^{T/2} x^2(t)\, dt = R_{xx}(0) \\
&= 1
\end{aligned}
$$

Example 2.6. Find the autocorrelation function and the power spectral density of a rectangular waveform with a period T_0, a peak amplitude of A, and an average value of $A/2$.

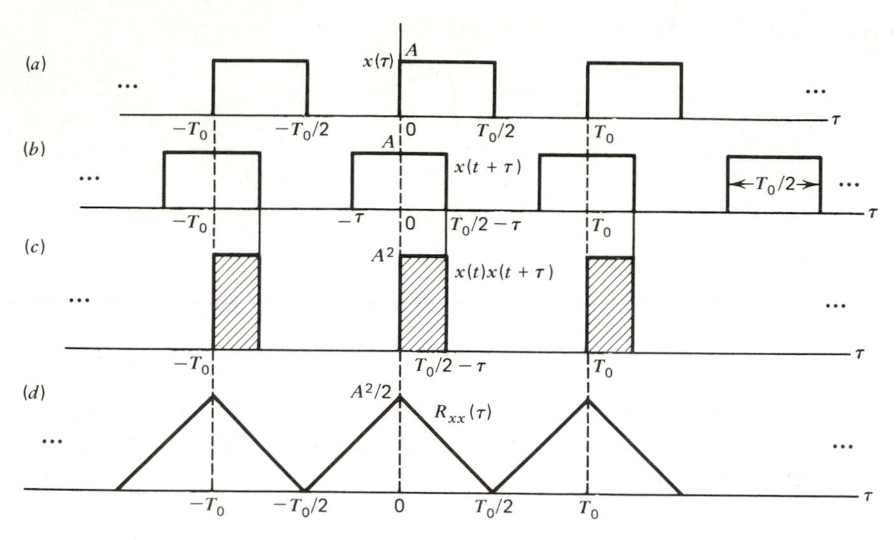

Figure 2.10 (a) Signal $x(t)$. (b) $x(t+\tau)$. (c) $x(t)x(t+\tau)$, which is integrated from $-T_0/2$ to $T_0/2$ to get $R_{xx}(\tau)$. (d) $R_{xx}(\tau)$.

Solution. Since $x(t)$ is periodic, we need to carry out the "time averaging" for autocorrelation over one period only, that is,

$$R_{xx}(\tau) = \frac{1}{T_0}\int_{-T_0/2}^{T_0/2} x(t)x(t+\tau)\,dt$$

Sketches of $x(t)$ and $x(t+\tau)$ are shown in Figure 2.10. For $0 < \tau < T_0/2$, the value of $R_{xx}(\tau)$ is equal to the shaded area shown in Figure 2.10*c*:

$$R_{xx}(\tau) = \frac{A^2}{T_0}\left(\frac{T_0}{2} - \tau\right) = A^2\left(\frac{1}{2} - \frac{\tau}{T_0}\right), \quad 0 < \tau < \frac{T_0}{2}$$

It can be easily verified that $R_{xx}(\tau)$ is an even function, and that it will be periodic with a period of T_0. A sketch of $R_{xx}(\tau)$ is shown in Figure 2.10*d*. The psd of $x(t)$ is given by

$$\begin{aligned}G_x(f) &= \int_{-\infty}^{\infty} R_{xx}(\tau)\exp(-j2\pi f\tau)\,d\tau \\ &= \int_{-\infty}^{\infty} \frac{A^2}{4}\left[1 + \sum_{\substack{n=-\infty \\ n \text{ odd}}}^{\infty}\left(\frac{4}{\pi^2 n^2}\right)\exp(j2\pi n f_0\tau)\right]\exp(-j2\pi f\tau)\,d\tau\end{aligned}$$

where the bracketed term within the integral is the Fourier series for $R_{xx}(\tau)$.

Completing the integration, we obtain

$$G_x(f) = \frac{A^2}{4}\left[\delta(f) + \sum_{\substack{n=-\infty \\ n \text{ odd}}}^{\infty} \frac{4}{\pi^2 n^2}\delta(f - nf_0)\right]$$

2.5 SYSTEM RESPONSE AND FILTERS

Having developed a set of models for signals, we now look at the characterization of systems and their effects on signals. We will develop both time domain and frequency domain models for the response of systems to an arbitrary input signal $x(t)$. We will focus our attention on linear, time-invariant, causal, and asymptotically stable systems. We will assume that there is no stored energy in the system when the input is applied.

2.5.1 Impulse Response, Step Response, and Time Domain Analysis

A linear time-invariant system is characterized in the time domain by an *impulse response* $h(t)$, which is defined to be the response $y(t)$ of the system to a unit impulse applied at $t = 0$. That is,

$$h(t) \stackrel{\Delta}{=} y(t) \quad \text{when } x(t) = \delta(t)$$

The response to an arbitrary input $x(t)$ is then found by convolving $x(t)$ with $h(t)$ in the time domain:

$$y(t) = x(t) * h(t) = \int_{-\infty}^{\infty} h(\lambda)x(t - \lambda)\, d\lambda \tag{2.49}$$

Since $h(t) = 0$ for $t < 0$ for causal systems, we can write $y(t)$ as

$$y(t) = \int_{0}^{\infty} h(\lambda)x(t - \lambda)\, d\lambda = \int_{-\infty}^{\infty} x(\lambda)h(t - \lambda)\, d\lambda \tag{2.50}$$

The relationships given in Equations (2.49) and (2.50) are called the superposition integrals.

The impulse response of a system is often obtained from the *step response* of the system $y_u(t)$, which is defined as

$$y_u(t) \stackrel{\Delta}{=} y(t) \quad \text{when } x(t) = u(t) = \begin{cases} 0 & \text{for } t < 0 \\ 1 & \text{for } t > 0 \end{cases}$$

Since $y_u(t) = h(t) * u(t)$, we have

$$\frac{d}{dt}[y_u(t)] = h(t) * \frac{d}{dt}[u(t)] = h(t) * \delta(t) = h(t)$$

Thus the impulse response of a system may be obtained by differentiating the step response of the system.

2.5.2 Transfer Function and Frequency Domain Analysis

In the frequency domain, a linear time-invariant system is characterized by a *transfer function* $H(f)$, which is defined as

$$H(f) \overset{\Delta}{=} \frac{y(t)}{x(t)}, \qquad \text{when } x(t) = \exp(j2\pi ft) \tag{2.51}$$

The transfer function $H(f)$ is a complex valued function. By applying the convolution integral to Equation (2.51), we have

$$\begin{aligned} y(t) &= h(t) * x(t) = h(t) * \exp(j2\pi ft) \\ &= \int_{-\infty}^{\infty} h(\lambda) \exp(j2\pi f(t-\lambda))\, d\lambda \\ &= \exp(2\pi jft) \left[\int_{-\infty}^{\infty} h(\lambda) \exp(-2\pi jf\lambda)\, d\lambda \right] \end{aligned}$$

The expression in brackets is the Fourier transform of $h(t)$. Therefore, the impulse response and the transfer function form a Fourier transform pair; That is,

$$h(t) \leftrightarrow H(f)$$

Furthermore, we have seen in section 2.3.3. that the convolution in time domain corresponds to the multiplication of Fourier transforms in the frequency domain. Thus, for a linear time-invariant system, the Fourier transforms of the input and output (when they exist) are related to each other by

$$Y(f) = H(f)X(f) \tag{2.52}$$

and

$$y(t) \leftrightarrow Y(f) = H(f)X(f)$$

We will use the relationship given in Equation (2.52) as the basis of our frequency domain analysis of linear systems. Since the transfer function is in general a complex quantity, it can be expressed in magnitude and angle form as

$$H(f) = |H(f)| \exp(j[\text{angle of } H(f)])$$

$|H(f)|$ is called the *amplitude response* of the system and angle of $H(f)$ is called the *phase response* of the system. The magnitude response is often expressed in *decibels* (dB) using the definition

$$|H(f)|_{dB} = 20 \log_{10}|H(f)|$$

In physical systems $h(t)$ is a real-valued function and hence $H(f)$ has conjugate symmetry in the frequency domain. That is,

$$H(f) = H^*(-f) \tag{2.53}$$

If the input to the system $x(t)$ is a periodic power signal, then the steady-state response $y(t)$ can be expressed as

$$y(t) = \sum_{n=-\infty}^{\infty} H(nf_0)C_x(nf_0) \exp(j2\pi nf_0 t) \tag{2.54}$$

where $C_x(nf_0)$ are the coefficients of the complex exponential Fourier series of the input signal $x(t)$.

2.5.3 Effect of Transfer Function on Spectral Densities

Since we will be using spectral densities as important signal models, it is worthwhile to investigate how the spectral density of an input signal is modified by the system. From Equation (2.52) it follows that the amplitude and phase spectra of the input and output signals are related to each other by

$$|Y(f)| = |H(f)||X(f)| \tag{2.55a}$$

$$\text{angle of } \{Y(f)\} = \text{angle of } \{X(f)\} + \text{angle of } \{H(f)\} \tag{2.55b}$$

If the signal is modeled by power spectral density, then

$$G_y(f) = |H(f)|^2 G_x(f) \tag{2.55c}$$

The preceding equation is derived from Equation (2.47), and we will use this relationship quite extensively in our study of communication systems.

2.5.4 Real and Ideal Filters

Equations (2.55*a*), (2.55*b*), and (2.55*c*) show that a linear system acts as a frequency selective device. Based on the functional dependence of $H(f)$ on f, certain frequency components are *amplified* (*or emphasized*) whereas other

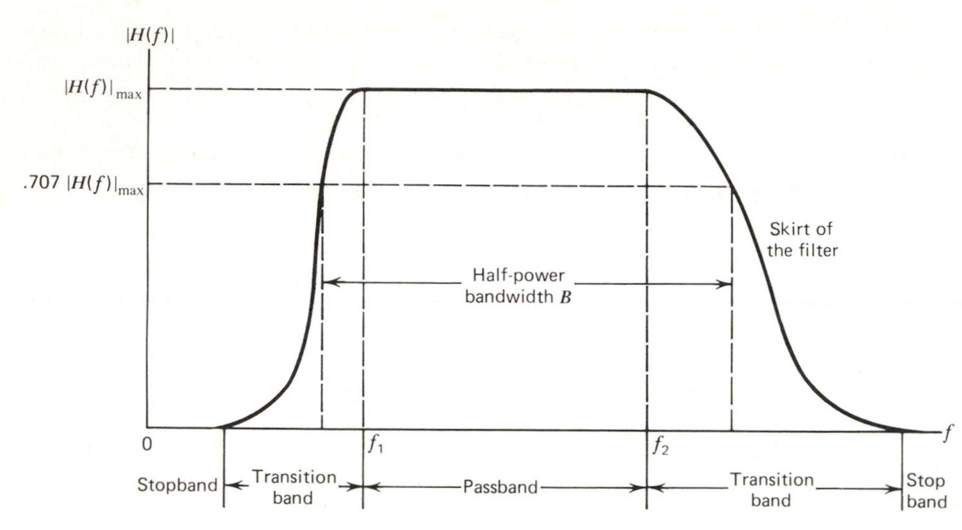

Figure 2.11 Parameters of a filter; $|H(f)| = |H(-f)|$.

components are *attenuated* (*or deemphasized*). We call this aspect of linear systems, namely, the frequency selectivity, as *filtering*.

From a filtering standpoint, a linear system is sometimes characterized by its *stopband, passband,* and *half-power bandwidth.* These parameters are identified in Figure 2.11 for a typical system.

It is often convenient to work with filters having idealized transfer functions with rectangular amplitude response functions and linear phase response functions within the passband. We will consider three general types of ideal filters: lowpass, highpass, and bandpass. Characteristics of these ideal filters are shown in Figure 2.12. These filters cannot be built with practical components; hence, they are called *unrealizable* filters. However, we will use them to approximate the characteristics of practical filters. Sharp-eyed readers might recognize that signals can also be classified into lowpass, bandpass, and highpass categories depending on their spectrum! Indeed, if we replace $|H(f)|$ by $|X(f)|$ or $G_x(f)$ in Figure 2.12, we have frequency domain plots of *lowpass, highpass,* and *bandpass signals.*

Example 2.7. The network shown in Figure 2.13 is often used in communication systems as a lowpass filter. Find the transfer function of the network and sketch the amplitude response. Discuss the possibility of approximating the amplitude response of this filter by that of an ideal lowpass filter with an appropriate cutoff frequency.

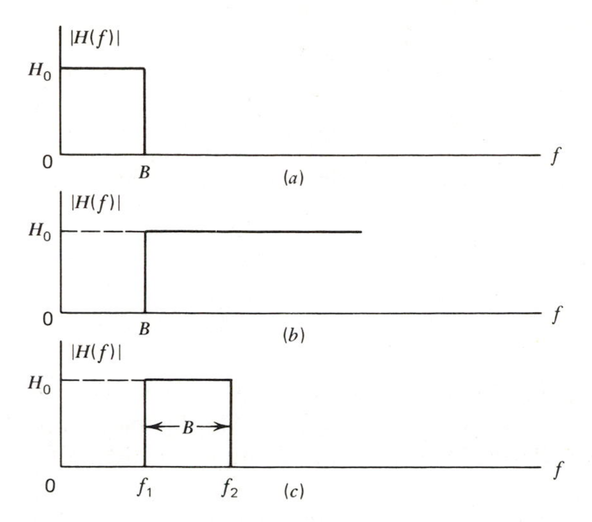

Figure 2.12 Amplitude response of idealized filters. Phase response is assumed to be linear over the passband.

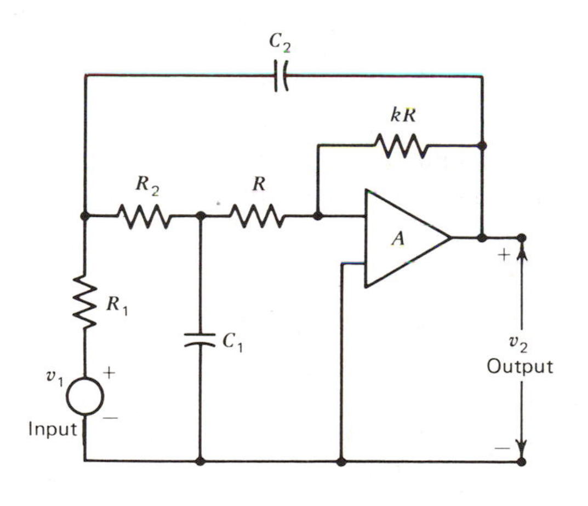

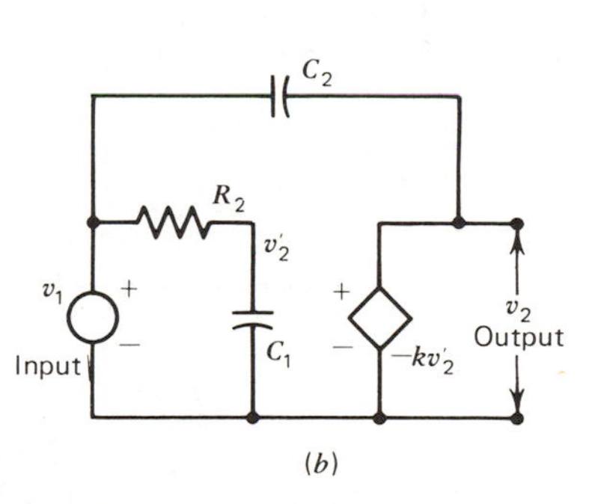

Figure 2.13 (*a*) Lowpass filter.
$C_1 = 0.01\ \mu f$,
$C_2 = 0.1\ \mu f$,
$R_1 = 1\ k\Omega$,
$R_2 = 10\ k\Omega$,
$k = 1.0$.
$A \gg 1$.

(*b*) Equivalent circuit of the filter.

Solution. The transfer function of the filter is given by

$$H(f) = \frac{-k}{(j2\pi f)^2 C_1 C_2 R_1 R_2 + (j2\pi f)(R_1 C_2 k + C_1 R_1 + C_2 R_1 + C_1 R_2) + 1}$$

Substituting element values and letting $2\pi f = \omega$, we have

$$H(\omega) = \frac{-1.0}{10^{-8}(j\omega)^2 + 10^{-4}(3.1)j\omega + 1}$$

$$\frac{|H(\omega)|}{|H(0)|} = \left[\sqrt{[1 - 10^{-8}\omega^2]^2 + [(3.1)(10^{-4})\omega]^2}\right]^{-1}$$

A plot of $|H(f)|/|H(0)|$ is shown in Figure 2.14. An ideal lowpass approximation is also shown in Figure 2.14. The lowpass approximation has a cutoff frequency $B \approx 600$ Hz, which is taken to be the half-power frequency of the actual filter.

For spectral analysis, we will mainly be interested in $|H(f)|^2$ rather than $H(f)$. Figure 2.14 shows that the actual $|H(f)|^2$ and the ideal lowpass approximation differ very little when $\log_{10}(2\pi f) < 3$, and also when $\log_{10}(2\pi f) > 5$. In the transition region, there are some differences between the actual response and the approximation. However, if the input signal does not have any significant spectral components beyond, say, 500 Hz, then the actual filter and the ideal lowpass filter (approximation) will yield the same output (power or energy) spectral distribution.

Thus, the validity of the approximation will depend not only on the response of the filter but also on the spectral characteristics of the input signal. We will use ideal filters as approximations to actual filters whenever possible since spectral analysis is greatly simplified if the filters are ideal.

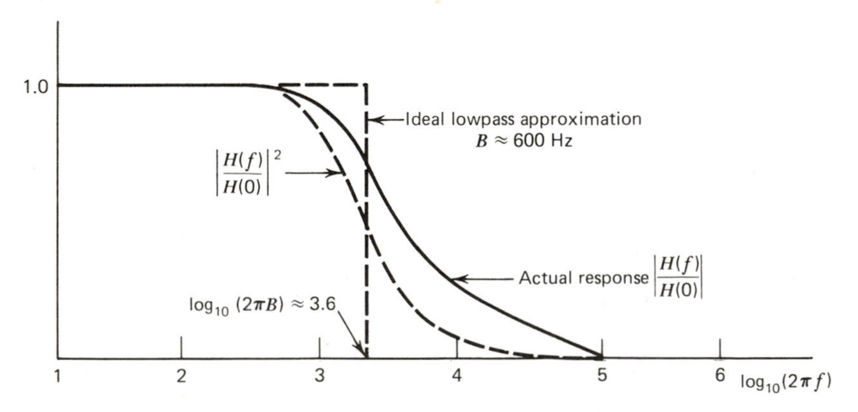

Figure 2.14 Ideal lowpass approximation.

2.6 SPECTRAL ANALYSIS OF MODULATION AND DEMODULATION OPERATIONS

Modulation and demodulation are two important signal processing operations that take place in a communication system. While we will take a detailed look at various types of modulation schemes in later chapters of the book, we present here an example to give the reader a preview of how we will go about analyzing modulation schemes. In particular, the example given below is intended to show the reader the importance of spectral analysis in the study of communication systems. We will use an analog modulation scheme as an example in this section.

In analog carrier wave modulation schemes, a parameter of a high-frequency carrier is varied proportionally to the message signal $x(t)$ such that a one-to-one correspondence exists between the parameter and the message signal. The carrier waveform is usually assumed to be sinusoidal and the modulated carrier has the mathematical form

$$x_c(t) = A(t)\cos(\omega_c t + \phi(t)), \quad \omega_c = 2\pi f_c \tag{2.56}$$

In Equation (2.56), f_c is referred to as the *carrier frequency*. $A(t)$ and $\phi(t)$ are the instantaneous amplitude and phase angle of the carrier, respectively. If $A(t)$ is varied in proportion to the message signal $x(t)$, the result is *linear modulation*. If $\phi(t)$ or its derivative is varied according to $x(t)$, then we have *angle modulation*. In either case, modulation changes the spectrum of the message signal in the frequency domain.

The process of generating $x_c(t)$ from $x(t)$ is called *modulation* and the reverse operation, namely, the extraction of $x(t)$ from $x_c(t)$, is called *demodulation* (see Figure 2.15).

2.6.1 Analysis of a Linear Modulation System

Suppose that we want to transmit a lowpass (baseband) signal $x(t)$ (with the amplitude spectrum shown in Figure 2.16*b*) over a bandpass channel with a response

$$H(f) = \begin{cases} 1, & |f - f_c| < B_c/2; \quad B_c \geq 2f_x \\ 0, & \text{elsewhere} \end{cases}$$

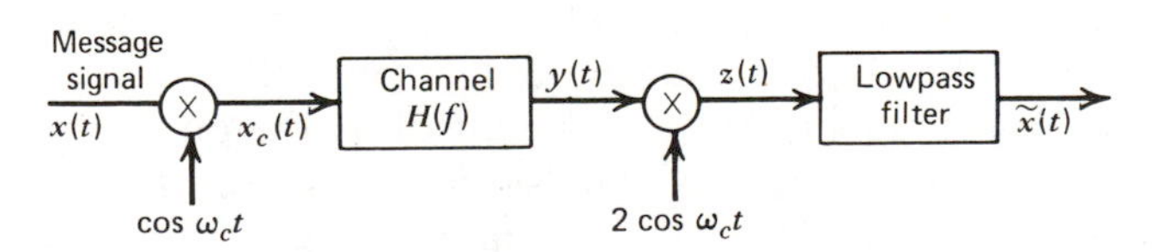

Figure 2.15 A linear modulation scheme.

where B_c is the bandwidth of the channel and f_x is the signal bandwidth. A linear modulation scheme can be used to translate the frequency content of the baseband signal to match the response of the channel. This modulation operation is represented in the time domain by (Figure 2.16*c*)

$$x_c(t) = x(t)\cos \omega_c t, \qquad f_c = \omega_c/2\pi > f_x$$

and in the frequency domain by (Figure 2.16*d*)

$$X_c(f) = \tfrac{1}{2}[X(f - f_c) + X(f + f_c)]$$

where $X_c(f)$ and $X(f)$ are the Fourier transforms of $x(t)$ and $x_c(t)$. As shown in Figure 2.16*d*, the modulated signal passes through the channel without suffering any changes, that is, without distortion. If the channel does not add noise, then the received signal is $y(t) = x_c(t)$, and the baseband signal can be recovered from $x_c(t)$ using the demodulation scheme shown in Figure 2.15.

In time domain, the first step in a possible demodulation operation consists of multiplying the channel output with a "local" carrier to obtain

$$z(t) = y(t)(2\cos \omega_c t)$$

or

$$\begin{aligned} Z(f) &= Y(f - f_c) + Y(f + f_c) \\ &= X(f) + \tfrac{1}{2}[X(f - 2f_c) + X(f + 2f_c)] \end{aligned} \tag{2.57}$$

The lowpass filter rejects $X(f - 2f_c)$ and $X(f + 2f_c)$ since they are made up of spectral components that lie outside the passband of the filter. The filter however passes $X(f)$ without any distortion. Thus the receiver output is equal to the transmitted message signal $x(t)$.

The preceding analysis could have been carried out entirely in time domain by expressing the filter response and the channel response in time domain. Such an analysis would have been tedious, and a true picture of what is happening in the system will not be obvious as is the case in frequency domain analysis.

In our study of communication systems, we will be emphasizing the frequency domain approach. Indeed, it is quite unlikely that many of the significant modulation techniques that are in use today could have been developed without the aid of frequency domain (spectral) analysis. However, the spectral approach should not be thought of as the only method used for analyzing communication systems. For some classes of signals, such as the ones we will encounter in digital communication systems, time domain techniques are quite convenient.

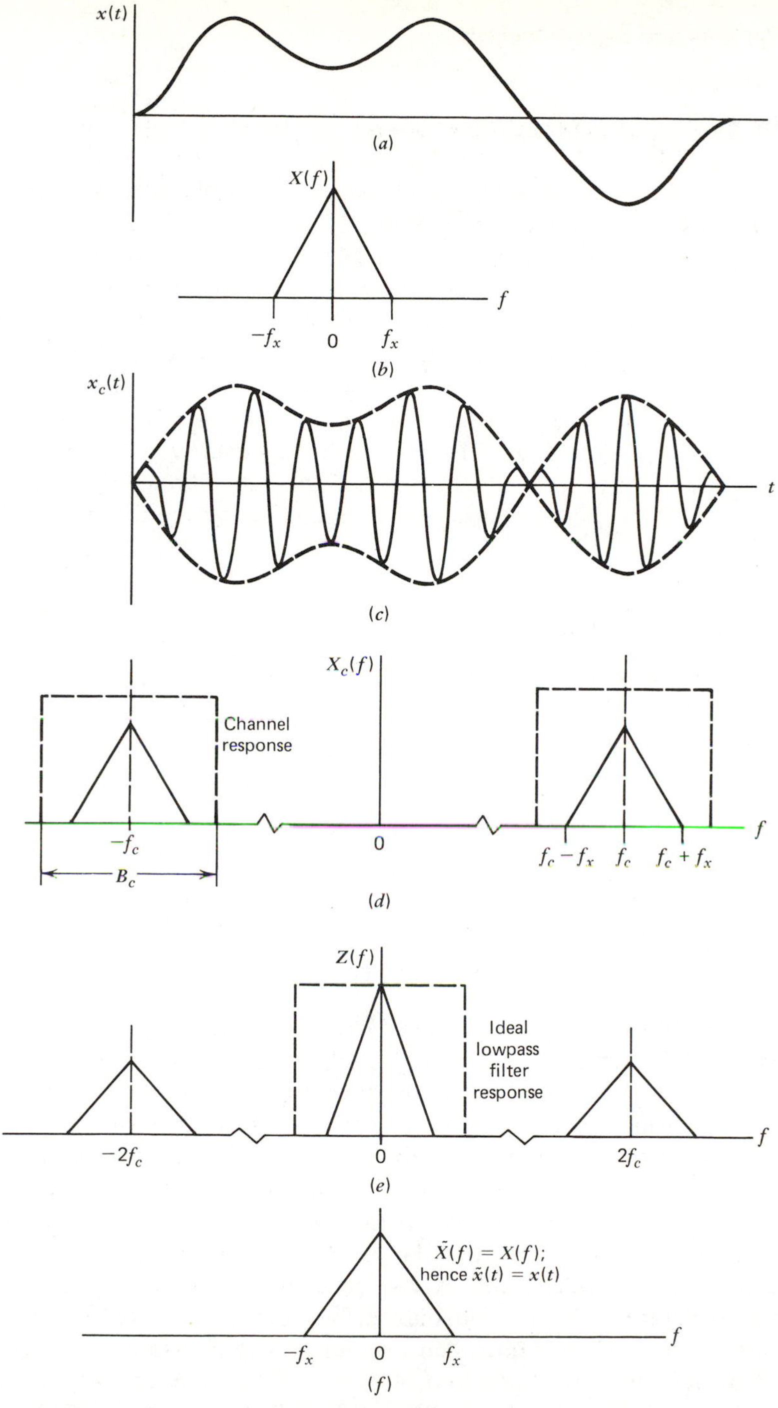

Figure 2.16 Spectral analysis of signals in the system shown in Figure 2.15. (*a*) Signal $x(t)$. (*b*) Signal spectrum $X(f)$. (*c*) Modulated carrier $x_c(t)$. (*d*) Spectrum of the modulated signal. (*e*) Input to the lowpass filter. (*f*) Output.

2.7 SPECTRAL MEASUREMENTS AND COMPUTATIONS

In the preceding sections we discussed several models for characterizing time domain signals in the frequency domain. These models consisted of several (spectral) functions of frequency. For signals that are described by "well-behaved" functions in the time domain, the signal spectrum can be obtained by carrying out the appropriate mathematical operations. In many practical applications, analytical models of signals are not always available. The signal is usually observed (or recorded) for a period of time and the spectrum is calculated either in real time while the signal is being observed (on-line) or from a recording of the signal (off-line). On-line measurements are often made using a device called the spectrum analyzer. Off-line measurements are usually obtained via digital or analog processing of the recorded signal.

2.7.1 Spectrum Analyzer

Suppose we want to measure the power spectral density of a signal $x(t)$ at frequency f_c. Now, if we pass $x(t)$ through a narrow bandpass filter centered at f_c, with a bandwidth of W, then the filter output $y(t)$ will have an average power

$$S_y = \lim_{T\to\infty} \frac{1}{T} \int_{-T/2}^{T/2} y^2(t)\, dt \approx 2K^2 G_x(f_c) W$$

where K^2 is the filter power gain and $G_x(f)$ is the power spectral density of the input signal $x(t)$. If $K^2 = 1/(2W)$, then

$$S_y \approx G_x(f_c)$$

The average value of the output power can be obtained by passing the filter output through a square-law device followed by an integrator. The entire spectrum of $x(t)$ can be "scanned" by changing the center frequency of the filter and measuring the average output power at various frequencies.

Scanning spectrum analyzers that measure power spectral densities use an electronically tunable filter (Figure 2.17). Here, the actual filter center frequency is fixed. Various portions of the signal spectrum are swept into the range of the filter by frequency translation that results when the input signal is multiplied by a sinusoidal signal whose frequency is slowly changing. With reference to Figure 2.17, this sinusoidal waveform is generated by applying a ramp voltage to a voltage controlled oscillator. The sinusoidal voltage multiplies $x(t)$ which shifts the spectrum of $x(t)$. As the ramp progresses, the net effect is to move the signal spectrum past the bandpass filter! The ramp voltage can also be used to sweep the beam of an oscilloscope, and the output

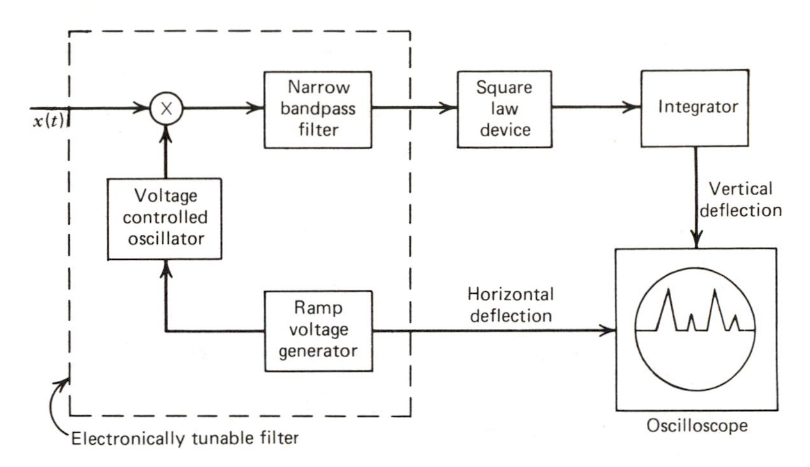

Figure 2.17 Scanning spectrum analyzer.

of the integrator which is proportional to the psd of $x(t)$ can be used to deflect the beam vertically. Any nonzero spectral component present results in a vertical deflection as its frequency is passed. Thus a display of $G_x(f)$ versus f will be traced on the oscilloscope.

The accuracy of a scanning spectrum analyzer will improve with decreasing filter bandwidth and slower ramp sweep rate, both of which will increase the measurement time. If the filter bandwidth is W and the sweep rate is $\Delta f/\Delta t$, then the sweep goes through W Hz in $W/(\Delta f/\Delta t)$ seconds. The filter rise time t_R will be of the order of $1/W$ seconds and in order to allow for the filter to respond we must have

$$\frac{W}{(\Delta f/\Delta t)} > \frac{1}{W} \quad \text{or} \quad \frac{\Delta f}{\Delta t} < W^2$$

that is, the rate of frequency change (in Hz/sec) can be no greater than W^2 Hz/sec.

2.7.2 Numerical Computation of Fourier Series Coefficients

The Fourier series coefficients may be calculated numerically when an analytical expression for $x(t)$ is not known and $x(t)$ is available only as numerical data points. To illustrate how this can be done, let us assume that the sampled values $x(kT_s)$ of a periodic power signal $x(t)$ are available and that we want to compute $C_x(nf_0)$ using these sample values.

The sampled version of $x(t)$ has the form (Figure 2.7 and Equation (2.39))

$$x_s(t) = \sum_{k=-\infty}^{\infty} x(kT_s)\delta(t - kT_s) \tag{2.58}$$

Let us assume that $x(t)$ has a complex exponential Fourier series of the form

$$x(t) = \sum_{n=-\infty}^{\infty} C_x(nf_0) \exp(j2\pi nf_0 t) \tag{2.59}$$

where

$$C_x(nf_0) = \frac{1}{T_0} \int_{-T_0/2}^{T_0/2} x(t) \exp(-j2\pi nf_0 t)\, dt, \qquad f_0 = \frac{1}{T_0} \tag{2.60}$$

For convenience, let us further assume that $T_0/T_s = M$ (see Figure 2.18), and let $N = M/2$.

The Fourier series coefficients of $x_s(t)$ are obtained from

$$\begin{aligned} C_{x_s}(nf_0) &= \frac{1}{T_0} \int_{-T_0/2}^{T_0/2} \sum_k x(kT_s)\delta(t - kT_s) \exp(-j2\pi nf_0 t)\, dt \\ &= \frac{1}{T_0} \sum_{k=-N}^{N} x(kT_s) \exp(-j2\pi nkT_s/T_0) \\ &= \frac{1}{T_0} \sum_{k=-N}^{N} x(kT_s) \exp(-j2\pi nk/M) \end{aligned} \tag{2.61}$$

Equation (2.61) shows that the Fourier series coefficients of $x_s(t)$ are obtained by forming a weighted sum of the samples $x(kT_s)$, an operation which can be carried out easily on a digital computer.

By virtue of the sampling theorem, $C_{x_s}(nf_0)$ is related to $C_x(nf_0)$ by

$$\begin{aligned} C_{x_s}(nf_0) = f_s[C_x(nf_0)] & \\ + f_s[C_x(nf_0 - f_s) + C_x(nf_0 + f_s)] & \\ + f_s[C_x(nf_0 - 2f_s) + C_x(nf_0 + 2f_s)] + \cdots & \end{aligned} \tag{2.62}$$

An example of this relationship is illustrated in Figure 2.18.

From Equation (2.62) and the example shown in Figure 2.18, we draw the following conclusions:

1. If aliasing error is negligible, then $C_{x_s}(nf_0) \approx f_s[C_x(nf_0)]$ for $n = 0, 1, \ldots, M/2$, that is, the first $M/2$ coefficients of the Fourier series representation of $x(t)$ can be computed using Equation (2.61).
2. The highest harmonic coefficient of $x(t)$ that can be determined uniquely from M data points is that one for which the harmonic number is less than or equal to $M/2$.

The presence of aliasing terms is the main source of inaccuracy in the computation of the Fourier series coefficients. The aliasing effect can be minimized by increasing the sampling rate although this increases the number of data points that have to be stored and processed.

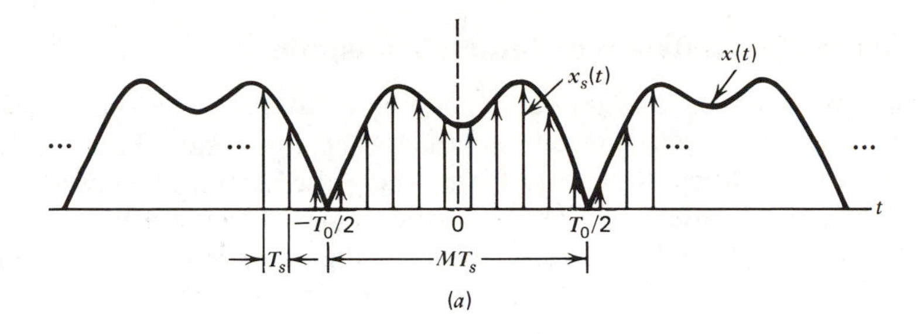

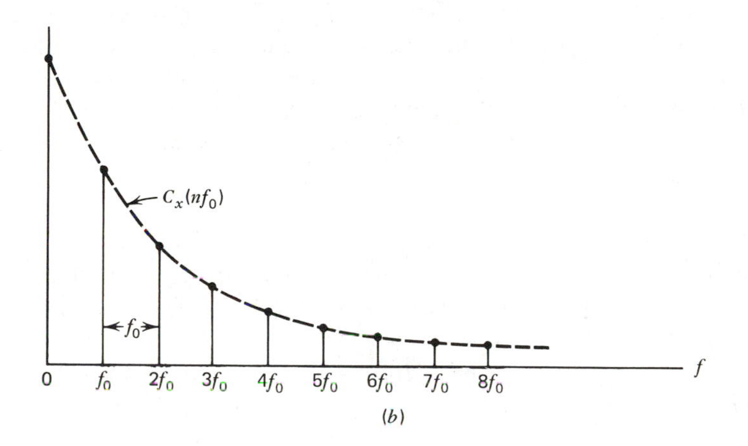

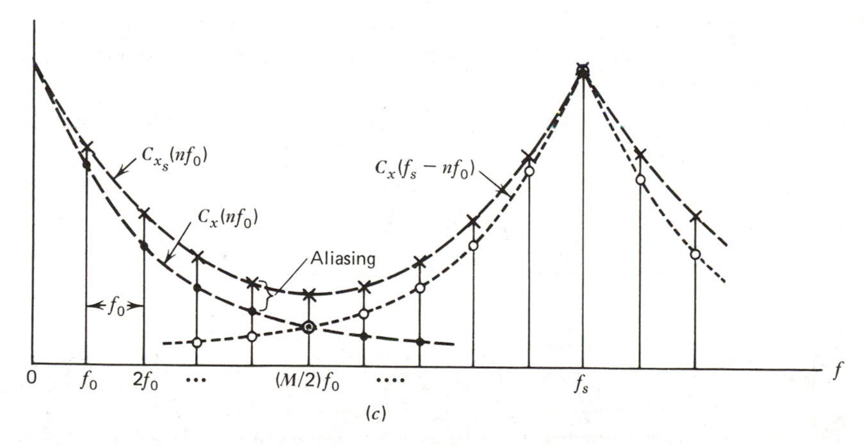

Figure 2.18 Discrete Fourier series coefficients. (*a*) Signal $x(t)$ and its samples. (*b*) Spectrum of $x(t)$. (*c*) Spectrum of $x_s(t)$; $f_s = Mf_0$.

2.7.3 Numerical Computation of Fourier Transforms

Suppose that we want to use numerical methods to compute the Fourier transform $X(f)$ of an energy signal $x(t)$ that lasts for T seconds (Figure 2.19). The first step in this computation is to form a periodic sampled version $x_{s_p}(t)$ of $x(t)$ as shown in Figure 2.19. The periodic sampled version is obtained from samples of $x(t)$ padded with zeros. Now, if the sampling rate $f_s \to \infty$ and $T_0 \to \infty$, then the transform of $x_{s_p}(t) \to X(f)$, except for a constant of proportionality.

Since $x_{s_p}(t)$ is periodic, its Fourier transform may be obtained from its Fourier series representation. The coefficients of the Fourier series representation for $x_{s_p}(t)$ can be computed using Equation (2.61). If we take T_0 to be large in comparison to T, then the Fourier series coefficients of $x_{s_p}(t)$ give us a good approximation to the sampled version of the Fourier transform $X(f)$ of $x(t)$. In using this approach the reader should be aware of the following limitations:

1. Aliasing errors may be present.
2. If T_0 is the period of the extended periodic version of $x(t)$, then the frequency resolution of the approximation will be $1/T_0$. The spectral components of $x(t)$ cannot be resolved any finer.
3. With M sample points, the highest frequency for which $X(f)$ can be determined is given by $f_{\max} \leq M/(2T_0)$.

This method of computing the Fourier transform can also be used to calculate the power spectral density of a power signal by transforming its autocorrelation function.

The Fourier transform obtained via Equation (2.61) is called the *discrete*

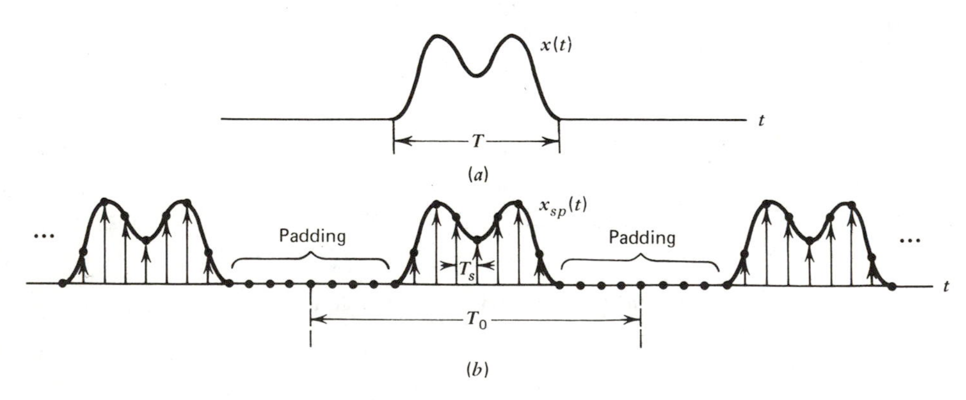

Figure 2.19 (*a*) Aperiodic energy signal $x(t)$. (*b*) Periodic sampled version of $x(t)$.

Fourier transform. Computation of the coefficients defined by Equation (2.61) requires M^2 multiplications and additions. When M is large, the storage requirements and processing time required to find the discrete Fourier transform may become excessive.

Recent advances in digital signal processing techniques have resulted in a class of fast algorithms known as the *Fast Fourier Transform* (*FFT*), which offers significant reductions in computer time. This algorithm is particularly efficient when $M = 2^r$, $r =$ any integer. The number of additions and multiplications required to compute the discrete Fourier transform using the FFT algorithm is of the order of $M \log_2 M$, which will be considerably less than the M^2 operations required to compute the discrete Fourier transform directly. The limitation of space prevents us from going into a detailed discussion of the FFT algorithm. The interested reader is referred to Brigham's book [6] or the IEEE press book on digital signal processing [7].

2.8 SUMMARY

We have considered time domain and frequency domain models for signals and systems. Since signal location in the frequency domain and signal bandwidth are important parameters of communication systems, frequency domain models were emphasized. The type of model used to represent a signal in the frequency domain depends on whether the signal is a periodic power signal, an aperiodic energy signal, or an aperiodic power signal. Regardless of the model used, a signal can be described in the frequency domain in terms of its spectral content. A summary of the important aspects of various signal models is given in Table 2.1.

The response of systems can also be modeled in the frequency domain using the transfer functions of the system and appropriate frequency domain signal models. Many time domain operations such as convolution, modulation, and sampling have simple frequency domain models.

The effect of a linear system on an input signal may be viewed as a filtering operation—modification of the spectral components of the input signal. Ideal models of filters are often used to approximate practical systems.

The spectral characteristics of a signal may be obtained experimentally using either a spectrum analyzer or by processing the sampled values of the signal on the digital computer. If the sampling rate is high and if the signal was observed for a long time, the discrete Fourier series and the discrete Fourier transform provide good approximations of the true spectrum.

In our analysis of communication systems, we will model the message signal as an arbitrary waveform $x(t)$ that is assumed to be bandlimited to f_x.

Table 2.1. Summary of Signal Models

Type of Signal	Applicable Models	Details of the Model[a]	Response of a Linear System[a]
Periodic power signal	Fourier series, amplitude, and phase spectra, power spectral density[b] (psd)	$x(t) = \sum_{n=-\infty}^{\infty} C_x(nf_0) \exp(j2\pi n f_0 t), \quad f_0 = 1/T_0$	$y(t) = \sum_{n=-\infty}^{\infty} C_y(nf_0) \exp(j2\pi n f_0 t)$
		$C_x(nf_0) = \frac{1}{T_0}\int_{-T_0/2}^{T_0/2} x(t) \exp(-j2\pi n f_0 t)\, dt$	$C_y(nf_0) = H(nf_0)C_x(nf_0)$
		$\text{psd} = G_x(f) = \sum_{n=-\infty}^{\infty} \lvert C_x(nf_0)\rvert^2 \delta(f - nf_0)$	$G_y(f) = \lvert H(f)\rvert^2 G_x(f)$
		$\frac{1}{T_0}\int_0^{T_0} \lvert x(t)\rvert^2\, dt = \sum_{n=-\infty}^{\infty} \lvert C_x(nf_0)\rvert^2$	
Aperiodic energy signal	Fourier transform, amplitude, and phase spectra, energy spectral density[c]	$x(t) = \int_{-\infty}^{\infty} X(f) \exp(j2\pi f t)\, df$	$y(t) = \int_{-\infty}^{\infty} Y(f) \exp(j2\pi f t)\, df$
		$X(f) = \int_{-\infty}^{\infty} x(t) \exp(-j2\pi f t)\, dt$	$Y(f) = H(f)X(f)$
		$\int_{-\infty}^{\infty} \lvert x(t)\rvert^2\, dt = \int_{-\infty}^{\infty} \lvert X(f)\rvert^2\, df$	
Aperiodic power signal	Power spectral density, autocorrelation function	$\text{psd} = G_x(f) = \lim_{T\to\infty} \frac{\lvert X_T(f)\rvert^2}{T} = F\{R_{xx}(\tau)\}$	
		$X_T(f) = \int_{-T/2}^{T/2} x(t) \exp(-j2\pi f t)\, dt$	
		$R_{xx}(\tau) = \lim_{T\to\infty} \frac{1}{T}\int_{-T/2}^{T/2} x(t)x(t+\tau)\, dt$	$G_y(f) = \lvert H(f)\rvert^2 G_x(f)$

[a] $x(t)$—input signal; $y(t)$—output signal; $h(t)$—impulse response; $H(f)$—transfer function. $G_x(f)$ and $G_y(f)$ are power spectral densities of $x(t)$ and $y(t)$.
[b] Spectra are defined only for discrete values of frequency.
[c] Spectra are defined for all values of frequencies.

That is,

$$\left.\begin{array}{c} G_x(f) = 0 \\ \text{or} \\ X(f) = 0 \end{array}\right\} \text{for } |f| > f_x$$

The Fourier transform is appropriate for energy signals and the power spectral density is appropriate for power signals (random or deterministic).

Occasionally, an analysis with arbitrary $x(t)$ will prove to be difficult if not impossible. For such situations we will assume the message signal $x(t)$ to be a simple sinusoid (or tone)

$$x(t) = A_m \cos 2\pi f_m t, \quad f_m < f_x$$

While this might seem to be an unduly simplified approach, it has definite advantages. In a complex system tones are often the only signals that are traceable; they facilitate the calculation of spectra, bandwidths, signal power, and so forth, that would otherwise be impossible to trace. Furthermore, if we can find the response of a communication system to a specific frequency in the message band, then we can infer the response for all frequencies in the message band.

REFERENCES

Basic treatments of Fourier theory and spectral analysis may be found in a number of texts. Introductory coverage of Fourier series, Fourier transforms, impulse response, and transfer functions is found in most undergraduate circuit-analysis texts; see, for example, the books by Cruz and Valkenberg (1974) and Cooper and McGillem (1974). An excellent treatment of signal and system models from a mathematical viewpoint may be found in Papoulis (1977). Spectral analysis as applied to communication systems is treated well in the book by Panter (1965). The last two books are written at a more advanced level than this text.

Spectral measurement techniques using analog and digital methods are covered well in the book by Bendat and Piersol (1971). Brigham's book (1974) and the IEEE Press book on digital signal processing deal with discrete Fourier transforms in great detail.

1. J. B. Cruz and M. E. Van Valkenberg. *Signals in Linear Circuits.* Houghton–Mifflin Boston (1974).
2. G. R. Cooper and C. D. McGillem. *Continuous and Discrete Signal and System Analysis.* Holt, Rinehart, and Winston, New York (1974).

3. A. Papoulis. *Signal Analysis.* McGraw–Hill, New York (1977).
4. P. F. Panter. *Modulation, Noise, and Spectral Analysis.* McGraw–Hill, New York (1965).
5. J. S. Bendat and A. G. Piersol. *Random Data: Analysis and Measurement Procedures.* Wiley, New York (1971).
6. E. O. Brigham. *The Fast Fourier Transform.* Prentice-Hall, Englewood Cliffs, N. J. (1974).
7. *Digital Signal Processing,* edited by L. R. Rabiner and C. R. Rader. IEEE Press (1973).

PROBLEMS

Section 2.1

2.1. Classify the following signals as energy signals or power signals. Also, find the normalized energy or normalized power of each signal.

(a) $x(t) = \begin{cases} A\exp(-\alpha t), & t > 0,\ \alpha > 0 \\ 0 & \text{elsewhere} \end{cases}$

(b) $x(t) = \cos t + 3\cos 2t, \quad -\infty < t < \infty$

(c) $x(t) = \exp(-2t^2), \ -\infty < t < \infty$

(d) $x(t) = \exp(-\alpha|t|), \ -\infty < t < \infty,\ \alpha > 0$

(e) $x(t) = \exp(-at)\begin{cases} a > 0 \\ -\infty < t < \infty \end{cases}$

2.2. Classify the following signals as periodic or aperiodic. For those that are periodic, find the period (all signals are defined over $-\infty < t < \infty$).

(a) $x(t) = \cos 2t + \cos 5t$

(b) $x(t) = \exp(-2t)\sin 10t$

(c) $x(t) = \cos 2t + \cos 4t$

(d) $x(t) = \exp(2j\pi t)$

2.3. The input–output relationships for five systems are given below. Classify the systems into one or more of the six categories discussed in Section 2.1.2 ($y(t)$ is the output signal and $x(t)$ is the input signal).

(a) $y(t) = 2tx^2(t)$

(b) $y(t) = 5tx(t)$

(c) $y(t) = \int_{-\infty}^{t} x(\tau)h(t-\tau)\,d\tau,\ h(\tau) = 0 \text{ for } \tau < 0$

(d) $y(t) = 4 + 5x(t)$

(e) $y(t) = \int_{-\infty}^{\infty} x^2(\tau)h(t-\tau)\,d\tau$

2.4. Evaluate the following integrals ($u(t)$ is the unit step function defined as $u(t) = 1$ for $t > 0$ and $u(t) = 0$ for $t < 0$):

(a) $\int_{-\infty}^{\infty} \delta(t-2)\cos 2\pi t\,dt$

(b) $\int_{-\infty}^{\infty} (t^2 + 3t)\delta(t-1)\, dt$

(c) $\int_{-\infty}^{\infty} \delta(t-2)\delta(t-1)\, dt$

(d) $\int_{-\infty}^{\infty} \delta(t-2)u(t)\, dt$

(e) $\int_{-\infty}^{\infty} \delta(t-2)u(t-3)\, dt$

Section 2.2

2.5 Define $\phi_n(t) = \exp(j2\pi n f_0 t)$, where $(1/f_0) = t_2 - t_1 > 0$. Show that

$$\frac{1}{t_2 - t_1}\int_{t_1}^{t_2} \phi_n(t)\phi_m^*(t)\, dt = \begin{cases} 1, & m = n \\ 0, & m \neq n \end{cases}$$

2.6. Using the result derived in Problem 2.5, show that

$$\frac{1}{t_2 - t_1}\int_{t_1}^{t_2} |x(t)|^2\, dt = \sum_{n=-\infty}^{\infty} |C_x(nf_0)|^2$$

where $C_x(nf_0)$ are the coefficients of the complex exponential Fourier series representation of $x(t)$ over the interval (t_1, t_2).

2.7. Consider a periodic function $x(t)$ that can be represented by a complex exponential Fourier series. Show that:
(a) If $x(t)$ is real, then $C_x(nf_0) = C_x^*(-nf_0)$.
(b) If $x(t)$ is real and even, then $C_x(nf_0)$ is real.
(c) If $x(t)$ is real and odd, then $C_x(nf_0)$ is imaginary.

2.8. $x(t)$ is a periodic power signal with a period T_0. $y(t)$ is a periodic signal related to $x(t)$ by

$$y(t) = a + bx(\alpha t)$$

Find the relationship between $C_y(nf_0')$ and $C_x(nf_0)$, where f_0' is the fundamental frequency of $y(t)$.

2.9. Find the complex exponential Fourier series representation for the rectangular waveform given in Figure 2.20. Define S_n as the average

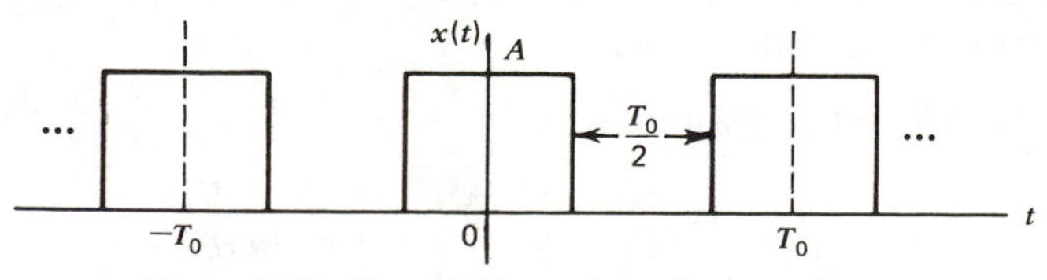

Figure 2.20 Symmetric rectangular waveform.

power content of the first n spectral components, that is,

$$S_n = |C_x(0)|^2 + 2\sum_{k=1}^{n} |C_x(kf_0)|^2$$

Plot (S_n/S_x) versus n, where S_x is the normalized average power of $x(t)$. (Use up to 10 components in your plot.)

2.10. Find the complex exponential Fourier series representation for the triangular waveform shown in Figure 2.21.

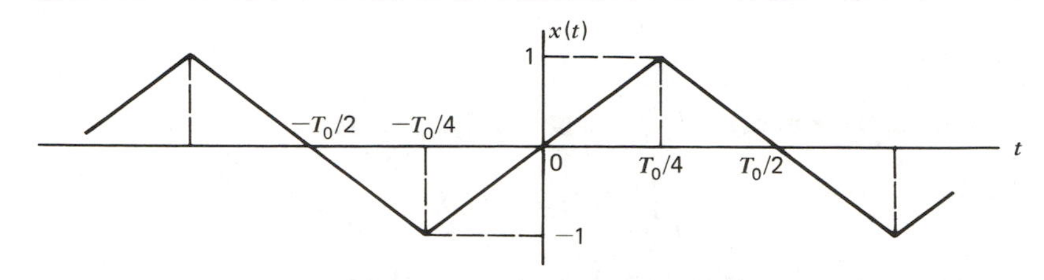

Figure 2.21 Triangular waveform.

2.11. (a) Sketch the amplitude spectrum and the psd of the waveform shown in Figure 2.22. Include at least 10 harmonics in your plot.
(b) Using the power spectral density, find the normalized average power of the signal that is contained in the frequency interval 1.5 MHz to 6.5 MHz.
(c) Plot (S_n/S_x) versus n for $n = 1, 2, 3, \ldots, 10$.

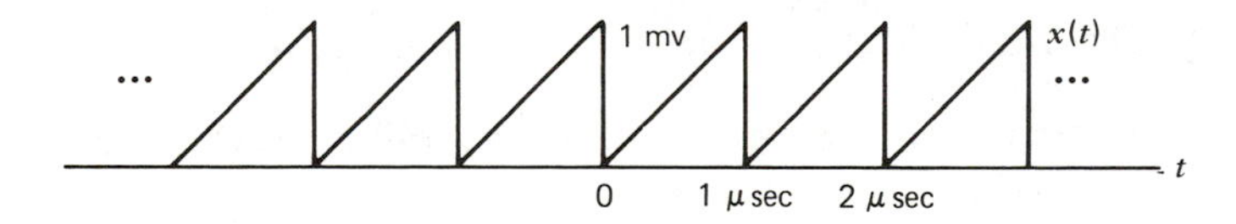

Figure 2.22 Sawtooth waveform.

Section 2.3

2.12. Find the Fourier transform of $x(t) = A\exp(-\pi(t/\tau)^2)$ for $\tau = 4$ and $\tau = 1$ and sketch $|X(f)|^2$ versus f. a) Verify that smaller values of τ spread the spectrum in the frequency domain. b) Calculate the time-bandwidth product using Equation 2.42.

2.13. Find the Fourier transform of

$$x(t) = \begin{cases} \cos(\pi t/\tau), & \text{for } |t| < \tau/2 \\ 0, & \text{elsewhere} \end{cases}$$

Sketch $|X(f)|$ versus f. On the same graph, sketch the amplitude spectrum of a unit amplitude rectangular pulse having a width of τ. Compare the two spectra and indicate which one has more energy in the higher frequencies.

2.14. A rectangular pulse of unit height, unit width, and centered at $t = 0$ has a Fourier transform (Example 2.3)

$$X(f) = \frac{\sin \pi f}{\pi f}$$

(a) Plot the magnitude of $X(f)$ versus f.
(b) Plot, on the same graph, the amplitude spectrum of a rectangular pulse train consisting of unit amplitude and unit width pulses with a repetition rate of one pulse every T seconds ($T \gg 1$).
(c) Based on the plot, comment about the possibility of finding the Fourier transform of an aperiodic function from the Fourier series coefficients of the periodic extension.

2.15. Suppose that $x(t)$ is a real-valued energy signal with a Fourier transform $X(f)$. Show that:
(a) $X(f) = X^*(-f)$.
(b) $\int_{-\infty}^{\infty} x^2(t)\, dt = \int_{-\infty}^{\infty} |X(f)|^2\, df$.
(c) If $x(t)$ is odd, then $X(f) = -2j \int_0^{\infty} x(t) \sin(2\pi f t)\, dt$.
(d) If $x(t)$ is even, then $X(f) = 2 \int_0^{\infty} x(t) \cos(2\pi f t)\, dt$.

2.16. Show that
(a) $\int_{-\infty}^{t} x(\tau)\, d\tau \leftrightarrow \frac{X(f)}{2\pi j f} + \frac{X(0)\delta(f)}{2}$.
(b) $\frac{d}{dt}[x(t)] \leftrightarrow 2\pi j f X(f)$.

2.17. Given the Fourier transform $X(f)$ of $x(t)$, how would you obtain the Fourier transform of $x^2(t)$?

2.18. The differentiation theorem stated in the Problem 2.16 can be generalized and written in the form

$$X(f) = \frac{1}{(2\pi j f)^n} F\left\{\frac{d^n x(t)}{dt^n}\right\}$$

Now, if the nth derivative of $x(t)$ consists entirely of simple functions (impulses and pulses), then the Fourier transform of the nth derivative of $x(t)$ can be easily found and $X(f)$ can be obtained from it. Using this method, find the transform of

$$x(t) = \begin{cases} t, & \text{for } |t| < \tau \\ 0, & \text{elsewhere} \end{cases}$$

2.19. Find the Fourier transform of the triangular pulse shown in Figure 2.23 using the principle stated in Problem 2.18.

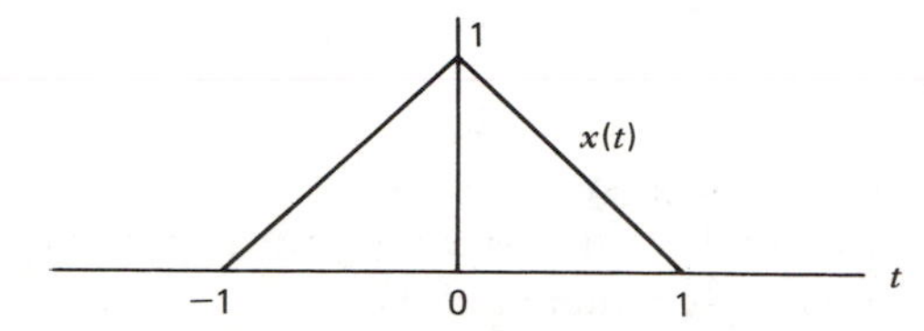

Figure 2.23 A triangular pulse.

2.20. The magnitude of the Fourier transform of a signal is shown in Figure 2.24.
(a) Find the normalized energy content of the signal.
(b) Calculate the frequency f_1 such that one half the normalized energy lies in the range $-f_1$ to f_1.

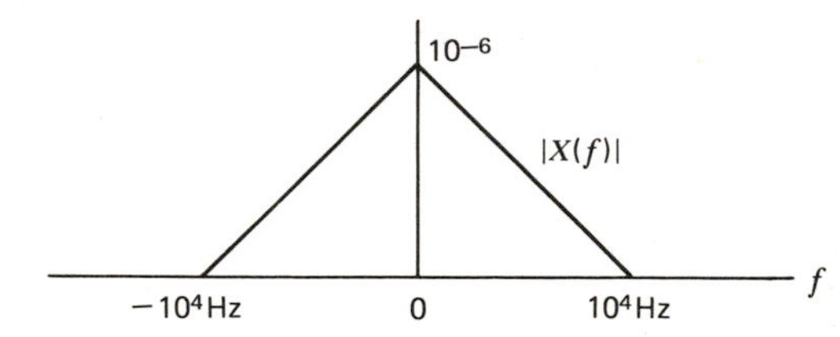

Figure 2.24 Amplitude spectrum for Problem 2.20

2.21. Define $E_x(B)$ as the portion of the normalized energy that lies in the frequency range $-B$ to B Hz, that is, define

$$E_x(B) = \frac{\int_{-B}^{B} |X(f)|^2 \, df}{\int_{-\infty}^{\infty} |X(f)|^2 \, df}$$

Sketch $E_x(B)$ versus B for the signals shown in Figure 2.25.

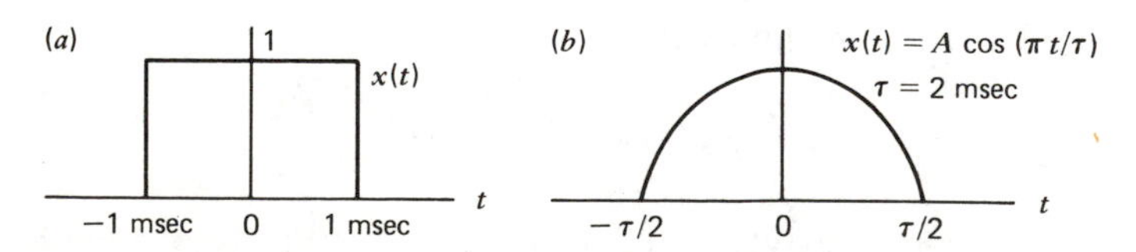

Figure 2.25 Pulse waveforms for Problem 2.21.

2.22. Define the bandwidth of a lowpass signal to be the smallest value of B for which

$$.95 \int_{-\infty}^{\infty} |X(f)|^2 \, df = \int_{-B}^{B} |X(f)|^2 \, df$$

Using this definition, find the bandwidth of the signals shown in Figure 2.25.

2.23. The input to a linear system along with the impulse response of the system are shown in Figure 2.26.
(a) Find the output $y(t)$ by convolution.
(b) Find the Fourier transform of $y(t)$ obtained in (a).
(c) Show that the Fourier transform obtained in (b) is equal to the product of the Fourier transforms of $x(t)$ and $h(t)$.

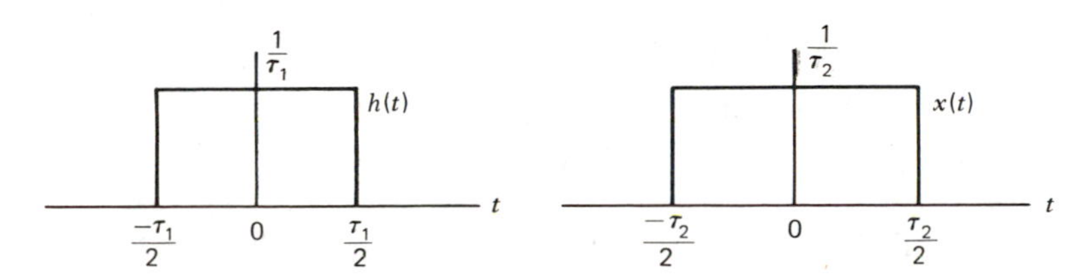

Figure 2.26 Impulse response $h(t)$ and input $x(t)$ for Problem 2.23.

2.24. Using the frequency translation theorem (Equation (2.38)) find the Fourier transform of

$$x(t) = \begin{cases} \cos 2\pi f_c t, & |t| < \tau/2 \\ 0 \text{ elsewhere}, & f_c \gg 1/\tau \end{cases}$$

2.25. The practical sampling scheme shown in Figure 2.27 uses a rectangular sampling function rather than an impulse sampling function. For this sampling scheme find the Fourier transform of the sampled signal and compare it with the result given in Equation (2.40). Assume that $x(t)$ has the spectrum shown in Figure 2.6a, and that $\tau/T_0 \ll 1$, $1/T_0 \gg f_x$.

Figure 2.27 Sampling using a rectangular waveform.

2.26. Find the Fourier transforms of:
(a) $x(t) = 2 \cos 10\pi t$.
(b) $x(t) = 10 \sin 10\pi t + 20 \cos 10\pi t$.
(c) $x(t) = 2 \cos 10\pi t + 10 \sin 20\pi t$.

Section 2.4

2.27. Find the power spectral density and autocorrelation function of

$$x(t) = \cos 10\pi t + \cos 20\pi t$$

Sketch the power spectral density function.

2.28. Establish the following properties of the autocorrelation function $R_{xx}(\tau)$ of a real-valued signal $x(t)$.
(a) $R_{xx}(0) = \int_{-\infty}^{\infty} G_x(f)\, df = \lim_{T\to\infty} \frac{1}{T} \int_{-T/2}^{T/2} x^2(t)\, dt$
(i.e., $R_{xx}(0)$ = normalized average power of $x(t)$).
(b) $R_{xx}(-\tau) = R_{xx}(\tau)$; that is, $R_{xx}(\tau)$ is even.
(c) $|R_{xx}(\tau)| \leq R_{xx}(0)$.
(d) Using the result derived in (b), show that $G_x(f) = F\{R_{xx}(\tau)\}$ is even and real.

2.29. Let $x(t)$ be a periodic power signal and let $C_x(nf_0)$ be the coefficient of the complex exponential Fourier series representation of $x(t)$.
(a) Find the autocorrelation function of $x(t)$.
(b) Find $G_x(nf_0)$ using the result obtained in (a).
(c) Compare the answer to (b) with the psd given in Equation (2.22).

2.30. The power spectral density of a signal is shown in Figure 2.28.
(a) Find the normalized average power of the signal.
(b) Find the amount of power contained in the frequency range 5 to 10 kHz.

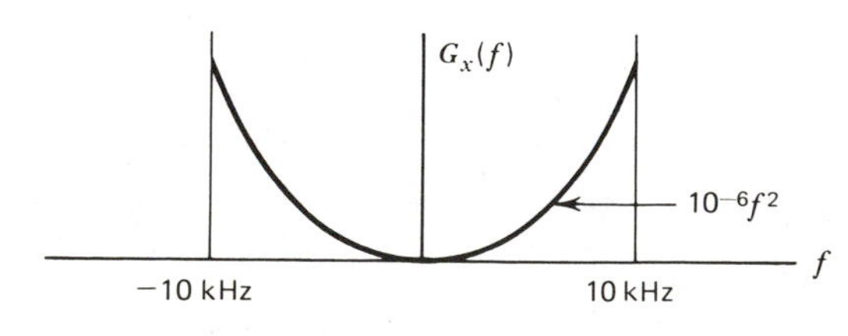

Figure 2.28 Psd for Problem 2.30.

Section 2.5

2.31. For the RC network shown in Figure 2.29, find:
(a) the step response (assume zero initial condition).
(b) the impulse response.
(c) Verify that the impulse response is the time derivative of the step response.

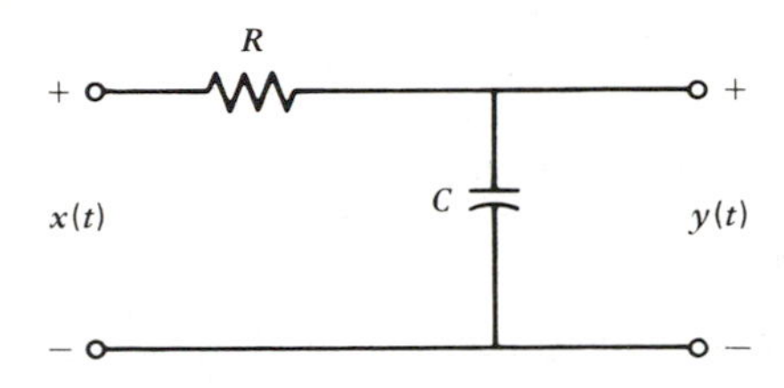

Figure 2.29 RC network for Problem 2.31.

2.32. Assume that the input to the RC network shown in Figure 2.29 is a single rectangular pulse of width 1 ms and amplitude 1 volt. Assume that the 3-dB (half-power) frequency of the network is 1 kHz.
(a) Find the normalized energy content of the input signal.
(b) Find the energy spectral density of the output.
(c) Find the normalized energy content of the output signal $y(t)$.

2.33. The input to the RC network shown in Figure 2.29 is

$$x(t) = 1 + 2\cos 2\pi f_0 t + .5 \sin 2\pi f_0 t$$

Assume that the 3-dB frequency of the network is $f_c = 2f_0$.
(a) Find $G_x(f)$.
(b) Find $G_y(f)$.
(c) Find the normalized average power content of $y(t)$.

2.34. For the network shown in Figure 2.29 assume that the 3-dB frequency is 1000 Hz and that the input voltage is a symmetric square wave with a period of 1 ms and a peak to peak amplitude of 10 volts.
(a) Calculate $G_x(f)$.
(b) Calculate $G_y(f)$.

2.35. A signal with a psd shown in Figure 2.28 is applied to an ideal lowpass filter with a bandwidth of 5 kHz. Find the psd of the filter output and the average normalized power content of the output signal.

2.36. Repeat Problem 2.35 with a highpass filter with a cutoff frequency of 5 kHz.

2.37. A symmetrical square wave of zero DC value, peak to peak amplitude of 10 volts, and period $10^{-4}/3$ is applied to an ideal lowpass filter with a transfer function

$$|H(f)| = \begin{cases} \frac{1}{2} & \text{for } |f| < 100 \text{ kHz} \\ 0 & \text{elsewhere} \end{cases}$$

(a) Plot the output psd.
(b) What is the normalized average power of the filter output?

Section 2.6

2.38. Consider the system shown in Figure 2.30. Show that

$$y(t) \approx kx(t)$$

where k is a scale factor, when the filter is ideal lowpass with a cut-off frequency f_x (i.e., show that it is possible to reconstruct a signal $x(t)$ from its sampled version by lowpass filtering).

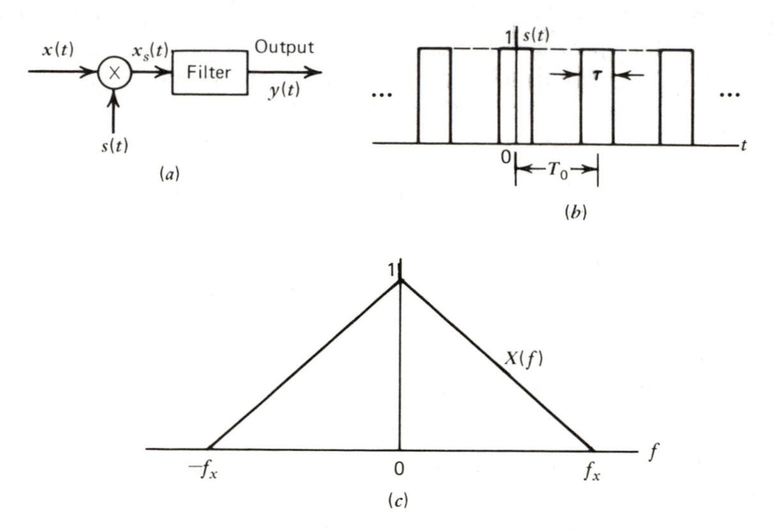

Figure 2.30 Signal and System models for Problems 2.38 and 2.39. $\tau/T_0 \ll 1$, and $1/T_0 \gg f_x$.

2.39. In Figure 2.30 assume that the sampling function consists of a sequence of impulses with a period T_0 and that the filter is ideal bandpass with a center frequency f_0 and a bandwidth of $2f_x$. Show that

$$y(t) \approx kx(t) \cos 2\pi f_0 t, \quad f_0 = 1/T_0$$

where k is a scale factor.

2.40. The signal $x(t) = \cos 200\pi t + 0.2 \cos 700\pi t$ is sampled (ideally) at a rate of 400 samples per second. The sampled waveform is then passed through an ideal lowpass filter with a bandwidth of 200 Hz. Write an expression for the filter output.

2.41. A signal $x(t)$ with a Fourier transform $X(f)$,

$$X(f) = \begin{cases} 10^{-5} & \text{for } |f| < 15 \text{ kHz} \\ 0, & \text{elsewhere} \end{cases}$$

is passed through a nonlinearity whose output $y(t)$ is given by

$$y(t) = x(t) + 2x^2(t)$$

Find $Y(f)$ and sketch it.

2.42. Let $y(t) = [1 + x(t)] \cos 2\pi f_0 t$, where $x(t)$ is a lowpass signal with a bandwidth f_x having a Fourier transform $X(f)$ shown in Figure 2.30*c*. Assuming $f_0 \gg f_x$, sketch the transform of $y(t)$.

2.43. Assume that $y(t)$ described in Problem 2.42 is such that

$$|x(t)| \ll 1$$

$y(t)$ is passed through a nonlinearity whose output $z(t)$ is given by

$$z(t) = 2y^2(t)$$

(a) Find $Z(f)$ (assume that $x^2(t) \ll |x(t)|$).
(b) If $z(t)$ is passed through an ideal lowpass filter with a bandwidth of f_x Hz, find the transform of the filter output.

Section 2.7

2.44. Write a computer program* to calculate the discrete exponential Fourier series (Equation (2.61)) of the waveform shown in Figure 2.20. Sketch the amplitude spectrum. (Use 64 sample points/period.) Compare the coefficient values with the values obtained in Problem 2.9.

2.45. Using a computer program*, calculate the discrete Fourier transform of a unit amplitude rectangular pulse of duration one second. Take 64 samples of the pulse and pad it with 96 zero-valued samples on either side so that a Fast Fourier Transform algorithm can be used. Sketch the amplitude of the transform and compare with the answer to Problem 2.44.

*Fast Fourier Transform programs written in Fortran may be found in References 6 and 7 listed on page 58.

3

RANDOM SIGNAL THEORY

3.1 INTRODUCTION

In Chapter 2, we developed time domain and frequency domain models for several classes of deterministic signals and systems. In the analysis of communication systems we often encounter random signals—signals whose behaviour cannot be predicted exactly. For deterministic signals we have implicitly assumed that it is possible to write an explicit time function. For random signals it is not possible to write such expressions. The main objective of this chapter is to develop probabilistic models for random signals.

The output of information sources, and noise, are random in nature. For, if the output of the information source is known in advance to the receiver, there is no point in transmitting the message. Thus from a receiving point of view, all meaningful information-bearing signals are random in nature. By way of an example, consider a communication system for transmitting the daily up/down movement of the stock market. For purposes of simplification, let us assume that one of three messages (upward movement, downward movement, or no change) are to be transmitted and that each message is transmitted using a distinct electrical waveform. The receiver knows that one of these three messages or waveforms will be transmitted. However, it does not know in advance which one of the three messages will be transmitted. Thus from a receiving point of view the information source *randomly* selects for transmission one of three messages according to some *probabilities*. Correspondingly, the signal waveforms transmitted over the channel are chosen from a collection of three waveforms. Once the message is selected, the waveform shape is completely specified.

Thus, all information carrying functions (messages) in the form of fluctuating quantities such as voltages, currents, temperatures, and pressures are characterized by not being subjected to precise prediction. Similarly, thermal noise due to random motion of electrons in a conducting media, shot noise, and other forms of disturbances in the transmission media have the same characteristics. Such random signals cannot be expressed as explicit functions of time. However, when examined over a long period, a random signal may exhibit certain regularities that can be described in terms of probabilities and statistical averages. Thus, although we lack an exact description, we may be able to model such a signal in terms of average values and the probability that the random signal will be in a given range at a specific time. Description and analysis of messages and noise using concepts that have been borrowed from the probability theory are very useful in communication theory.

We begin our development of random signal theory with a brief review of the important concepts of probability theory. Definitions of random experiments, outcomes of random experiments, random events, and probabilities of combinations of random events are introduced first. Following these definitions, we introduce the concept of a random variable that maps the space of outcomes of a random experiment to points on the real line. Descriptions of random variables in terms of probability distribution functions and statistical averages are discussed followed by the Chebyshev's inequality which gives bounds on probabilities in terms of statistical averages. We conclude our discussion of random variables by deriving the statistical properties of the output of a memoryless system whose input is a random variable.

The concept of a random process is introduced next as a model for describing a collection of functions of time. The properties of a special class of random process, the ergodic random process, are discussed in detail. The concepts of autocorrelation function and power spectral density are introduced next and the response of linear time-invariant systems to a random input signal is derived. It is shown that the power spectral density function provides a meaningful frequency domain description of random signals. Finally, we develop models for random noise encountered in communication systems.

The material presented in this chapter is intended to serve as a review of probability theory and random processes. Only those aspects of probability theory and random processes used in later chapters are covered. The presentation in this chapter relies heavily on intuitive reasoning rather than mathematical rigor. A bulk of the proofs of statements and theorems are left as exercises for the reader to complete. Those wishing a more detailed treatment of the subject are referred to several well-written texts listed at the end of this chapter.

3.2 INTRODUCTION TO PROBABILITIES

In this section we outline mathematical techniques for dealing with the results of an experiment whose outcomes are not known in advance. Such an experiment is called a random experiment. The mathematical approach used for studying the results of random experiments and random phenomena is called probability theory. We begin our study of probability theory with some basic definitions and axioms.

3.2.1 Definitions

(i) *Outcome*: The end result of an experiment.

(ii) *Random experiment*: An experiment whose outcome is not known in advance.

(iii) *Random event*: A random event A is an outcome or a set (or collection) of outcomes of a random experiment.

(iv) *Sample space*: The sample space S of a random experiment is the set of all possible outcomes of the random experiment.

(v) *Mutually exclusive events*: Events A and B are said to be mutually exclusive if they have no common elements (or outcomes).

(vi) *Union of events*: The union of two events A and B, denoted by $A \cup B$ (sometimes also denoted by "A or B"), is the set of all outcomes which belong to A or B.

(vii) *Intersection of events*: The intersection of two events A and B, denoted by $A \cap B$ (also denoted by "A and B" or AB), is the set of all outcomes which belong to both A and B.

(viii) *Occurrence*: Event A of a random experiment is said to have occurred if the experiment terminates in an outcome that belongs to event A.

3.2.2 Probabilities of Random Events

Using the simple definitions given above, we now proceed to define the probabilities of occurrence of random events. There are several definitions for probabilities of random events. Two of the most popular definitions are the relative frequency definition and the classical definition.

Relative Frequency Definitions. Suppose that a random experiment is repeated n times. If the event A occurs n_A times, then its probability $P(A)$ is defined as the limit of the relative frequency n_A/n of the occurrence of A. That is,

$$P(A) \triangleq \lim_{n\to\infty} \frac{n_A}{n} \tag{3.1}$$

For example, if a coin (fair or not) is tossed n times and heads show up n_h times, then the probability of heads equals the limiting value of n_h/n.

Classical Definition. In this definition, the probability $P(A)$ of an event A is found without experimentation. This is done by counting the total number N of the possible outcomes of the experiment (i.e., the number of outcomes in S). If N_A of these outcomes belong to event A, then $P(A)$ is defined to be

$$P(A) \triangleq \frac{N_A}{N} \tag{3.2}$$

If we use this definition to get the probability of getting a tail when a coin is tossed, we will get an answer of $\frac{1}{2}$. This answer is correct when we have a fair coin. If the coin is not fair, then the classical definition will lead to wrong values for probabilities. We can take this possibility into account and modify the definition as: the probability of an event A consisting of N_A outcomes equals the ratio N_A/N provided the outcomes are *equally likely* to occur.

Axioms of Probability Theory. In order to arrive at a satisfactory theory of probability (a theory that does not depend on the actual definition of probability) we require the probability of an event A to be a number $P(A)$ that obeys the following postulates (or axioms):

1. $P(A) \geq 0$. (3.3a)
2. $P(S) = 1$. (3.3b)
3. If $A \cap B = 0$, then $P(A \cup B) = P(A) + P(B)$. (3.3c)

The reader can verify that the two definitions of probabilities given in the preceding paragraphs indeed satisfy these axioms.

Useful Laws of Probability. Using any of the many definitions of probability that satisfies the axioms given in Equations (3.3a), (3.3b), and (3.3c), we can establish the following relationships:

1. $P(A \cup B) = P(A) + P(B) - P(AB)$ (3.4)

2. If $A_1, A_2, \ldots, A_n$ are n random events such that

$$\begin{aligned} A_i \cap A_j &= 0 \quad \text{for } i \neq j \\ A_1 \cup A_2 \cup \cdots \cup A_n &= S, \end{aligned} \tag{3.5}$$

then

$$\begin{aligned} P(A) = P(A \cap S) &= P[A \cap (A_1 \cup A_2 \cup \cdots \cup A_n)] \\ &= P[(A \cap A_1) \cup (A \cap A_2) \cup \cdots \cup (A \cap A_n)] \\ &= P(A \cap A_1) + P(A \cap A_2) + \cdots + P(A \cap A_n) \end{aligned} \tag{3.6}$$

The sets $A_1, A_2, \ldots, A_n$ are said to be *mutually exclusive and exhaustive* if Equation (3.5) is satisfied.

4. If $A \cup \bar{A} = S$ and $A \cap \bar{A} = 0$, then $\bar{A}$ is called the *complement* of A and

$$P(\bar{A}) = 1 - P(A) \tag{3.7}$$

3.2.3 Joint and Conditional Probabilities

In many engineering applications, including communication systems, we often perform experiments that consist of many subexperiments. A simple example is the simultaneous observation of the input and output digits of a binary communication system. Suppose we have a random experiment E consisting of two subexperiments E_1 and E_2. Then the sample space of the experiment E consists of outcomes of the subexperiments E_1 and E_2. If events $A_1, A_2, \ldots, A_n$ are defined for the first subexperiment E_1 and the events $B_1, B_2, \ldots, B_m$ are defined for the second subexperiment E_2, then event A_iB_j is an event of the total experiment.

Joint Probability. The probability of an event such as A_iB_j that is the intersection of events from subexperiments is called the joint probability of the event and is denoted by $P(A_iB_j)$.

Marginal Probability. If the events $A_1, A_2, \ldots, A_n$ associated with subexperiment E_1 are mutually exclusive and exhaustive, then

$$\begin{aligned} P(B_j) = P(B_jS) &= P[B_j(A_1 \cup A_2 \cup \cdots \cup A_n)] \\ &= \sum_{i=1}^{n} P(A_iB_j) \end{aligned} \tag{3.8}$$

Since B_j is an event associated with subexperiment E_2, $P(B_j)$ is called a marginal probability.

Conditional Probability. Quite often, the probability of occurrence of event B_j may depend on the occurrence of a related event A_i. For example, imagine a box containing six resistors and one capacitor. Suppose we draw a component from the box. Then, without replacing the first component, we draw a second component. Now, the probability of getting a capacitor on the second draw depends on the outcome of the first draw. For, if we had drawn a capacitor on the first draw, then the probability of getting a capacitor on the second draw is zero since there is no capacitor left in the box! Thus we have a situation where the occurrence of event B_j (getting a capacitor on the second draw) on the second subexperiment is conditional on the occurrence of event A_i (the component drawn first) on the first subexperiment. We denote the probability of event B_j occurring given that event A_i is known to have occurred by the *conditional probability* $P(B_j|A_i)$.

An expression for the conditional probability $P(B|A)$ in terms of the joint probability $P(AB)$ and the marginal probabilities $P(A)$, $P(B)$ can be obtained as follows using the classical definition of probability. Let N_A, N_B, and N_{AB} be the number of outcomes belonging to events A, B, and AB, respectively, and let N be the total number of outcomes in the sample space. Then,

$$P(AB) = \frac{N_{AB}}{N}$$

$$P(A) = \frac{N_A}{N} \tag{3.9}$$

Given that the event A has occurred, we know that the outcome is in A. There are N_A outcomes in A. Now, for B to occur given that A has occurred, the outcome should belong to A and B. There are N_{AB} outcomes in AB. Thus the probability of occurrence of B given A has occurred is

$$P(B|A) = \frac{N_{AB}}{N_A} = \frac{N_{AB}/N}{N_A/N} = \frac{P(AB)}{P(A)} \tag{3.10}$$

The implicit assumption here is that $P(A) \neq 0$.

Relationships Involving Joint, Marginal, and Conditional Probabilities. The reader could use the results given in Equations (3.9) and (3.10) to establish the following useful relationships.

1. $P(AB) = P(A|B)P(B) = P(B|A)P(A)$ (3.11)
2. $P(A|B) = \dfrac{P(B|A)P(A)}{P(B)}$ (*Bayes Rule*; $P(B) \neq 0$) (3.12)
3. $P(ABC) = P(A)P(B|A)P(C|AB)$ (*Chain Rule*) (3.13)
4. If $B_1, B_2, \ldots, B_m$ are a set of mutually exclusive and exhaustive set of events, then

$$P(A) = \sum_{j=1}^{m} P(A|B_j)P(B_j) \tag{3.14}$$

and

$$P(B_j|A) = \frac{P(A|B_j)P(B_j)}{\sum_{j=1}^{m} P(A|B_j)P(B_j)} \tag{3.15}$$

Statistical Independence. Suppose that A_i and B_j are events associated with the outcomes of two experiments. Suppose that event B_j is independent of A_i so that the occurrence of A_i does not influence the occurrence of B_j and vice versa. Then we say that the events are statistically independent. More precisely, we say that two events A_i and B_j are *statistically independent* if

$$P(A_iB_j) = P(A_i)P(B_j) \tag{3.16}$$

or when

$$P(A_i|B_j) = P(A_i) \tag{3.17}$$

Observe that statistical independence is quite different from mutual exclusiveness. Indeed, if A_i and B_j are mutually exclusive, then $P(A_iB_j) = 0$ by definition.

3.3 DISCRETE RANDOM VARIABLES

It is often useful to describe the outcome of an experiment by a number (for example, the value of the output signal in a communication system, the number of telephone calls arriving at a central switching station, or the time of failure of a component in a system). The rule or *functional relationship*, which assigns real numbers $X(\lambda)$ to each possible outcome λ of a random experiment, is called a *random variable.* The numbers $X(\lambda)$ are called values of the random variable. We will use the notation X to denote the random variable and $x = X(\lambda)$ to denote a particular value of the random variable.

A random variable may be discrete or continuous. A *discrete random variable* can take on only a countable number of distinct values. A *continuous random variable* can assume any value within one or more intervals on the real line. Examples of discrete random variables are the number of telephone calls arriving at an office in a finite interval of time, or a student's numerical score on an examination. The exact time of arrival of a telephone call is an example of a continuous random variable.

3.3.1 Probability Mass Functions

A discrete random variable X is characterized by a set of allowed values $x_1, x_2, \ldots, x_n$ and the probabilities of the random variable taking on one of these values based on the outcome of the underlying random experiment. The probability that $X = x_i = X(\lambda_i)$ is denoted by $P(X = x_i)$ and is called the *probability mass function.* We will assume that $P(X = x_i) > 0$ for $i = 1, 2, \ldots, n$.

The probability mass function of a random variable has the following important properties:

1. $\sum_{i=1}^{n} P(X = x_i) = 1$ (3.18)

2. $P(X \leq x) = \sum_{\text{all } x_i \leq x} P(X = x_i)$ (3.19)

$P(X \leq x)$ is called the *cumulative probability distribution function* of X and is denoted by $F_X(x)$.

It is of course possible to define two or more random variables on the sample space of a single random experiment or on the combined sample spaces of many random experiments. If these variables are all discrete, then they are characterized by a joint probability mass function. Let us take the example of two random variables X and Y that take on the values $x_1, x_2, \ldots, x_n$ and $y_1, y_2, \ldots, y_m$. These two variables can be characterized by a joint probability mass function $P(X = x_i, Y = y_j)$, which gives the probability that $X = x_i$ and $Y = y_j$.

Using the results stated in the preceding sections, we can derive the following relationships involving joint, marginal, and conditional probability mass functions:

1. $P(X \leq x, Y \leq y) = \sum_{x_i \leq x} \sum_{y_j \leq y} P(X = x_i, Y = y_j)$ (3.20)

2. $$P(X = x_i) = \sum_{j=1}^{m} P(X = x_i, Y = y_j) = \sum_{j=1}^{m} P(X = x_i | Y = y_j) P(Y = y_j) \quad (3.21)$$

3. $P(X = x_i | Y = y_j) = \dfrac{P(X = x_i, Y = y_j)}{P(Y = y_j)}$ (assume $P(Y = y_j) \neq 0$) (3.22)

$$= \frac{P(Y = y_j | X = x_i) P(X = x_i)}{\sum_{i=1}^{n} P(Y = y_j | X = x_i) P(X = x_i)} \quad \text{(Bayes Rule)} \quad (3.23)$$

4. Random variables X and Y are *statistically independent* if

$$P(X = x_i, Y = y_j) = P(X = x_i) P(Y = y_j) \quad (i = 1, 2, \ldots, n; j = 1, 2, \ldots, m) \quad (3.24)$$

3.3.2 Statistical Averages

The probability mass function provides as complete a description as possible for a discrete random variable. For many purposes this description is often too detailed. It is sometimes simpler and more convenient to describe a random variable by a few characteristic *numbers* that are representative of its probability mass function. These numbers are the various statistical averages or expected values. The *expected value* or *statistical average* of a function $g(X)$ of a discrete random variable X is defined as

$$E\{g(X)\} \triangleq \sum_{i=1}^{n} g(x_i)P(X = x_i) \tag{3.25}$$

Two statistical averages that are most commonly used for characterizing a random variable X are its *mean* μ_X and its *variance* σ_X^2. The mean and variance are defined as

$$E\{X\} = \mu_X = \sum_{i=1}^{n} x_i P(X = x_i) \tag{3.26}$$

$$E\{[X - \mu_X]^2\} = \sigma_X^2 = \sum_{i=1}^{n} (x_i - \mu_X)^2 P(X = x_i) \tag{3.27}$$

The square-root of variance is called the *standard deviation.* When the probability mass function is not known, then the mean and variance can be used to arrive at bounds on probabilities via the *Chebyshev's inequality,* which is stated as

$$P\{|X - \mu_X| > K\} \leq \sigma_X^2/K^2 \tag{3.28}$$

The mean of the random variable is its average value and the variance of the random variable is a measure of the "spread" of the values of the random variable. The Chebyshev's inequality can be used to obtain bounds on the probability of finding X outside of an interval $\mu_X \pm K$.

The expected value of a function of two random variables is defined as

$$E\{g(X, Y)\} = \sum_{i=1}^{n} \sum_{j=1}^{m} g(x_i, y_j)P(X = x_i, Y = y_j) \tag{3.29}$$

A useful statistical average that gives a measure of dependence between two random variables X and Y is the *correlation coefficient* defined as

$$\rho_{XY} = \frac{E\{(X - \mu_X)(Y - \mu_Y)\}}{\sigma_X \sigma_Y} \tag{3.30}$$

The numerator of the right-hand side of Equation (3.30) is called the *covariance* of X and Y. The reader can verify that if X and Y are statistically independent, then $\rho_{XY} = 0$ and that in the case when X and Y are linearly dependent (i.e., when $Y = kX$), then $|\rho_{XY}| = 1$. Observe that $\rho_{XY} = 0$ does not imply statistical independence.

3.3.3 Examples of Probability Mass Functions

The probability mass functions of some random variables have nice analytical forms. Two examples are presented below. We will encounter these probability mass functions very often in our analysis of digital communication systems.

The Uniform Probability Mass Function. A random variable X is said to have a uniform probability mass function (or distribution) when

$$P(X = x_i) = 1/n, \quad i = 1, 2, 3, \ldots, n \tag{3.31}$$

The Binomial Probability Mass Function. Let p be the probability of an event A of a random experiment E. If the experiment is repeated n times, then the probability that event A occurs k times is given by the binomial probability mass function defined as

$$P(X = k) = \binom{n}{k} p^k (1 - p)^{n-k}, \quad k = 0, 1, 2, \ldots, n \tag{3.32}$$

where

$$\binom{n}{k} = \frac{n!}{k!(n-k)!}, \quad \text{and} \quad m! = m(m-1)(m-2)\cdots(3)(2)(1)$$

The reader can verify that the mean and variance of the binomial random variable are given by (see Problem 3.13)

$$\mu_X = np \tag{3.33}$$

$$\sigma_X^2 = np(1-p) \tag{3.34}$$

Before we proceed to describe continuous random variables, let us look at two examples that illustrate the concepts described in the preceding sections.

Example 3.1. The input to a binary communication system, denoted by a random variable X, takes on one of two values 0 or 1 with probabilities $\frac{3}{4}$ and $\frac{1}{4}$, respectively. Due to errors caused by noise in the system, the output Y differs from the input X occasionally. The behavior of the communication system is modeled by the conditional probabilities

$$P(Y = 1 | X = 1) = \tfrac{3}{4} \quad \text{and} \quad P(Y = 0 | X = 0) = \tfrac{7}{8}$$

(a) Find $P(Y = 1)$ and $P(Y = 0)$. (b) Find $P(X = 1 | Y = 1)$.

Solution

(a) Using Equation (3.21), we have

$$P(Y=1)=P(Y=1|X=0)P(X=0)+P(Y=1|X=1)P(X=1)$$
$$=(1-\tfrac{7}{8})(\tfrac{3}{4})+(\tfrac{3}{4})(\tfrac{1}{4})=\tfrac{9}{32}$$
$$P(Y=0)=1-P(Y=1)=\tfrac{23}{32}$$

(b) Using Bayes Rule, we obtain

$$P(X=1|Y=1)=\frac{P(Y=1|X=1)P(X=1)}{P(Y=1)}$$
$$=\frac{(\frac{1}{4})\frac{3}{4}}{\frac{9}{32}}=\frac{2}{3}$$

($P(X=1|Y=1)$ is the probability that the input to the system is indeed 1 when the output is 1.)

Example 3.2. Binary data are transmitted over a noisy communication channel in blocks of 16 binary digits. The probability that a received binary digit is in error due to channel noise is 0.1. Assume that the occurrence of an error in a particular digit does not influence the probability of occurrence of an error in any other digit within the block (i.e., errors occur in various digit positions within a block in a statistically independent fashion).

(a) Find the average (or expected) number of errors per block.
(b) Find the variance of the number of errors per block.
(c) Find the probability that the number of errors per block is greater than or equal to 5.

Solution

(a) Let X be the random variable representing the number of errors per block. Then, X has a binomial distribution

$$P(X=k)=\binom{16}{k}(.1)^k(.9)^{16-k}, \quad k=0,1,\ldots,16$$
$$E\{X\}=np=(16)(.1)=1.6$$

(b) Variance of $X=\sigma_X^2=np(1-p)=(16)(.1)(.9)=1.44$.

(c) $P(X\geq 5)=1-P(X\leq 4)=.0170$. We can obtain a bound on $P(X\geq 5)$ using the Chebyshev's inequality as follows:

$$P(X\geq 5)=P(|X-\mu_X|>3.3)$$
$$P(|X-\mu_X|>3.3)\leq\frac{1.44}{(3.3)^2}=0.13$$

Hence

$$P(X \geq 5) \leq 0.13$$

The Chebyshev's bound is not a tight bound as illustrated by this example. But remember that this bound is valid for *all* distributions.

3.4 CONTINUOUS RANDOM VARIABLES

3.4.1 Probability Density Functions and Statistical Averages

A continuous random variable can take on more than a countable number of values in one or more intervals on the real line. A continuous random variable X is characterized by a probability density function $f_X(x)$, which has the following properties:

1. $f_X(x) \geq 0, \quad -\infty < x < \infty$ (3.35)
2. $\int_{-\infty}^{\infty} f_X(x)\,dx = 1$ (3.36)
3. $P(X \leq a) = F_X(a) = \int_{-\infty}^{a} f_X(x)\,dx$ (3.37)

$F_X(x)$ is called the distribution function of X, and

$$\frac{dF_X(x)}{dx} = f_X(x)$$

The probability density function (pdf) plays a role in describing a continuous random variable similar to the role played by the probability mass function in describing discrete random variables.

Some random processes involve two or more continuous random variables. In such cases we use the joint probability density function to describe the variables. For example, if there are two random variables X and Y, they may be characterized by a joint probability density function $f_{X,Y}(x, y)$. From the joint probability density function one can obtain marginal probability density functions $f_X(x)$, $f_Y(y)$, and conditional probability density functions $f_{X|Y}(x|y)$ and $f_{Y|X}(y|x)$ as follows:

$$f_X(x) = \int_{-\infty}^{\infty} f_{X,Y}(x, y)\,dy \tag{3.38}$$

$$f_Y(y) = \int_{-\infty}^{\infty} f_{X,Y}(x, y)\,dx \tag{3.39}$$

$$f_{X|Y}(x|y) = \frac{f_{X,Y}(x, y)}{f_Y(y)}, \quad f_Y(y) > 0 \tag{3.40}$$

$$f_{Y|X}(y|x) = \frac{f_{X,Y}(x, y)}{f_X(x)}, \quad f_X(x) > 0 \tag{3.41}$$

$$f_{Y|X}(y|x) = \frac{f_{X|Y}(x|y)f_Y(y)}{\int_{-\infty}^{\infty} f_{X|Y}(x|\alpha)f_Y(\alpha)\, d\alpha} \quad \text{(Bayes Rule; } f_X(x) \neq 0) \tag{3.42}$$

Finally, random variables X and Y are said to be statistically independent if

$$f_{X,Y}(x, y) = f_X(x)f_Y(y), \quad -\infty < x,\ y < \infty \tag{3.43}$$

As in the case of discrete random variables, continuous random variables can also be described by statistical averages or expected values. The expected values of functions of continuous random variables are defined by

$$E\{g(X, Y)\} = \int_{-\infty}^{\infty} \int_{-\infty}^{\infty} g(x, y)f_{X,Y}(x, y)\, dx\, dy \tag{3.44}$$

$$\mu_X = E\{X\} = \int_{-\infty}^{\infty} xf_X(x)\, dx \tag{3.45}$$

$$\sigma_X^2 = E\{(X - \mu_X)^2\} = \int_{-\infty}^{\infty} (x - \mu_X)^2 f_X(x)\, dx \tag{3.46}$$

and

$$\rho_{XY} = \frac{E\{(X - \mu_X)(Y - \mu_Y)\}}{\sigma_X \sigma_Y} \tag{3.47}$$

The Chebyshev's inequality for a continuous random variable has the same form as given in Equation (3.28).

3.4.2 Examples of Probability Density Functions

We now look at three useful models for continuous random variables that we will be encountering in our later work.

Uniform Probability Density Functions. A random variable X is said to have a uniform pdf if

$$f_X(x) = \begin{cases} 1/(b - a), & a \leq x \leq b \\ 0, & \text{elsewhere} \end{cases} \tag{3.48}$$

The mean and variance of a uniform random variable can be shown to be

(Problem 3.18)

$$\mu_X = \frac{b+a}{2} \tag{3.49}$$

$$\sigma_X^2 = \frac{(b-a)^2}{12} \tag{3.50}$$

Gaussian Probability Density Function. One of the most widely used pdf's is the Gaussian or *normal* probability density function. This pdf occurs in so many applications because of a remarkable phenomenon called the *central limit theorem*. The central limit theorem implies that a random variable which is determined by a large number of independent causes tends to have a Gaussian probability distribution. Several versions of this theorem have been proven by statisticians and verified experimentally from data by engineers and physicists.

Our interest in studying about the Gaussian pdf is from the viewpoint of using it to model random electrical noise. Electrical noise in communication systems is often due to the cumulative effects of a large number of randomly moving charged particles and hence the instantaneous value of the noise will have a Gaussian distribution—a fact that can be verified experimentally. (The reader is cautioned at this point that there are examples of noise that cannot be modeled by Gaussian pdf's. Such examples include pulse-type disturbances on a telephone line and the electrical noise from nearby lightning discharges.)

The Gaussian pdf has the form (see Figure 3.1)

$$f_X(x) = \frac{1}{\sqrt{2\pi\sigma_X^2}} \exp\left(-\frac{(x-\mu_X)^2}{2\sigma_X^2}\right), \quad -\infty < x < \infty \tag{3.51}$$

where $\exp(y)$ denotes e^y. The family of Gaussian pdf's is characterized by two parameters μ_X and σ_X^2, which are the mean and variance of the random variable X. In our studies on the effect of Gaussian noise on digital signal transmission, we will often be interested in probabilities such as

$$P(X > a) = \int_a^\infty \frac{1}{\sqrt{2\pi\sigma_X^2}} \exp\left(-\frac{(x-\mu_X)^2}{2\sigma_X^2}\right) dx$$

By making a change of variable $z = (x - \mu_X)/\sigma_X$, the preceding integral can be reduced to

$$P(X > a) = \int_{(a-\mu_X)/\sigma_X}^\infty \frac{1}{\sqrt{2\pi}} \exp(-z^2/2)\, dz$$

Unfortunately, this integral cannot be solved in closed form and requires numerical evaluation. Several versions of the integral are well tabulated and

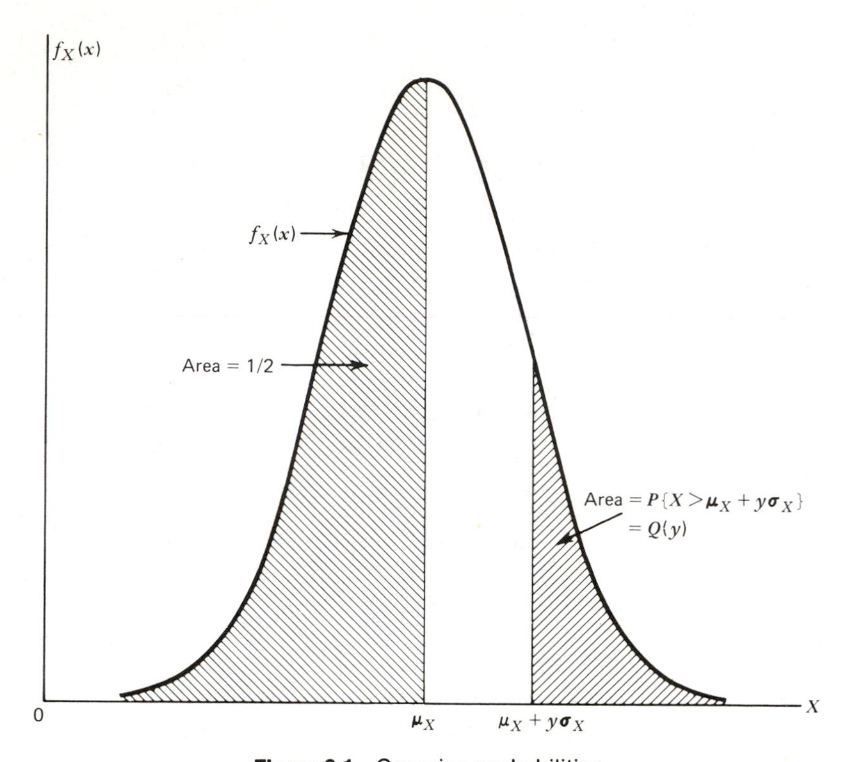

Figure 3.1 Gaussian probabilities.

we will use tabulated values (Appendix D) of the *Marcum Q function*, which is defined as

$$Q(y) = \frac{1}{\sqrt{2\pi}} \int_y^\infty \exp(-z^2/2)\, dz;\ y > 0 \tag{3.52}$$

In terms of the values of the Q function we can write $P(X > a)$ as

$$P(X > a) = Q\left(\frac{a - \mu_X}{\sigma_X}\right) \tag{3.53}$$

Bivariate Gaussian pdf. Occasionally we will encounter the situation when the instantaneous amplitude of the input signal to a linear system has a Gaussian pdf and we might be interested in the joint pdf of the amplitude of the input and the output signals. The bivariate Gaussian pdf is a valid model for describing such situations. The bivariate Gaussian pdf has the form

$$f_{X,Y}(x, y) = \frac{1}{2\pi\sigma_X\sigma_Y\sqrt{1-\rho^2}} \exp\left\{\frac{-1}{2(1-\rho^2)}\left[\left(\frac{x-\mu_X}{\sigma_X}\right)^2 + \left(\frac{y-\mu_Y}{\sigma_Y}\right)^2 - \frac{2\rho(x-\mu_X)(y-\mu_Y)}{\sigma_X\sigma_Y}\right]\right\}, \quad -\infty < x, y < -\infty \tag{3.54}$$

The reader can verify that the marginal pdf's of X and Y are Gaussian with means μ_X, μ_Y, and variances σ_X^2, σ_Y^2, respectively, and

$$\rho = \rho_{XY} = \frac{E\{(X-\mu_X)(Y-\mu_Y)\}}{\sigma_X\sigma_Y}$$

(see Problem 3.22).

3.4.3 Transformations of Random Variables

In many engineering problems we will be given a mathematical model of the system of interest and the input signal, and will be asked to find the characteristics of the output signal. As a simple example, consider the problem of finding the pdf of the output Y of a rectifier given that the input X has a Gaussian pdf and that the rectifier transfer characteristic is $Y = X^2$. We give below two theorems (without proof) that will enable us to solve problems of this type. For proofs of these theorems, the reader is referred to Papoulis' book (Chapters 5 and 6).

Theorem 3.1

Let X be a continuous random variable whose pdf $f_X(x)$ is given, and let $Z = g(X)$ be a given transformation of X. To determine the pdf of Z, we solve the equation $z = g(x)$ for x in terms of z. If $x_1, x_2, \ldots, x_n$ are all real solutions of $z = g(x)$, then

$$f_Z(z) = \sum_{i=1}^{n} f_X(x_i)\left|\frac{dx_i}{dz}\right| \tag{3.55}$$

Theorem 3.2

Let X and Y be continuous random variables whose joint pdf $f_{X,Y}(x, y)$ is given. Let $Z = g(X, Y)$ and $W = h(X, Y)$. To determine $f_{Z,W}(z, w)$, we solve the equations $g(x, y) = z$ and $h(x, y) = w$ for x and y in terms of z and w. If $(x_1, y_1), \ldots, (x_n, y_n)$ are all real solutions of these equations (that is, if $g(x_i, y_i) = z$ and $h(x_i, y_i) = w$ for $i = 1, 2, \ldots, n$), then

$$f_{Z,W}(z, w) = \sum_{i=1}^{n} f_{X,Y}(x_i, y_i)|J_i| \tag{3.56a}$$

where J_i is the *Jacobian* of the transformation defined as

$$J_i = \begin{vmatrix} \frac{\partial x_i}{\partial z} & \frac{\partial x_i}{\partial w} \\ \frac{\partial y_i}{\partial z} & \frac{\partial y_i}{\partial w} \end{vmatrix} \tag{3.56b}$$

Let us now look at some examples that illustrate the use of Theorems 3.1 and 3.2.

Example 3.3. X and Y are two independent random variables, each having a Gaussian pdf with a mean of zero and a variance of one.

(a) Find $P(|X|>3)$ using the tabulated values of $Q(y)$, and also obtain an upper bound using the Chebyshev's inequality.
(b) Find the joint pdf of

$$Z = \sqrt{X^2 + Y^2}$$

and

$$W = \tan^{-1}(Y/X)$$

(c) Find $P(Z>3)$.

Solution
(a) $P(|X|>3) = 2P(X>3)$, since the pdf of X is an even function. Hence, from the tabulated values of $Q(y)$, we obtain

$$P(|X|>3) = 2Q(3) = 2(.0013) = .0026$$

Using Chebyshev's inequality, we have

$$P(|X|>3) \leq 1/3^2 = .111$$

(b) Since X and Y are given to be independent, the joint pdf of X and Y is equal to the product of the pdf's of X and Y. Hence,

$$f_{X,Y}(x, y) = \frac{1}{2\pi} \exp(-(x^2 + y^2)/2), \quad -\infty < x, y < \infty$$

Solving for x and y in terms of z and w, we have the unique solution

$$x_1 = z \cos w$$
$$y_1 = z \sin w$$

Hence,

$$J_1 = \begin{vmatrix} \dfrac{\partial x_1}{\partial z} & \dfrac{\partial x_1}{\partial w} \\ \dfrac{\partial y_1}{\partial z} & \dfrac{\partial y_1}{\partial w} \end{vmatrix} = \begin{vmatrix} \cos w & -z \sin w \\ \sin w & z \cos w \end{vmatrix} = z$$

and $|J_1| = z$ since $z > 0$. The joint pdf of Z and W is obtained using Equation (3.56) as

$$f_{Z,W}(z, w) = \frac{z}{2\pi} \exp(-z^2/2), \quad 0 < z < \infty, -\pi < w < \pi$$

The reader can easily verify that

$$f_Z(z) = z \exp(-z^2/2), \qquad f_W(w) = \frac{1}{2\pi}, \quad z > 0, |w| < \pi$$

and $f_{Z,W}(z, w) = f_Z(z) f_W(w)$. Hence, Z and W are independent.

(c) $P(Z > 3) = \int_3^\infty z \exp(-z^2/2)\, dz = -\exp(-z^2/2)\Big|_3^\infty$

$= \exp(-4.5) = .011$

Example 3.4. The input to a noisy communication channel is a binary random variable X with $P(X = 0) = P(X = 1) = \frac{1}{2}$. The output of the channel Z is given by $Z = X + Y$, where Y is the additive noise introduced by the channel. Assume that the channel noise Y has the pdf

$$f_Y(y) = \frac{1}{\sqrt{2\pi}} \exp(-y^2/2), \quad -\infty < y < \infty$$

and that X and Y are statistically independent.

(a) Find the pdf of the output Z.
(b) Find the conditional pdf $f_{Z|X}(z|X = 0)$ and $f_{Z|X}(z|X = 1)$.

Solution

(a) To find $f_Z(z)$, let us first find $F_Z(z) = P(Z \leq z)$. We start with

$$P(Z \leq z) = P(Z \leq z|X = 0)P(X = 0) + P(Z \leq z|X = 1)P(X = 1)$$

Since

$$Z = X + Y$$

$$P(Z \leq z|X = 0) = P(X + Y \leq z|X = 0) = P(Y \leq z) = F_Y(z)$$

Similarly,

$$P(Z \leq z|X = 1) = P(Y + 1 \leq z) = P(Y \leq z - 1) = F_Y(z - 1)$$

Hence,

$$P(Z \leq z) = F_Z(z) = \tfrac{1}{2}F_Y(z) + \tfrac{1}{2}F_Y(z - 1)$$

and

$$\begin{aligned} f_Z(z) &= \frac{d}{dz}[F_Z(z)] = \frac{1}{2}\frac{d}{dz}[F_Y(z) + F_Y(z - 1)] \\ &= \tfrac{1}{2}f_Y(z) + \tfrac{1}{2}f_Y(z - 1) \\ &= \frac{1}{2}\left(\frac{1}{\sqrt{2\pi}}\exp(-z^2/2) + \frac{1}{\sqrt{2\pi}}\exp(-(z - 1)^2/2)\right) \end{aligned}$$

(b) When X is given to be zero, $Z = Y$ since $Z = X + Y$. Thus

$$f_{Z|X=0}(z|X = 0) = f_Y(z) = \frac{1}{\sqrt{2\pi}}\exp(-z^2/2), \quad -\infty < z < \infty$$

similarly,

$$f_{Z|X=1}(z|X = 1) = \frac{1}{\sqrt{2\pi}}\exp(-(z - 1)^2/2), \quad -\infty < z < \infty$$

Thus

$$f_Z(z) = f_{Z|X}(z|X = 0)P(X = 0) + f_{Z|X}(z|X = 1)P(X = 1), \quad -\infty < z < \infty$$

3.5 RANDOM PROCESSES

The main objective of an electrical communication system is the transfer of messages over a communication channel using electrical waveforms. These waveforms are in many instances unpredictable. For, if the waveform is exactly predictable (i.e., deterministic), then its transmission will be unnecessary and the entire communication system would serve no purpose. In this sense, particularly from the receiving viewpoint, all meaningful message signals are random. The randomness arises due to the fact that the receiver does not know a priori which one of the many possible message waveforms will be transmitted. The number of message waveforms emitted by an information source may be very large. For analog information sources, such as a microphone actuated by a speech signal, there may be an uncountably infinite number of message waveforms. Obviously, we need a unified mathe-

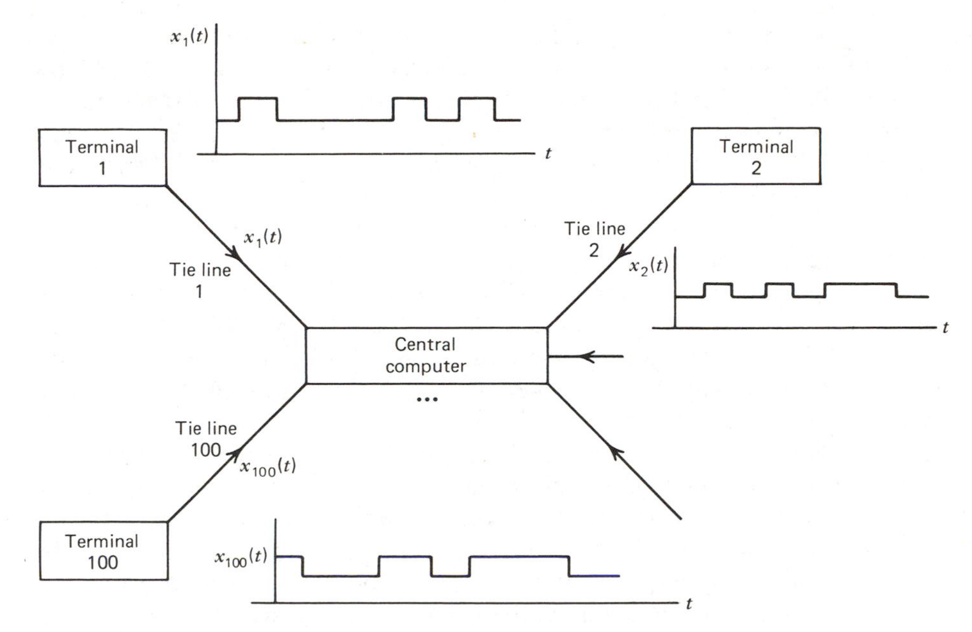

Figure 3.2 Example of message waveforms in a data communication system.

matical model to describe this collection of waveforms for purposes of analysis and design of communication systems.

By way of elaborating on the need for such a model, consider the example shown in Figure 3.2, which depicts a data communication system consisting of 100 remote data terminals connected to a central computer. For purposes of discussion let us assume that the terminals are transmitting binary data at a fixed rate. Typical waveforms are shown in this figure. Now, suppose that we are working on the design of this system and that we are asked to specify the bandwidth requirements of the tie lines that connect the terminals to the computer.

We may attempt to answer the bandwidth question by using one of the following approaches:

1. Observe one of the waveforms, say $x_1(t)$, for a reasonably long time, obtain a spectral description of $x_1(t)$, compute its bandwidth and use it to specify the bandwidth of *all* the tie lines.
2. Repeat (1) for *all* the waveforms in the collection and specify the bandwidth of ith tie line based on the properties of $x_i(t)$.
3. *Average* the bandwidth obtained in (2) and specify the average value as the bandwidth of *all* the tie lines.

Obviously we are going to get different answers for the bandwidth and thus we are faced with the following questions: Which is the correct answer? How do we go about judging the validity of the answers? Under what conditions are the answers nearly the same? We could answer these questions if we had a model for describing the common properties of the collection of signals that we are dealing with. It turns out that we can indeed develop such a model in the form of a probabilistic description for a collection of functions of time.

The concept of random variables discussed in the preceding section provides us a probabilistic description of the numerical values of a variable. A similar model, called a random process (or a stochastic process), can be used as a probabilistic description of functions of time. Random processes are used to model message waveforms as well as noise waveforms that are encountered in communication systems.

In the following sections, we describe several classes of random processes. We start with basic definitions and classifications of random processes and develop the important concepts of correlation and power spectral density. We will show that, under some mild assumptions, a collection of functions can be adequately characterized using the concept of power spectral density. The power spectral density function can be used to analyze signal location in the frequency domain, and for computing bandwidth, and signal power.

3.5.1 Definitions and Notations

Consider a random experiment E with outcomes λ and a sample space S. To every outcome $\lambda \in S$ we assign, according to a certain rule, a real valued* function of time $X(t, \lambda)$. We have thus created a family of functions, one function for each outcome λ. This family is called a random process.

A random process $X(t, \lambda)$ can be viewed as a function of two variables, $\lambda \in S$ and $t \in (-\infty, \infty)$. For a specific outcome, say λ_i, we have a single time function $X(t, \lambda_i) = x_i(t)$. This time function is called a *member function*, a *sample function*, or a *realization of the random process*. The totality of all sample functions is called an *ensemble*. For a specific time t_0, $X(t_0, \lambda)$ is a random variable whose value depends on the outcome λ. Finally for $\lambda = \lambda_i$ and $t = t_0$, $X(t_0, \lambda_i)$ is merely a number.

We will use the notation $X(t)$ to represent the random process. From the discussion in the preceding paragraph it is clear that $X(t)$ may represent a family of functions, a single time function, a random variable, or a single number. The specific interpretation will be obvious from the context.

*We could define a complex-valued random process similarly. However, unless otherwise specified, we will assume $X(t, \lambda)$ to be real valued.

Random processes can be described by referring to the physical phenomenon (or the underlying random experiment) that generates the family of functions. By modeling the physical phenomenon, we can arrive at an appropriate random process model for a given family of functions. In addition, random processes can also be described in terms of joint probability density (or probability mass) functions, in terms of analytic functions of random variables, or in terms of statistical averages.

Joint Distribution Function. A random process can be described by its joint probability distribution function (subjected to some very reasonable symmetry and compatibility conditions):

$$P\{X(t_1) \leq a_1, X(t_2) \leq a_2, \ldots, X(t_n) \leq a_n\}$$

for any n and $t_1, t_2, \ldots, t_n$.

Analytical Description. A random process can also be defined as a function of one or more random variables as

$$X(t) = g(Y_1, Y_2, \ldots, Y_n; t)$$

where $Y_1, Y_2, \ldots, Y_n$ are n random variables and g is an ordinary function.

Statistical Averages. As in the case of random variables, random processes are often described using statistical averages (also referred to as ensemble averages). Two statistical averages that are most often used in the description of a random process are the *mean* $\mu_X(t)$ and the *autocorrelation function* $R_{XX}(t_1, t_2)$, defined as

$$\mu_X(t) = E\{X(t)\} \tag{3.57}$$

$$R_{XX}(t_1, t_2) = E\{X(t_1)X(t_2)\} \tag{3.58}$$

The expected values are taken with respect to the appropriate pdf's.

Example 3.5. A random process $X(t)$ is defined by

$$X(t) = 2\cos(2\pi t + Y)$$

where Y is a discrete random variable with $P(Y = 0) = \frac{1}{2}$ and $P(Y = \pi/2) = \frac{1}{2}$. Find $\mu_X(1)$ and $R_{XX}(0, 1)$.

Solution. This is an example of a random process that has an analytical description. $X(1)$ is a random variable with values $2\cos(2\pi)$ and $2\cos(2\pi + \pi/2)$ with probabilities $\frac{1}{2}$ each, that is, $P[X(1) = 2] = \frac{1}{2}$, and $P[X(1) = 0] = \frac{1}{2}$. Hence

$$\mu_X(1) = E\{X(1)\} = 2 \cdot \tfrac{1}{2} + 0 \cdot \tfrac{1}{2} = 1$$

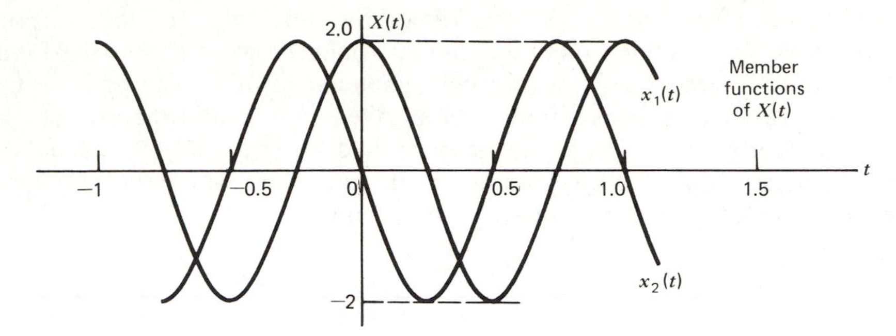

Figure 3.3 Random process $X(t)$.

The joint probability mass function of $X(0)$ and $X(1)$ can be obtained (by inspection) from Figure 3.3 as

$$P(X(0)=2, X(1)=2)=\tfrac{1}{2}$$
$$P(X(0)=0, X(1)=0)=\tfrac{1}{2}$$

Hence

$$\begin{aligned} R_{XX}(0,1) &= E\{X(0)X(1)\} \\ &= (2)(2)\tfrac{1}{2}+(0)(0)\tfrac{1}{2}=2 \end{aligned}$$

3.5.2 Stationarity, Time Averages, and Ergodicity

A random process $X(t)$ is said to be *stationary, in the strict sense,* if its statistics are not affected by a shift in the time origin. This means that $X(t)$ and $X(t+\epsilon)$, where ϵ is an arbitrary time shift, have the same statistical properties. Another idea of stationarity is that a time translation of a sample function results in another sample function of the random process having the same probability.

In the case when we are describing a random process by its mean and autocorrelation function, it seems reasonable to call the random process *stationary, in the wide sense,* if the mean and autocorrelation function do not vary with a shift in the time origin. Thus we call a process to be wide sense stationary if

$$\begin{aligned} E\{X(t)\} &= K = \text{const}, \qquad t \in (-\infty, \infty) \\ R_{XX}(t_1, t_2) &= R_{XX}(t_1+t, t_2+t), \qquad t, t_1, t_2 \in (-\infty, \infty) \\ &= R_{XX}(|t_1-t_2|) \end{aligned} \tag{3.59}$$

Strict sense stationarity implies wide sense stationarity, but not vice versa.

The autocorrelation function of a stationary process is a function of the time difference, and hence it is often defined as

$$R_{XX}(\tau) \triangleq E\{X(t)X(t+\tau)\} \tag{3.60}$$

Whether a process is stationary or not can be verified easily if an analytical description of the process is available. Lacking such a description, one has to collect data and analyze it to check whether the process is stationary. For details of procedures to be used for collecting, analyzing, and testing data for stationarity the reader is referred to the book by Bendat and Piersol on the measurement and analysis of random data.

Time Averages. The mean and autocorrelation function of a random process defined in Equations (3.57) and (3.58) are obtained by *ensemble averaging*. For example,

$$\mu_X(t_0) = \int_{-\infty}^{\infty} x f_{X(t_0)}(x)\, dx$$

where $f_{X(t_0)}(x)$ is the pdf of the random variable $X(t_0)$. Similarly, $R_{XX}(t_1, t_2)$ is obtained by averaging with respect to the joint pdf $f_{X(t_1),X(t_2)}(x_1, x_2)$.

It is possible, however, to define the autocorrelation function using *time averaging* (similar to the definition of autocorrelation function for deterministic power signals given in Chapter 2) as

$$\langle R_{XX}(\tau)\rangle = \lim_{T\to\infty} \frac{1}{T}\int_{-T/2}^{T/2} X(t)X(t+\tau)\, dt \tag{3.61}$$

Similarly, we can define the time-averaged mean as

$$\langle \mu_X \rangle = \lim_{T\to\infty} \frac{1}{T}\int_{-T/2}^{T/2} X(t)\, dt \tag{3.62}$$

Observe that $\langle R_{XX}(\tau)\rangle$ and $\langle \mu_X \rangle$ are random variables; their values depend on which member function of $X(t)$ is used in the time averaging process. In contrast, $R_{XX}(\tau)$, is an ordinary function of τ and μ_X is a constant.

To compute $R_{XX}(\tau)$ and μ_X by ensemble averaging, we need to average across all the member functions. This requires complete knowledge of the first- and second-order joint pdf's of the process. In many practical applications these pdf's are not known. About the only thing that may be available is the recording of one or more (but definitely a small number) of the member functions of the random process. From this data we could compute some time averages and attempt to use them in place of the ensemble averages for characterizing the random process. Before interchanging time averages and ensemble averages we should make sure that they are equal (at least in a mean squared sense*). Unfortunately, ensemble averages and time averages are not in general equal except for a very special class of random processes.

Ergodicity. Ergodicity deals with the problem of determining the properties of a random process, such as the mean and autocorrelation function, from a single member function. A random process is said to be ergodic in the most general form if time averages equal ensemble averages. We are usually not interested in all the ensemble averages of a random process but in only certain averages such as the mean and the autocorrelation function. We can define ergodicity with respect to these parameters. For example, we say that a process is ergodic in the mean if $E\{\langle \mu_X \rangle\} = E\{X(t)\}$, and the variance of $\langle \mu_X \rangle \to 0$ as $T \to \infty$. Similarly, we say that a process is ergodic in the autocorrelation function if $E\{\langle R_{XX}(\tau) \rangle\} = R_{XX}(\tau)$ and the variance of $\langle R_{XX}(\tau) \rangle \to 0$ as $T \to \infty$.

In practice it is very hard to decide on the basis of data if a random process is ergodic. One has to decide about ergodicity based on reasoning about the physical phenomenon involved. In order to be ergodic, a random process must be stationary and randomness must be evident in time variation as well as in the selection of a sample function. In addition, the time averages must not depend on which sample function is selected for computing the time-averaged values (see Problem 3.33).

3.5.3 Power Spectral Density of Stationary Random Processes

In Section 2.4 we showed that the Fourier transform of the autocorrelation function of a deterministic signal is its power spectral density function. The power spectral density of a stationary random process is defined similarly. If the autocorrelation function $R_{XX}(\tau)$ of a stationary random process is such that

$$\int_{-\infty}^{\infty} |R_{XX}(\tau)| \, d\tau < \infty$$

then its Fourier transform $G_X(f)$ given by

$$G_X(f) = \int_{-\infty}^{\infty} R_{XX}(\tau) \exp(-2\pi j f \tau) \, d\tau \tag{3.63}$$

is called the *power density spectrum* or the *power spectral density function* of $X(t)$.

*A random variable X is said to be equal to a constant b in a mean squared sense if $E\{X\} = b$ and $E\{(X - b)^2\} = 0$.

The name power spectral density (psd) comes from the fact that $E\{X^2(t)\}$ may be thought of as the average power dissipated by the random process across a 1 ohm resistor, and

$$E\{X^2(t)\} = R_{XX}(0) = \int_{-\infty}^{\infty} G_X(f)\, df$$

In Section 2.4, we defined the psd of a deterministic signal $x(t)$ as

$$G_x(f) = \lim_{T\to\infty} \frac{\left| \int_{-T/2}^{T/2} x(t) \exp(-2\pi jft)\, dt \right|^2}{T} \tag{3.64}$$

Now, we can give a similar definition for the time-averaged version of the psd of an ergodic random process as

$$\langle G_X(f) \rangle = \lim_{T\to\infty} \frac{\left| \int_{-T/2}^{T/2} X(t) \exp(-2\pi jft)\, dt \right|^2}{T} \tag{3.65}$$

For an ergodic random process $\langle G_X(f)\rangle = G_X(f)$, where the equality is in a mean squared sense (i.e., $E\{\langle G_X(f)\rangle\} = G_X(f)$, and variance of $\langle G_X(f)\rangle \to 0$ as $T \to \infty$).

Thus, the concept of power spectral density carries the same meaning for deterministic power signals and for ergodic random processes. In the remainder of this book we will deal almost exclusively with random signals that are sample functions of ergodic processes. These signals are power signals, and we will use their psd functions for determining such things as the locations of these signals in the frequency domain, bandwidths, and so forth.

It is useful at this point to emphasize the meaning of various time averages for an ergodic random process.

1. The mean $\langle X(t)\rangle$ is the *DC component.*
2. The mean squared value $\langle X^2(t)\rangle$ is the *total average power.*
3. The square of the mean value $\langle X(t)\rangle^2$ is the *DC power.*
4. The variance $\langle X^2(t)\rangle - \langle X(t)\rangle^2$ is the *AC power.*
5. The standard deviation is the *rms value.*

The time averages mentioned above are familiar quantities for an electrical engineer. For an ergodic random process, we can replace ensemble or statistical averages by time averages and by finite time approximation to these time averages measured in the laboratory.

Example 3.6—Randomly Phased Sinusoids. Consider the random process

$$X(t) = A\cos(\omega_c t + \theta)$$

where A and ω_c are constants while θ is a random variable with an uniform pdf

$$f_\theta(\theta) = 1/2\pi, \quad -\pi < \theta < \pi$$

(a) Find the mean, the autocorrelation function, and the psd of $X(t)$. (Show that $X(t)$ is wide sense stationary before finding the psd.)

(b) Find the autocorrelation function by time averaging and show that $\langle R_{XX}(\tau)\rangle = R_{XX}(\tau)$.

Solution

(a) $E\{X(t)\} = E\{A\cos(\omega_c t + \theta)\}$, where the expected value is taken with respect to the pdf of θ:

$$\begin{aligned} E\{X(t)\} &= \int_{-\pi}^{\pi} \frac{1}{2\pi}[A\cos(\omega_c t + \theta)]\, d\theta \\ &= \frac{A\cos\omega_c t}{2\pi}\int_{-\pi}^{\pi}\cos\theta\, d\theta - \frac{A\sin\omega_c t}{2\pi}\int_{-\pi}^{\pi}\sin\theta\, d\theta \\ &= 0 \quad \text{(since the two integrals are zero)} \end{aligned}$$

Using a similar approach, the reader can verify that

$$\begin{aligned} R_{XX}(t, t+\tau) &= E\{X(t)X(t+\tau)\} \\ &= \frac{A^2}{2}\cos\omega_c\tau + \frac{A^2}{2}E\{\cos(2\omega_c t + \omega_c\tau + 2\theta)\} \\ &= \frac{A^2}{2}\cos\omega_c\tau = R_{XX}(\tau) \end{aligned}$$

The process $X(t)$ is stationary since $\mu_X(t)$ and $R_{XX}(t, t+\tau)$ do not depend on t. The power spectral density of $X(t)$ is given by

$$G_X(f) = F\{R_{XX}(\tau)\} = \frac{A^2}{4}[\delta(f - f_c) + \delta(f + f_c)]$$

Observe that $R_{XX}(0) = E\{X^2(t)\} = A^2/2$ which is the same as the average power of a deterministic sinusoidal waveform with an amplitude A.

(b) Suppose that we use the member function $x_0(t) = A\cos(\omega_c t + \theta_0)$, where θ_0 is a particular value in the interval $-\pi$ to π, to perform time averaging.

Then,

$$\langle R_{XX}(\tau)\rangle = \lim_{T\to\infty}\frac{1}{T}\int_{-T/2}^{T/2} A^2\cos(\omega_c t+\theta_0)\cos(\omega_c t+\omega_c\tau+\theta_0)\,dt$$

$$= \lim_{T\to\infty}\frac{1}{T}\int_{-T/2}^{T/2}\frac{A^2}{2}\cos(\omega_c\tau)\,dt + \lim_{T\to\infty}\frac{1}{T}\int_{-T/2}^{T/2}\frac{A^2}{2}\cos(2\omega_c t+\omega_c\tau+2\theta_0)\,dt$$

The average value of the first term is $(A^2/2)\cos\omega_c\tau$, and the average value of the second term is zero. Hence $\langle R_{XX}(\tau)\rangle = (A^2/2)\cos\omega_c\tau$. The autocorrelation function obtained by time averaging does not depend on which member function was used to obtain the time-averaged value, and hence

$$\langle R_{XX}(\tau)\rangle = R_{XX}(\tau)$$

We will use the randomly phased sinusoids for developing time domain models for narrowband noise.

Example 3.7. A class of modulated signals is modeled by

$$X_c(t) = AX(t)\cos(\omega_c t+\theta) \tag{3.66}$$

where $X(t)$ is the message signal and $A\cos(\omega_c t+\theta)$ is the carrier. The message signal $X(t)$ is modeled as a zero mean stationary random process with the autocorrelation function $R_{XX}(\tau)$ and the psd $G_X(f)$. The carrier amplitude A and the frequency ω_c are assumed to be constants and the initial carrier phase θ is assumed to be a random variable uniformly distributed in the interval $[-\pi, \pi]$. Furthermore, $X(t)$ and θ are assumed to be independent.

(a) Show that $X_c(t)$ is stationary (in the wide sense).
(b) Find the psd of $X_c(t)$.

Solution

(a)
$$E\{X_c(t)\} = E\{AX(t)\cos(\omega_c t+\theta)\}$$
$$= E\{AX(t)\}E\{\cos(\omega_c t+\theta)\} = 0$$

since $X(t)$ and θ are independent and $E\{X(t)\}=0$,

$$R_{X_cX_c}(t, t+\tau) = E\{X_c(t)X_c(t+\tau)\}$$
$$= E\{A^2X(t)X(t+\tau)\cos(\omega_c t+\theta)\cos(\omega_c t+\omega_c\tau+\theta)\}$$
$$= \frac{A^2}{2}E\{X(t)X(t+\tau)\}E\{\cos\omega_c\tau+\cos(2\omega_c t+\omega_c\tau+2\theta)\}$$
$$= \frac{A^2}{2}R_{XX}(\tau)\cos\omega_c\tau,\quad \text{since } E\{\cos(2\omega_c t+\omega_c\tau+2\theta)\}=0$$

Since $\mu_{X_c}(t)$ and $R_{X_cX_c}(t, t+\tau)$ do not depend on t, the process $X_c(t)$ is stationary in the wide sense.

(b)
$$G_{X_c}(f) = F\{R_{X_cX_c}(\tau)\} = \frac{A^2}{2}F\{R_{XX}(\tau)\cos\omega_c\tau\}$$

$$= \frac{A^2}{2}\{G_X(f) * \tfrac{1}{2}[\delta(f-f_c) + \delta(f+f_c)]\}$$

$$= \frac{A^2}{4}\{G_X(f-f_c) + G_X(f+f_c)\} \tag{3.67}$$

Example 3.8—Random Binary Waveform. Consider a random binary waveform that consists of a sequence of pulses with the following properties:

1. Each pulse is of duration T_b.
2. Pulses are equally likely to be ± 1.
3. All pulses (pulse amplitudes) are statistically independent.
4. The pulses are not synchronized; that is, the starting time T of the first pulse is equally likely to be anywhere between 0 to T_b.

(a) Find the autocorrelation and power spectral density function of $X(t)$. (Show that $R_{XX}(t_1, t_2)$ is a function of $|t_1 - t_2|$ before proceeding to find $G_X(f)$.)

(b) Define the bandwidth of the signal to be B, where B is such that

$$\int_{-B}^{B} G_X(f)\,df = 0.95\int_{-\infty}^{\infty} G_X(f)\,df$$

(i.e., 95% of the signal power is to be contained in the interval $-B$ to B). Find B.

Solution

(a) A member function of the random process is shown in Figure 3.4*a*. The reader can easily verify that

$$E\{X(t)\} = 0$$

To find the autocorrelation function, let t_1 and t_2 be two arbitrary values of time t, and assume that $0 < t_1 < t_2 < T_b$ and $|t_1 - t_2| < T_b$. Then, depending on the value of T, $X(t_1)$ and $X(t_2)$ may or may not be in the same pulse interval (Figure 3.4*b*). Let $A(t_1, t_2)$ be the random event that the value of T is such that t_1 and t_2 occur during the same pulse. For this situation, depicted in

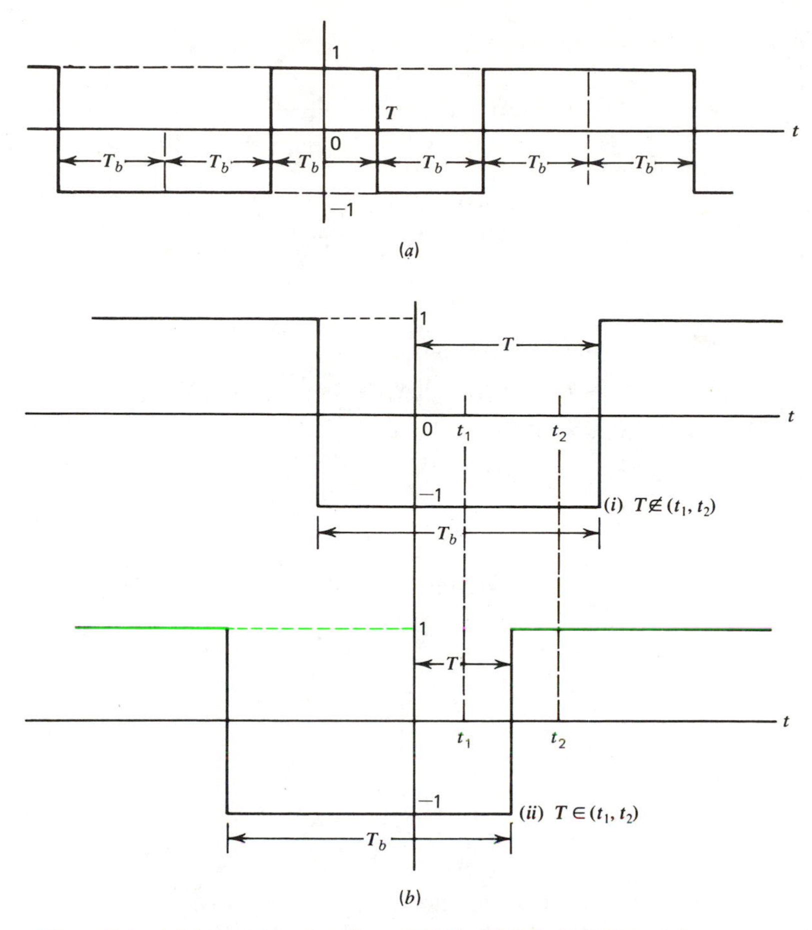

Figure 3.4 (*a*) A member function of $X(t)$. (*b*) Detailed timing diagram for calculating $R_{XX}(t_1, t_2)$. It is assumed that $t_1 < t_2 < T_b$.

Figure 3.4*b*, we have

$$P[A(t_1, t_2)] = P[T < t_1 \text{ or } T > t_2]$$

$$= \frac{t_1}{T_b} + \frac{T_b - t_2}{T_b}$$

$$= 1 - \frac{(t_2 - t_1)}{T_b}$$

$$P[\bar{A}(t_1, t_2)] = P[t_1 < T < t_2]$$

$$= \frac{t_2 - t_1}{T_b}$$

If t_2 is assumed to be $<t_1$, then

$$P[A(t_1, t_2)] = 1 - \frac{(t_1 - t_2)}{T_b}$$

and in general we have

$$P[A(t_1, t_2)] = 1 - \frac{|t_1 - t_2|}{T_b}$$

Now,

$$\begin{aligned} R_{XX}(t_1, t_2) &= E\{X(t_1)X(t_2)\} \\ &= E\{X(t_1)X(t_2)|A\}P(A) + E\{X(t_1)X(t_2)|\bar{A}\}P(\bar{A})^* \\ E\{X(t_1)X(t_2)|A\} &= \tfrac{1}{2}(1)^2 + \tfrac{1}{2}(-1)^2 = 1 \\ E\{X(t_1)X(t_2)|\bar{A}\} &= \tfrac{1}{4}(1)(1) + \tfrac{1}{4}(-1)(-1) + \tfrac{1}{4}(1)(-1) + \tfrac{1}{4}(-1)(1) = 0 \end{aligned}$$

Hence,

$$R_{XX}(t_1, t_2) = 1 - \frac{|t_1 - t_2|}{T_b}$$

Letting $\tau = |t_1 - t_2|$, we have

$$R_{XX}(\tau) = 1 - \frac{|\tau|}{T_b}, \qquad |\tau| < T_b$$

When $|t_1 - t_2| > T_b$, $P[A(t_1, t_2)] = 0$ and $R_{XX}(\tau) = 0$. Thus,

$$R_{XX}(\tau) = \begin{cases} 1 - |\tau|/T_b, & \text{for } |\tau| < T_b \\ 0, & \text{for } |\tau| > T_b \end{cases} \tag{3.68}$$

It can be shown that $R_{XX}(\tau)$ will be the same for any arbitrary choice of $t_1 = t$ and $t_2 = t + \tau$. By taking the Fourier transform of $R_{XX}(\tau)$, we obtain the psd of $X(t)$ as

$$G_X(f) = T_b\left(\frac{\sin(\pi f T_b)}{(\pi f T_b)}\right)^2 = \frac{|P(f)|^2}{T_b} \tag{3.69}$$

where $P(f)$ is the Fourier transform of a unit amplitude pulse of width T_b.

(b) By numerical integration, we obtain

$$B \approx 1.5/T_b$$

The random binary waveform is quite useful in modeling the signals encountered in digital communication systems. The reader is now asked to compare

*To show that $E\{Y\} = E\{Y|A\}P(A) + E\{Y|\bar{A}\}P(\bar{A})$ is left as an exercise for the reader (Problem 3.17).

the psd of the random binary waveform with the psd of a deterministic rectangular waveform derived in Example 2.6.

The reader can verify that the random binary waveform can be written as

$$X(t) = \sum_{k=-\infty}^{\infty} a_k p(t - kT_b - T) \tag{3.70}$$

where $\{a_k\}$ is a sequence of independent random variables with $P(a_k = 1) = P(a_k = -1) = \frac{1}{2}$, and $p(t)$ is a unit amplitude pulse of width T_b; T is a random variable having a uniform pdf in the interval $[0, T_b]$.

Example 3.9. $X(t)$ is a random binary waveform with $P(X(t) = 1) = P(X(t) = -1) = \frac{1}{2}$. Let $Y(t)$ be a random process derived from $X(t)$,

$$Y(t) = g_0 X(t) + g_1 X(t - T_b) + g_2 X(t - 2T_b)$$

where T_b is the bit duration of $X(t)$ and g_0, g_1, g_2 are constants.

(a) Show that $Y(t)$ can be expressed as

$$Y(t) = \sum_k A_k p(t - kT_b - T)$$

where $\{A_k\}$ is a sequence of random amplitudes and T is a random delay with a uniform distribution in the interval $[0, T_b]$.

(b) Show that the psd of $Y(t)$ can be expressed as

$$G_Y(f) = \frac{|P(f)|^2}{T_b}\left(R(0) + 2\sum_{k=1}^{2} R(k)\cos 2\pi k f T_b\right)$$

where $P(f)$ is the Fourier transform of $p(t)$ and

$$R(k) = E\{A_j A_{j+k}\}, \quad k = 0, 1, 2$$

Solution

(a) Since $Y(t) = g_0 X(t) + g_1 X(t - T_b) + g_2 X(t - 2T_b)$, and

$$X(t) = \sum_k a_k p(t - kT_b - T)$$

we can express $Y(t)$ as

$$Y(t) = \sum_k A_k p(t - kT_b - T)$$

where

$$A_k = g_0 a_k + g_1 a_{k-1} + g_2 a_{k-2}$$

The amplitude sequence $\{a_k\}$ of the random binary waveform has the properties

$$E\{a_k\} = 0, \qquad E\{a_i a_j\} = \begin{cases} 0, & i \neq j \\ 1, & i = j \end{cases}$$

(b) $$R_{YY}(t, t+\tau) = E\{Y(t)Y(t+\tau)\}$$
$$= E\{[g_0 X(t) + g_1 X(t - T_b) + g_2 X(t - 2T_b)] \times [g_0 X(t+\tau) + g_1 X(t + \tau - T_b) + g_2 X(t + \tau - 2T_b)]\}$$

or

$$R_{YY}(\tau) = [g_0^2 + g_1^2 + g_2^2] R_{XX}(\tau) + [g_0 g_1 + g_1 g_2][R_{XX}(\tau - T_b) + R_{XX}(\tau + T_b)] + [g_0 g_2][R_{XX}(\tau - 2T_b) + R_{XX}(\tau + 2T_b)]$$

The reader can verify that

$$g_0^2 + g_1^2 + g_2^2 = E\{A_k^2\} = R(0)$$
$$g_0 g_1 + g_1 g_2 = E\{A_k A_{k+1}\} = R(1)$$
$$g_0 g_2 = E\{A_k A_{k+2}\} = R(2)$$

Hence

$$R_{YY}(\tau) = R(0) R_{XX}(\tau) + R(1)[R_{XX}(\tau - T_b) + R_{XX}(\tau + T_b)] + R(2)[R_{XX}(\tau - 2T_b) + R_{XX}(\tau + 2T_b)]$$

Now, taking the Fourier transform of $R_{YY}(\tau)$ we get

$$G_Y(f) = G_X(f) R(0) + 2R(1)[G_X(f) \cos 2\pi f T_b] + 2R(2)[G_X(f) \cos 4\pi f T_b]$$

since

$$F\{R_{XX}(\tau - kT_b) + R_{XX}(\tau + kT_b)\} = G_X(f) 2 \cos 2\pi k f T_b$$

Substituting $G_X(f) = |P(f)|^2/T_b$, we can simplify $G_Y(f)$ to

$$G_Y(f) = \frac{|P(f)|^2}{T_b}\left(R(0) + 2\sum_{k=1}^{2} R(k) \cos 2\pi k f T_b\right)$$

where

$$R(0) = E\{A_k^2\} \quad \text{and} \quad R(k) = E\{A_j A_{j+k}\}$$

We can generalize the result derived in the preceding example as follows. If

$$Y(t) = \sum_{k=-\infty}^{\infty} A_k p(t - kT_b - T) \tag{3.71}$$

where $\{A_k\}$ is a stationary sequence of random variables, $p(t)$ is a pulse of duration T_b, and T is a random delay with a uniform pdf in the interval

$[0, T_b]$, then the psd of $Y(t)$ is given by

$$G_Y(f) = \frac{|P(f)|^2}{T_b}\left(R(0) + 2\sum_{k=1}^{\infty} [R(k)\cos 2\pi k f T_b]\right) \tag{3.72a}$$

where

$$R(k) = E\{A_j A_{j+k}\} \quad \text{and} \quad E\{A_k\} = 0 \tag{3.72b}$$

We will use this result quite often in our analysis of the spectra of pulse waveforms used in digital communication systems.

3.5.4 Special Classes of Random Processes

Gaussian Random Process. In the study of noise problems in communication systems, we encounter a special type of random process that is characterized by the fact that all the basic probability distribution functions of the process are Gaussian. We describe below the important properties of the Gaussian random process.

A random process $X(t)$ is said to be Gaussian if for any $n \geq 1$ and $t_1, t_2, \ldots, t_n \in (-\infty, \infty)$, the joint distribution of $X(t_1), X(t_2), \ldots, X(t_n)$ is Gaussian. That is,

$$f_{X_1, X_2, \ldots, X_n}(x_1, x_2, \ldots, x_n) = \frac{1}{|C|^{1/2}(2\pi)^{n/2}} \exp\left(-\frac{1}{2}\sum_{i=1}^{n}\sum_{j=1}^{n} [(x_i - \mu_i) d_{ij} (x_j - \mu_j)]\right) \tag{3.73}$$

where $X_i = X(t_i)$, $c_{ij} = E\{(X_i - \mu_i)(X_j - \mu_j)\}$, $\mu_j = E\{X_j\}$, $C = \text{matrix}[c_{ij}]$, and $D = \text{matrix}[d_{ij}] = C^{-1}$.

A Gaussian random process is completely described when μ_j and c_{ij} are specified for all $t_i, t_j \in (-\infty, \infty)$. A special case of the Gaussian random process is the zero mean stationary Gaussian random process that is completely described by the autocorrelation function $R_{XX}(\tau)$. The first- and second-order pdf's of a stationary zero mean Gaussian process are given by

$$\begin{aligned} f_{X(t)}(x) &= \frac{1}{\sqrt{2\pi R_{XX}(0)}} \exp\left(-\frac{x^2}{2R_{XX}(0)}\right), \quad -\infty < x < \infty \\ f_{X(t), X(t+\tau)}(x_1, x_2) &= \frac{\exp[-(x_1^2 - 2\rho x_1 x_2 + x_2^2)/2(1-\rho^2) R_{XX}(0)]}{2\pi R_{XX}(0)\sqrt{(1-\rho^2)}} \end{aligned} \tag{3.74}$$

where $\rho = R_{XX}(\tau)/R_{XX}(0)$.

For a Gaussian random process, the nth-order distribution functions are completely specified by $R_{XX}(t_i, t_j)$, and $\mu_X(t_i)$, $t_i, t_j \in (-\infty, \infty)$. If $E\{X(t_i)\} = \mu_X$ and $R_{XX}(t_i, t_j) = R_{XX}(|t_i - t_j|)$, that is, if the process is wide sense stationary, then the nth-order joint distribution depends only on time differences.

Hence, wide sense stationarity implies strict sense stationarity for Gaussian random processes.

Markoff Sequences. Consider a random process $X(t)$ that is defined only at a countable number of discrete values of time $t_1, t_2, t_3 \ldots$. Such a random process can be represented by a sequence of random variables $X_1, X_2, X_3, \ldots$, where X_n denotes $X(t_n)$. Furthermore, if we assume $X(t)$ to have only a finite set of permissible values, then the random process $X(t)$ is called a *discrete time, discrete amplitude* random process. It can be represented by a sequence of discrete random variables $\{X_n\}$. Discrete time discrete amplitude random processes are used as models for the output of discrete information sources such as teletypes and A/D converters. (The output of these sources are sequences of symbols or letters.)

A special class of discrete time discrete amplitude random processes that we will use in our later work is the Markoff sequence. Suppose that the random variables X_n are of the discrete type taking the values $a_1, a_2, \ldots, a_N$. Then the sequence $\{X_n\}$ is called a Markoff sequence (or chain) if

$$P\{X_n = a_{i_n} | X_{n-1} = a_{i_{n-1}}, X_{n-2} = a_{i_{n-2}}, \ldots, X_1 = a_{i_1}\} = P\{X_n = a_{i_n} | X_{n-1} = a_{i_{n-1}}\} \tag{3.75}$$

where $a_{i_k} = a_j$, j being an arbitary integer $\in [1, N]$
The preceding equation says that $\{X_n\}$ is a Markoff sequence if the past values of the sequence have no influence on the statistics of the future values under the condition that the present value is known.

If we use the notation $p_i(n) = P(X_n = a_i)$ and $p_{ij}(s, n) = P(X_n = j | X_s = i)$ $(n > s)$, then

$$p_j(n) = \sum_{i=1}^{N} p_i(s) p_{ij}(s, n) \tag{3.76}$$

If $p_{ij}(n, s) = p_{ij}(n - s)$, that is, if the transition probabilities depend only on the time difference, then the sequence $\{X_n\}$ is called *homogeneous*. Such a chain is completely characterized by a set of initial probabilities $p_i(1)$ $(i = 1, 2, \ldots, N)$ and a set of transition probabilities p_{ij} $(i, j = 1, 2, \ldots, N)$. If we let $P(k)$ be a column vector whose ith entry is $p_i(k)$ and let ϕ be an $N \times N$ matrix whose (i, j)-th entry is $p_{ij}(1)$, then we have

$$P(k+1) = \phi^T P(k) \tag{3.77a}$$

or

$$P(k+1) = (\phi^T)^k P(1) \tag{3.77b}$$

The matrix ϕ is called the probability transition matrix of the Markoff process. The Markoff process is called a *stationary Markoff process* if $P(k+1) = P(k)$.

3.6 SYSTEMS AND RANDOM SIGNALS

In Chapter 2 we derived input–output relationships for systems responding to deterministic input signals. We now turn our attention to the problem of obtaining some description of the output $Y(t)$ of a system responding to a random process $X(t)$ at its input. We will consider the system to be fixed (i.e., all parameters of the system are constants) and will assume that the statistical characteristics of the input random process $X(t)$, and the transfer characteristics of the system are known (Figure 3.5).

Now, if the input to the system is a particular member function $x(t)$ of $X(t)$, then the output $y(t)$ can be calculated using deterministic procedures. However, probabilities enter the picture in the selection of the member function $x(t)$. For different member functions $x(t)$ of $X(t)$, it is reasonable to expect different responses $y(t)$. Thus the output is a collection of functions of time, that is, the output is a random process. We are interested in obtaining a random process model for $Y(t)$ given the model of the input signal and the model of the system. Two special cases are described in the following sections.

3.6.1 Response of Memoryless Systems

Let us first consider systems in which the instantaneous value of the output signal at time t_0 depends only on the instantaneous value of the input at t_0. That is, the system is described by

$$Y(t_0) = g[X(t_0)] \tag{3.78}$$

Systems that are modeled by equations of this form include linear devices such as resistive networks as well as nonlinear devices such as square law devices, voltage limiters, and so forth. Given the model stated in Equation (3.78) and the distribution of $X(t)$, the distribution of $Y(t)$ can be derived using the techniques described in Section 3.4.3. For example, if $f_{X(t_1)}$ is given, then $Y(t_1) = g[X(t_1)]$ and the pdf of $f_{Y(t_1)}(y_1)$ can be obtained using the result stated in Theorem 3.1. Higher order distributions such as $f_{Y(t_1),Y(t_2),\ldots,Y(t_n)}$ can be obtained using a similar technique if $f_{X(t_1),X(t_2),\ldots,X(t_n)}$ is given. However, the calculations can become formidable.

In some cases only the mean and autocorrelation function of $Y(t)$ are needed. If this is the case, μ_Y and $R_{YY}(\tau)$ may be found easily. On many occasions, we can do this without ever finding the joint pdf of $Y(t)$ and $Y(t+\tau)$. The example given below illustrates this point.

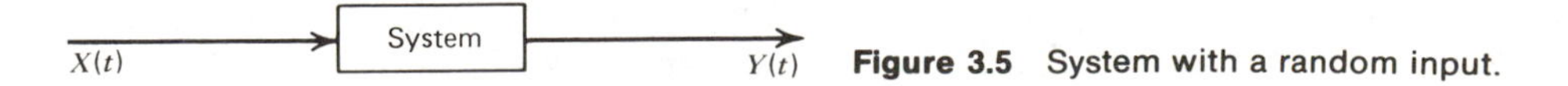

Figure 3.5 System with a random input.

Example 3.10. The input $X(t)$ to a diode with a transfer characteristic $Y = X^2$ is a zero mean stationary Gaussian random process with an autocorrelation function $R_{XX}(\tau) = \exp(-|\tau|)$. Find the mean $\mu_Y(t)$ and $R_{YY}(t_1, t_2)$.

Solution

$$\mu_Y = E\{Y(t)\} = E\{X^2(t)\} = R_{XX}(0) = 1$$
$$R_{YY}(t_1, t_2) = E\{Y(t_1)Y(t_2)\} = E\{X^2(t_1)X^2(t_2)\}$$

To calculate $E\{X^2(t_1)X^2(t_2)\}$ we use a general relationship for Gaussian random variables with zero means. This result is (see Problem 3.37)

$$E\{X^2(t_1)X^2(t_2)\} = E\{X^2(t_1)\}\, E\{X^2(t_2)\} + 2[E\{X(t_1)X(t_2)\}]^2 \tag{3.79}$$

Now,

$$E\{X^2(t_1)\} = E\{X^2(t_2)\} = R_{XX}(0)$$
$$E\{X(t_1)X(t_2)\} = R_{XX}(|t_1 - t_2|)$$

since $X(t)$ is stationary, and hence

$$R_{YY}(t_1, t_2) = [R_{XX}(0)]^2 + 2[R_{XX}(|t_1 - t_2|)]^2$$

or

$$R_{YY}(\tau) = [R_{XX}(0)]^2 + 2[R_{XX}(\tau)]^2 = 1 + 2\exp(-2|\tau|)$$

From $R_{YY}(\tau)$, we can obtain the psd $G_Y(f)$ as

$$G_Y(f) = [R_{XX}(0)]^2\delta(f) + 2G_X(f) * G_X(f)$$
$$= F\{1 + 2\exp(-2|\tau|)\}$$

3.6.2 Response of Linear Time-Invariant Systems

We now turn our attention to the problem of finding the statistical properties of the response $Y(t)$ of a linear time-invariant causal system when the input is a stationary random process $X(t)$. The output $Y(t)$ is given by the convolution integral

$$Y(t) = \int_{-\infty}^{\infty} X(t - \alpha)h(\alpha)\, d\alpha$$
$$= \int_{-\infty}^{\infty} X(\alpha)h(t - \alpha)\, d\alpha$$

where $h(t)$ is the impulse response of the system. We can obtain the mean or the expected value of the output as

$$E\{Y(t)\} = E\left\{\int_{-\infty}^{\infty} X(\alpha)h(t-\alpha)\,d\alpha\right\}$$

$$= \int_{-\infty}^{\infty} E\{X(\alpha)\}h(t-\alpha)\,d\alpha$$

The last step follows from the fact that the expected value operation is a linear operation and hence the expected value and integration operations can be interchanged (subjected to some mild continuity requirements). Since $E\{X(\alpha)\} = \mu_X$, we can write $E\{Y(t)\}$ as

$$E\{Y(t)\} = \mu_X \int_{-\infty}^{\infty} h(t-\alpha)\,d\alpha = \mu_X H(0) \tag{3.80}$$

where $H(f)$ is the Fourier transform of $h(t)$. Equation (3.80) shows that the expected value of $Y(t)$ does not depend on t.

To determine the autocorrelation function of $Y(t)$, let us first find the *cross correlation* between $X(t)$ and $Y(t)$ defined as

$$R_{YX}(\tau) \triangleq E\{Y(t)X(t-\tau)\} = E\left\{X(t-\tau)\int_{-\infty}^{\infty} X(t-\alpha)h(\alpha)\,d\alpha\right\}$$

$$= \int_{-\infty}^{\infty} R_{XX}(\tau-\alpha)h(\alpha)\,d\alpha$$

The right-hand side of the preceding equation is independent of t and it equals the convolution of $R_{XX}(\tau)$ with $h(\tau)$. That is,

$$R_{YX}(\tau) = R_{XX}(\tau) * h(\tau) \tag{3.81}$$

Now,

$$R_{YY}(\tau) = E\{Y(t+\tau)Y(t)\}$$

$$= E\left\{Y(t+\tau)\int_{-\infty}^{\infty} X(t-\alpha)h(\alpha)\,d\alpha\right\}$$

$$= \int_{-\infty}^{\infty} R_{YX}(\tau+\alpha)h(\alpha)\,d\alpha = R_{YX}(\tau) * h(-\tau) \tag{3.82}$$

Combining Equations (3.81) and (3.82) we obtain

$$R_{YY}(\tau) = R_{XX}(\tau) * h(\tau) * h(-\tau) \tag{3.83}$$

Equations (3.80) and (3.83) indicate that $Y(t)$ is wide sense stationary if $X(t)$ is wide sense stationary.

Finally, taking the Fourier transform of Equations (3.81) and (3.82), we get the following results:

$$G_{YX}(f) = F\{R_{YX}(\tau)\} = F\{R_{XX}(\tau) * h(\tau)\} = G_X(f)H(f)$$

$$\begin{aligned} G_Y(f) &= F\{R_{YX}(\tau) * h(-\tau)\} = G_{YX}(f)H^*(f) \\ &= G_X(f)|H(f)|^2 \end{aligned} \tag{3.84}$$

Equation (3.84) shows that the psd of the output $Y(t)$ is the product of the psd of the input signal and the square of the amplitude response of the system. This result is identical to Equation (2.55), which gives the relationship between the input and output psd functions for deterministic signals.

Example 3.11. The input to an RC lowpass network is a zero mean stationary Gaussian random process $X(t)$ with $R_{XX}(\tau) = \exp(-\alpha|\tau|)$. Find the mean, variance, and psd of the output $Y(t)$.

Solution. The transfer function of an RC lowpass network is

$$H(f) = \frac{1}{1 + 2\pi jfRC} = \frac{\beta}{\beta + j2\pi f}, \quad \beta = \frac{1}{RC}$$

From Equation (3.80) we obtain

$$E\{Y(t)\} = H(0)E\{X(t)\} = 0$$

The psd of the output $G_Y(f)$ is given by

$$\begin{aligned} G_Y(f) &= G_X(f)|H(f)|^2 \\ &= F\{R_{XX}(\tau)\}|H(f)|^2 \\ &= \left(\frac{2\alpha}{\alpha^2 + (2\pi f)^2}\right)\left(\frac{\beta^2}{\beta^2 + (2\pi f)^2}\right) \end{aligned}$$

$$\begin{aligned} R_{YY}(\tau) &= F^{-1}\{G_Y(f)\} \\ &= F^{-1}\left(\frac{2\alpha\beta^2}{[\alpha^2 + (2\pi f)^2][\beta^2 + (2\pi f)^2]}\right) \\ &= F^{-1}\left(\frac{A}{\alpha^2 + (2\pi f)^2} + \frac{B}{\beta^2 + (2\pi f)^2}\right) \end{aligned}$$

where

$$A = \frac{2\alpha\beta^2}{\beta^2 - \alpha^2}, \qquad B = \frac{2\alpha\beta^2}{\alpha^2 - \beta^2}$$

Using the transform pairs given in Appendix C, we obtain

$$R_{YY}(\tau) = \left(\frac{\beta^2}{\beta^2 - \alpha^2}\right)\exp(-\alpha|\tau|) + \frac{\alpha\beta}{\alpha^2 - \beta^2}\exp(-\beta|\tau|)$$

and

$$E\{Y^2(t)\} = R_{YY}(0) = \frac{\beta}{\alpha + \beta}$$

3.7 NOISE IN COMMUNICATION SYSTEMS

Successful electrical communication depends on how accurately the receiver can determine the transmitted signal. Perfect signal identification might be possible in the absence of "noise" and other contaminations, but noise is always present in electrical systems. The presence of noise superimposed on a signal limits the receiver's ability to correctly identify the intended signal and thereby limits the rate of information transmission.

The term "noise" is used in electrical communication systems to refer to unwanted electrical signals that accompany the message signals. These unwanted signals arise from a variety of sources and can be classified as man-made or naturally occurring. Man-made noise includes such things as electromagnetic pickup of other radiating signals, 60 cycle hum due to inadequate power supply filtering, switching transients, and so forth. Natural noise-producing phenomena include atmospheric disturbances, extra-terrestrial radiation, and internal circuit noise. The effects of many noise sources can be reduced or eliminated completely by careful engineering design. A simple example is the use of a shielded cable to minimize the amount of ignition noise picked up by the antenna cable of a TV receiver.

In spite of careful engineering design, there are some unavoidable causes of electrical noise. One such unavoidable cause is the thermal motion of electrons in conducting media—wires, resistors, and so forth. As long as electrical communication systems use such materials, *thermal noise* is inevitable. Thermal noise corrupts the desired signal in an *additive* fashion and its effects can be minimized by appropriate modulation techniques. We now develop statistical models for thermal noise that will be useful in the analysis and design of modulation schemes.

The properties of thermal noise have been derived by physicists using the quantum mechanics approach. We will simply quote the appropriate results and derive useful time domain and frequency domain models for thermal noise. We will confine our interest for the most part to noise which may be described by an ergodic, Gaussian random process. In many communication systems and in a wide variety of other applications, the assumptions of a Gaussian density and ergodicity are justifiable. On the other hand, it needs to be noted that such an assumption is hardly universally valid. For example, the noise encountered in a system due to a nearby thunderstorm is nonstationary and impulsive in nature due to pulse-type disturbances generated by lightning discharges. The assumption of Gaussian densities and ergodicity cannot be justified in such cases. However, noise from far away thunderstorms on a global basis may be assumed to have a Gaussian amplitude distribution.

3.7.1 Thermal Noise

Thermal noise is produced by the random motion of charged particles (usually electrons) in conducting media. This random motion at the atomic level is a universal characteristic of matter. It accounts for the ability of matter to store energy (supplied through the flow of heat) in the form of random agitation. The conductivity of a resistor or, for that matter, the flow of current through any medium results from the availability of electrons or other charged particles that are free to move. The random movement of these electrons produces the resistor noise. Resistor noise as well as other noise of similar origin is called *thermal noise.*

It has been determined experimentally (and derived theoretically from thermodynamic and quantum mechanical considerations) that the noise voltage $V(t)$ which appears across the terminals of a resistor of value R ohms has a Gaussian distribution with $\mu_V = 0$ and

$$E\{V^2\} = \frac{2(\pi kT)^2}{3h} R \text{ (volts squared or volt}^2\text{)} \tag{3.85}$$

where

k = Boltzman constant = $(1.37)(10^{-23})$ Joule/deg
h = Plank's constant = $(6.62)(10^{-34})$ Joule sec.

and T is the temperature of the resistor measured in degrees Kelvin (°K = °C + 273). Furthermore, it has been shown that the power spectral density of thermal noise is

$$G_V(f) = \frac{2Rh|f|}{\exp(h|f|/kT) - 1} \text{ volt}^2/\text{Hz} \tag{3.86}$$

A plot of $G_V(f)$ versus f for $f > 0$ is shown in Figure 3.6. Figure 3.6 shows that $G_V(f)$ is essentially flat over the range $|f| < 0.1(kT/h)$. At room temperature $G_V(f)$ is flat over the range of frequencies $|f| < 10^{12}$ Hz. The upper limit 10^{12} Hz is well outside the range of frequencies used in conventional electrical communication systems. Hence, for purposes of modeling, the psd of thermal noise in electrical communication systems can be assumed to be

$$G_V(f) = 2RkT \text{ volt}^2/\text{Hz} \tag{3.87}$$

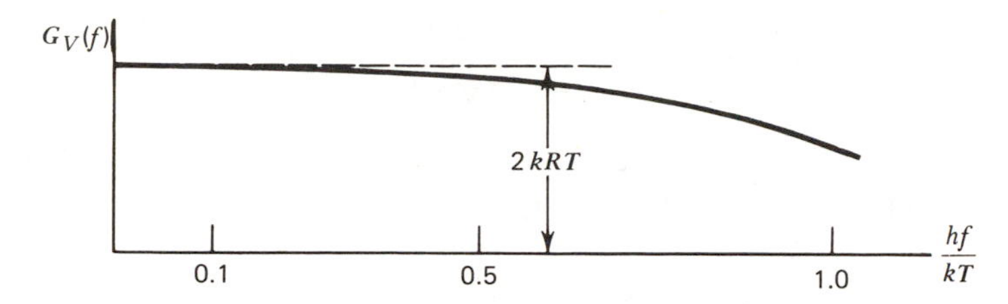

Figure 3.6 Psd of thermal noise.

Equation (3.87) seems to imply that

$$E\{V^2(t)\} = \int_{-\infty}^{\infty} G_V(f)\, df = \infty$$

However, one never deals with infinite bandwidth systems. $V(t)$ is always subjected to some filtering, and $E\{V^2(t)\}$ over a finite bandwidth is always finite.

Rather than dealing with the mean squared value of noise voltage, it is often more convenient to describe thermal noise using available noise power. *Available power* is the maximum power that can be delivered to a load from a source having a fixed but non-zero resistance. If we represent a noisy resistor by a voltage source $V(t)$ in series with a noiseless resistor R, then the maximum power is delivered to the load when the load resistance R_L matches the source resistance R, that is, when $R_L = R$. If the load is matched to the source, then the maximum power delivered is given by

$$P_{\max} = \frac{\langle V^2(t)\rangle}{4R} = \frac{E\{V^2(t)\}}{4R}$$

Extending this concept to the thermal resistor viewed as a noise source, the available power spectral density at the matched load becomes

$$G_V(f) = kT/2 \text{ watts/Hz} \tag{3.88}$$

Equation (3.88) implies that a noisy resistor delivers a maximum of $kT/2$ watts/Hz power to a matched load at all frequencies, regardless of the value of R.

Besides thermal noise many other types of noise sources are Gaussian (by virtue of the central limit theorem) and have spectral densities that are flat over a wide range of frequencies. Examples of such noise sources include solar thermal noise, cosmic noise from our own galaxy and from extragalactic sources. A noise signal having a flat power spectral density over a wide range of frequencies is called *white noise* by analogy to white light. The power spectral density of white noise is denoted by

$$G_V(f) = \eta/2 \text{ watts/Hz} \tag{3.89}$$

where the factor of 2 is included to indicate that $G_V(f)$ is a two-sided psd. The autocorrelation function of the white noise is given by

$$R_{VV}(\tau) = F^{-1}\{G_V(f)\} = \frac{\eta}{2}\delta(\tau) \tag{3.90}$$

Equation (3.90) shows that any two different samples of a zero mean Gaussian white noise are uncorrelated and hence independent.

We will use zero mean stationary Gaussian bandlimited white noise as the model for the noise that accompanies the signal at the receiver input. The reason for concentrating on the noise at the receiver input is due to the fact that it is at this point in the system that the signal component is weakest. Any attempt to increase signal power by amplification at the receiver will amplify the noise power also. Thus the noise that accompanies the signal at the receiver input will have considerable influence on the quality of the output signal. The bandwidth of the noise will be assumed to be the same as the bandwidth of the signal $X_c(t)$.

3.7.2 Time Domain Representation of Narrowband Noise

The received signal $X_c(t)$ in many analog communication systems has the general form

$$X_c(t) = A(t)\cos(\omega_c t + \phi(t)), \quad \omega_c = 2\pi f_c \tag{3.91}$$

where $A(t)$ and $\phi(t)$ are lowpass power signals. The receiver performs filtering and demodulation operations on $X_c(t)$. These operations are well defined and are easily tracked either in the time domain or in the frequency domain. If the signal component is tracked through the receiver using frequency domain analysis, then the noise component can be tracked through the receiver using the same frequency domain analysis. If the signal is tracked using time domain analysis, then we can track the noise also using a similar approach if we have a time domain representation for the filtered version of the noise $Y(t)$ (Figure 3.7). The filtering that takes place before demodulation is bandpass filtering and hence we are interested in time domain representation for bandpass noise. That is, we are interested in a representation for $Y(t)$, the filtered version of $V(t)$, of the form

$$Y(t) = R(t)\cos[2\pi f_c t + \theta(t)] \tag{3.92}$$

where f_c is the center frequency of the filter. To elaborate further, we are

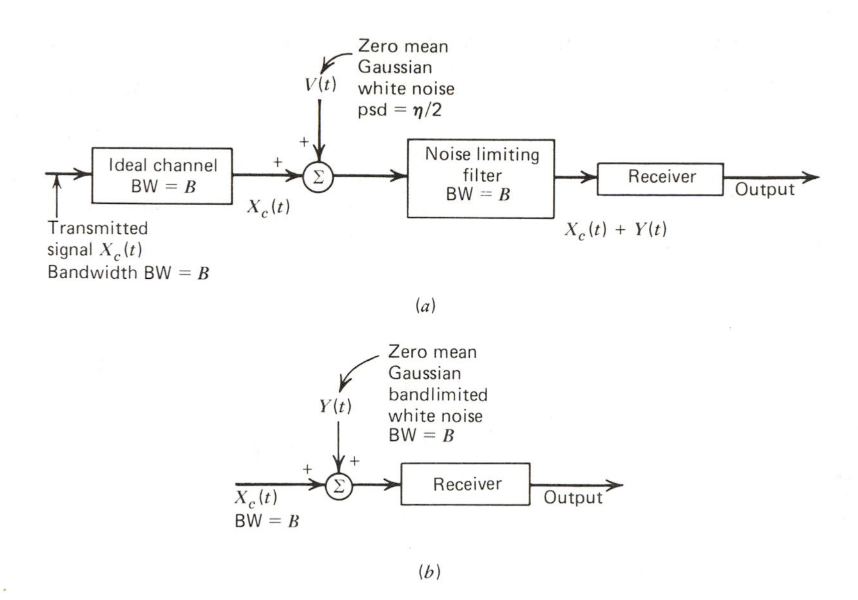

Figure 3.7 (*a*) Model of a noisy communication system. (*b*) Simplified model of the system.

interested in deriving the properties of $R(t)$ and $\theta(t)$ given the frequency domain description of $Y(t)$.

Filtered White Noise. Let $Y(t)$ be the output of an ideal bandpass filter with a center frequency f_c and bandwidth Δf whose input is a zero mean stationary Gaussian white noise $V(t)$ with a psd = $\eta/2$. The output of the filter will be a zero mean stationary Gaussian process (Problem 3.36) with a psd shown in Figure 3.8. Let us now attempt to obtain a time domain representation of $Y(t)$ given its frequency domain representation. The autocorrelation function of $Y(t)$ is given by

$$\begin{aligned} R_{YY}(\tau) &= F^{-1}\{G_Y(f)\} \\ &= 2[G_Y(f_c)\Delta f](\cos \omega_c\tau)(\text{sinc } \Delta f\tau) \end{aligned} \tag{3.93}$$

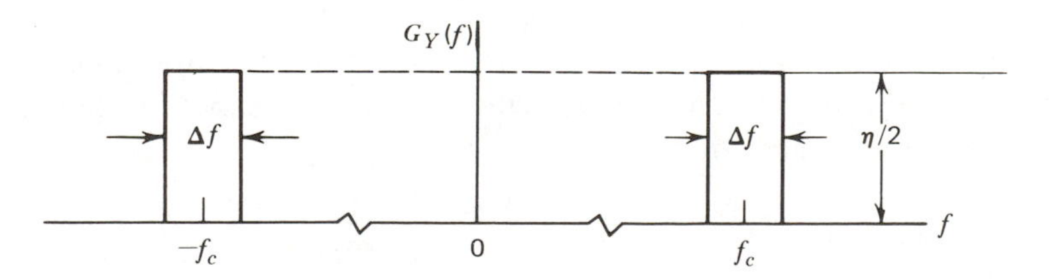

Figure 3.8 Psd of filtered white noise. Assume Δf to be very small.

Now if we restrict our attention to relatively small values of τ ($|\tau| \ll 1/\Delta f$), then sinc $\Delta f\tau \approx 1$ and

$$R_{YY}(\tau) \approx 2G_Y(f_c)\Delta f \cos \omega_c\tau, \qquad |\tau| \ll 1/\Delta f \tag{3.94}$$

If $\Delta f \to 0$, the preceding expression is valid for all values of τ. Comparing Equation (3.94) with autocorrelation function of the randomly phased sinusoids (Example 3.6), we find that we can approximate $Y(t)$ by

$$Y(t) \approx \tilde{Y}(t) = C \cos(\omega_c t + \theta) \tag{3.95a}$$

$$= A \cos \omega_c t + B \sin \omega_c t \tag{3.95b}$$

where A, B, C, and θ are random variables.

Now, we have to choose appropriate probability distribution for A and B (or C and θ) such that $\tilde{Y}(t)$ is a stationary zero mean Gaussian random process. The reader can verify that (Problem 3.42) if A and B are chosen to be Gaussian random variables with

$$E\{A\} = E\{B\} = 0 \tag{3.96}$$

$$E\{A^2\} = E\{B^2\} = 2G_y(f_c)\Delta f \tag{3.97}$$

and

$$E\{AB\} = 0 \tag{3.98}$$

then $\tilde{Y}(t)$ is a zero mean stationary Gaussian random process with an autocorrelation function

$$R_{\tilde{Y}\tilde{Y}}(\tau) = 2G_Y(f_c)\Delta f \cos \omega_c\tau \approx R_{YY}(\tau)$$

Thus $\tilde{Y}(t)$ is a reasonable approximation of $Y(t)$.

The time domain representation given in Equation (3.95) is a valid approximation as long as Δf is small and A, B are Gaussian variables with the properties given in Equations (3.96), (3.97), and (3.98).

Time Domain Representation of Narrowband Noise. We now turn our attention to the problem of obtaining time domain representations for narrowband noise using the results derived in the preceding section. To begin with, let $Y(t)$ be the output of a bandpass filter whose input is a zero mean stationary Gaussian random process $X(t)$. Assume that the filter response $H(f)$ is zero for $|f - f_c| > B/2$. Then $Y(t)$ will be a zero mean Gaussian random process (Problem 3.36) with a psd

$$G_Y(f) = \begin{cases} G_X(f)|H(f)|^2, & |f - f_c| < B/2 \\ 0, & \text{elsewhere} \end{cases}$$

An example of the psd of $G_Y(f)$ is shown in Figure 3.9.

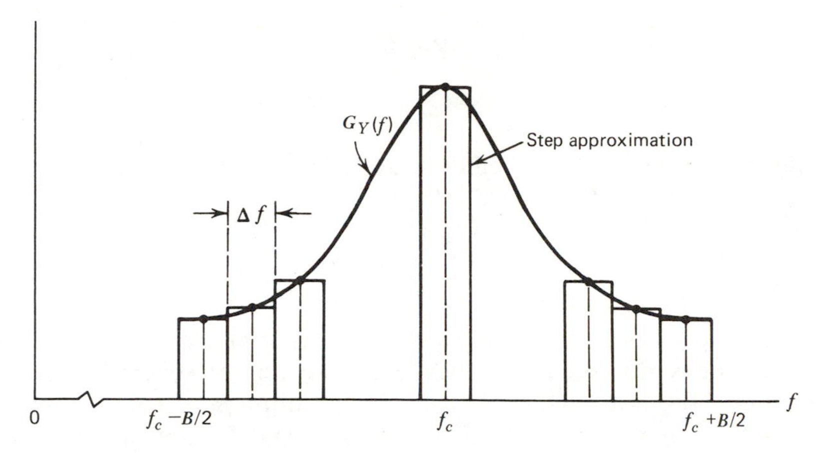

Figure 3.9 Power spectral density of narrowband noise. $G_Y(f)$ is shown for positive values of f only. It is assumed that $B \ll f_c$.

We can divide the spectrum of $G_Y(f)$ into $2n+1$ very narrow bands of width Δf centered at $f_c \pm m\Delta f$ $(m = 0, \pm 1, \ldots, \pm n)$. A spectral pair of $Y(t)$ at $(f_c + m\Delta f)$ and $-(f_c + m\Delta f)$ can be approximated by

$$\tilde{Y}_m(t) = A_m \cos(2\pi f_m t) + B_m \sin(2\pi f_m t)$$

where $f_m = f_c + m\Delta f$. A_m and B_m are uncorrelated, zero mean Gaussian random variables with

$$\begin{aligned} &E\{A_m\} = E\{B_m\} = 0 \\ &E\{A_m^2\} = E\{B_m^2\} = 2G_Y(f_m)\Delta f \\ &E\{A_m B_m\} = 0 \end{aligned} \tag{3.99}$$

Using a similar representation for other nonoverlapping spectral components of $Y(t)$, we can approximate $Y(t)$ by

$$\tilde{Y}(t) = \sum_{m=-n}^{n} A_m \cos(2\pi f_m t) + B_m \sin(2\pi f_m t) \tag{3.100}$$

Now for $\tilde{Y}(t)$ to be stationary, we need the additional constraints (see Problem 3.43) that

$$E\{A_i A_j\} = E\{A_i B_j\} = E\{B_i B_j\} = 0, \quad i \neq j \tag{3.101}$$

With these constraints we have

$$R_{\tilde{Y}\tilde{Y}}(\tau) = \tfrac{1}{2} \sum_{i=-n}^{n} E\{(A_i^2 + B_i^2)\} \cos(2\pi f_i \tau)$$

and

$$E\{\tilde{Y}^2(t)\} = R_{\tilde{Y}\tilde{Y}}(0)$$
$$= \sum_{i=-n}^{n} 2G_Y(f_i)\Delta f$$

The right-hand side of the preceding equation is a stepwise approximation to the area under the psd function $G_Y(f)$. That is,

$$\sum_{i=-n}^{n} 2G_Y(f_i)\Delta f \approx 2\int_{f_c-B/2}^{f_c+B/2} G_Y(f)\, df = E\{Y^2(t)\}$$

In summary, we find that a narrowband stationary Gaussian random process can be represented as a linear superposition of spectral components of the form given in Equation (3.100). This description is a reasonably good approximation when $\Delta f \to 0$. The coefficients A_k and B_k are Gaussian random variables with zero mean and equal variances. The coefficients A_k and B_k are uncorrelated with each other and are uncorrelated also with the coefficients of a spectral component at a different frequency.

Quadrature Representation of Narrowband Noise. It is often convenient to express narrowband noise in a quadrature form

$$\tilde{Y}(t) = Y_c(t)\cos 2\pi f_c t - Y_s(t)\sin 2\pi f_c t \tag{3.102}$$

We may readily transform Equation (3.100) into the form (3.102) and in so doing arrive at explicit expressions for $Y_c(t)$ and $Y_s(t)$ as follows:

$$\begin{aligned}\tilde{Y}(t) &= \sum_{m=-n}^{n} \{A_m \cos[2\pi(f_c + m\Delta f)t] + B_m \sin[2\pi(f_c + m\Delta f)t]\}\\ &= \cos 2\pi f_c t \sum_{m=-n}^{n} [A_m \cos(2\pi m\Delta f t) + B_m \sin(2\pi m\Delta f t)]\\ &\quad -\sin 2\pi f_c t \sum_{m=-n}^{n} [A_m \sin(2\pi m\Delta f t) - B_m \cos(2\pi m\Delta f t)]\end{aligned} \tag{3.103}$$

From Equation (3.103), $Y_c(t)$ and $Y_s(t)$ are readily identified as

$$Y_c(t) = \sum_{m=-n}^{n} \sqrt{A_m^2 + B_m^2}\cos(2\pi m\Delta f t - \theta_m) \tag{3.104}$$

$$Y_s(t) = \sum_{m=-n}^{n} \sqrt{A_m^2 + B_m^2}\sin(2\pi m\Delta f t - \theta_m) \tag{3.105}$$

where $\theta_m = \tan^{-1}(B_m/A_m)$.

Using the properties given in Equations (3.99) and (3.100), the reader can verify that $Y_c(t)$ and $Y_s(t)$ are zero mean Gaussian random processes and that

$Y_c(t)$ and $Y_s(t)$ are *uncorrelated*, that is, $E\{Y_c(t_1)Y_s(t_2)\} = 0$ for all values of $t_1, t_2 \in (-\infty, \infty)$ (Problems 3.44 and 3.45).

The power spectral density of $Y_c(t)$ and $Y_s(t)$ can be obtained from the power spectral density of $Y(t)$ as follows: Let $[G_{Y_c}(m\Delta f)]2\Delta f$ be the total average power contained in the spectral components of $Y_c(t)$ at frequencies $\pm m\Delta f$. From Equation (3.104), the power contained in these components may be obtained as

$$[G_{Y_c}(m\Delta f)]2\Delta f = E\left(\frac{A_m^2 + B_m^2}{2} + \frac{A_{-m}^2 + B_{-m}^2}{2}\right)$$

Substituting for $E\{A_m^2\}$, $E\{B_m^2\}$, $E\{A_{-m}^2\}$, and $E\{B_{-m}^2\}$ from Equation (3.99), we have

$$\begin{aligned}[G_{Y_c}(m\Delta f)]2\Delta f &= 2\Delta f[G_Y(f_m) + G_Y(f_{-m})] \\ &= 2\Delta f[G_Y(f_c + m\Delta f) + G_Y(f_c - m\Delta f)]\end{aligned}$$

or

$$G_{Y_c}(m\Delta f) = G_Y(f_c + m\Delta f) + G_Y(f_c - m\Delta f)$$

Letting $\Delta f \to 0$ and $m\Delta f = f$, we have

$$G_{Y_c}(f) = \begin{cases} G_Y(f_c + f) + G_Y(f_c - f), & |f| < B/2 \\ 0, & \text{elsewhere} \end{cases} \tag{3.106}$$

The power spectral density of $G_{Y_s}(f)$ has the same form as the psd of $G_{Y_c}(f)$.

Equation (3.106) allows us to construct the spectra $G_{Y_c}(f)$ and $G_{Y_s}(f)$ in the following manner:

1. Move the positive frequency portion of the plot of $G_Y(f)$ to the left by f_c so that the portion of the plot originally located at f_c is now moved to the origin on the frequency axis.
2. Move the negative frequency portion of $G_Y(f)$ to the right by f_c.
3. Add the two shifted plots.

Since $G_{Y_c}(f) = G_{Y_s}(f)$, it follows from Equation (3.106) that (Problem 3.45)

$$E\{Y_c^2(t)\} = E\{Y_s^2(t)\} = E\{Y^2(t)\} = N_0 \tag{3.107}$$

Envelope and Phase Representation of Narrowband Noise. We may represent the narrowband noise $\tilde{Y}(t)$ in an envelope and phase form as

$$\tilde{Y}(t) = R(t)\cos[2\pi f_c t + \theta(t)] \tag{3.108}$$

where the envelope $R(t)$ and the phase $\theta(t)$ are given by

$$R(t) = \sqrt{[Y_c(t)]^2 + [Y_s(t)]^2} \tag{3.109a}$$

and

$$\theta(t) = \tan^{-1}\left(\frac{Y_s(t)}{Y_c(t)}\right) \tag{3.109b}$$

Using the results of Example 3.3 we can show that the pdf's of R and θ are

$$f_R(r) = \frac{r}{N_0}\exp(-(r^2/2N_0)), \quad N_0 = E\{Y_c^2(t)\} = E\{Y_s^2(t)\}, \quad r > 0 \tag{3.110a}$$

$$f_\theta(\theta) = \frac{1}{2\pi}, \quad -\pi < \theta < \pi \tag{3.110b}$$

The pdf of R given in Equation (3.110) is called a *Rayleigh pdf*.

We will use both the quadrature representation and the envelope and phase form when we analyze the effects of noise in communication systems.

Envelope of Narrowband Noise Plus Sine Wave. The message signal in many communication systems is modeled as a sinusoidal wave $A\cos(2\pi f_c t)$, where A and f_c are constants. When the message signal arrives at the receiver, it is often accompanied by additive narrowband Gaussian noise. To analyze how this additive Gaussian noise affects the signal amplitude, we need to determine the pdf of the envelope of the sinusoidal signal plus noise. That is, we need to find the pdf of the envelope of

$$Z(t) = A\cos(2\pi f_c t) + n(t) \tag{3.111}$$

where $n(t)$ is a stationary zero mean Gaussian random process with a bandpass psd $G_n(f)$ centered at f_c. We will assume that the bandwidth of the noise is very small compared to f_c.

We can make use of the quadrature components of $n(t)$ to express $Z(t)$ as

$$\begin{aligned} Z(t) &= A\cos(2\pi f_c t) + n_c(t)\cos(2\pi f_c t) - n_s(t)\sin(2\pi f_c t) \\ &= [A + n_c(t)]\cos(2\pi f_c t) - n_s(t)\sin(2\pi f_c t) \end{aligned}$$

The envelope of $Z(t)$ is given by

$$R(t) = \sqrt{[A + n_c(t)]^2 + [n_s(t)]^2}$$

To obtain the pdf or $R(t)$, let us use the following notation:

$$X_1 = A + n_c(t)$$
$$X_2 = n_s(t)$$
$$R = R(t)$$

Then, X_1 and X_2 are two independant Gaussian random variables with

$$E\{X_1\} = A, \qquad E\{X_2\} = 0$$

and

$$\sigma_{X_1}^2 = \sigma_{X_2}^2 = E\{n^2(t)\} = N_0$$

Hence, the joint pfd of X_1 and X_2 is

$$f_{X_1,X_2}(x_1, x_2) = \frac{1}{2\pi N_0} \exp\left(-\frac{[(x_1 - A)^2 + x_2^2]}{2N_0}\right)$$

To calculate the pdf of R, let us define another variable θ as

$$\theta = \tan^{-1}(X_2/X_1)$$

so that we can write the joint pdf of R and θ as

$$f_{R,\theta}(r, \theta) = \frac{r}{2\pi N_0} \exp\left(-\frac{[r^2 + A^2 - 2Ar\cos\theta]}{2N_0}\right); \quad r > 0, -\pi < \theta < \pi$$

The pdf of R is obtained by integrating out θ as (see Example 3.3)

$$\begin{aligned} f_R(r) &= \int_{-\pi}^{\pi} f_{R,\theta}(r, \theta)\, d\theta \\ &= \frac{r}{N_0} \exp\left(-\frac{(A^2 + r^2)}{2N_0}\right)\left[\frac{1}{2\pi}\int_{-\pi}^{\pi} \exp\left(-\frac{Ar\cos\theta}{N_0}\right) d\theta\right] \end{aligned}$$

The bracketed integral in the preceding equation is the *modified Bessel function* of the first kind of zero order defined by

$$I_0(a) = \frac{1}{2\pi}\int_{-\pi}^{\pi} \exp(a\cos(u))\, du = \frac{1}{2\pi}\int_{0}^{2\pi} \exp(a\cos(u))\, du$$

Using the Bessel function defined above, we can write the pdf of the envelope of a sine wave plus narrowband noise as

$$f_R(r) = \frac{r}{N_0} I_0\left(\frac{Ar}{N_0}\right) \exp\left(-\frac{(r^2 + A^2)}{2N_0}\right), \quad r > 0 \tag{3.112}$$

The reader should note that $f_{R,\theta}(r, \theta) \neq f_R(r) f_\theta(\theta)$ because of the presence of the term $2Ar\cos\theta$ in the joint pdf.

The pdf given in Equation (3.112) is known as the Rice–Nagakami pdf or simply the Rician pdf. The Rician pdf reduces to the Rayleigh pdf when $A = 0$.

3.7.3 Signal-to-Noise Ratio and Probability of Error

The presence of noise degrades the performance of analog and digital communication systems. The extent to which noise affects the performance of communication systems is measured by the output signal to noise power ratio and the probability of error. The signal-to-noise ratio is used to measure the performance in analog communication systems, whereas the probability of error is used as a performance measure in digital communication systems.

Signal-to-Noise Ratios. Consider an analog communication system that is designed to transmit the output of an analog information source. Assume that the source output is modeled by the random process $X(t)$, and that the channel introduces no distortion other than additive random noise. Then, the output of the system can be represented as

$$Y(t) = KX(t - t_d) + n_0(t)$$

where $KX(t - t_d)$ is the delayed version of the input $X(t)$, and $n_0(t)$ is the noise that accompanies the output signal.

The average signal power at the receiver output is given by

$$S_0 = E\{[KX(t - t_d)]^2\} = K^2 \int_{-\infty}^{\infty} G_X(f)\, df \tag{3.113a}$$

and the average noise power at the output is

$$N_0 = E\{n_0^2(t)\} \tag{3.113b}$$

The output signal quality is measured in terms of the output *signal-to-noise ratio* $(S/N)_0$ defined as

$$\left(\frac{S}{N}\right)_0 = \left(\frac{S_0}{N_0}\right) = \frac{K^2 E\{X^2(t)\}}{E\{n_0^2(t)\}} \tag{3.114}$$

The ratio defined in Equation (3.114) is a dimensionless quantity and is often expressed in decibels as

$$\left(\frac{S}{N}\right)_0 \text{ in dB} = 10 \log_{10}\left(\frac{S_0}{N_0}\right)$$

We will use the output signal-to-noise ratio for comparing the performance of analog communication systems. The output signal-to-noise ratio will range from 60 dB for superior quality audio systems to 10 dB for barely intelligible voice communication.

Probability of Error. Consider a digital communication system whose input is a sequence of symbols $\{S_k\}$. The output of the system will be a sequence of symbols $\{\hat{S}_k\}$. In an ideal system $\{\hat{S}_k\}$ will be the same as $\{S_k\}$. However, in practical systems the input and output sequences will differ occasionally due to errors caused by the channel noise. The overall performance of a digital communication system is measured in terms of the probability of symbol errors P_e defined as

$$P_e = P(\hat{S}_k \neq S_k) \tag{3.115}$$

We will use P_e for comparing the performance of digital communication systems. The values of P_e range from 10^{-4} to 10^{-7} in practical digital communication systems.

3.7.4 Noise Equivalent Bandwidth, Effective Noise Temperature, and Noise Figure

Two of the important signal processing operations performed in a communication receiver are filtering and amplification. In practical systems, these operations are performed by two-port devices (with one input port and one output port) consisting of active devices and passive components. The signal at the input to these devices in the receiver will be accompanied by noise. If the two-port device amplifies the signal, the accompanying noise also will be amplified. Furthermore, the resistors and active devices in a two-port network act as noise sources and this internal noise degrades the signal as it passes through the network.

In this section we develop appropriate measures for describing the extent to which a signal is degraded in passing through a two-port network. The signal quality at the input is characterized by the input signal-to-noise ratio. Because of the noise sources within the two-port, the signal-to-noise ratio at the output will be lower than at the input. This degradation in the signal quality is characterized in terms of noise equivalent bandwidth, effective noise temperature, and noise figure.

Noise Equivalent Bandwidth. In the absence of internal noise sources, the noise power at the output of a two-port will depend on the input noise psd and the transfer function of the two-port. If the noise at the input is white, then the output noise power N_0 is given by

$$N_0 = \int_{-\infty}^{\infty} \left(\frac{\eta}{2}\right) |H(f)|^2 \, df = \eta \int_0^{\infty} |H(f)|^2 \, df$$

Since the integral in the preceding equation depends only on the filter transfer function, we can simplify the discussion of noise power by defining a *noise equivalent bandwidth* B_N as (see Figure 3.10)

$$B_N \stackrel{\Delta}{=} \frac{1}{g_a} \int_0^{\infty} |H(f)|^2 \, df \tag{3.116}$$

where $g_a = |H(f)|^2_{\max}$ is the *maximum available power gain.* The output noise power can be expressed in terms of g_a, η, and B_N as

$$N_0 = g_a \eta B_N \tag{3.117}$$

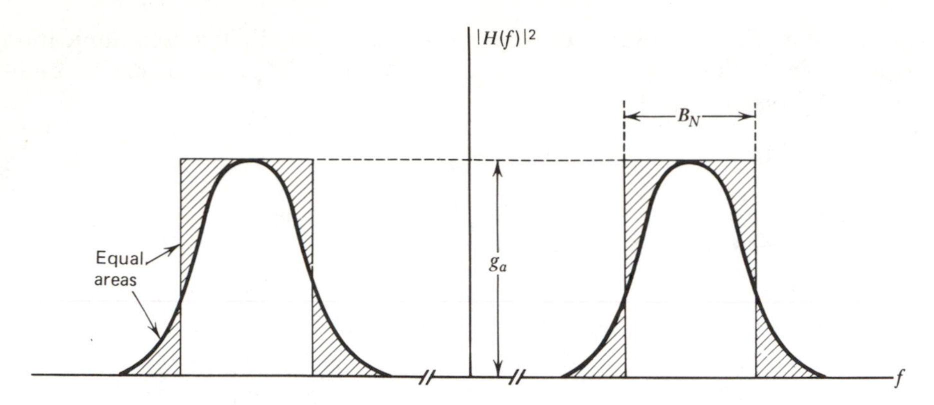

Figure 3.10 Noise equivalent bandwidth of a bandpass filter.

In most practical filters, g_a represents the midband power gain.

Equation (3.117) states, in a simple form, that the noise power at the output of a filter is characterized by the parameters g_a and B_N of the filter and the input noise psd. Furthermore, B_N is equal to the bandwidth of an ideal rectangular filter with a power gain g_a that would pass as much white noise power as the filter in question. The noise bandwidth of an ideal filter is the actual filter bandwidth. However, for practical filters, B_N will be somewhat greater than the half-power (3-dB) bandwidth. It is left as an exercise for the reader to verify that an RC lowpass filter with a 3-dB bandwidth of B has a noise equivalent bandwidth $B_N = \pi B/2$.

Effective Noise Temperature. With reference to Figure 3.11, let us assume that the two-port network is driven by a matched source (source impedance = two-port input impedance) and the filter output drives a matched load (load impedance = two-port output impedance). With matched impedances we can use available noise and signal powers. Thus, if the source delivers its available power S_{ai} to the two-port, then the output signal power is

$$S_{a0} = g_a S_{ai}$$

where g_a is the available power gain. Similarly, the output noise power is

$$N_{a0} = g_a N_{ai} + N_{ax}$$

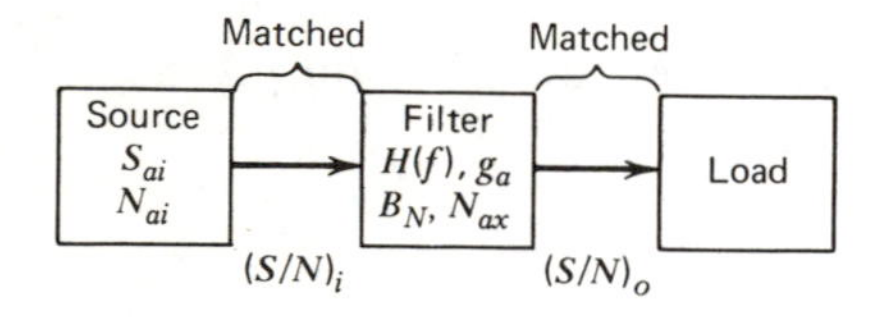

Figure 3.11 A two-port device with a matched source and a matched load.

where N_{ax} is the excess noise power at the output due to internal noise generated within the two-port network. If the input noise is white and represented by an equivalent noise temperature T_i, then

$$N_{ai} = kT_iB_N$$

and

$$\begin{aligned} N_{a0} &= g_akB_N\left(T_i + \frac{N_{ax}}{g_akB_N}\right) \\ &= g_akB_N[T_i + T_e] \end{aligned} \tag{3.118}$$

where

$$T_e = \frac{N_{ax}}{g_akB_N} \tag{3.119}$$

Equation (3.118) shows that the output noise power appears to originate from a thermal noise source at a temperature of $T_i + T_e$. Hence, we might model the two-port itself as a noise free device, and account for the increased noise by assigning to the input noise source a new temperature higher than T_i by T_e. The temperature T_e is called the *effective (input) noise temperature* of the two-port. Typical effective noise temperature of some practical devices are given in Table 3.1. Some of these devices are cooled by liquid nitrogen or helium to achieve low effective noise temperatures ($T_e = 0$ for a noise free two-port device).

The signal-to-noise ratios at the input and output of a two-port device can be shown to be

$$\left(\frac{S}{N}\right)_0 = \left(\frac{S}{N}\right)_i \frac{1}{1 + (T_e/T_i)} \tag{3.120}$$

Table 3.1. Typical Noise Figures and Effective Noise Temperatures

Device	Noise Figure F (dB)	T_e (°K)	g_a (dB)	Frequency
Maser	0.16	11	20–30	6 GHz
Parametric Amplifier				
Uncooled	1.61	130	10–20	9 GHz
Liquid N_2 Cooled	0.69	50		3 &6 GHz
Liquid He Cooled	0.13	9		4 GHz
Traveling Wave Tube				
(TWT) Amplifier	2.70	250	20–30	3 GHz
Integrated Circuit (IC)				
Amplifier	7.0	1163	50	≤70 MHz

As $T_e/T_i \to 0$, the output signal quality is the same as the signal quality at the input as measured by the signal-to-noise ratio. Thus T_e is a measure of noisiness of the two-port referred to the input temperature T_i.

Noise Figure. Another measure of the excess internal noise generated by a two-port is the integrated *noise figure* F that is defined as the actual output noise power divided by the output noise power if the two-port were noiseless, *assuming that the noise source at the input is at room temperature* T_0 (290°K or 17°C). The noise figure is a power ratio written as

$$F \triangleq \frac{N_{a0}}{g_a k T_0 B_N} \tag{3.121a}$$

$$= 1 + \frac{T_e}{T_0} \tag{3.121b}$$

Equation (3.121b) shows that in general the noise figure $F \geqslant 1$ and that $F = 1$ for a noiseless two-port.

Finally, if the input noise source is at room temperature, then it can be shown that

$$\left(\frac{S}{N}\right)_0 = \frac{1}{F}\left(\frac{S}{N}\right)_i \tag{3.122}$$

Two-ports are most commonly characterized in terms of noise figure when the driving noise source is at or near T_0, while the effective noise temperature T_e is more convenient when the input noise temperature is not near T_0.

The concept of noise figure also applies to devices such as attenuator and other devices that are characterized by power loss rather than gain. Now, if such devices consist entirely of resistive elements in thermal equilibrium at temperature T_0, then available noise power at the output is

$$N_{a0} = kT_0B_N \tag{3.123a}$$

From Equation (3.118), $N_{a0} = g_a k B_N (T_i + T_e)$, and with $T_i = T_0$ we have

$$N_{a0} = kB_N(T_0 + T_e)/L \tag{3.123b}$$

where $g_a = 1/L$ and L is the attenuator power loss. From (3.123a), (3.123b), and (3.121b) we obtain

$$F = L \tag{3.123c}$$

for a lossy two-port comprised of resistors in thermal equilibrium at T_0 (= 290°K).

Example 3.12. Consider two two-ports in cascade each with the same noise bandwidth but with different available power gains and noise figures (Figure

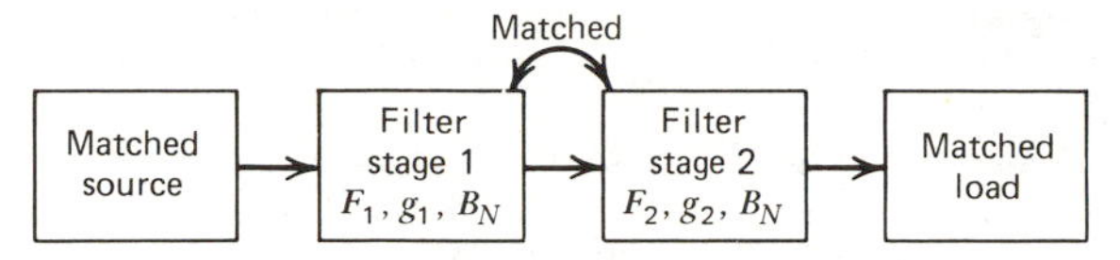

Figure 3.12 Multistage filtering.

3.12). Find the overall noise figure of the two circuits in cascade. (Assume perfect impedance matching.)

Solution

$$\begin{aligned} N_{a0} &= \text{noise from stage 1 at the output of stage 2} \\ &\quad + \text{noise } \Delta N_2 \text{ introduced by network 2.} \\ &= (kT_0B_N)F_1g_1g_2 + \Delta N_2 \\ &= (kT_0B_N)F_1g_1g_2 + \underbrace{(F_2-1)T_0}_{T_{e2}}(kB_Ng_2) \\ &= (kT_0B_N)Fg_1g_2 \end{aligned}$$

where F is the overall noise figure. Hence,

$$F = F_1 + \frac{F_2 - 1}{g_1}$$

Generalizing this result to n stages in cascade, we have

$$F = F_1 + \frac{F_2-1}{g_1} + \frac{F_3-1}{g_1g_2} + \cdots + \frac{F_n-1}{g_1g_2\cdots g_{n-1}} \tag{3.124}$$

Equation (3.124) is known as the *Friis' formula* and it implies that the overall noise figure is highly dependent on the noise figure of the first stage. Hence the first stage or the "front end" should have a low noise figure and high gain. With a good "front end," the remaining stages provide additional gain, amplifying both signal and noise without appreciably changing the signal-to-noise ratio.

Example 3.13. In TV receivers, the antenna is often mounted on a tall mast and a long lossy cable is used to connect the antenna and the receiver. To overcome the effect of the lossy cable, a preamplifier is mounted on the antenna as shown in Figure 3.13. Typical values of the parameters are also shown in the figure.

(a) Find the overall noise figure of the system.

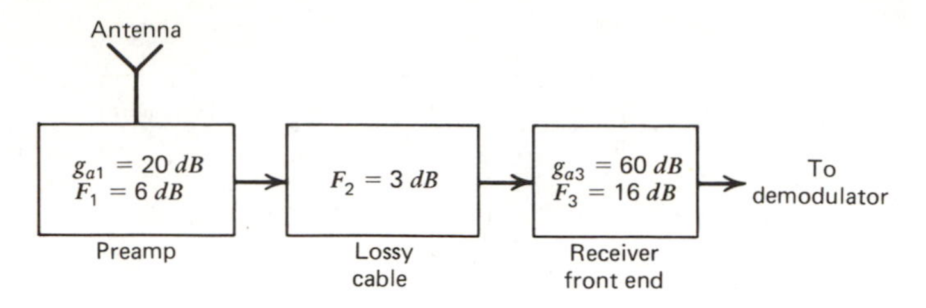

Figure 3.13 Receiver front end with preamplifier.

(b) Find the overall noise figure of the system if the preamplifier is omitted and the gain of the front end is increased by 20 dB (assume $F = 16$ dB for the front end).

Solution

(a) Converting the gain and noise figures from decibels to ordinary power ratios, we have

$$(F)_{\text{overall}} = 4 + \frac{2-1}{100} + \frac{40-1}{(100)\frac{1}{2}} = 4.79 = 6.8 \text{ dB}$$

Note that we have made use of the fact that $L_2 = 1/g_{a2} = F_2$.

(b) Without the preamplifier,

$$F = F_2 + \frac{(F_3 - 1)}{(1/L_2)}$$
$$= 2 + (40 - 1)2 = 80 = 19 \text{ dB}$$

Example 3.14. A ground station for receiving signals from a satellite has an antenna with a gain of 40 dB and a noise temperature of 60°K (i.e., the antenna acts like a noisy resistor at 60°K). The antenna feeds a preamplifier with an effective input noise temperature of 125°K and a gain of 20 dB. The preamplifier is followed by an amplifier with a noise figure of 10 dB and a gain of 80 dB. The transmission bandwidth is 1 MHz. The satellite has an antenna with a power gain of 6 dB and the total path losses are 190 dB.

(a) Calculate the average thermal noise power at the receiver output.

(b) Calculate the minimum satellite transmitter power required tc achieve a 20-dB signal-to-noise ratio at the receiver output.

Solution

(a) Source temperature of noise at the receiver input = 60°K. Effective input noise temperature of the preamp = T_{e1} = 125°K. Available gain of the preamp $= g_{a1} = 100$. Effective noise temperature of the second amplifier $=$ T_{e2}, $T_{e2} = (F_2 - 1)T_0$, $F_2 = 10$, and $T_0 = 290°\text{K}$. Hence, $T_{e2} = 2610°\text{K}$. Effective input noise temperature of the receiver $= T_e$,

$$T_e = T_{e1} + \frac{T_{e2}}{g_{a1}} = 125 + \frac{2610}{100} = 151°\text{K}$$

Total available gain $= g_a = g_{a1}g_{a2} = 10^{10}$. Noise power at the receiver output $= k(T_i + T_e)Bg_a = 1.38(10^{-23})211(10^6)(10^{10}) = \underline{2.91\ (10^{-5})}$ watt.

(b) For $S/N = 20$ dB, signal power at the output $= 100$ (noise power) $=$ 2.91 (10^{-3}) watt. Total power loss $= 190 - 6 - 40 - 20 - 80 = -44$ dB $=$ 3.98 (10^{-5}). Hence, signal power at the transmitter $= 2.91(10^{-3})/3.98(10^{-5}) =$ $\underline{73.12 \text{ watts}}$.

3.8 SUMMARY

In this chapter, we developed models for random message signals and noise encountered in communication systems. Random signals cannot be explicitly described, prior to their occurrence, by deterministic functions of time. However, when examined over a long period, a random signal may exhibit certain regularities that can be described in terms of probabilities and statistical averages. Such a model, in the form of a probabilistic description of a collection of time functions, is called a random process. The random process model can be used to obtain, for example, the probability that a random signal is in a given range at a given time. Furthermore, under some mild assumptions, the random process model can be used to develop frequency domain descriptions of random message signals and noise. The description and analysis of random signals in the frequency domain using concepts borrowed from probability theory are very useful for the analysis and design of communication systems.

Probability theory deals with the concept of probabilities of random events and the concept of random variables. A random variable is a functional relationship that assigns numerical values to the outcomes of a random experiment. If the probabilities of the outcomes are known, then the probability that the random variable will have a value within a specified interval on the real line is known. Thus the random variable provides us with a probabilistic description of the numerical values associated with the outcomes of a random experiment. If all of the message waveforms that an information source can emit and the probabilities of their occurrences are known, then we can describe the collection of message waveforms emitted by the source using the concepts of probability theory. Such a description is called a random process.

A random process can be described in the time domain by joint probability distribution functions or by statistical averages. Two statistical averages that are most often used are the mean and autocorrelation function of a random process. If the mean and autocorrelation functions of a process are invariant under a translation of the time axis, then the random process is said to be wide sense stationary. Stationary random processes can be described in the frequency domain using power spectral density functions that are useful for determining signal locations in the frequency domain and bandwidths. A special subclass of stationary random processes called ergodic random processes has the property that time averages of the process are equal to statistical averages. For an ergodic process we can obtain a complete description based on time averages.

Random process models can also be used for characterizing internal and external noise in communication systems. The additive thermal noise in communication systems is usually modeled as a zero mean stationary Gaussian random process with a constant power spectral density over a wide range of frequencies. When such a noise is filtered, the filtered version can be represented as a sum of sinusoids with random amplitudes and phase angles. This representation leads to a quadrature model for narrowband noise that will be useful in our analysis of the performance of communication systems in the presence of additive noise.

Finally, when a random process is the input to a linear time-invariant system, the output is also a random process. If the input is stationary and Gaussian, so is the output. The relationships between the mean and autocorrelation function of the input, and the mean and autocorrelation function of the output are quite simple.

In the following chapters, we will use the zero mean, stationary Gaussian random process as the model for noise in communication systems. If the performance of the communication system is to be evaluated in the *absence of noise*, then we will use deterministic models and methods of analysis described in Chapter 2. However, if the *presence of noise* is to be treated, then we will use random process models for signals and noise.

REFERENCES

There are several well-written books that deal with topics in probability, random variables, and random processes. The books by Breipohl (1970) and Papoulis (1965) are excellent in their coverage and are recommended reading. Hogg and Craig's book (1965) is an introductory level treatment of the mathematical theory of probability and statistics. This book contains a large number of worked examples and is very readable. An advanced level treatment of topics in probability theory may be found in Feller's books (1957, 1967).

Davenport and Root's book (1958) and the books by Van Trees (1968, 1971) are good reference texts that deal with the subject matter covered in this chapter at a more advanced level. Finally, the book by Bendat and Piersol (1971) covers several useful topics dealing with the measurement and analysis of random data.

1. R. V. Hogg and A. T. Craig. *Introduction to Mathematical Statistics*. Macmillan, New York (1965).
2. W. Feller. *Introduction to Probability Theory and Applications*, vols. I–II. Wiley, New York (1957, 1967).
3. A. M. Breipohl. *Probabilistic Systems Analysis*. Wiley, New York (1970).
4. W. B. Davenport and W. L. Root. *Random Signals and Noise*. McGraw-Hill, New York (1958).
5. A. Papoulis. *Probability, Random Variables, and Stochastic Processes*. McGraw-Hill, New York (1965).
6. H. L. Van Trees, *Detection, Estimation, and Modulation Theory*, vols. I–III. Wiley, (1968, 1971).
7. J. S. Bendat and A. G. Piersol. *Random Data: Analysis and Measurement Procedures*. Wiley, New York (1971).

PROBLEMS

Section 3.2

3.1. A random experiment consists of tossing three fair coins and observing the faces showing up.
(a) List all the outcomes of this random experiment.
(b) Find the probability of the outcome head, tail, head.
(c) Find the probability of two heads in three tosses by counting the number of outcomes that have two heads.

3.2. A card is drawn from an ordinary deck of cards. The following events are associated with this random experiment:

$$A = \{\text{aces}\}$$
$$B = \{\text{cards belonging to the spade suit}\}$$
$$C = \{\text{cards belonging to the club suit}\}$$
$$D = \{\text{cards belonging to either diamond or heart suits}\}$$

(a) Find $P(A)$, $P(B)$, $P(C)$, and $P(D)$.
(b) Are events C and D mutually exclusive and exhaustive?

(c) Are events B and C mutually exclusive and exhaustive?
(d) Find $P(B \cup C)$.
(e) Find $P(\bar{A})$.

3.3. A box contains three 100 ohm resistors labeled R_1, R_2, and R_3 and two 1000 ohm resistors labeled R_4 and R_5. Two resistors are drawn from this box without replacement (and without looking into the box!).
(a) List all the outcomes of this random experiment. (A typical outcome may be listed as (R_1, R_5) to represent that R_1 was drawn first followed by R_5.)
(b) Find the probability that both resistors are 100 ohm resistors.
(c) Find the probability of drawing one 100 ohm resistor and one 1000 ohm resistor.
(d) Find the probability of drawing a 100 ohm resistor on the first draw and a 1000 ohm resistor on the second draw.
Work parts (b), (c), and (d) by counting the outcomes that belong to the appropriate events.

3.4. With reference to the random experiment described in Problem 3.3, define the following outcomes:

A_1 = 100 ohm resistor on the first draw
A_2 = 1000 ohm resistor on the first draw
B_1 = 100 ohm resistor on the second draw
B_2 = 1000 ohm resistor on the second draw

(a) Find $P(A_1B_1)$, $P(A_2B_1)$, $P(A_1B_2)$, and $P(A_2B_2)$.
(b) Find $P(A_1)$, $P(A_2)$, $P(B_1|A_1)$, and $P(B_1|A_2)$. Verify that $P(B_1) = P(B_1|A_1)P(A_1) + P(B_1|A_2)P(A_2)$.

3.5. Show that:
(a) $P(A \cup B \cup C) = P(A) + P(B) + P(C) - P(AB) - P(BC) - P(CA) + P(ABC)$.
(b) $P(A|B) = P(A)$ implies $P(B|A) = P(B)$.
(c) $P(ABC) = P(A)P(B|A)P(C|AB)$.

3.6. Suppose that four cards are drawn from an ordinary deck of cards without replacement. Find the probability of obtaining four aces using the chain rule.

3.7. A_1, A_2, and A_3 are three mutually exclusive and exhaustive sets of events associated with a random experiment E_1. Events B_1, B_2, and B_3 are mutually exclusive and exhaustive sets of events associated with a random experiment E_2. The joint probabilities of occurrence of these events, and some marginal probabilities are listed in the table given below.

A_i \ B_j	B_1	B_2	B_3
A_1	3/36	*	5/36
A_2	5/36	4/36	5/36
A_3	*	6/36	*
$P(B_j)$	12/36	14/36	*

(a) Find the missing probabilities (*) in the table.
(b) Find $P(B_3|A_1)$ and $P(A_1|B_3)$.
(c) Are events A_1 and B_1 statistically independent?

3.8. The probability that a student passes a certain exam is 0.9, given that he studied. The probability that he passes the exam without studying is 0.2. Assume that the probability that the student studies for an exam is 0.75 (somewhat a lazy student). Given that the student passed the exam, what is the probability that he studied?

Section 3.3

3.9. A fair coin is tossed four times and the faces showing up are observed.
(a) List all the outcomes of this random experiment.
(b) If X is the number of heads in each of the outcomes of this experiment, find the probability mass function of X.

3.10. Two dies are tossed. Let X be the sum of the numbers showing up. Find the probability mass function of X.

3.11. A random experiment E can terminate in one of three events A, B, or C with probabilities $\frac{1}{2}$, $\frac{1}{4}$, and $\frac{1}{4}$, respectively. The experiment is repeated three times. Find the probability that events A, B, and C each occur exactly one time.

3.12. Consider a random experiment that can terminate in three events A_1, A_2, or A_3 with probabilities p_1, p_2, and p_3 ($p_1+p_2+p_3=1$). Let n be the number of times the experiment is repeated. Let N_1, N_2, and N_3 be three random variables denoting the number of times events A_1, A_2, and A_3 occur, respectively, in n trials of the experiment. Then, $P(N_1=n_1$, $N_2=n_2$, and $N_3=n_3)$, $n_1+n_2+n_3=n$ is given by the *trinomial* probability mass function

$$P(N_1=n_1, N_2=n_2, N_3=n_3)=\frac{n!}{n_1!n_2!n_3!}(p_1)^{n_1}(p_2)^{n_2}(p_3)^{n_3}$$

Verify your answer to Problem 3.11 using the trinomial probability mass function.

3.13. Show that the mean and variance of a binomial random vairable X are $\mu_X = np$ and $\sigma_X^2 = npq$, where $q = 1 - p$.

3.14. When $n \gg 1$ and $p \ll 1$, the binomial probability mass function can be approximated by the *Poisson* probability mass function:

$$P(X = k) = \frac{\lambda^k}{k!}\exp(-\lambda), \quad \lambda = np, k = 0, 1, 2, \ldots$$

Use this model to find the probability of no transmission errors when a sequence of 100 binary digits is sent over a communication system that has a bit error probability of 10^{-3}. (Assume that occurrence of error in a particular bit position does not influence the probability of occurrence of an error in any other bit position.)

3.15. Suppose that you are trying to market a digital transmission system (modem) that has a bit error probability of 10^{-4}. The buyer will test your modem by sending a known message of 10^4 digits and checking the received message. If more than two errors occur, your modem will be rejected. Find the probability that the customer will buy your modem. (Use the approximation given in Problem 3.14.)

3.16. The input to a communication channel is a random variable X and the output is another random variable Y. The joint probability mass functions of X and Y are listed below:

Y \ X	−1	0	1
−1	$\frac{1}{4}$	$\frac{1}{8}$	0
0	0	$\frac{1}{4}$	0
1	0	$\frac{1}{8}$	$\frac{1}{4}$

(a) Find $P(Y = 1|X = 1)$.
(b) Find $P(X = 1|Y = 1)$.
(c) Find ρ_{XY}.

3.17. X and Y are discrete random variables with values $x_1, x_2, \ldots, x_n$ and $y_1, y_2, \ldots, y_m$. Define the *conditional expected value* of X given $Y = y_j$ as

$$E\{X|Y = y_j\} = \sum_{i=1}^{n} x_i P(X = x_i|Y = y_j)$$

Show that

$$E\{X\} = \sum_{j=1}^{m} E\{X|Y = y_j\}P(Y = y_j)$$

Section 3.4

3.18. Show that the mean and variance of a random variable X having a uniform distribution in the interval $[a, b]$ are $\mu_X = (a+b)/2$ and $\sigma_X^2 = (a-b)^2/12$.

3.19. X is a Gaussian random variable with $\mu_X = 2$ and $\sigma_X^2 = 9$. Find $P(-4 < X \leq 5)$.

3.20. X is a zero mean Gaussian random variable with a variance of σ_X^2. Show that

$$E\{X^n\} = \begin{cases} (\sigma_X)^n \, 1 \cdot 3 \cdot 5 \cdots (n-1), & n \text{ even} \\ 0, & n \text{ odd} \end{cases}$$

3.21. The joint pdf of random variables X and Y is

$$f_{X,Y}(x, y) = \tfrac{1}{2}, \quad 0 \leq x \leq y, 0 \leq y \leq 2$$

(a) Find the marginal pdf's $f_X(x)$ and $f_Y(y)$.
(b) Find the conditional pdf's $f_{X|Y}(x|y)$ and $f_{Y|X}(y|x)$.
(c) Find $E\{X|Y = 1\}$ and $E\{X|Y = 0.5\}$.
(d) Are X and Y statistically independent?

3.22. X and Y have a bivariate Gaussian pdf given in Equation (3.54).
(a) Show that the marginals are Gaussian pdf's.
(b) Find the conditional pdf $f_{X|Y}(x|y)$. Show that this conditional pdf has a mean

$$E\{X|Y = y\} = \mu_X + \rho\frac{\sigma_X}{\sigma_Y}(y - \mu_Y)$$

and a variance

$$\sigma_X^2(1 - \rho^2)$$

3.23. Let $Z = X + Y - c$, where X and Y are independent random variables with variance σ_X^2 and σ_Y^2 and c is a constant. Find the variance of Z in terms of σ_X^2, σ_Y^2, and c.

3.24. X and Y are independent zero mean Gaussian random variables with variances σ_X^2 and σ_Y^2. Let

$$Z = \tfrac{1}{2}(X + Y) \quad \text{and} \quad W = \tfrac{1}{2}(X - Y)$$

(a) Find the joint pdf $f_{Z,W}(z, w)$.
(b) Find the marginal pdf $f_Z(z)$.

3.25. $X_1, X_2, \ldots, X_n$ are n independent zero mean Gaussian random variables with equal variances, $\sigma_{X_i}^2 = \sigma^2$. Show that

$$Z = \frac{1}{n}[X_1 + X_2 + \cdots + X_n]$$

is a Gaussian random variable with $\mu_Z = 0$ and $\sigma_Z^2 = \sigma^2/n$. (Use the result derived in Problem 3.24b.)

3.26. $X(t)$ is a Gaussian random variable with mean $= 0$ and variance $= \sigma_X^2$. Find the pdf of Y if:
(a) $Y = X^2$
(b) $Y = |X|$
(c) $Y = \frac{1}{2}[X + |X|]$
(d) $Y = \begin{cases} 1 & \text{if } X > \sigma_X \\ X & \text{if } |X| \leq \sigma_X \\ -1 & \text{if } X < -\sigma_X \end{cases}$

3.27. X is a Gaussian random variable with mean 0 and variance 4. Find $P(X^2 > 10)$.

Section 3.5

3.28. Define the random process $X(t)$ based on the outcome k of tossing a die as

$$X(t) = \begin{cases} -2, & k = 1 \\ -1, & k = 2 \\ +1, & k = 3 \\ +2, & k = 4 \\ +t, & k = 5 \\ -t, & k = 6 \end{cases} \qquad -\infty < t < \infty$$

(a) Find the probability mass functions of $X(0)$ and $X(1)$.
(b) Find the joint probability mass function of $X(0)$ and $X(1)$.
(c) Find $E\{X(0)\}$, $E\{X(1)\}$, and $R_{XX}(0, 1)$.

3.29. The member functions of two random processes are given in Figure 3.14. Assume that the member functions have equal probabilities of occurrence.
(a) Find $\mu_X(t)$ and $R_{XX}(t, t+\tau)$. Is $X(t)$ stationary in the wide sense?
(b) Find $\mu_Y(t)$ and $R_{YY}(t, t+\tau)$. Is $Y(t)$ stationary in the wide sense?

3.30. $Z(t) = X \cos \omega_0 t - Y \sin \omega_0 t$ is a random process. X and Y are independent Gaussian random variables, each with zero mean and variance 1.

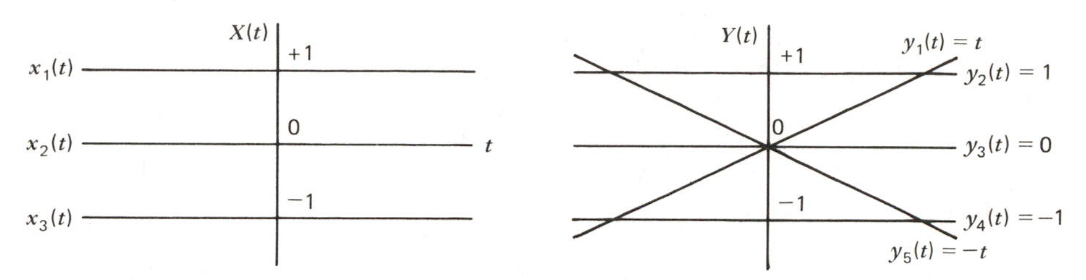

Figure 3.14 Member functions of $X(t)$ and $Y(t)$.

(a) Find $\mu_Z(t)$ and $R_{ZZ}(t, t+\tau)$. Is $Z(t)$ stationary in the wide sense?
(b) Find $f_Z(z)$. (*Hint*: Show that Z is Gaussian, then find μ_Z and σ_Z^2.)

3.31. Let $Z(t) = X(t)\cos(\omega_0 t + Y)$, where $X(t)$ is a zero mean stationary Gaussian random process with $E\{X^2(t)\} = \sigma_X^2$.
(a) If Y is a constant, say zero, find $E\{Z^2(t)\}$. Is $Z(t)$ stationary?
(b) If Y is a random variable with a uniform pdf in the interval $(-\pi, \pi)$, find $R_{ZZ}(t, t+\tau)$. Is $Z(t)$ wide sense stationary? If so, find the psd of $Z(t)$. (Assume X and Y to be independent.)

3.32. Show that the autocorrelation function $R_{XX}(\tau)$ of a stationary random process has the following properties:
(a) $R_{XX}(0) = E\{X^2(t)\}$
(b) $R_{XX}(\tau) = R_{XX}(-\tau)$
(c) $R_{XX}(0) \geq |R_{XX}(\tau)|$
(d) If $X(t)$ and $X(t+\tau)$ are independent for large τ, then $\lim_{\tau\to\infty} R_{XX}(\tau) = (\mu_X)^2$.

3.33. With reference to the random process $X(t)$ shown in Figure 3.14, compute $\langle\mu_X\rangle$ and $\langle R_{XX}(\tau)\rangle$ by the time averaging for each of the member functions of the process.
(a) Is $X(t)$ ergodic in the mean?
(b) Find $E\{\langle R_{XX}(\tau)\rangle\}$ and variance of $\langle R_{XX}(\tau)\rangle$. Is $E\{\langle R_{XX}(\tau)\rangle\} = R_{XX}(\tau)$? Is $X(t)$ ergodic in the autocorrelation function?

3.34. $X(t)$ is a zero mean stationary random process with an autocorrelation function $R_{XX}(\tau)$. By integrating $X(t)$ we form a random variable Y as

$$Y = \frac{1}{2T}\int_{-T}^{T} X(t)\,dt$$

(a) Show that

$$E\{Y\} = \mu_X = 0 \quad \text{and} \quad \sigma_Y^2 = \frac{1}{T}\int_0^{2T}\left(1 - \frac{\tau}{2T}\right)R_{XX}(\tau)\,d\tau$$

(b) Y is the time averaged mean of the random process $X(t)$, that is, $Y = \langle \mu_X \rangle$. Show that, if

$$\lim_{T\to\infty} \frac{1}{T} \int_0^{2T} \left(1 - \frac{\tau}{2T}\right) R_{XX}(\tau)\, d\tau = 0,$$

then $X(t)$ is ergodic in the mean.

3.35. For a real-valued stationary random process $X(t)$, show that its power spectral density $G_X(f)$ is *even*, *nonnegative* and that

$$R_{XX}(\tau) = 2 \int_0^{\infty} G_X(f) \cos(2\pi f \tau)\, df$$

Section 3.6

3.36. The input to a linear time-invariant system is a random process $X(t)$ that is a zero mean stationary Gaussian random process. Show that the output is a zero mean stationary Gaussian random process.

3.37. X and Y have a bivariate Gaussian pdf with $E\{X\} = E\{Y\} = 0$. Show that

$$E\{X^2Y^2\} = E\{X^2\}E\{Y^2\} + 2[E\{XY\}]^2$$

(*Hint*: Use conditional expected values.)

3.38. The output of a linear system is $Y(t) = X(t+a) - X(t-a)$, where the input $X(t)$ is a stationary random process. Show that

$$R_{YY}(\tau) = 2R_{XX}(\tau) - R_{XX}(\tau + 2a) - R_{XX}(\tau - 2a)$$

and

$$G_Y(f) = 4G_X(f) \sin^2(2\pi a f)$$

3.39. Find the autocorrelation function and the psd of

$$Y(t) = \frac{d}{dt}[X(t)] = \lim_{a\to 0} \frac{x(t+a) - x(t)}{a}$$

where $X(t)$ is a stationary random process with an autocorrelation function $R_{XX}(\tau)$. Assume that $R_{XX}(\tau)$ is twice differentiable at $\tau = 0$.

3.40. The input voltage to an RLC series circuit is a stationary random process $X(t)$ with $E\{X(t)\} = 2$ and $R_{XX}(\tau) = 4 + \exp(-2|\tau|)$. Let $Y(t)$ be the voltage across the capacitor. Find $E\{Y(t)\}$, and $G_Y(f)$.

Section 3.7

3.41. A random noise process $V(t)$ has no dc component and an rms value of 2 mvolt. Measurements indicate that $V(t)$ and $V(t+\tau)$ have a joint

Gaussian distribution and that $V(t)$ and $V(t+\tau)$ are independent for $|\tau| > 4$ msec while $R_{VV}(\tau)$ decreases linearly with $|\tau|$ for $|\tau| < 4$ msec.

(a) Sketch $R_{VV}(\tau)$ versus τ.

(b) Write down the pdf of $V(t)$ and the joint pdf of $V(t)$ and $V(t+0.001)$.

3.42. Show that A and B have to be Gaussian random variables with $E\{A\} = E\{B\} = 0$, $E\{A^2\} = E\{B^2\} = \sigma^2$, and $E\{AB\} = 0$ if $\tilde{Y}(t) = A\cos\omega_c t + B\sin\omega_c t$ is to be a zero mean stationary Gaussian random process.

3.43. Consider two spectral components of a zero mean stationary Gaussian random process:

$$Y_k(t) = A_k \cos\omega_k t + B_k \sin\omega_k t$$
$$Y_j(t) = A_j \cos\omega_j t + B_j \sin\omega_j t$$

Let

$$Y_{k+j}(t) = Y_k(t) + Y_j(t)$$

Show that $Y_{k+j}(t)$ is stationary (wide sense, and hence strict sense for a Gaussian process) if A_k's and B_k's are Gaussian random variables with

$$E\{A_k\} = E\{B_k\} = E\{A_j\} = E\{B_j\} = 0$$

and

$$E\{A_k^2\} = E\{B_k^2\} = E\{A_j^2\} = E\{B_j^2\} = \sigma^2$$

and

$$E\{A_kB_j\} = E\{A_jB_k\} = E\{A_kB_k\} = E\{A_kA_j\} = E\{B_jB_k\} = 0; \quad k \neq j$$

3.44. Show that $Y_c(t)$ and $Y_s(t)$ given in Equations (3.104) and (3.105) are zero mean Gaussian random processes.

3.45. Verify that $Y_c(t)$ and $Y_s(t)$ are uncorrelated and that $E\{Y_c^2(t)\} = E\{Y_s^2(t)\} = E\{Y^2(t)\}$.

3.46. A zero mean stationary Gaussian noise $Y(t)$ has the psd $G_Y(f)$ shown in Figure 3.15. Find $G_{Y_c}(f)$ and $G_{Y_s}(f)$.

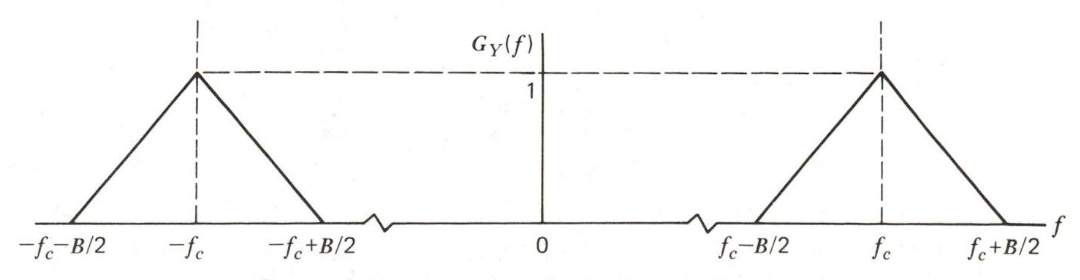

Figure 3.15 Psd of $Y(t)$ (Problem 3.46).

3.47. A zero mean Gaussian white noise with a psd $\eta/2$ is passed through an ideal lowpass filter with a bandwidth B.
(a) Find the autocorrelation function of the output $Y(t)$.
(b) Write down the joint pdf of $Y(t)$ and $Y(t+\tau_0)$, where $\tau_0 = 1/2B$. Are $Y(t)$ and $Y(t+\tau_0)$ independent?

3.48. An amplifier with $g_a = 40$ dB and $B_N = 20$ kHz is found to have $N_{a0} = 10^9 kT_0$ when $T_i = T_0$. Find T_e and the noise figure.

3.49. A simple laboratory procedure for measuring the noise figure of a two-port consists of the following steps: First, connect a resistor at room temperature T_0 to the input and measure the output power. Second, heat the resistor until the output power reading has doubled and record the temperature in degrees Kelvin. Let the temperature of the resistor be T_R. Assuming that the resistor and the power meter are impedance matched to the two-port, show that the noise figure is given by

$$F = T_R/T_0 - 1$$

(Note that g_a and B_N need not be known.)

3.50. Show that the effective noise temperature of n networks in cascade is given by

$$T_e = T_{e_1} + \frac{T_{e_2}}{g_1} + \frac{T_{e_3}}{g_1 g_2} + \cdots + \frac{T_{e_n}}{g_1 g_2 \cdots g_{n-1}}$$

3.51. A low noise receiver for a satellite ground station is shown in Figure 3.16.
(a) Calculate the effective noise temperature of the receiver.
(b) If $(S/N)_d > 30$ dB, find the (available) signal power required at the antenna.

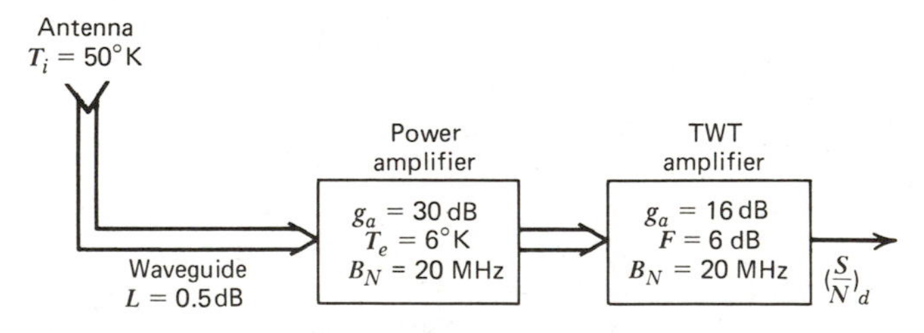

Figure 3.16 A low noise receiver.

3.52. An amplifier has input and output impedances of 75 ohm, 60 dB power gain, and a noise equivalent bandwidth of 15 kHz. When a 75 ohm resistor at 290°K is connected to the input, the output rms noise voltage is 75 μvolt. Determine the effective noise temperature of the amplifier assuming that the meter is impedance matched to the amplifier.

3.53. A proposed ground station for receiving signals from a satellite has a receiver with a cooled parametric amplifier with a 0.3 dB noise figure, a bandwidth of 1 MHz, and a center frequency of 3 GHz. The receiving antenna has a power gain of 50 dB and the satellite antenna has a power gain of 6 dB. Path loss in decibels is $a = 40 + 20 \log_{10}(f) + 20 \log_{10}(d)$, where f is the center frequency in MHz and d is the distance in miles. If the transmitted power is 15 watt and a minimum 15 dB signal-to-noise ratio is to be maintained at the receiver output, estimate the distance over which communication can be maintained.

4

INFORMATION AND CHANNEL CAPACITY

4.1 INTRODUCTION

In this chapter, we attempt to answer two basic questions that arise in the analysis and design of communication systems: (1) Given an information source, how do we evaluate the "rate" at which the source is emitting information? (2) Given a noisy communication channel, how do we evaluate the maximum "rate" at which "reliable" information transmission can take place over the channel? We develop answers to these questions based on probabilistic models for information sources and communication channels.

Information sources can be classified into two categories: analog (or continuous-valued) and discrete. Analog information sources, such as a microphone actuated by a voice signal, emit a continuous-amplitude, continuous-time electrical waveform. The output of a discrete information source such as a teletype consists of sequences of letters or symbols. Analog information sources can be transformed into discrete information sources through the process of sampling and quantizing. In the first part of this chapter we will deal with models for discrete information sources as a prelude to our study of digital communication systems.

A discrete information source consists of a discrete set of letters or *alphabet of symbols*. In general, any *message* emitted by the source consists of a string or sequence of symbols. The symbols in the string or sequence are emitted at *discrete moments*, usually at a fixed time rate, and each symbol emitted is chosen from the source alphabet. Every message coming out of the source contains some information, but some messages convey more information than others. In order to quantify the information content of messages and the average information content of symbols in messages, we will define a measure of information in this chapter. Using the average information content of symbols and the symbol rate, we will then define an average information rate for the source.

If the units of information are taken to be binary digits or bits, then the average information rate represents the minimum average number of bits per second needed to represent the output of the source as a binary sequence. In order to achieve this rate, we need a functional block in the system that will replace strings of symbols by strings of binary digits. We will discuss a procedure for designing this functional block, called the source encoder, in the first part of this chapter.

In the second part of this chapter, we will develop statistical models that adequately represent the basic properties of communication channels. For modeling purposes, we will divide communication channels into two categories: analog channels and discrete channels. An analog channel accepts a continuous-amplitude continuous-time electrical waveform as its input and produces at its output a noisy smeared version of the input waveform. A discrete channel accepts a sequence of symbols as its input and produces an output sequence that is a replica of the input sequence, except for occasional errors. We will first develop models for discrete communication channels and derive the concept of "capacity" of a communication channel. The channel capacity is one of the most important parameters of a data communication system and it represents the maximum rate at which data can be transferred over the channel with an arbitrarily small probability of error.

While expressions for the capacity of discrete channels are easily derived, such is not the case when we deal with the continuous portion of the channel. For this portion we will simply state the model of the channel and then the expression for the capacity of the channel. For the case of a bandlimited channel with bandwidth B, Shannon has shown that the capacity C is equal to $B \log_2(1 + S/N)$, where S is the average signal power at the output of the channel and N is the average power of the bandlimited Gaussian noise that accompanies the signal. While we will not attempt to derive this expression for channel capacity, we will consider in detail the implications of $C = B \log_2(1 + S/N)$ in the design of communication systems.

The material contained in this chapter is based on the pioneering work of Shannon. In 1948, he published a treatise on the mathematical theory of communication in which he established basic theoretical bounds for the performances of communication systems. Shannon's theory is based on probabilistic models for information sources and communication channels. We present here some of the important aspects of his work.

4.2 MEASURE OF INFORMATION

4.2.1 Information Content of a Message

The output of a discrete information source is a message that consists of a sequence of symbols. The actual message that is emitted by the source during a message interval is selected at random from a set of possible messages. The communication system is designed to reproduce at the receiver either exactly or approximately the message emitted by the source.

As mentioned earlier, some messages produced by an information source contain more information than other messages. The question we ask ourselves now is, how can we measure the information content of a message quantitatively so that we can compare the information content of various messages produced by the source? In order to answer this question, let us first review our intuitive concept of the amount of information in the context of the following example.

Suppose you are planning a trip to Miami, Florida from Minneapolis in the winter time. To determine the weather in Miami, you telephone the Miami weather bureau and receive one of the following forecasts:

1. mild and sunny day,
2. cold day,
3. possible snow flurries.

The amount of information received is obviously different for these messages. The first message contains very little information since the weather in Miami is mild and sunny most of the time. The forecast of a cold day contains more information since it is not an event that occurs often. In comparison, the forecast of snow flurries conveys even more information since the occurrence of snow in Miami is a rare event. *Thus on an intuitive basis the amount of information received from the knowledge of occurrence of an event is related to the probability or the likelihood of occurrence of the event.* The message associated with an event least likely to occur contains most information. The above conjecture applies to messages related to any uncertain event, such as the behavior of the stock market. The amount of information in a message depends only on the uncertainty of the underlying event rather than its actual content. We can now formalize this concept in terms of probabilities as follows:

Suppose an information source emits one of q possible messages $m_1, m_2, \ldots, m_q$ with probabilities of occurrence $p_1, p_2, \ldots, p_q$; $p_1 + p_2 + \cdots + p_q = 1$. According to our intuition, the information content or the amount of information in the kth message, denoted by $I(m_k)$, must be inversely related to p_k. Also, to satisfy our intuitive concept of information, $I(m_k)$ must approach 0 as p_k approaches 1. For example, if the forecast in the preceding example said the sun will rise in the east, this does not convey any information since the sun will rise in the east with probability 1. Furthermore, the information content $I(m_k)$ must be nonnegative since each message contains some information. At worst, $I(m_k)$ may be equal to zero. Summarizing these requirements, $I(m_k)$ must satisfy:

$$I(m_k) > I(m_j) \quad \text{if } p_k < p_j \tag{4.1}$$

$$I(m_k) \to 0 \quad \text{as } p_k \to 1 \tag{4.2}$$

$$I(m_k) \geq 0 \quad \text{when } 0 \leq p_k \leq 1 \tag{4.3}$$

Before we start searching for a measure of information that satisfies the above constraints, let us impose one more requirement. Namely, when two *independent* messages are received the total information content is the sum of the information conveyed by each of the two messages. For example, suppose that you read in the newspaper two items of news: (1) scientists have discovered a cure for the common cold and (2) a NASA space probe has found evidence of life on planet Mars. It is reasonable to assume that the two events mentioned in the news items are independent, and that the total information received from the two messages is the same as the sum of the information contained in each of the two news items.

We can apply the same concept to independent messages coming from the same source. For example, the information received in the message, "It will be sunny today and cloudy tomorrow," is the same as the sum of information received in the two messages, "It will be sunny today" and "It will be cloudy tomorrow" (assuming that weather today does not affect weather tomorrow). Mathematically, we can state this requirement by

$$I(m_k \text{ and } m_j) \stackrel{\Delta}{=} I(m_k m_j) = I(m_k) + I(m_j) \tag{4.4}$$

where m_k and m_j are two independent messages.

A continuous function of p_k that satisfies the constraints specified in Equations (4.1)–(4.4) is the logarithmic function and we can define a measure of information as

$$I(m_k) = \log(1/p_k) \tag{4.5}$$

The base for the logarithm in (4.5) determines the unit assigned to the information content. If the natural logarithm base is used, then the unit is *nat*, and if the base is 10, then the unit is *Hartley* or *decit*. When the base is 2, then the unit of information is the familiar *bit*, an abbreviation for binary digit. Using the binary digit as the unit of information is based on the fact that if two possible binary digits occur with equal probabilities ($p_1 = p_2 = \frac{1}{2}$), then the correct identification of the binary digit conveys an amount of information $I(m_1) = I(m_2) = -\log_2(\frac{1}{2}) = 1$ bit. Unless otherwise specified, we will use the base 2 in our definition of information content.

Example 4.1. A source puts out one of five possible messages during each message interval. The probabilities of these messages are

$$p_1 = \tfrac{1}{2}, \quad p_2 = \tfrac{1}{4}, \quad p_3 = \tfrac{1}{8}, \quad p_4 = \tfrac{1}{16}, \quad p_5 = \tfrac{1}{16}$$

Find the information content of each of these messages. (Observe that the actual meaning or interpretation of the message does not enter into our computation of information content.)

Solution

$$I(m_1) = -\log_2(\tfrac{1}{2}) = 1 \text{ bit}$$
$$I(m_2) = -\log_2(\tfrac{1}{4}) = 2 \text{ bits}$$
$$I(m_3) = -\log_2(\tfrac{1}{8}) = 3 \text{ bits}$$
$$I(m_4) = -\log_2(\tfrac{1}{16}) = 4 \text{ bits}$$
$$I(m_5) = -\log_2(\tfrac{1}{16}) = 4 \text{ bits}$$

4.2.2 Average Information Content (Entropy) of Symbols in Long Independent Sequences

Messages produced by information sources consist of sequences of symbols. While the receiver of a message may interpret the entire message as a single unit, communication systems often have to deal with individual symbols. For example, if we are sending messages in the English language using a teletype, the user at the receiving end is interested mainly in words, phrases, and sentences, whereas the communication system has to deal with individual letters or symbols. Hence, from the point of view of communication systems that have to deal with symbols, we need to define the information content of symbols.

When we attempt to define the information content of symbols, we need to keep the following two factors in mind: First, the instantaneous flow of information in a system may fluctuate widely due to the randomness involved in the symbol selection. Hence we need to talk about *average information content* of symbols in a long message. Secondly, the statistical dependence of symbols in a message sequence will alter the average information content of symbols. For example, the presence of the letter U following Q in an English word carries less information than the presence of the same letter U following the letter T. We will first define the average information content of symbols assuming the source selects or emits symbols in a statistically independent sequence, with the probabilities of occurrence of various symbols being invariant with respect to time. Later in the chapter we will deal with sources emitting symbols in statistically dependent sequences.

Suppose we have a source that emits one of M possible symbols $s_1, s_2, \ldots, s_M$ in a statistically independent sequence. That is, the probability of occurrence of a particular symbol during a symbol time interval does not depend on the symbols emitted by the source during the preceding symbol intervals. Let $p_1, p_2, \ldots, p_M$ be the probabilities of occurrence of the M symbols, respectively. Now, in a long message containing N symbols, the symbol s_1 will occur on the average p_1N times, the symbol s_2 will occur p_2N times, and in general the symbol s_i will occur p_iN times. Treating individual symbols as messages of length one, we can define the information content of the ith symbol as $\log_2(1/p_i)$ bits. Hence the p_iN occurrences of s_i contributes an information content of $p_iN \log_2(1/p_i)$ bits. The total information content of the message is then the sum of the contribution due to each of the M symbols of the source alphabet and is given by

$$I_{\text{total}} = \sum_{i=1}^{M} Np_i \log_2(1/p_i) \text{ bits}$$

We obtain the *average information per symbol* by dividing the total information content of the message by the number of symbols in the message, as

$$H = \frac{I_{\text{total}}}{N} = \sum_{i=1}^{M} p_i \log_2(1/p_i) \text{ bits/symbol} \tag{4.6}$$

Observe that the definition of H given in Equation (4.6) is based on "time averaging." In order for this definition to be valid for ensemble averages, the source has to be ergodic (see Section 3.5.2).

The expression given in Equation (4.6) was used by Shannon as the starting point in his original presentation of the mathematical theory of communication.

The average information content per symbol is also called the *source entropy* since the expression in (4.6) is similar to the expression for entropy in statistical mechanics. A simple interpretation of the source entropy is the following: On the average, we can expect to get H bits of information per symbol in long messages from the information source even though we cannot say in advance what symbol sequences will occur in these messages.

Example 4.2. Find the entropy of a source that emits one of three symbols A, B, and C in a statistically independent sequence with probabilities $\frac{1}{2}$, $\frac{1}{4}$, and $\frac{1}{4}$, respectively.

Solution. We are given $s_1 = A$, $s_2 = B$, and $s_3 = C$, with $p_1 = \frac{1}{2}$ and $p_2 = p_3 = \frac{1}{4}$. The information content of the symbols are one bit for A, two bits for B, and two bits for C. The average information content per symbol or the source entropy is given by

$$\begin{aligned} H &= \tfrac{1}{2}\log_2(2) + \tfrac{1}{4}\log_2(4) + \tfrac{1}{4}\log_2(4) \\ &= 1.5 \text{ bits/symbol} \end{aligned}$$

(A typical message or symbol sequence from this source could be: *ABCCABAABCABABAACAAB*.)

To explore the dependence of H on the symbol probabilities, let us consider a source emitting two symbols with probabilities p and $1-p$, respectively $(0 < p < 1)$. The entropy for this source is given by

$$H = p \log_2\left(\frac{1}{p}\right) + (1-p)\log_2\left(\frac{1}{1-p}\right) \text{ bits/symbol}$$

It is easy to verify that the maximum value of H is reached when $p = \frac{1}{2}$ ($dH/dp = 0$ requires $\log((1-p)/p) = 0$ and hence $p = \frac{1}{2}$), and $H_{\max}$ is 1 bit/symbol. In general, for a source with an alphabet of M symbols, the *maximum entropy* is attained when the symbol probabilities are equal, that is, when $p_1 = p_2 = \cdots = p_M = 1/M$, and $H_{\max}$ is given by

$$H_{\max} = \log_2 M \text{ bits/symbol} \tag{4.7}$$

It was mentioned earlier that symbols are emitted by the source at a fixed time rate, say r_s symbols/sec. We can bring this time element into the picture and define the average *source information rate* R in bits per second as the product of the average information content per symbol and the symbol rate r_s.

$$R = r_s H \text{ bits/sec} \tag{4.8}$$

The abbreviation BPS is often used to denote bits per second.

Example 4.3. A discrete source emits one of five symbols once every millisecond. The symbol probabilities are $\frac{1}{2}$, $\frac{1}{4}$, $\frac{1}{8}$, $\frac{1}{16}$, and $\frac{1}{16}$, respectively. Find the source entropy and information rate.

Solution

$$H = \sum_{i=1}^{5} p_i \log_2\left(\frac{1}{p_i}\right) \text{ bits/symbol}$$
$$= \tfrac{1}{2}\log_2(2) + \tfrac{1}{4}\log_2(4) + \tfrac{1}{8}\log_2(8) + \tfrac{1}{16}\log_2(16) + \tfrac{1}{16}\log_2(16)$$
$$= 0.5 + 0.5 + 0.375 + 0.25 + 0.25$$
$$= 1.875 \text{ bits/symbol}$$

Information rate $R = r_s H$ bits/sec $= (1000)(1.875) = 1875$ bits/sec.

4.2.3 Average Information Content of Symbols in Long Dependent Sequences

The entropy or average information per symbol and the source information rate defined in Equations (4.6) and (4.8) apply to sources that emit symbols in statistically independent sequences. That is, the occurrence of a particular symbol during a symbol interval does not alter the probability of occurrences of symbols during any other symbol intervals. However, nearly all practical sources emit sequences of symbols that are statistically dependent. In telegraphy, for example, the messages to be transmitted consist of a sequence of letters, numerals, and special characters. These sequences, however, are not completely random. In general, they form sentences and have a *statistical structure* of, say, the English language. For example, the letter E occurs more frequently than letter Q; occurrence of letter Q implies that the following letter will most probably be the letter U; the occurrence of a consonant implies that the following letter will most probably be a vowel, and so on. This statistical dependence or structure reduces the amount of information coming out of such a source compared to the amount of information coming from a source emitting the same set of symbols in independent sequences. The problem we address now is one of calculating the information rate for discrete sources that emit dependent sequences of symbols or messages. We will first develop a statistical model for sources emitting symbols in dependent sequences, and then use this model to define the entropy and the information rate for the source.

4.2.4 Markoff Statistical Model for Information Sources

For the purpose of analysis, we will assume that the discrete information source emits a symbol once every T_s seconds. The source puts out symbols belonging to a finite alphabet according to certain probabilities depending, in general, on *preceding symbols* as well as the *particular symbol in question.* A physical system or statistical model of a system that produces such a sequence of symbols governed by a set of probabilities is known as a stochastic or random process. We may consider a discrete source, therefore, to be represented by a random process. Conversely, any random process that produces a discrete sequence of symbols chosen from a finite set may be considered a discrete source. This will include for example, natural written languages such as English and German, and also continuous information sources that have been rendered discrete by sampling and quantization.

We can statistically model the symbol sequences emitted by the discrete source by a class of random processes known as discrete stationary Markoff processes (see Section 3.5.4). The general case can be described as follows:

1. The source is in one of n possible states, $1, 2, \ldots, n$ at the beginning of each symbol interval. The source changes state once during each symbol interval from say i to j. The probability of this transition is p_{ij}, which depends only on the initial state i and the final state j, but does not depend on the states during any of the preceding symbol intervals. The transition probabilities p_{ij} $(i, j = 1, 2, \ldots, n;\ \Sigma_{j=1}^{n} p_{ij} = 1)$ remain constant as the process progresses in time.
2. As the source changes state from i to j it emits a symbol. The particular symbol emitted depends on the initial state i and the transition $i \to j$.
3. Let $s_1, s_2, \ldots, s_M$ be the symbols of the alphabet, and let $X_1, X_2, \ldots, X_k, \ldots$ be a sequence of random variables where X_k represents the kth symbol in a sequence emitted by the source. Then, the probability that the kth symbol emitted is s_q will depend on the previous symbols $X_1, X_2, \ldots, X_{k-1}$ emitted by the source, that is, s_q is emitted by the source with the conditional probability,

$$P(X_k = s_q | X_1, X_2, \ldots, X_{k-1})$$

4. The residual influence of $X_1, X_2, \ldots, X_{k-1}$ on X_k is represented by the state of the system at the beginning of the kth symbol interval. That is, the probability of occurrence of a particular symbol during the kth symbol interval depends only on the state of the system at the beginning of the symbol interval or

$$P(X_k = s_q | X_1, X_2, \ldots, X_{k-1}) = P(X_k = s_q | S_k) \tag{4.9}$$

where S_k is a discrete random variable representing the state of the system at the beginning of the kth interval. (We use the "states"* to "remember"

*In general, a discrete stationary source with M letters in the alphabet emitting a symbol sequence with a residual influence lasting q symbols can be represented by n states, where $n \leq (M)^q$.

past history or residual influence in the same context as the use of state variables in systems theory, and states in sequential logic networks.)

5. At the beginning of the first symbol interval, the system is in one of the n possible states $1, 2, \ldots, n$ with probabilities $P_1(1), P_2(1), \ldots, P_n(1)$, respectively ($\sum_{i=1}^{n} P_i(1) = 1$).
6. If the probability that the system is in state j at the beginning of the kth symbol interval is $P_j(k)$, then we can represent the transitions of the system as

$$P_j(k+1) = \sum_{i=1}^{n} P_i(k) p_{ij} \tag{4.10}$$

If we let $P(k)$ be an $n \times 1$ column vector whose ith entry is $P_i(k)$ and let ϕ to be an $n \times n$ matrix whose (i, j)th entry is p_{ij}, then we can rewrite Equation (4.10) in matrix form as

$$P(k+1) = \phi^T P(k)$$

The matrix ϕ is called the probability transition matrix of the Markoff process. The process is called a stationary Markoff process if $P(k) = \phi^T P(k)$ for $k = 1$.

Information sources whose outputs can be modeled by discrete stationary Markoff processes are called discrete stationary Markoff sources.

Discrete stationary Markoff sources are often represented in a graph form where the states are represented by nodes of the graph, and the transition between states is represented by a directed line from the initial to the final state. The transition probabilities and the symbols emitted corresponding to various transitions are usually shown marked along the lines of the graph. An example is shown in Figure 4.1. This example corresponds to a source emitting one of three symbols, A, B, and C. *The probability of occurrence of a symbol depends on the particular symbol in question and the symbol immediately preceding it*, that is, the residual or past influence lasts only for a duration of one symbol. Since the last symbol emitted by the source can be A or B or C, the past history can be represented by three states, one for each of the three symbols of the alphabet. If the system is in state one, then the last symbol emitted by the source was A, and the source now emits letter A with probability $\frac{1}{2}$ and returns to state one, or it emits letter B with probability $\frac{1}{4}$ and goes to state three, or it emits letter C and goes to state two with probability $\frac{1}{4}$. The state transition and symbol generation can also be illustrated using a "tree" diagram. A tree diagram is a planar graph where the nodes correspond to states and branches correspond to transitions. Transition between states occurs once every T_s seconds, where $1/T_s$ is the number of symbols per second emitted by the source. Transition probabilities and symbols emitted

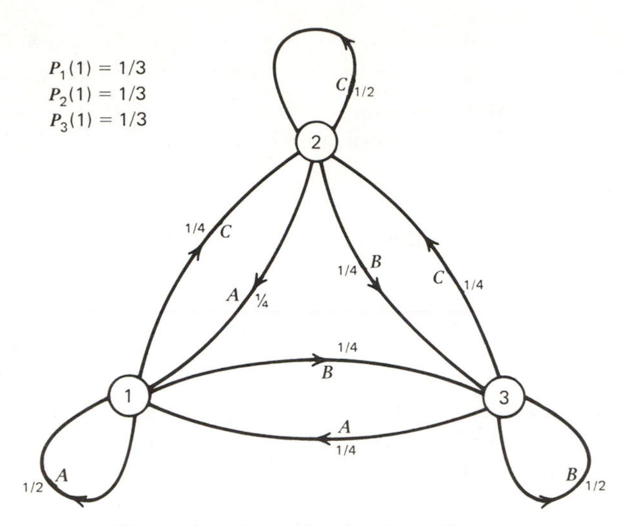

Figure 4.1 Example of a Markoff source.

are shown along the branches. A tree diagram for the source shown in Figure 4.1 is shown in Figure 4.2.

The tree diagram can be used to obtain the probabilities of generating various symbol sequences. For example, the symbol sequence AB can be generated by either one of the following transitions: $1 \to 1 \to 3$ or $2 \to 1 \to 3$ or $3 \to 1 \to 3$. Hence the probability of the source emitting the two-symbol sequence AB is given by

$$\begin{aligned} P(AB) = P(S_1 = 1,\ S_2 = 1,\ S_3 = 3 \\ \text{or} \quad S_1 = 2,\ S_2 = 1,\ S_3 = 3) \\ \text{or} \quad S_1 = 3,\ S_2 = 1,\ S_3 = 3 \end{aligned} \tag{4.11}$$

Since the three transition paths are disjoint, we get

$$\begin{aligned} P(AB) = P(S_1 = 1,\ S_2 = 1,\ S_3 = 3) \\ + P(S_1 = 2,\ S_2 = 1,\ S_3 = 3) \\ + P(S_1 = 3,\ S_2 = 1,\ S_3 = 3) \end{aligned} \tag{4.12}$$

Using the chain rule of probability we can rewrite the first term on the right-hand side of Equation (4.12) as

$$\begin{aligned} P\{S_1 = 1,\ S_2 = 1,\ S_3 = 3\} \\ = P(S_1 = 1)P(S_2 = 1|S_1 = 1)P(S_3 = 3|S_1 = 1,\ S_2 = 1) \\ = P(S_1 = 1)P(S_2 = 1|S_1 = 1)P(S_3 = 3|S_2 = 1) \end{aligned} \tag{4.13}$$

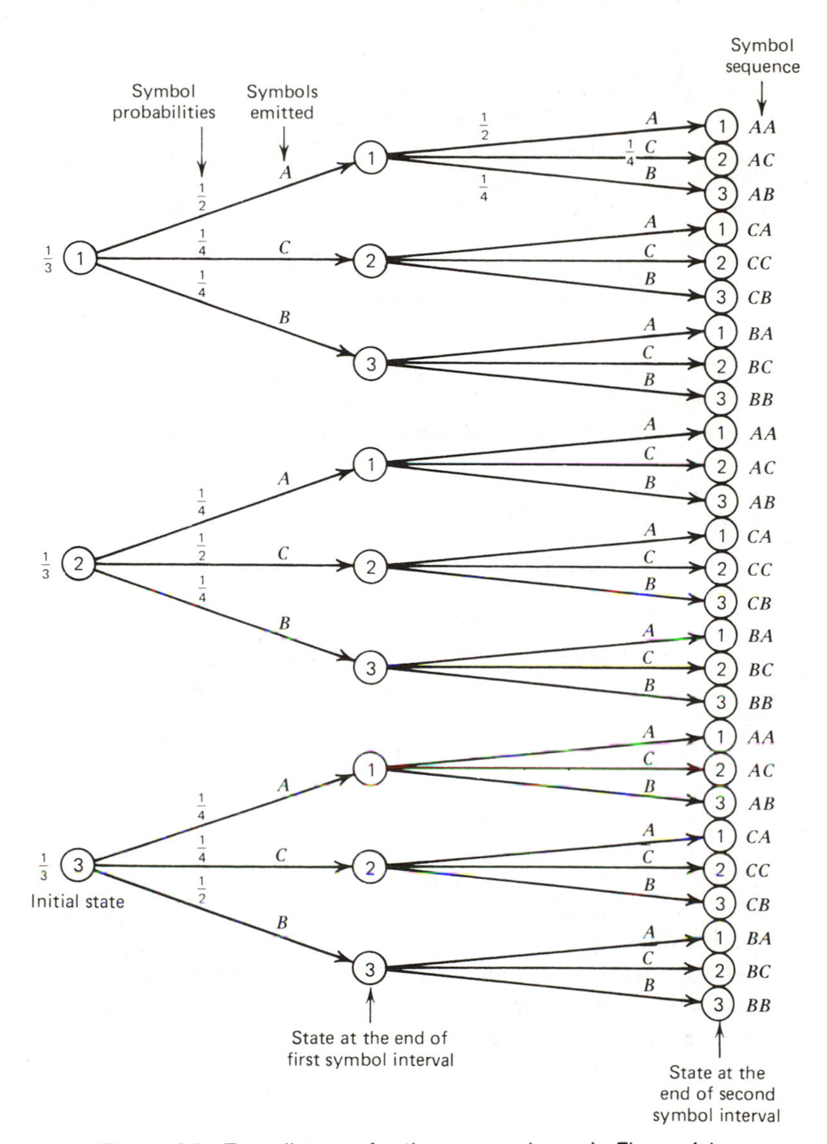

Figure 4.2 Tree diagram for the source shown in Figure 4.1.

The last step is based on the fact that the transition probability to S_3 depends on S_2 but not on how the system got to S_2 (i.e., the Markoff property).

The right-hand side of the previous equation is the product of probabilities shown along the branches representing the transition path and the probability of being at state 1 at the starting point. Other terms on the right-hand side of

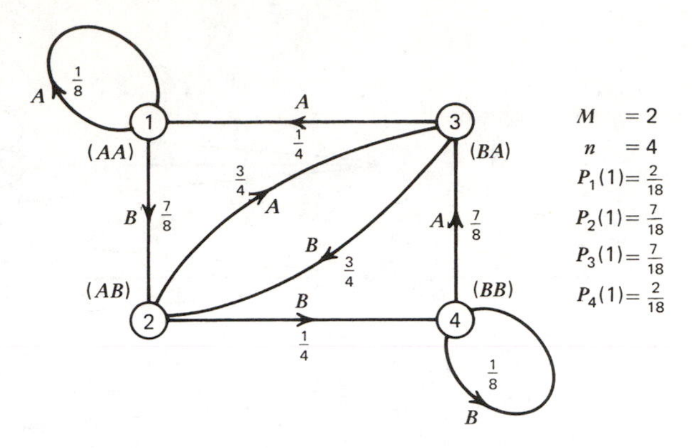

Figure 4.3 Another example of a Markoff source.

Equation (4.12) can be similarly evaluated, and $P(AB)$ is given by

$$P(AB) = (\tfrac{1}{3})(\tfrac{1}{2})(\tfrac{1}{4}) + (\tfrac{1}{3})(\tfrac{1}{4})(\tfrac{1}{4}) + (\tfrac{1}{3})(\tfrac{1}{4})(\tfrac{1}{4})$$
$$= \tfrac{1}{12}$$

Similarly, the probabilities of occurrence of other symbol sequences can be computed. In general, the probability of the source emitting a particular symbol sequence can be computed by summing the product of probabilities in the tree diagram along all the paths that yield the particular sequence of interest.

The model shown in Figure 4.1 corresponds to a source in which the residual influence lasts over one symbol interval. Shown in Figure 4.3 is a source where the probability of occurrence of a symbol depends not only on the particular symbol in question, but also on the two symbols preceding it. It is easy to verify that if the system is in state two, at the beginning of a symbol interval, then the last two symbols emitted by the source were AB, and so on.

It is left as an exercise for the reader to draw the tree diagram for this source and to calculate the probabilities of some typical symbol sequences (Problem 4.11).

4.2.5 Entropy and Information Rate of Markoff Sources

In this section we will define the entropy and information rate of Markoff sources. We will assume that the source can be modeled as a discrete finite-state Markoff process. Furthermore, we will assume the process to be *ergodic* (Chapter 3, Section 3.4) so that time averages can be applied. The ergodic assumption implies that the process is stationary, and hence $P_i(k) =$

$P_i(k+j)$ for any values of k and j. In other words, the probability of being in state i at the beginning of the first symbol interval is the same as the probability of being in state i at the beginning of the second symbol interval, and so on. The probability of going from state i to j also does not depend on time.

We define the entropy of the source as a weighted average of the entropy of the symbols emitted from each state, where the entropy of state i, denoted by H_i, is defined as the average information content of the symbols emitted from the i-th state:

$$H_i = \sum_{j=1}^{n} p_{ij} \log_2\left(\frac{1}{p_{ij}}\right) \text{ bits/symbol} \tag{4.14}$$

The entropy of the source is then the average of the entropy of each state. That is,

$$\begin{aligned} H &= \sum_{i=1}^{n} P_i H_i \\ &= \sum_{i=1}^{n} P_i \left[\sum_{j=1}^{n} p_{ij} \log_2\left(\frac{1}{p_{ij}}\right) \right] \text{ bits/symbol} \end{aligned} \tag{4.15}$$

where P_i is the probability that the source is in state i. The average information rate R for the source is defined as

$$R = r_s H \text{ bits/sec} \tag{4.16}$$

where r_s is the number of state transitions per second or the symbol rate of the source.

The entropy H defined in Equation (4.15) carries the same meaning as H defined in Equation (4.6). In both cases, we can expect the source output to convey H bits of information per symbol in long messages. This is precisely stated in the following theorem.

Theorem 4.1

Let $P(m_i)$ be the probability of a sequence m_i of N symbols from the source. Let

$$G_N = -\frac{1}{N} \sum_i P(m_i) \log_2 P(m_i) \tag{4.17}$$

where the sum is over all sequences m_i containing N symbols. Then G_N is a monotonic decreasing function of N and

$$\lim_{N \to \infty} G_N = H \text{ bits/symbol} \tag{4.18}$$

A detailed proof of this theorem can be found in Reference 1. We will illustrate the concept stated in Theorem 4.1 by the following example.

Example 4.4. Consider an information source modeled by a discrete ergodic Markoff random process whose graph is shown in Figure 4.4. Find the source entropy H and the average information content per symbol in messages containing one, two, and three symbols, that is, find G_1, G_2, and G_3.

Solution. The source shown above emits one of three symbols A, B, and C. The tree diagram for the output of the source is shown in Figure 4.5 and Table 4.1 lists the various symbol sequences and their probabilities. To illustrate how these messages and their probabilities are generated, let us consider the sequence CCC. There are two paths on the graph that terminate in CCC corresponding to the transition sequences $1\to 2\to 1\to 2$ and $2\to 1\to 2\to 1$. The probability of the path $1\to 2\to 1\to 2$ is given by the product of the probability that the system is in state one initially, and the probabilities of the transitions $1\to 2$, $2\to 1$, and $1\to 2$. These probabilities are $\frac{1}{2}$, $\frac{1}{4}$, $\frac{1}{4}$, and $\frac{1}{4}$, respectively, and hence the path probability is 1/128. Similarly, the probability of the second path can be calculated as 1/128. Hence, the probability of the sequence CCC is given by the sum of the two paths as 2/128.

From the definition of H_i (Equation (4.14)) we get

$$H_1 = \tfrac{1}{4}\log_2(4) + \tfrac{3}{4}\log_2(\tfrac{4}{3}) = 0.8113$$
$$H_2 = \tfrac{1}{4}\log_2(4) + \tfrac{3}{4}\log_2(\tfrac{4}{3}) = 0.8113$$

and using Equation (4.15) we obtain the source entropy as

$$H = (\tfrac{1}{2})(0.8113) + (\tfrac{1}{2})(0.8113) = 0.8113 \text{ bits/symbol}$$

Let us now calculate the average information content per symbol in messages containing two symbols. There are seven messages. The information contents of these messages are $I(AA) = I(BB) = 1.83$, $I(BC) = I(AC) = I(CB) =$

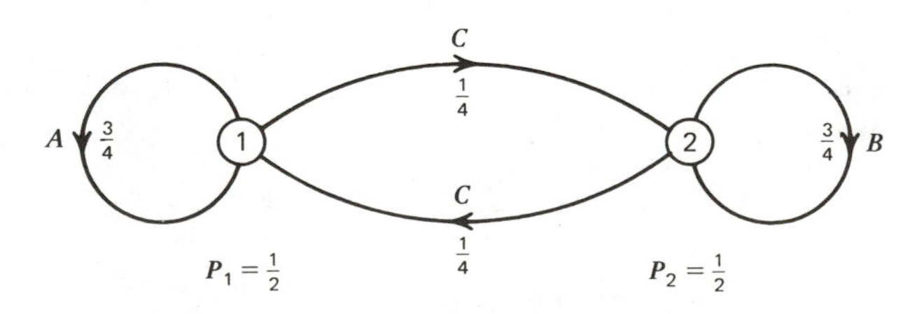

Figure 4.4 Source model for Example 4.4.

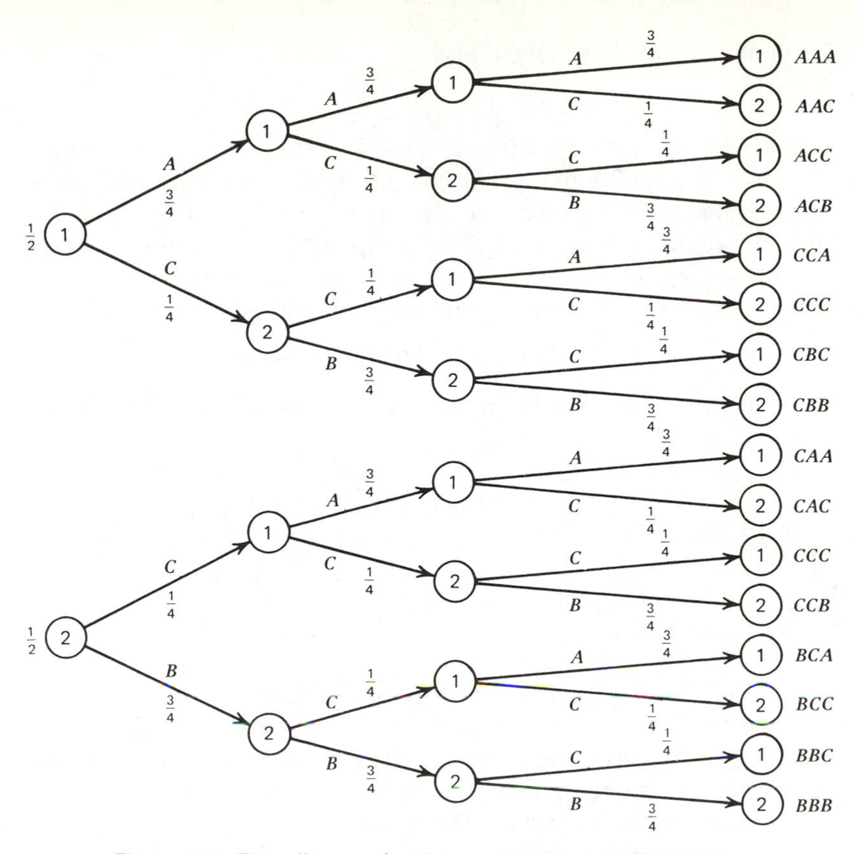

Figure 4.5 Tree diagram for the source shown in Figure 4.4.

Table 4.1. Messages of length 1, 2, and 3 and their probabilities

Messages of length 1	Messages of length 2	Messages of length 3
A (3/8)	*AA* (9/32)	*AAA* (27/128)
B (3/8)	*AC* (3/32)	*AAC* (9/128)
C (1/4)	*CB* (3/32)	*ACC* (3/128)
	CC (2/32)	*ACB* (9/128)
	BB (9/32)	*BBB* (27/128)
	BC (3/32)	*BBC* (9/128)
	CA (3/32)	*BCC* (3/128)
		BCA (9/128)
		CCA (3/128)
		CCB (3/128)
		CCC (2/128)
		CBC (3/128)
		CAC (3/128)
		CBB (9/128)
		CAA (9/128)

$I(CA) = 3.4150$, and $I(CC) = 4.0$ bits. The average information content of these messages is given by the sum of the products of the information content of each message and its respective probability. This can be computed as 2.5598 bits. Now we can obtain the average information content per symbol in messages containing two symbols by dividing the average information content of the messages by the number of symbols in these messages, that is,

$$G_2 = 2.5598/2 = 1.2799 \text{ bits/symbol}$$

In a similar fashion, we can obtain the values of G_1 and G_3. The reader can easily verify that

$$G_1 = 1.5612 \text{ bits/symbol}$$

$$G_3 = 1.0970 \text{ bits/symbol}$$

Thus,

$$G_1 \geqslant G_2 \geqslant G_3 \geqslant H$$

as stated in Theorem 4.1.

The preceding example illustrates the basic concept that the average information content per symbol from a source emitting symbols in a dependent sequence decreases as the message length increases. Alternatively, the average number of bits per symbol needed to represent a message decreases as the message length increases. The decrease in entropy is due to the structure of the messages—messages that are highly structured usually convey less information per symbol than messages containing the same number of symbols when the symbols are chosen independently.

In the next section we will discuss a source coding technique that takes advantage of the statistical structure of the source to reduce the average number of bits per symbol needed to represent the output of an information source. But, before we discuss source coding techniques, let us take a brief look at how the statistical model of a source is constructed.

The development of a model for an information source consists of two parts: (1) the development of the model structure and (2) estimation of the parameters of the model. The structure of the model is usually derived from knowledge about the physical nature of the source. Parameter values are obtained through the use of statistical estimation procedures. In some cases, both the structure and the parameters of the source model can be derived from test data using estimation techniques.

The test data used for estimating the parameters of a source model can be derived either from the simultaneous recordings of the outputs of a number of identical sources for a short time or from the recording of the output of a single source for a long time period. Estimates based on data from a large number of sources are called ensemble estimates, while estimates based on the output of a single source are called time-averaged estimates. These two estimates will be equal if the source is ergodic. An excellent treatment of how to collect and analyze random data and to estimate parameter values can be found in Reference 2. We will illustrate through an example the principle involved in such procedures.

Example 4.5. We want to design a system to report the heading of a collection of 400 cars. The heading is to be quantized into three levels: heading straight (S), turning left (L), and turning right (R). This information is to be transmitted every second. Based on the test data given below, construct a model for the source and calculate the source entropy.

1. On the average, during a given reporting interval, 200 cars were heading straight, 100 were turning left, and 100 cars were turning right.
2. Out of 200 cars that reported heading straight during a reporting period, 100 of them (on the average) reported going straight during the next reporting period, 50 of them reported turning left during the next period, and 50 of them reported turning right during the next period.
3. On the average, out of 100 cars that reported as turning during a signaling period, 50 of them continued their turn during the next period and the remaining headed straight during the next reporting period.
4. The dynamics of the cars did not allow them to change their heading from left to right or right to left during subsequent reporting periods.

Solution. The source model for this process can be developed as follows:

1. Since the past history or residual influence lasts one reporting interval, we need only three states to "remember" the last symbol emitted by the source. The state probabilities are given as (statement one)

$$P_1 = \tfrac{1}{4}, \quad P_2 = \tfrac{1}{2}, \quad P_3 = \tfrac{1}{4}$$

2. The transition probabilities are

$$P\left(\begin{matrix}\text{present}\\ \text{heading}\end{matrix} = S \,\middle|\, \begin{matrix}\text{previous}\\ \text{heading}\end{matrix} = S\right) = P(S|S) = 0.5$$

$$P(S|S) = P(R|R) = P(L|L) = 0.5$$

$$P(L|S) = P(R|S) = 0.25$$

$$P(S|L) = P(S|R) = 0.5$$

$$P(L|R) = P(R|L) = 0$$

3. The state diagram is shown in Figure 4.6.

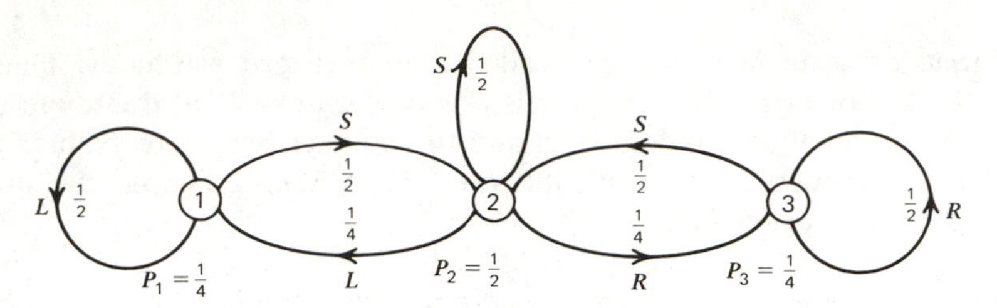

Figure 4.6 Source model for Example 4.5.

The data discussed in the preceding example could have been obtained by monitoring the heading of a single car over an extended period of time instead of monitoring 400 cars. We can obtain good estimates of statistical parameters from a single time history record if the process is ergodic. Here, we will estimate the probability of a car heading straight by the portion of time the car was heading straight during a long monitoring period (as opposed to the ratio of 400 cars heading straight at a given time in the previous method).

The information rate for this source can be computed as follows:

$$
\begin{aligned}
H_1 &= \text{entropy of state one} \\
&= -\sum p_{1j} \log_2(p_{1j}), \text{ where the sum is over all} \\
&\quad \text{symbols emitted from state one} \\
&= \tfrac{1}{2}\log_2 2 + \tfrac{1}{2}\log_2 2 = 1 \text{ bit/symbol} \\
H_2 &= \tfrac{1}{2}\log_2 2 + \tfrac{1}{4}\log_2 4 + \tfrac{1}{4}\log_2 4 \\
&= 1.5 \text{ bits/symbol} \\
H_3 &= 1 \text{ bit/symbol, and hence} \\
H &= P_1H_1 + P_2H_2 + P_3H_3 \\
&= \tfrac{1}{4}(1) + \tfrac{1}{2}(1.5) + \tfrac{1}{4}(1) \\
&= 1.25 \text{ bits/symbol, and} \\
R &= 1.25 \text{ bits/sec (per car)}
\end{aligned}
$$

The values of probabilities obtained from data are estimated values and are not the actual values. The sample size and the type of estimator used will determine the bias (if any) and the variance of the estimator for each of the parameters.

4.3 ENCODING OF THE SOURCE OUTPUT

Source encoding is the process by which the output of an information source is converted into a binary sequence. The functional block that performs this task in a communication system is called the *source encoder*. The input to the source encoder is the symbol sequence emitted by the information source. The source encoder assigns variable length binary code words to blocks of symbols and produces a binary sequence as its output. If the encoder operates on blocks of N symbols in an optimum way, it will produce an average bit rate of G_N bits/symbol, where G_N is defined in Theorem 4.1. In general, the average bit rate of the encoder will be greater than G_N due to practical constraints. As the block length N is increased, the average output bit rate per symbol will decrease and in the limiting case as $N \to \infty$, the bit rate per symbol will approach G_N and G_N will approach H. Thus, with a large block size the output of the information source can be encoded into a binary sequence with an average bit rate approaching R, the source information rate.

The performance of the encoder is usually measured in terms of the *coding efficiency* that is defined as the ratio of the source information rate and the average output bit rate of the encoder. There are many algorithms available for designing a source encoder. The following section deals with a simple, yet most powerful, source coding algorithm given by Shannon.

4.3.1 Shannon's Encoding Algorithm

The design of the source encoder can be formulated as follows:

The input to the source encoder consists of one of q possible messages, each message containing N symbols. Let $p_1, p_2, \ldots, p_q$ be the probabilities of these q messages. We would like to code (or replace) the ith message m_i using a unique binary code word c_i of length n_i bits, where n_i is an integer. Our objective is to find n_i and c_i for $i = 1, 2, \ldots, q$ such that the average number of bits per symbol $\hat{H}_N$ used in the coding scheme is as close to G_N as possible. In other words, we want

$$\hat{H}_N = \frac{1}{N} \sum_{i=1}^{q} n_i p_i \to \frac{1}{N} \sum_{i=1}^{q} p_i \log_2\left(\frac{1}{p_i}\right)$$

Several solutions have been proposed to the above problem, and the algorithm given by Shannon (and Fano) is stated below.

Suppose the q messages $m_1, m_2, \ldots, m_q$ are arranged in order of decreasing probability such that $p_1 \geq p_2 \geq \cdots \geq p_q$. Let $F_i = \sum_{k=1}^{i-1} p_k$, with $F_1 = 0$. Let n_i be an integer such that

$$\log_2(1/p_i) \leq n_i < 1 + \log_2(1/p_i) \tag{4.19}$$

Then, the code word for the message m_i is the binary expansion* of the fraction F_i up to n_i bits, that is,

$$c_i = (F_i)_{\text{binary } n_i \text{ bits}}$$

This algorithm yields a source encoding procedure that has the following properties:

1. Messages of high probability are represented by short code words and those of low probability are represented by long code words. This can be easily verified using the inequality stated in (4.19).
2. The code word for m_i will differ from all succeeding code words in one or more places and hence it is possible to decode messages uniquely from their code words. We can prove this by rewriting inequality (4.19) as

$$\frac{1}{2^{n_i}} \leq p_i < \frac{1}{2^{n_i-1}}$$

   Hence the binary expansion of F_i will differ from all succeeding ones in one or more places. For example, F_i and F_{i+1} will differ in the n_ith bit since $p_i \geq 1/2^{n_i}$. Hence the code word for m_{i+1} will differ from m_i in at least one bit position or more.
3. The average number of bits per symbol used by the encoder is bounded by

$$G_N \leq \hat{H}_N < G_N + 1/N \tag{4.20}$$

This bound can be easily verified as follows:
From (4.19) we have

$$\log_2(1/p_i) \leq n_i < 1 + \log_2(1/p_i)$$

Multiplying throughout by p_i and summing over i, we obtain

$$\sum_{i=1}^{q} p_i \log_2(1/p_i) \leq \sum_{i=1}^{q} n_i p_i < 1 + \sum_{i=1}^{q} p_i \log_2(1/p_i)$$

or

$$\frac{1}{N}\sum_{i=1}^{q} p_i \log_2\left(\frac{1}{p_i}\right) \leq \frac{1}{N}\sum_{i=1}^{q} p_i n_i < \frac{1}{N} + \frac{1}{N}\sum_{i=1}^{q} \log_2\left(\frac{1}{p_i}\right)$$

or

$$G_N \leq \hat{H}_N < 1/N + G_N$$

Hence as $N \to \infty$, $G_N \to H$ and $\hat{H}_N \to H$.

*Binary fraction

$$.b_1 b_2 b_3 \ldots b_k = \frac{b_1}{2} + \frac{b_2}{2^2} + \frac{b_3}{2^3} + \cdots + \frac{b_k}{2^k}$$

where $b_i = 0$ or 1.

The *rate efficiency* e of the encoder using blocks of N symbols is defined as

$$e = H/\hat{H}_N \tag{4.21}$$

The following example illustrates the concepts involved in the design of a source encoder.

Example 4.6. Design a source encoder for the information source given in Example 4.4. Compare the average output bit rate and efficiency of the coder for $N = 1$, 2, and 3.

Solution. Let us first design the encoder with a block size $N = 3$. From Example 4.4 we know that the source emits 15 distinct three-symbol messages. These messages and their probabilities are shown in columns 1 and 2 of Table 4.2; the messages are arranged in column 1 according to decreasing order of probabilities. The number of bits n_1 assigned to message m_1 is bounded by

$$\log_2\left(\frac{128}{27}\right) \leq n_1 < 1 + \log_2\left(\frac{128}{27}\right)$$

or

$$2.245 \leq n_1 < 3.245$$

Table 4.2. Source encoder design for Example 4.6

Messages m_i	p_i	n_i	F_i	Binary expansion of F_i	Code word c_i
AAA	27/128	3	0	.0000000	000
BBB	27/128	3	27/128	.0011011	001
CAA	9/128	4	54/128	.0110110	0110
CBB	9/128	4	63/128	.0111111	0111
BCA	9/128	4	72/128	.1001000	1001
BBC	9/128	4	81/128	.1010001	1010
AAC	9/128	4	90/128	.1011010	1011
ACB	9/128	4	99/128	.1100011	1100
CBC	3/128	6	108/128	.1101100	110110
CAC	3/128	6	111/128	.1101111	110111
CCB	3/128	6	114/128	.1110010	111001
CCA	3/128	6	117/128	.1110101	111010
BCC	3/128	6	120/128	.1111000	111100
ACC	3/128	6	123/128	.1111011	111101
CCC	2/128	6	126/128	.1111110	111111

$\Sigma\, p_i n_i = 3.89 \quad \hat{H}_3 = \dfrac{3.89}{3} \approx 1.30$ bits/symbol

Since n_1 has to be an integer, the above inequality yields $n_1 = 3$ bits. The code word c_1 is generated from $F_1 \triangleq 0$. Hence, $c_1 = 000$. For m_2, it is easy to verify that $n_2 = 3$ bits and $F_2 = \Sigma_{i=1}^{1} p_i$ or $F_2 = 27/128$. The binary expansion of 27/128 is .0011011. Truncating this expansion to 3 bits, we obtain the code word 001 for m_2. The complete design of the encoder for $N = 3$ is shown in Table 4.2.

It can be easily verified that the average number of bits per symbol used by the encoder is 1.30 bits/symbol. Table 4.3 summarizes the characteristics of the encoder for $N = 1$ and 2.

The performance of the encoder is summarized in Table 4.4. The numbers in Table 4.4 show that the average output bit rate of the encoder decreases as N increases, and hence the efficiency of the encoder increases as N increases. Also we can verify that

$$\hat{H}_N < G_N + 1/N \quad \text{for } N = 1, 2, 3$$

To illustrate how the encoder operates, let us consider a symbol string $ACBBCAAACBBB$ at the encoder input. If the encoder uses a block size $N = 3$, then the output of the encoder can be obtained by replacing successive

Table 4.3. Encoder for $N = 1$ and $N = 2$

Message	p_i	n_i	C_i
	$N = 1$		
A	3/8	2	00
B	3/8	2	01
C	1/4	2	11
	$\hat{H}_1 = 2$ bits/symbol		
	$N = 2$		
AA	9/32	2	00
BB	9/32	2	01
AC	3/32	4	1001
CB	3/32	4	1010
BC	3/32	4	1100
CA	3/32	4	1101
CC	2/32	4	1111
	$\hat{H}_2 = 1.44$ bits/symbol		

Table 4.4. Performance of the encoder for Example 4.6

Average number of bits per symbol used	$N = 1$	$N = 2$	$N = 3$
$\hat{H}_N$	2	1.44	1.30
G_N	1.561	1.279	1.097
$G_N + 1/N$	2.561	1.779	1.430
Efficiency $= \dfrac{H}{\hat{H}_N}$	40.56%	56.34%	62.40%
$H = .8113$			

groups of three input symbols by the code words shown in Table 4.2 as

$$\begin{array}{cccc} ACB & BCA & AAC & BBB \\ 1100 & 1001 & 1011 & 001 \end{array}$$

The same symbol string will be encoded as 10010111010010101001 if the encoder operates on two symbols at a time with code words listed in Table 4.3. The decoding is accomplished by starting at the left-most bit and matching groups of bits with the code words listed in the code table.

For the $N = 3$ example, we take the first 3-bit group 110 (this is the shortest code word) and check for a matching word in Table 4.2. Finding none, we try the first 4-bit group 1100, find a match, and decode this group as ACB. Then the procedure is repeated beginning with the fifth bit to decode the rest of the symbol groups. The reader can verify that the decoding can be done easily by knowing the code word lengths a priori if no errors occur in the bit string in the transmission process.

Bit errors in transmission lead to serious decoding problems. For example, if the bit string 1100100110111001 ($N = 3$) was received at the decoder input with one bit error as 110$\underline{1}$100110111001, the message will be decoded as $CBCCAACCB$ instead of $ACBBCAAACBCA$. This type of error is a major disadvantage of an encoder using variable length code words. Another disadvantage lies in the fact that output data rates measured over short time periods will fluctuate widely. To avoid this problem, buffers of large length will be needed at both the encoder and the decoder to store the variable rate bit stream if a fixed output rate is to be maintained.

Some of the above difficulties can be resolved by using *fixed length* code words at the expense of a slight increase in data rate. For example, with $N = 3$ we can encode the output of the source discussed in the preceding example using 4-bit code words $0000, 0001, \ldots, 1110$. The output data rate now is *fixed* at 1.333 bits/symbol compared to an *average output* data rate of 1.30 bits/symbol for the variable length scheme discussed before. The encoder/decoder structure using fixed length code words will be very simple compared to the complexity of an encoder/decoder using variable length code words. Also single bit errors lead to single block errors when fixed length code words are used. These two advantages more than make up for the slight increase in data rate from 1.30 bits/symbol to 1.33 bits/symbol.

Another important parameter in the design of encoders is the delay involved in decoding a symbol. With large block sizes, the first symbol in the block cannot be decoded until the bit string for the entire block is received by the decoder. The average delay will be $N/2$ symbols for a block size of N symbols. The time delay $(N/2)T_s$ seconds, where $1/T_s$ is the number of symbols emitted by a source, may be unacceptable in some real time applications.

It must be pointed out here that the encoding algorithm presented in the preceding pages is only one of many encoding algorithms for representing a source output. Other encoding procedures such as the one given by Huffman yield the shortest average word length. These schemes are more difficult to implement and the interested reader is referred to Abramson's book on information theory [2] for details.

It is also possible to represent the output of an information source using code words selected from an alphabet containing more than two letters. The design of source encoder using nonbinary code words is rather involved. The interested reader is referred to the books of Abramson, Wozencraft and Jacobs, and Gallager [2–4].

Having developed the concept of information rate for sources, we now turn our attention to the problem of evaluating the maximum rate at which reliable information transmission can take place over a noisy channel.

4.4 COMMUNICATION CHANNELS

We can divide a practical communication system into a transmitter, physical channel, or transmission medium, and a receiver. The transmitter consists of an encoder and a modulator, while the receiver consists of a demodulator and a decoder. The term "communication channel" carries different meanings and characterizations depending on its terminal points and functionality. Between points c and g in the system shown in Figure 4.7 we have a discrete channel, often referred to as a *coding channel*, that accepts a sequence of symbols at its input and produces a sequence of symbols at its output. This channel is completely characterized by a set of transition probabilities p_{ij}, where p_{ij} is the probability that the channel output is the jth symbol of the alphabet when the channel input is the ith symbol. These probabilities will depend on the parameters of the modulator, transmission media, noise, and demodulator.

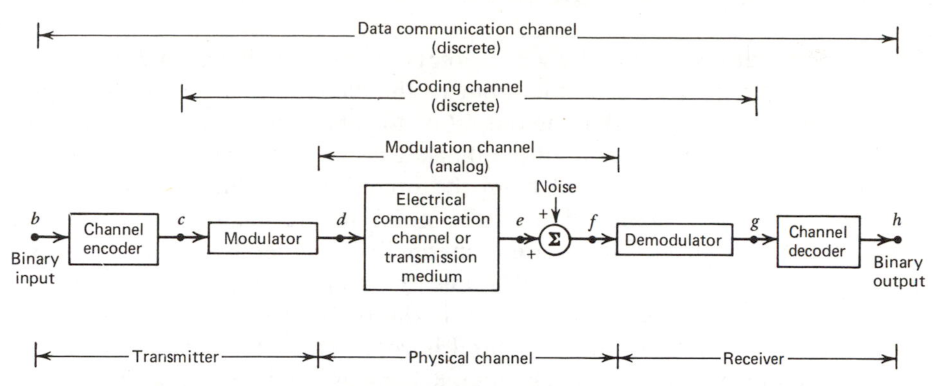

Figure 4.7 Characterization of a binary communication channel.

However, this dependence is transparent to a system designer who is concerned with the design of the digital encoder and decoder.

The communication channel between points d and f in the system provides the electrical connection between the transmitter and the receiver. The input and output are analog electrical waveforms. This portion of the channel is often called a continuous or *modulation channel.* Examples of analog electrical communication channels are voiceband and wideband telephone systems, high frequency radio systems, and troposcatter systems. These channels are subject to several varieties of impairments. Some are due to amplitude and frequency response variations of the channel within its passband. Other channel impairments are due to variations of channel characteristics with time and nonlinearities in the channel. All of these result in the channel modifying the input signal in a deterministic (although not necessarily a known) fashion. In addition, the channel can also corrupt the signal statistically due to various types of additive and multiplicative noise and *fades* (random attenuation changes within the transmission medium). All of these impairments introduce errors in data transmission and limit the maximum rate at which data can be transferred over the channel.

In the following sections we will develop simple mathematical models for discrete communication channels and develop the concept of capacity of a discrete communication channel. The channel capacity is one of the most important parameters of a data communication system since it represents the maximum rate at which data can be transferred between two points in the system, with an arbitrarily small probability of error. After we deal with discrete channels, we will discuss the Shannon–Hartley theorem, which defines the capacity of certain types of continuous channels.

4.5 DISCRETE COMMUNICATION CHANNELS

The communication channel between points c and g in Figure 4.7 is discrete in nature. In the general case, the input to the channel is a symbol belonging to an alphabet of M symbols. The output of the channel is a symbol belonging to the same alphabet of M input symbols. Due to errors in the channel, the output symbol may be different from the input symbol during some symbol intervals. Errors are mainly due to the noise in the analog portion of the communication channel. The discrete channel is completely modeled by a set of probabilities p_i^t $(i = 1, 2, \ldots, M)$ and p_{ij} $(i, j = 1, 2, \ldots, M)$. p_i^t is the probability that the input to the channel is the ith symbol of the alphabet and p_{ij} is the probability that the ith symbol is received as the jth symbol of the alphabet at the output of the channel. Channels designed to transmit and receive one of M possible symbols are called discrete M-ary channels $(M > 2)$. In the binary case we can statistically model the digital channel as shown in Figure 4.8.

The input to the channel is a binary valued discrete random variable X, and the two nodes on the left-hand side of the graph in Figure 4.8 represent the values 0 and 1 of the random variable X. The output of the channel is also a binary valued random variable Y and its values are shown marked at the nodes on the right-hand side of the graph. Four paths connect the input nodes to the output nodes. The path on the top portion of the graph represents an input 0 and a correct output 0. The diagonal path from 0 to 1 represents an input bit 0 appearing incorrectly as 1 at the output of the channel due to noise. Errors occur in a random fashion and we can statistically model the occurrence of errors by assigning probabilities to the paths shown in Figure 4.8. To simplify the analysis, we will assume that the occurrence of an error during a bit interval does not affect the behavior of the system during other bit intervals (i.e., we will assume the channel to be *memoryless*).

Letting $P(X = 0) = p_0^t$, $P(X = 1) = p_1^t$, $P(Y = 0) = p_0^r$, $P(Y = 1) = p_1^r$, we

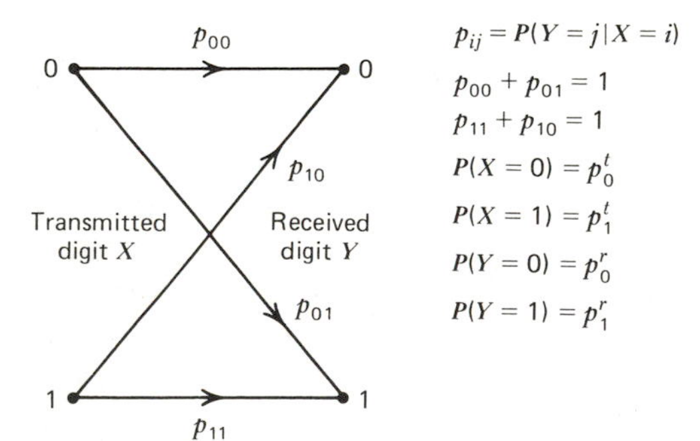

Figure 4.8 Model of a discrete channel.

have the following relationships:

$$P(\text{error}) = P_e = P(X \neq Y) = P(X = 0, Y = 1) + P(X = 1, Y = 0)$$
$$= P(Y = 1|X = 0)P(X = 0) + P(Y = 0|X = 1)P(X = 1)$$

or

$$P_e = p_0^t p_{01} + p_1^t p_{10} \tag{4.22}$$

Also, p_0^r and p_1^r can be expressed as

$$p_0^r = p_0^t p_{00} + p_1^t p_{10}$$
$$p_1^r = p_0^t p_{01} + p_1^t p_{11} \tag{4.23}$$

The channel is called a *binary symmetric channel* (BSC) if $p_{00} = p_{11} = p$. The only parameter needed to characterize a BSC is p.

We can extend our model to the general case where the channel input X can assume M values ($M > 2$). There are commercial modems available today where up to eight distinct levels or waveforms are transmitted over the channel. Figure 4.9 shows a model for the general case. Analysis of this channel is similar to the analysis of the binary channel discussed before. For example,

$$p_j^r = \sum_{i=1}^{M} p_i^t p_{ij}$$

and

$$P(\text{error}) = P_e = \sum_{i=1}^{M} p_i^t \left[\sum_{\substack{j=1 \\ j \neq i}}^{M} p_{ij} \right] \tag{4.24}$$

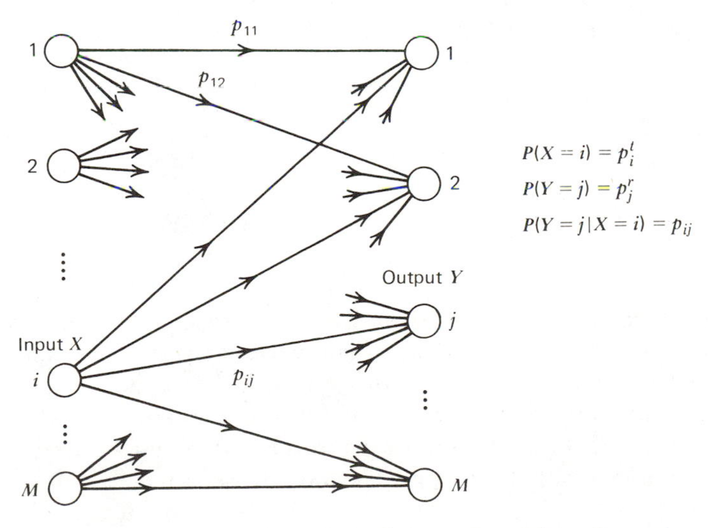

Figure 4.9 Model of an M-ary discrete memoryless channel.

In a discrete memoryless channel such as the one shown in Figure 4.9, there are two statistical processes at work: the input to the channel and the noise. Thus there are a number of entropies or information contents that we can calculate. First, we have the entropy of the input X defined as

$$H(X) = -\sum_{i=1}^{M} p_i^t \log_2(p_i^t) \text{ bits/symbol} \tag{4.25}$$

where p_i^t is the probability that the ith symbol of the alphabet is transmitted. Similarly, we can also define the entropy of the output Y as

$$H(Y) = -\sum_{i=1}^{M} p_i^r \log_2(p_i^r) \text{ bits/symbol} \tag{4.26}$$

where p_i^r denotes the probability that the output of the channel is the ith symbol of the alphabet. $H(Y)$ represents the average number of bits per symbol needed to encode the output of the channel. We can also define a conditional entropy $H(X|Y)$, called *equivocation*, as

$$H(X|Y) = -\sum_{i=1}^{M}\sum_{j=1}^{M} P(X = i, Y = j) \log_2(P(X = i|Y = j)) \tag{4.27}$$

and a joint entropy $H(X, Y)$ as

$$H(X, Y) = -\sum_{i=1}^{M}\sum_{j=1}^{M} P(X = i, Y = j) \log_2 P(X = i, Y = j) \tag{4.28}$$

The conditional entropy $H(X|Y)$ represents how uncertain we are of X, on the average, when we know Y. The reader can verify the following relationships between $H(X)$, $H(Y)$, $H(X|Y)$, $H(Y|X)$, and $H(X, Y)$:

$$\begin{aligned} H(X, Y) &= H(X|Y) + H(Y) \\ &= H(Y|X) + H(X) \end{aligned} \tag{4.29}$$

where

$$H(Y|X) = -\sum_{i=1}^{M}\sum_{j=1}^{M} P(X = i, Y = j) \log_2(P(Y = j|X = i))$$

For a BSC, $P(X = i|Y = i)$ $(i = 0, 1)$ measures the uncertainty about the transmitted bit based on the received bit. The uncertainty is minimum when $P(X = i|Y = i) = 1$ for $i = 0, 1$, that is, an errorless channel. The uncertainty is maximum when $P(X = i|Y = i) = \frac{1}{2}$ for $i = 0, 1$. If we define the uncertainty as $-\log_2[P(X = i|Y = i)]$, then we have one bit of uncertainty when the output is independent of the input. When we have one bit of uncertainty associated with each received bit, the received value of the bit does not convey any information!

The conditional entropy $H(X|Y)$ is an *average measure of uncertainty* about X when we know Y. In one extreme we can have Y and X related in a one-to-one manner such as $Y = X$. For this case, there is no uncertainty about X when we know Y; $P(X = i|Y = j) = \delta_{ij}$, where δ_{ij} is the delta function that is 0 for $i \neq j$ and 1 for $i = j$. We can easily verify that $H(X|Y) = 0$ when $Y = X$. In the context of a communication channel $Y = X$ represents an errorless channel, and there is no uncertainty about the input when the output is known. Alternatively, we can say that no information is lost in the channel since the output is uniquely related to the input. As another example, let us consider a communication channel that is so noisy that the output is statistically independent of the input. In this case we can easily verify that $H(X, Y) = H(X) + H(Y)$, and $H(X|Y) = H(X)$, that is, Y does not contain any information about X (see Problem 4.19).

4.5.1 Rate of Information Transmission Over a Discrete Channel

In the case of an M-ary discrete memoryless channel accepting symbols at the rate of r_s symbols/sec, the average amount of information per symbol going into the channel is given by the entropy of the input random variable X as

$$H(X) = -\sum_{i=1}^{M} p_i^t \log_2 p_i^t \tag{4.30}$$

In Equation (4.30) we have assumed that the symbols in the sequence at the input to the channel occur in a statistically independent fashion. The average rate at which information is going into the channel is given by

$$D_{in} = H(X) r_s \text{ bits/sec} \tag{4.31}$$

Due to errors, it is not in general possible to reconstruct the input symbol sequence with certainty by operating on the received sequence. Hence we can say that some information is lost due to errors. Before we attempt to define the amount of information that is "lost", let us consider the following example.

Suppose there are two possible symbols 0 and 1 that are transmitted at a rate of 1000 symbols or bits per second with $p_0^t = \frac{1}{2}$ and $p_1^t = \frac{1}{2}$. The source information rate and D_{in} at the input to the channel are 1000 bits/sec. Let us assume that the channel is symmetric with the probability of errorless transmission p equal to 0.95. Now, let us ask ourselves the question, what is the rate of transmission of information? It is certainly less than 1000 bits/sec since on the average 50 out of every 1000 bits are incorrect. Our first impulse might be to say that the rate is 950 bits/sec by subtracting the number of errors from the data rate at the channel input. However, this is not satisfactory since the receiver does not know exactly which bits are in error, even though it knows that on the average 50 out of 1000 bits are incorrect. To further illustrate the difficulty in defining the amount of information transmitted as discussed above, let us consider the extreme case where the channel noise is so great that the probability of receiving a 1 or 0 is $\frac{1}{2}$ irrespective of what was transmitted. In such a case about $\frac{1}{2}$ of the received symbols are correct due to chance alone and we will give the system credit for transmitting 500 bits/sec, whereas no information is actually being transmitted. Indeed we can completely dispense with the channel and decide on the transmitted bit by flipping a coin at the receiving point, and correctly determine one half of the bits transmitted.

The inconsistency in defining information transmitted over a channel as the difference between input data rate and the error rate can be removed by making use of the information "lost" in the channel due to errors. In the preceding section we defined the conditional entropy of the input given the output $H(X|Y)$ as a measure of how uncertain we are of the input X given the output Y. For an ideal errorless channel we have no uncertainty about the input given the output and $H(X|Y)$ is equal to zero, that is, no information is lost. Knowing that $H(X|Y)=0$ for the ideal case wherein no information is lost, we may attempt to use $H(X|Y)$ to represent the amount of information lost in the channel. Accordingly we can define the amount of information transmitted over a channel by subtracting the information lost from the amount of information going into the channel. That is, we may define the *average rate of information transmission D_t* as

$$D_t \triangleq [H(X) - H(X|Y)]r_s \text{ bits/sec} \tag{4.32}$$

This definition takes care of the case when the channel is so noisy that the output is statistically independent of the input. When Y and X are independent, $H(X|Y) = H(X)$, and hence all the information going into the channel is lost and no information is transmitted over the channel. Let us illustrate these concepts by an example.

Example 4.7. A binary symmetric channel is shown in Figure 4.10. Find the rate of information transmission over this channel when $p = 0.9$, 0.8, and 0.6; assume that the symbol (or bit) rate is 1000/sec.

Solution

$$H(X) = \tfrac{1}{2}\log_2 2 + \tfrac{1}{2}\log_2 2 = 1 \text{ bit/symbol}$$

$$D_{in} = r_s H(X) = 1000 \text{ bits/sec}$$

To find D_t, we need the conditional probabilities $P(X|Y)$, $X, Y = 0, 1$. These

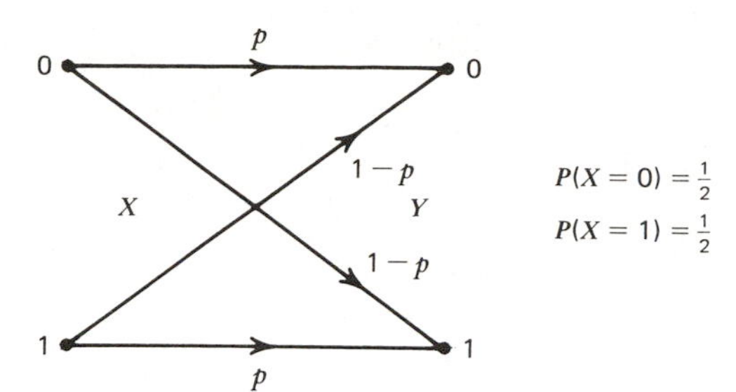

Figure 4.10 Binary symmetric channel.

conditional probabilities may be calculated as

$$P(X=0|Y=0)=\frac{P(Y=0|X=0)P(X=0)}{P(Y=0)}$$

and

$$\begin{aligned}P(Y=0)&=P(Y=0|X=0)P(X=0)+P(Y=0|X=1)P(X=1)\\&=p(\tfrac{1}{2})+(1-p)\tfrac{1}{2}=\tfrac{1}{2}\end{aligned}$$

Hence,

$$P(X=0|Y=0)=p$$

Similarly,

$$\begin{aligned}P(X=1|Y=0)&=1-p\\P(X=1|Y=1)&=p\\P(X=0|Y=1)&=1-p\end{aligned}$$

Hence,

$$\begin{aligned}H(X|Y)=&-P(X=0,\,Y=0)\log_2 P(X=0|Y=0)\\&-P(X=0,\,Y=1)\log_2 P(X=0|Y=1)\\&-P(X=1,\,Y=0)\log_2 P(X=1|Y=0)\\&-P(X=1,\,Y=1)\log_2 P(X=1|Y=1)\\=&-[\tfrac{1}{2}p\log_2 p+\tfrac{1}{2}(1-p)\log_2(1-p)\\&+\tfrac{1}{2}p\log_2 p+\tfrac{1}{2}(1-p)\log_2(1-p)]\\=&-[p\log_2 p+(1-p)\log_2(1-p)]\end{aligned}$$

and the rate of information transmission over the channel is given by

$$D_t=[H(X)-H(X|Y)]r_s \text{ bits/sec}$$

Values of D_t for $p = 0.9$, 0.8, and 0.6 are given in Table 4.5.
The values shown in Table 4.5 clearly indicate that the rate of information transmission over the channel decreases very rapidly as the probability of error $1-p$ approaches $\frac{1}{2}$.

Table 4.5. Rate of information transmission versus values of p

p	0.9	0.8	0.6
D_t	531 bits/sec	278 bits/sec	29 bits/sec

The reader should be aware of the fact that the *data rate* and *information rate* are two distinctly different quantities. With reference to Example 4.7, we often refer to the bit transition rate r_b at the channel input as the input data rate, or simply the bit rate. This is different from the information rate D_{in} at the channel input. D_{in} depends on r_b and the symbol (bit) probabilities. Furthermore, the rate of information transmission over the channel (D_t) depends not only on D_{in} but also on the channel symbol transition probabilities p_{ij}.

4.5.2 Capacity of a Discrete Memoryless Channel

The capacity of a noisy (discrete, memoryless) channel is defined as the maximum possible rate of information transmission over the channel. The maximum rate of transmission occurs when the source is "matched" to the channel. We define the *channel capacity* C as

$$\begin{aligned} C &\triangleq \max_{P(X)} \{D_t\} \\ &= \max_{P(X)} [H(X) - H(X|Y)]r_s \end{aligned} \tag{4.33}$$

where the maximum is with respect to all possible information sources; that is, the maximum is taken with respect to all possible probability distributions for the discrete random variable X.

Example 4.8. Calculate the capacity of the discrete channel shown in Figure 4.11. Assume $r_s = 1$ symbol/sec.

Solution. Let $\alpha = -[p \log p + q \log q]$ and let $P(X = 0) = P(X = 3) = P$ and $P(X = 1) = P(X = 2) = Q$ (these probabilities being equal from consideration of symmetry). Then, from the definition of channel capacity,

$$C = \max_{P,Q} [H(X) - H(X|Y)]$$

subject to the constraint $2P + 2Q = 1$ (why?). From the definition of $H(X)$

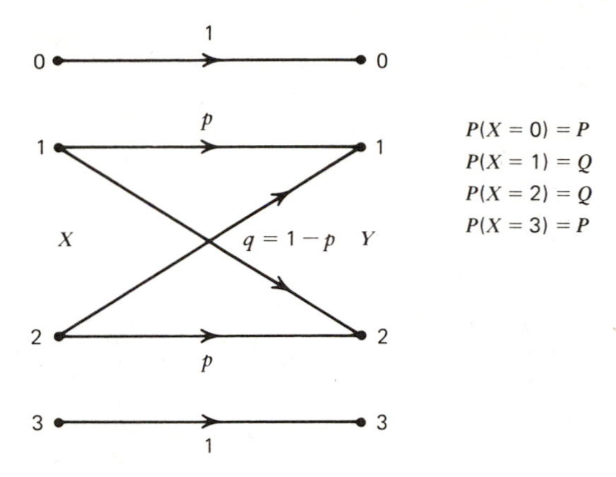

Figure 4.11 Channel model for Example 4.8.

and $H(X|Y)$, we obtain

$$H(X) = -2P \log_2 P - 2Q \log_2 Q$$
$$H(X|Y) = -2Q(p \log_2 p + q \log_2 q) = 2Q\alpha$$

Hence,

$$D_t = -2P \log_2 P - 2Q \log_2 Q - 2Q\alpha$$

We want to maximize D_t with respect to P and Q, subject to $2P + 2Q = 1$ (or $Q = \frac{1}{2} - P$).

Substituting $Q = \frac{1}{2} - P$, we have

$$D_t = -2P \log_2 P - 2(\tfrac{1}{2} - P) \log_2(\tfrac{1}{2} - P) - 2(\tfrac{1}{2} - P)\alpha$$

To find the value of P that maximizes D_t, we set

$$\frac{dD_t}{dP} = 0$$

or

$$\begin{aligned} 0 &= -\log_2 e - \log_2 P + \log_2 e + \log_2(\tfrac{1}{2} - P) + \alpha \\ &= -\log_2 P + \log_2 Q + \alpha \end{aligned}$$

Solving for P, we get

$$\begin{aligned} P &= Q2^{\alpha} \\ &= Q\beta \end{aligned}$$

where

$$\beta = 2^{\alpha}$$

Substituting $P = Q\beta$ in $2P + 2Q = 1$, we can obtain the optimum values of

P and Q as

$$P = \frac{\beta}{2(1+\beta)}$$

$$Q = \frac{1}{2(1+\beta)}$$

The channel capacity is then,

$$\begin{aligned} C &= -2(P \log_2 P + Q \log_2 Q + Q\alpha) \\ &= -2\left[\frac{\beta}{2(1+\beta)} \log_2\left(\frac{\beta}{2(1+\beta)}\right) + \frac{1}{2(1+\beta)} \log_2\left(\frac{1}{2(1+\beta)}\right) + \frac{1}{2(1+\beta)} \log_2 \beta\right] \\ &= \log_2\left(\frac{2(\beta+1)}{\beta}\right) \text{ bits/sec} \end{aligned}$$

A check with extreme values of $p = 1$ and $p = 0$ reveals the following: With $p = 1$, we have an errorless channel and the maximum rate of information transmission occurs when the input symbols occur with equal probability. The channel capacity for this ideal channel is 2 bits/symbol or 2 bits/sec with a symbol rate of 1 symbol/sec. For the noisy case with $p = \frac{1}{2}$, the capacity of the channel is $C = \log_2 3$. Here the first and fourth symbol are used more often than the other two because of their freedom from noise. Also the second and third symbols could not be distinguished at all and act together like one symbol. Hence, the capacity $\log_2 3$ seems to be a reasonable answer. For other values of p, the channel capacity will lie between $\log_2 3$ and $\log_2 4$ bits/sec.

The justification for defining a capacity for the noisy channel when we know that we can never send information without errors over such a channel is based on the fact that we can definitely reduce the probability of errors by repeating messages many times and studying the different received versions of the message. By increasing the redundancy of the encoding we can make the probability of error approach zero. This result is stated below as a theorem.

Theorem 4.2

Let C be the capacity of a discrete memoryless channel, and let H be the entropy of a discrete information source emitting r_s symbols per second. If $r_s H \leq C$, then there exists a coding scheme such that the output of the source can be transmitted over the channel with an arbitrarily small probability of error. It is not possible to transmit information at a rate exceeding C without a positive frequency of errors.

While a proof of this theorem is mathematically formidable, we will look at encoding schemes that will accomplish the task mentioned in the theorem in a later chapter when we look at the design of channel encoders. For now, it suffices to say that if the information rate of the source is less than the capacity of the channel, then we can design a channel encoder/decoder that will allow us to transmit the output of the source over the channel with an arbitrarily small probability of error.

4.5.3 Discrete Channels with Memory

In the preceding sections we looked at channels that have no memory, that is, channels in which the occurrence of error during a particular symbol interval does not influence the occurrence of errors during succeeding symbol intervals. However, in many channels, errors do not occur as independent random events, but tend to occur in bursts. Such channels are said to have memory. Telephone channels that are affected by switching transients and dropouts, and microwave radio links that are subjected to fading are examples of channels with memory. In these channels, impulse noise occasionally dominates the Gaussian noise and errors occur in infrequent long bursts. Because of the complex physical phenomena involved, detailed characterization of channels with memory is very difficult.

A model that has been moderately successful in characterizing error bursts in channels is the Gilbert model. Here the channel is modeled as a discrete memoryless BSC, where the probability of error is a time varying parameter. The changes in probability of error is modeled by a Markoff process shown in Figure 4.12. The error generating mechanism in the channel occupies one of

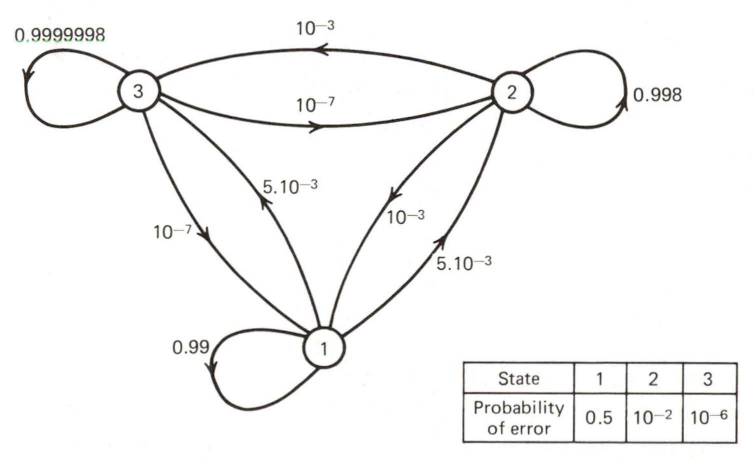

State	1	2	3
Probability of error	0.5	10^{-2}	10^{-6}

Figure 4.12 A three-state Gilbert model for communication channels.

three states, and transition from one state to another is modeled by a discrete, stationary Markoff process. When the channel is in state 2 for example, bit error probability during a bit interval is 10^{-2} and the channel stays in this state during the succeeding bit interval with a probability 0.998. However, the channel may go to state 1 wherein the bit error probability is 0.5. Since the system stays in this state with probability 0.99, errors tend to occur in bursts (or groups). State 3 represents a low bit error rate, and errors in this state are produced by Gaussian noise. Errors very rarely occur in bursts while the channel is in this state. Other details of the model are shown in Figure 4.12. The maximum rate at which data can be transmitted over the channel can be computed for each state of the channel using the BSC model of the channel corresponding to each of the three states. Other characteristic parameters of the channel such as the mean time between error bursts, and mean duration of the error bursts can be calculated from the model.

4.6 CONTINUOUS CHANNELS

The communication channel between points d and f in Figure 4.7 is analog or continuous in nature. In this portion of the channel, the input signals are continuous functions of time, and the function of the channel is to produce at its output the electrical waveform presented at its input. A real channel accomplishes this only approximately. First, the channel modifies the waveform in a deterministic fashion, and this effect can be adequately modeled by treating the channel as a linear system. The channel also modifies the input waveform in a random fashion due to additive and multiplicative noise. Throughout this book we will deal with additive noise only since it occurs more often than multiplicative noise. Additive noise can be Gaussian or impulsive in nature. Gaussian noise includes thermal and shot noise from equipment and radiation picked up by the receiving antenna. According to the central limit theorem the noise that results from the summed effects of many sources tends to have a Gaussian distribution. Because of this omnipresence, Gaussian noise is most often used to characterize the analog portion of communication channels. Modulation and demodulation techniques are designed with the primary objective of reducing the effects of Gaussian noise.

A second type of noise, impulse noise, is also encountered in the channel. Impulse noise is characterized by long quiet intervals followed by bursts of high amplitude noise pulses. This type of noise is due to switching transients, lightning discharges, and accidental hits during maintenance work, and so forth. The characterization of impulse noise is much more difficult than Gaussian noise. Also, analog modulation techniques are not as suitable as

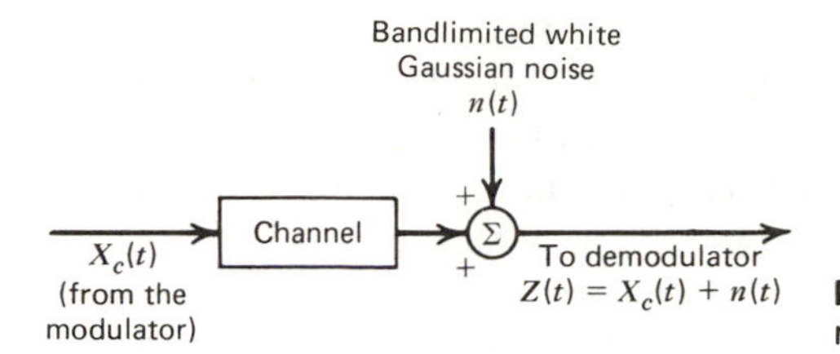

Figure 4.13 Analog portion of the communication channel.

digital coding methods for dealing with impulse noise phenomena. For these reasons the effects of impulse noise are often included in the model of the discrete portion of the channel, and only Gaussian noise is included in the model of the analog portion of the channel.

The analog portion of the communication channel can be modeled as shown in Figure 4.13 (see Section 3.7 and Figure 3.7). The input to the channel is a random process $X_c(t)$, which consists of the collection of all the waveforms generated by the modulator. The bandwidth of $X_c(t)$ and the channel is assumed to be B Hz (for convenience, let us assume $X_c(t)$ and the channel to be lowpass). The additive noise at the channel output is zero mean, bandlimited Gaussian white noise $n(t)$. The capacity of this portion of the channel is found by maximizing the rate of information transmission with respect to the distribution of $X_c(t)$. While the formulation of this problem is similar to the one we used for discrete channels in terms of $H(X_c)$ and $(H(X_c/Z)$, the optimization is very involved. However, the result has a very simple form; we state the result as a theorem (for a direct proof see Shannon's book*) and discuss how the result can be used in the design of communication systems.

4.6.1 Shannon–Hartley Theorem and Its Implications

Theorem 4.3

The capacity of a channel with bandwidth B and additive Gaussian bandlimited white noise is

$$C = B \log_2(1 + S/N) \text{ bits/sec} \tag{4.34}$$

where S and N are the average signal power and noise power, respectively, at the output of the channel. ($N = \eta B$ if the two-sided power spectral density of the noise is $\eta/2$ watts/Hz.)

*We give an indirect proof of Shannon's theorem in Section 8.7.3. Also, see Problem 4.27 in which the reader is asked to derive a relationship similar to the one given in Equation (4.34).

This theorem, referred to as the Shannon–Hartley theorem, is of fundamental importance and has two important implications for communication systems engineers. First, it gives us the upper limit that can be reached in the way of reliable data transmission rate over Gaussian channels. Thus a system designer always tries to optimize his system to have a data rate as close to C given in Equation (4.34) as possible with an acceptable error rate.

The second implication of the Shannon–Hartley theorem has to do with the exchange of signal-to-noise ratio for bandwidth. To illustrate this aspect of the theorem, suppose that we want to transmit data at a rate of 10,000 bits/sec over a channel having a bandwidth $B = 3000$ Hz. To transmit data at a rate of 10,000 bits/sec, we need a channel with a capacity of at least 10,000 bits/sec. If the channel capacity is less than the data rate, then errorless transmission is not possible. So, with $C = 10{,}000$ bits/sec we can obtain the (S/N) requirement of the channel as

$$\begin{aligned}(S/N) &= 2^{(C/B)} - 1\\ &= 2^{3.333} - 1 \approx 9\end{aligned}$$

For the same problem if we have a channel with a bandwidth of 10,000 Hz, then we need a S/N ratio of 1. Thus a bandwidth reduction from 10,000 Hz to 3000 Hz results in an increase in signal power from 1 to 9.

Another interesting aspect of the Shannon–Hartley theorem has to do with *bandwidth compression*. To illustrate this aspect let us ask ourselves the question, is it possible to quantize and transmit a signal whose spectral range extends up to a frequency f_m over a channel having a bandwidth less than f_m? The answer is yes and we can justify the answer as follows. Suppose we sample the analog signal at a rate of $3f_m$ samples/sec (i.e., at 1.5 times the Nyquist rate for example) and quantize the signal value into one of M possible levels. Then the data rate of the quantized signal is $3f_m \log_2 M$ bits/sec. If the bandwidth of the channel is B, then by an appropriate choice of signal power we can achieve a capacity C greater than $3f_m \log_2 M$. For example, with $M = 64$ and a channel bandwidth equal to half of the signal bandwidth, we would need a S/N ratio of about 109 dB to be able to transmit the quantized signal with a small probability of error. Thus a bandwidth compression by a factor of 2 is possible if we can maintain a S/N ratio of 109 dB (an impractical value) at the output of the channel. We are assuming that signal distortion due to sampling and quantizing is negligible.

The Shannon–Hartley theorem indicates that a *noiseless channel has an infinite capacity*. However, when noise is present the channel capacity does not approach infinity as the bandwidth is increased because the noise power increases as the bandwidth increases. The channel capacity reaches a *finite* upper limit with *increasing bandwidth* if the signal power is fixed. We can calculate this limit as follows. With $N = \eta B$, where $\eta/2$ is the noise power spectral density, we have

$$
\begin{aligned}
C &= B \log_2\left(1 + \frac{S}{\eta B}\right) \\
&= \left(\frac{S}{\eta}\right)\left(\frac{\eta B}{S}\right) \log_2\left(1 + \frac{S}{\eta B}\right) \\
&= \left(\frac{S}{\eta}\right) \log_2\left(1 + \frac{S}{\eta B}\right)^{\eta B/S}
\end{aligned} \tag{4.35}
$$

Recalling that $\lim_{x\to 0} (1 + x)^{1/x} = e$ and letting $x = S/\eta B$ in (4.35), we have

$$
\begin{aligned}
\lim_{B\to\infty} C &= \frac{S}{\eta} \log_2 e \\
&= 1.44\left(\frac{S}{\eta}\right)
\end{aligned} \tag{4.36}
$$

A communication system capable of transmitting information at a rate of $B \log_2(1 + S/N)$ is called an ideal system. Shannon proposed the following idea for such a system. Let us assume that the source puts out M equiprobable messages of length T seconds. The ideal communication system observes the source output for T seconds and the message is represented (encoded) by a channel signal chosen from a collection of M sample functions of white noise of duration T. At the output of the channel, the received signal plus noise is compared with stored versions of the channel signals. The channel signal that "best matches" the signal plus noise is presumed to have been transmitted and the corresponding message is decoded. The total amount of time delay involved in observing the message signal, transmitting, and decoding at the receiver is at best T seconds.

The ideal signalling scheme using noiselike signals can convey information at a rate approaching the channel capacity only when $T \to \infty$. Only in the limiting case do we have all the conditions satisfied. Under this limiting condition, the ideal system has the following characteristics:

1. The information rate $\to B \log_2(1 + S/N)$.
2. The error rate $\to 0$.
3. The transmitted and received signals have the characteristics of bandlimited Gaussian white noise.
4. As $T \to \infty$, the number of signals $M \to \infty$ and the coding delay also $\to \infty$.

It must be obvious from the preceding discussion that an ideal system cannot be realized in practice. Rather than trying to design a system using a large number of analog signals, we use a small number of analog signals in practical systems. This leads to a nonzero probability of error P_e. The data rate and the error probability define a discrete channel whose capacity C' will be less than $B \log_2(1 + S/N)$. Through this digital channel we try to achieve a data rate approaching C' with a probability of error approaching zero using digital error control encoding. Thus in practical systems we seldom try to achieve the maximum theoretical rate of information transmission over the analog portion of the channel. We keep this portion of the system reasonably simple. In the digital portion of the system, we try to achieve a rate approaching the capacity of the discrete portion of the channel since digital encoding is easier to implement.

In the following chapters we will discuss signaling schemes for transmitting digital information through an analog communication channel. For each type of signaling scheme, we will derive expressions for the error probability in terms of the bandwidth required, output S/N, and the data rate. These relationships define the parameters of the discrete portion of the channel. In Chapter 9 we will look at methods of error control coding that will enable us to transmit information over the discrete channel at a rate approaching its capacity with a small probability of error.

Before we conclude our discussion of the Shannon–Hartley theorem, it must be pointed out that the result given in Equation (4.34) is for the Gaussian channel. This limitation does not in any way diminish the importance and usefulness of the Shannon–Hartley law for the following reasons: First, most physical channels are generally at least approximately Gaussian. Second, it has been shown that the result obtained for the Gaussian channel provides a *lower bound* on the performance of a system operating over a non-Gaussian channel. That is, if a particular encoder/decoder yields an error probability P_e over the Gaussian channel, another encoder/decoder can be designed for a non-Gaussian channel to yield a smaller probability of error. Detailed expressions for the channel capacity have been derived for several non-Gaussian channels.

We now present several examples that illustrate the use of channel capacity in the design of communication systems.

Example 4.9. Calculate the capacity of a lowpass channel with a usable bandwidth of 3000 Hz and $S/N = 10^3$ at the channel output. Assume the channel noise to be Gaussian and white.

Solution. The capacity C is given by Equation (4.34) as

$$\begin{aligned} C &= B \log_2\left(1 + \frac{S}{N}\right) \\ &= (3000) \log_2(1 + 1000) \\ &\approx 30{,}000 \text{ bits/sec} \end{aligned}$$

The parameter values used in this example are typical of standard voice grade telephone lines. The maximum data rate achievable now on these channels is 9600 bits/sec. Rates higher than this require very complex modulation and demodulation schemes.

Example 4.10. An ideal lowpass channel of bandwidth B Hz with additive Gaussian white noise is used for transmitting digital information. (a) Plot C/B versus (S/N) in dB for an ideal system using this channel. (b) A practical signaling scheme on this channel uses one of two waveforms of duration T_b seconds to transmit binary information. The signaling scheme transmits data at a rate of $2B$ bits/sec, and the probability of error is given by*

$$P(\text{error}|1\text{ sent}) = P(\text{error}|0\text{ sent}) = P_e = Q(\sqrt{S/N})$$

where

$$Q(z) = \int_z^\infty \frac{1}{\sqrt{2\pi}} \exp(-x^2/2)\, dx$$

Using the tabulated values of $Q(z)$ given in Appendix D, plot the rate of information transmission versus (S/N) in dB for this scheme.

Solution

(a) For the ideal scheme, we have

$$\frac{C}{B} = \log_2\left(1 + \frac{S}{N}\right)$$

When $S/N \gg 1$, $C/B \approx \log_2(S/N)$.

(b) The binary signaling scheme corresponds to a discrete binary symmetric channel with $P_e = Q(\sqrt{S/N})$. The information rate over this channel is given by (see Example 4.7)

$$\begin{aligned} D_t &= 2B[1 - P_e \log_2(1/P_e) - (1 - P_e)\log_2(1/1 - P_e)] \\ &= 2B[1 + P_e \log_2 P_e + (1 - P_e)\log_2(1 - P_e)] \end{aligned}$$

For large S/N ratios, $P_e \approx 0$ and $D_t = 2B$. It also represents the capacity or the maximum rate at which we can transmit information using the binary signaling scheme. Values of C/B and D_t/B for various values of S/N are shown in Table 4.6 and shown plotted in Figure 4.14. These results show that for high S/N ratios, the binary signaling scheme is very inefficient. With high S/N ratios we can transmit and decode correctly a large

*Expressions for probabilities of error for various signaling schemes will be derived in Chapters 5 and 8.

Table 4.6. C/B and D_t/B for various values of S/N

$(S/N)_{dB}$	0	6	10	20
$(C/B)_{ideal}$	1	2.32	3.46	6.65
P_e	0.159	0.028	0.0008	≈ 0
$(D_t/B)_{binary}$	0.7236	1.6314	1.98	2.0

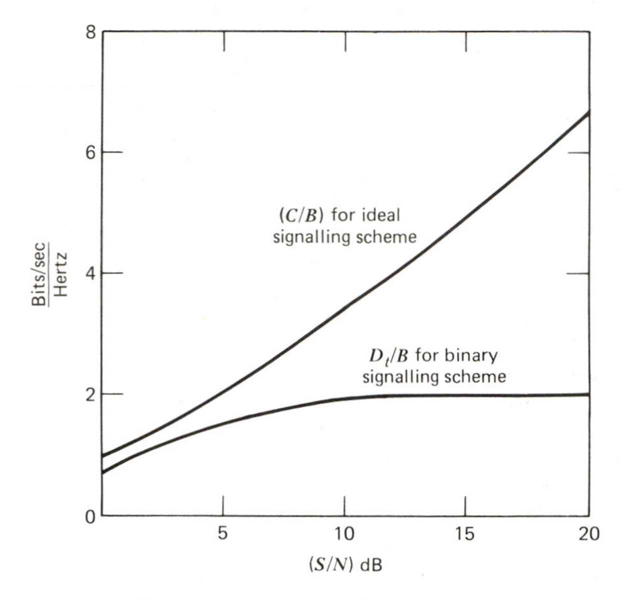

Figure 4.14 Plots of C/B and D_t/B.

number of waveforms and hence an M-ary signaling scheme, $M > 2$ should be used (we will discuss M-ary signaling schemes in Chapters 5 and 8).

The following example illustrates how we can use the concept of information rate and channel capacity in the design of communication systems.

Example 4.11. A CRT terminal is used to enter alphanumeric data into a computer. The CRT is connected to the computer through a voice grade telephone line having a usable bandwidth of 3000 Hz and an output S/N of 10 dB. Assume that the terminal has 128 characters and that the data sent from the terminal consist of independent sequences of equiprobable characters.

(a) Find the capacity of the channel.
(b) Find the maximum (theoretical) rate at which data can be transmitted from the terminal to the computer without errors.

Solution
(a) The capacity is given by:

$$C = B \log_2(1 + S/N)$$
$$= (3000) \log_2(11) = 10{,}378 \text{ bits/sec}$$

(b) Average information content/character:

$$H = \log_2(128) = 7 \text{ bits/character}$$

and the average information rate of the source $R = r_s H$. For errorless transmission, we need $R = r_s H < C$ or

$$7r_s < 10{,}378$$
$$r_s < 1482$$

Hence the maximum rate at which data can be transmitted without errors is 1482 characters/sec.

4.7 SUMMARY

A probabilistic model for discrete information sources was developed and the entropy of the source and the average information rate of the source were defined. The source entropy has the units of bits per symbol and it represents the average number of bits per symbol needed to encode long sequences of symbols emitted by the source. The average information rate represents the average number of bits per second needed to encode the source output. The functional block that maps the symbol sequence emitted by the source into a binary data stream is the source encoder. A procedure for designing an encoder using the algorithm given by Shannon was presented. The effect of design parameters such as block length and code word lengths on the complexity of the encoder, time delay in decoding and the efficiency of the encoder were discussed.

Mathematical models for discrete and continuous channels were discussed. The capacity of a channel represents the maximum rate at which data can be transmitted over the channel with an arbitrarily small probability of error. It was pointed out that the maximum rate of data transmission over a channel can be accomplished only by using signals of large dimensionality.

Several examples were presented to illustrate the concepts involved in modeling and analyzing discrete information sources and communication channels.

REFERENCES

Most of the material presented in this chapter is based on Shannon's work, which was first published in 1948. This classical work is very readable and quite interesting. Treatment of concepts in information theory and coding theory at an advanced level may be found in the books of Abramson (1963), and Wozencraft and Jacobs (1965) and Gallager (1968). Undergraduate texts such as the ones by Carlson (1975), and Taub and Schilling (1971) provide a brief treatment of information theory.

In order to understand and appreciate the concepts of source and channel models, the reader should be familiar with the theory of random variables and random processes. Books by Breipohl (1970—written for undergraduates) and by Papoulis (1965—beginning graduate level) provide thorough and easily readable treatment of the theory of random variables and random processes. Topics in measurement and analysis of random data are treated well in the book by Bendat and Piersol (1971).

1. C. E. Shannon. *Mathematical Theory of Communication.* Univ. of Illinois Press (1963) (original work was published in the Bell System Technical Journal, vol. 27, 379–423, 623–656, 1948).
2. N. Abramson. *Information Theory and Coding.* McGraw-Hill, New York (1963).
3. J. M. Wozencraft and I. M. Jacobs. *Principles of Communication Engineering.* Wiley, New York (1965).
4. R. G. Gallager. *Information Theory and Reliable Communication.* Wiley, New York (1968).
5. H. Taub and D. L. Schilling. *Principles of Communication Systems.* McGraw-Hill, New York (1971).
6. A. B. Carlson. *Communication Systems.* McGraw-Hill, New York (1975).
7. A. M. Breipohl. *Probabilistic Systems Analysis.* Wiley, New York (1970).
8. A. Papoulis. *Probability, Random Variables, and Stochastic Processes.* McGraw-Hill, New York (1965).
9. J. S. Bendat and A. G. Piersol. *Measurement and Analysis of Random Data.* Wiley, New York (1971).

PROBLEMS

Section 4.2

4.1. A source emits one of four possible messages m_1, m_2, m_3, and m_4 with probabilities $\frac{1}{2}$, $\frac{1}{4}$, $\frac{1}{8}$, and $\frac{1}{8}$, respectively. Calculate the information content of each message and the average information content per message.

4.2. A card is drawn from a deck of playing cards. (a) You are informed that the card you draw is a spade. How much information did you receive (in bits)? (b) How much information do you receive if you are told that the card that you drew is an ace? (c) How much information do you receive if you are told that the card you drew is an ace of spades? Is the information content of the message "ace of spades" the sum of the information contents of the messages "spade" and "ace"?

4.3. A source emits an independent sequence of symbols from an alphabet consisting of five symbols A, B, C, D, and E with symbol probabilities $\frac{1}{4}$, $\frac{1}{8}$, $\frac{1}{8}$, $\frac{3}{16}$, $\frac{5}{16}$, respectively. Find the entropy of the source.

4.4. A binary source is emitting an independent sequence of 0's and 1's with probabilities p and $1-p$, respectively. Plot the entropy of this source versus p $(0 < p < 1)$.

4.5. For a source emitting symbols in independent sequences, show that the source entropy is maximum when the symbols occur with equal probabilities.

4.6. The international Morse code uses a sequence of dots and dashes to transmit letters of the English alphabet. The dash is represented by a current pulse that has a duration of 3 units and the dot has a duration of 1 unit. The probability of occurrence of a dash is $\frac{1}{3}$ of the probability of occurrence of a dot.

(a) Calculate the information content of a dot and a dash.

(b) Calculate the average information in the dot–dash code.

(c) Assume that the dot lasts 1 msec, which is the same time interval as the pause between symbols. Find the average rate of information transmission.

4.7. The probability of occurrence of the various letters of the English alphabet are given below:

A 0.081	J 0.001	S 0.066
B 0.016	K 0.005	T 0.096
C 0.032	L 0.040	U 0.031
D 0.037	M 0.022	V 0.009
E 0.124	N 0.072	W 0.020
F 0.023	O 0.079	X 0.002
G 0.016	P 0.023	Y 0.019
H 0.051	Q 0.002	Z 0.001
I 0.072	R 0.060	

(a) What letter conveys the maximum amount of information?

(b) What letter conveys the minimum amount of information?

(c) What is the entropy of English text if you can assume that letters are chosen independently to form words and sentences (not a realistic assumption!).

(d) If I am thinking of a word and tell you the first letter of the word, which will be a more helpful clue, T or X? Why?

4.8. A black and white TV picture consists of 525 lines of picture information. Assume that each line consists of 525 picture elements and that each element can have 256 brightness levels. Pictures are repeated at the rate of 30/sec. Calculate the average rate of information conveyed by a TV set to a viewer.

4.9. The output of an information source consists of 128 symbols, 16 of which occur with a probability of 1/32 and the remaining 112 occur with a probability of 1/224. The source emits 1000 symbols/sec. Assuming that the symbols are chosen independently, find the average information rate of this source.

4.10. The state diagram of a stationary Markoff source is shown in Figure 4.15.

(a) Find the entropy of each state H_i $(i = 1, 2, 3)$.

(b) Find the entropy of the source H.

(c) Find G_1, G_2, and G_3 and verify that $G_1 \geq G_2 \geq G_3 \geq H$.

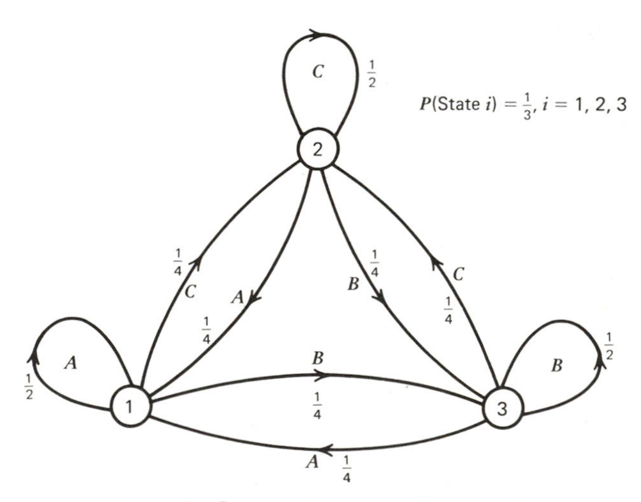

Figure 4.15 Source diagram for Problem 4.10.

4.11. Re-work the previous problem for the source shown in Figure 4.3.

Section 4.3

4.12. For the source described in Example 4.5:

(a) Design a source encoding scheme using a block size of two symbols and variable length code words. Calculate the actual number of bits per symbol $\hat{H}_2$ used by the encoder and verify that

$$\hat{H}_2 \leq G_2 + \tfrac{1}{2}$$

(b) Design a source encoding scheme using fixed length code words and a block size of four symbols. Compute the actual number of bits per symbol used.

(c) If the source is emitting symbols at a rate of 1000 symbols/sec, compute the output bit rate of the encoders (a) and (b).

4.13. Another technique used in constructing a source encoder consists of arranging the messages in decreasing order of probability and dividing the message into two almost equally probable groups. The messages in the first group are given the bit 0 and the messages in the second group are given the bit 1. The procedure is now applied again for each group separately, and continued until no further division is possible. Using this algorithm, find the code words for six messages occurring with probabilities $\frac{1}{3}, \frac{1}{3}, \frac{1}{6}, \frac{1}{12}, \frac{1}{24}, \frac{1}{24}$.

4.14. Another way of generating binary code words for messages consists of arranging the messages in decreasing order of probability and dividing code words as follows: The code word for the first message is "0". The code word for the ith message consists of $(i-1)$ bits of "1's" followed by a "0." The code word for the last message consists of all "1's"; the number of bits in the code word for the last message is equal to the total number of messages that are to be encoded.

(a) Find the code words and the average number of bits per message used if the source emits one of five messages with probabilities $\frac{1}{2}, \frac{1}{4}, \frac{1}{8}, \frac{1}{16}$, and $\frac{1}{16}$.

(b) Is this code uniquely decipherable? That is, for every possible sequence of bits, is there only one way of interpreting the messages?

4.15. A source emits *independent* sequences of symbols from a source alphabet containing five symbols with probabilities 0.4, 0.2, 0.2, 0.1 and 0.1.

(a) Compute the entropy of the source.

(b) Design a source encoder with a block size $n = 2$.

Section 4.5

4.16. A nonsymmetric binary channel is shown in Figure 4.16.

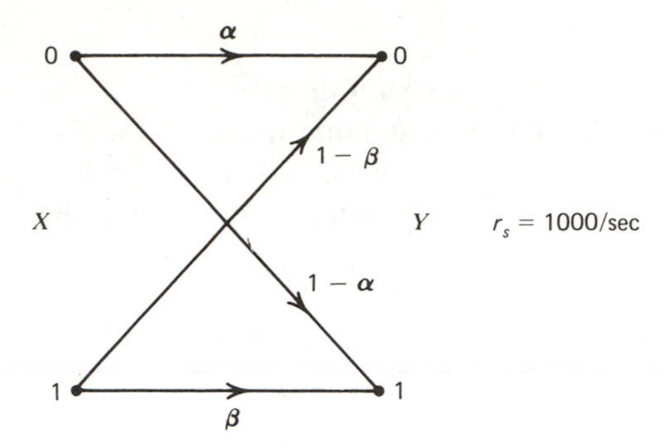

Figure 4.16 Binary channel model for Problem 4.17.

(a) Find $H(X)$, $H(Y)$, $H(X|Y)$, and $H(Y|X)$ when $P(X=0)=\frac{1}{4}$, $P(X=1)=\frac{3}{4}$, $\alpha=0.75$, and $\beta=0.9$.
(b) Find the capacity of the channel for $\alpha=0.75$ and $\beta=0.9$.
(c) Find the capacity of the binary symmetric channel ($\alpha=\beta$).

4.17. Show that $H(X, Y)=H(X)+H(Y|X)=H(Y)+H(X|Y)$.

4.18. Find the capacity of the discrete channel shown in Figure 4.17.

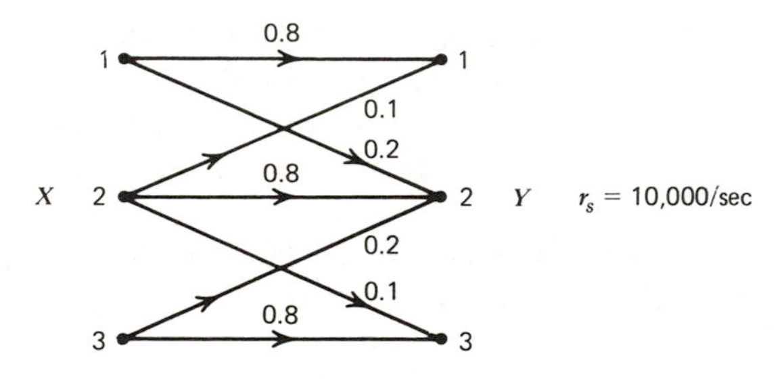

Figure 4.17 Channel model for Problem 4.18.

4.19. Show that (a) $H(X|Y)=H(X)$ when X and Y are statistically independent, and (b) $H(X|Y)=0$ when $X=Y$.

4.20. A discrete channel accepts as its input a binary sequence with a bit rate of r_b bits/sec. The channel signals are selected from a set of eight possible waveforms, each having a duration $3/r_b$ seconds. Thus, each waveform may convey up to three bits of information. The channel noise is such that when the received waveform is decoded, each block of three input bits is received with no errors, or with exactly one error in the first, second, or third bit position. Assuming that these four outcomes are equally likely to occur:

(a) Find the capacity of the discrete channel.

(b) Suppose that you want to transmit the output of an information source having a rate $R = r_b/3$ over this channel. How would you encode the data so that errorless transmission is possible?

4.21. The state model of a discrete channel with memory is shown in Figure 4.18. In state 1, the channel corresponds to a BSC with an error-probability of 0.001. At state 2, the channel is again a BSC with an error probability of 0.5. The state and transitional probabilities are shown in the diagram. Assume that the bit rate at the input to the channel is 1000 bits/sec and the transition rate of the state of the channel is also 1000/sec.

(a) Find the capacity of the channel for state 1 and state 2.

(b) Find the average capacity of the channel.

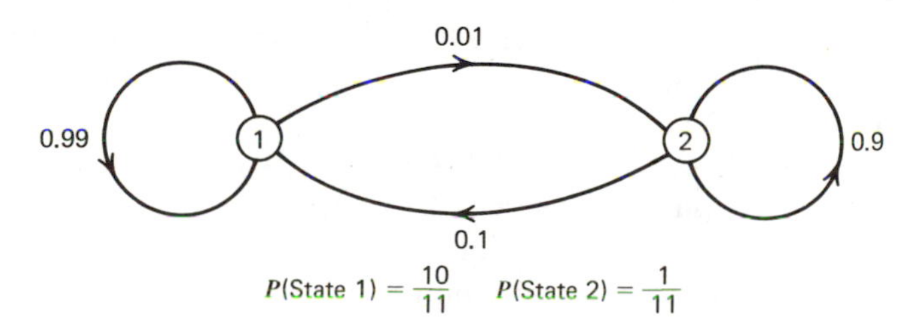

Figure 4.18 Channel state model for Problem 4.12.

Section 4.6

4.22. Calculate the capacity of a Gaussian channel with a bandwidth of 1 MHz and S/N ratio of 30 dB.

4.23. How long will it take to transmit one million ASCII characters over the channel in Problem 4.22? (In ASCII code, each character is coded as an 8-bit binary word; ignore start and stop bits.)

4.24. A Gaussian channel has a bandwidth of 4 kHz and a two-sided noise power spectral density $\eta/2$ of 10^{-14} watt/Hz. The signal power at the receiver has to be maintained at a level less than or equal to 1/10 of a milliwatt. Calculate the capacity of this channel.

4.25. An analog signal has a 4 kHz bandwidth. The signal is sampled at 2.5 times the Nyquist rate and each sample is quantized into one of 256 equally likely levels. Assume that the successive samples are statistically independent.

(a) What is the information rate of this source?
(b) Can the output of this source be transmitted without errors over a Gaussian channel with a bandwidth of 50 kHz and S/N ratio of 23 dB?
(c) What will be the bandwidth requirements of an analog channel for transmitting the output of the source without errors if the S/N ratio is 10 dB?

4.26. A friend of yours says that he can design a system for transmitting the output of a minicomputer to a line printer operating at a speed of 30 lines/minute over a voice grade telephone line with a bandwidth of 3.5 kHz, and $S/N = 30$ dB. Assume that the line printer needs eight bits of data per character and prints out 80 characters per line. Would you believe him?

4.27. The waveform shown in Figure 4.19 is used for transmitting digital information over a channel having a bandwidth $B \approx 1/2T$. Assume that the M levels are equally likely to occur and that they occur as an independent sequence.
(a) Find $E\{X^2(t)\}$.
(b) Find S/N (note: $N = \sigma^2$, and $S = E\{X^2(t)\}$).
(c) Assume λ is large enough so that errors occur with a probability $P_e \to 0$. Find the rate of information conveyed by the signal. Compare your result with Equation (4.34).

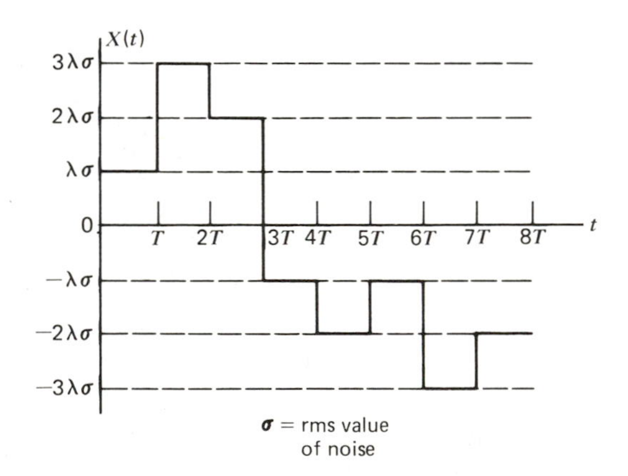

Figure 4.19 Signal waveform; sequences of levels convey messages.

5

BASEBAND DATA TRANSMISSION

5.1 INTRODUCTION

In the previous chapter, we discussed the theoretical limitations on the rate of information transmission over noisy channels. We pointed out that the maximum errorless rate of data transmission over a noisy channel could be achieved only by using signal sets of large dimensionality whose statistical characteristics match the noise characteristics. In a practical system, the large dimensionality of signals is realized by the digital encoding of a small set of basic waveforms generated by the modulator. The number of analog waveforms generated by commercial digital modulators range from 2 (binary) to a maximum of 8 or 16. There are many types of modulators corresponding to the many possible selections of modulator waveforms. In this chapter we take a detailed look at the analysis and design of discrete pulse modulation techniques that can be used for transmitting the output of a discrete source over a *baseband* channel. In a later chapter we will discuss discrete carrier modulation schemes that are used for transmitting digital information over *bandpass* channels.

In discrete pulse modulation, the amplitude, duration or position of the transmitted pulses is varied according to the digital information to be transmitted. These pulse modulation schemes are referred to as pulse amplitude (PAM), pulse duration (PDM), and pulse position modulation (PPM) schemes.

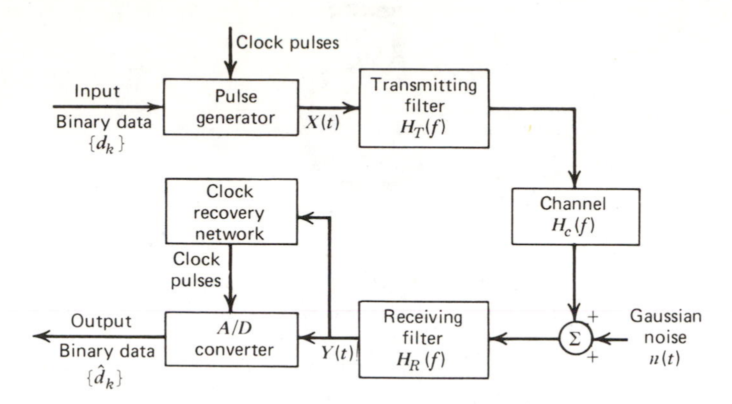

Figure 5.1 Baseband binary data transmission system.

Of these three methods, PAM systems are most efficient in terms of power and bandwidth utilization. This chapter is devoted to the study of PAM systems.

The elements of a baseband binary PAM system are shown in Figure 5.1. The input to the system is a binary data sequence with a bit rate of r_b and bit duration of T_b. The pulse generator output is a pulse waveform

$$X(t) = \sum_{k=-\infty}^{\infty} a_k p_g(t - kT_b) \tag{5.1a}$$

where $p_g(t)$ is the basic pulse whose amplitude a_k depends on the kth input bit. For convenience we will assume that $p_g(t)$ is normalized such that

$$p_g(0) = 1 \tag{5.1b}$$

and

$$a_k = \begin{cases} a & \text{if } k\text{th input bit is 1} \\ -a & \text{if } k\text{th input bit is 0} \end{cases} \tag{5.1c}$$

The PAM signal $X(t)$ passes through a transmitting filter $H_T(f)$, and then through the channel that adds random noise in addition to modifying the signal in a deterministic fashion. The noisy signal then goes through the receiving filter $H_R(f)$, and the output $Y(t)$ of the receiving filter is sampled by the analog-to-digital (A/D) converter. The transmitted bit stream is regenerated by the A/D converter based on the sampled values of $Y(t)$. The sampling instant is determined by the clock or timing signal that is usually generated from $Y(t)$ itself. A set of typical waveforms that occur at various points in the system is shown in Figure 5.2.

The A/D converter input $Y(t)$ can be written as

$$Y(t) = \sum_k A_k p_r(t - t_d - kT_b) + n_0(t) \tag{5.1d}$$

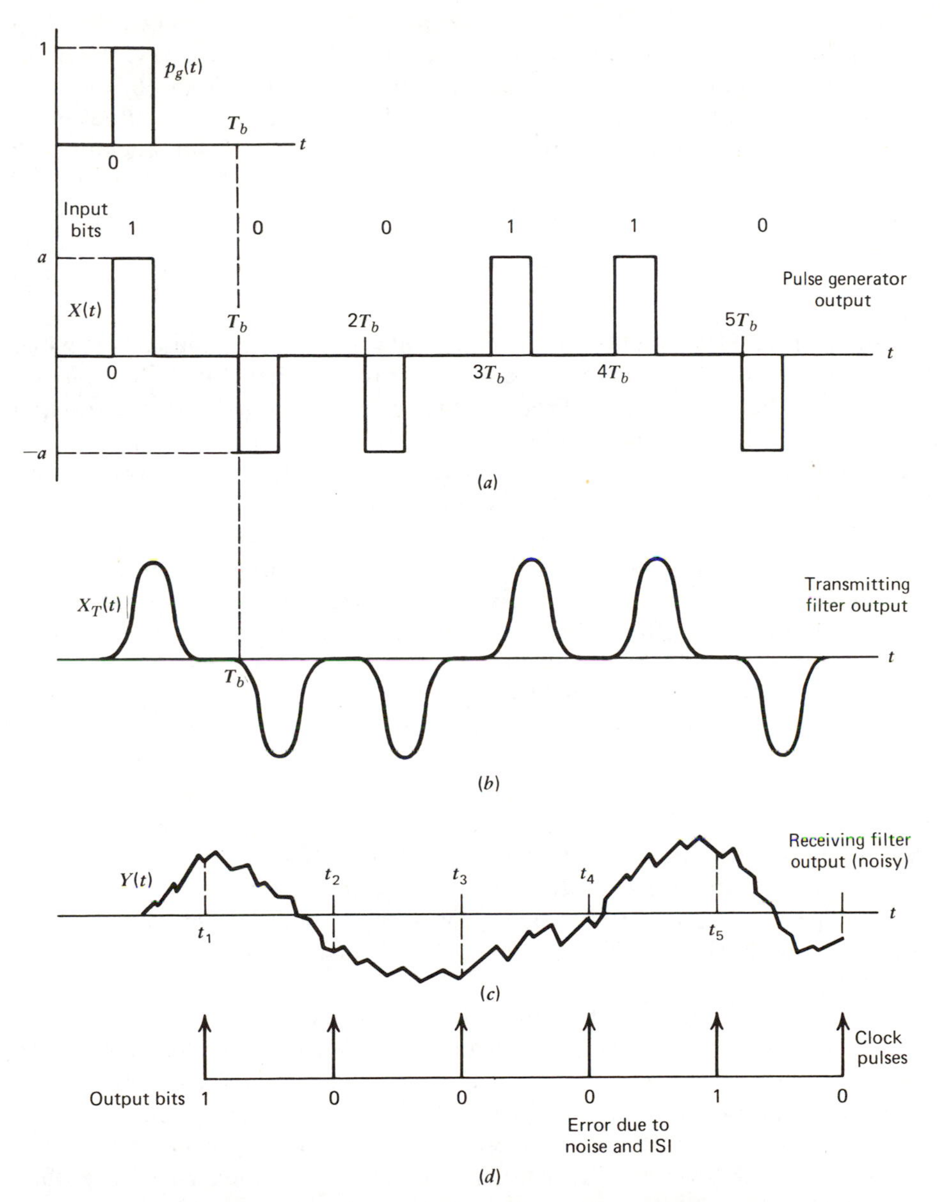

Figure 5.2 Example of typical waveforms in a binary PAM system. (*a*) Pulse generator output. (*b*) Transmitting filter output. (*c*) Receiving filter output (noisy). (*d*) Clock pulses.

where $A_k = K_c a_k$ and $K_c p_r(t - t_d)$ is the response of the system when the input is $p_g(t)$. In Equation (5.1d), t_d is an arbitrary time delay and $n_0(t)$ is the noise at the receiver output. K_c is a normalizing constant that yields $p_r(0) = 1$.

The A/D converter samples $Y(t)$ at $t_m = mT_b + t_d$ and the mth output bit is generated by comparing $Y(t_m)$ with a threshold (which is 0 for a symmetric binary PAM data transmission system). The input to the A/D converter at the sampling instant $t_m = mT_b + t_d$ is

$$Y(t_m) = A_m + \sum_{k \neq m} A_k p_r[(m - k)T_b] + n_0(t_m) \tag{5.2}$$

In Equation (5.2), the first term A_m represents the mth transmitted bit while the second term represents the residual effect of all other transmitted bits on the mth bit being decoded. This residual effect is called the *intersymbol interference* (ISI). The last term in (5.2) represents the noise.

In the absence of noise, and in the absence of ISI, the mth transmitted bit can be decoded correctly based on $Y(t_m)$ since $Y(t_m) = K_c a_m$ and a_m is uniquely related to the mth input bit. Noise and ISI introduce errors in the output. The major objectives of baseband PAM system design are to choose the transmitting and receiving filters to minimize the effects of noise and eliminate or minimize ISI. In addition, for a given transmitted power it may be desirable to maximize the signaling rate r_b for a given bandwidth B or minimize the bandwidth required for a given signaling rate.

We deal with the design and analysis of optimum baseband PAM system in this chapter. The design of the system starts with the assumption that the physical characteristics of the channel and the statistical characteristics of the noise and the input bit stream are known. The pulse shapes $p_g(t)$ and $p_r(t)$ and the transfer functions of the filters $H_T(f)$ and $H_R(f)$ are to be chosen to optimize the performance of the system, keeping the bit error probability below a specified value. The bit error probability, which is commonly used as a measure of performance for binary PAM systems, is defined as $P_e = P[\hat{d}_k \neq d_k]$ (see Figure 5.1).

The design and analysis of a binary baseband PAM system is treated first and the M-ary PAM schemes are treated next. We will see that the M-ary schemes require a smaller bandwidth but more power than a binary scheme for a given data rate and bit error probability. Special signaling schemes, such as the duobinary scheme, are discussed and the effects of precoding the input bit stream on the spectral characteristics of the transmitted signal are illustrated.

The design criteria are aimed at an overall pulse shaping that would yield zero ISI. However, in practical systems some amount of residual ISI will inevitably occur due to imperfect filter realization, incomplete knowledge of channel characteristics, and changes in channel characteristics. Hence, an equalizing filter is often inserted between the receiving filter and the A/D converter to compensate for changes in the parameters of the channel. We will look at procedures for designing zero forcing equalizers that will reduce the ISI.

In the PAM method of data transmission a clock signal must be recovered at the receiving end to set the sampling rate and sampling times. The clock information must somehow be carried in the transmitted signal. Methods of carrying the clock information and recovering it vary with the pulse shapes and coding methods used. We will discuss several methods of clock recovery at the receiver.

It must be pointed out that the important parameters involved in the design of a PAM system are data rate, error rate, transmitted power, noise power spectral density, and system complexity. These parameters are interrelated and the design procedure will involve trade-offs between the parameters to arrive at a system that meets the specified performance requirements and the given constraints.

5.2 BASEBAND BINARY PAM SYSTEMS

In this section we deal with the design of optimum baseband binary data transmission systems. Data rates in binary systems may range from a low rate of 100 bits/sec (BPS) in applications involving electromechanical devices such as a teletype to a high rate of up to tens of megabits per second in applications involving data transfer between computers. The rate typically is from 300 to 4800 bits/sec over voice grade telephone links to several hundred megabits per second over wideband microwave radio links. The acceptable bit error rate varies over a wide range depending on the application. Error probabilities in the range of 10^{-4} to 10^{-6} are representative and suitable to many applications. For design purposes we will assume that the input data rate and overall bit error probability are specified. Furthermore, we will assume that the characteristics of the channel are given, and that the channel noise can be represented by a zero mean Gaussian random process with a known power spectral density $G_n(f)$. The source that generates the input bit stream will be assumed to be ergodic and the source output will be assumed to be independent sequences of equiprobable bits.

The design of a baseband binary PAM system consists of specifying the pulse shapes $p_g(t)$ and $p_r(t)$ and the filters $H_R(f)$ and $H_T(f)$ to minimize the combined effects of intersymbol interference and noise in order to achieve a minimum probability of error for given data rate and power levels in the system.

5.2.1 Baseband Pulse Shaping

The intersymbol interference given by the second term in Equation (5.2) can be eliminated by proper choice of the received pulse shape $p_r(t)$. An inspection of Equation (5.2) reveals that for zero ISI, $p_r(t)$ should satisfy

$$p_r(nT_b) = \begin{cases} 1 & \text{for n} = 0 \\ 0 & \text{for n} \neq 0 \end{cases} \tag{5.3}$$

The constraint stated in (5.3) does not uniquely specify $p_r(t)$ for all values of t. To meet the constraint given in Equation (5.3), the Fourier transform $P_r(f)$ of $p_r(t)$ needs to satisfy a simple condition stated below.

Theorem 5.1

If $P_r(f)$ satisfies

$$\sum_{k=-\infty}^{\infty} P_r\left(f + \frac{k}{T_b}\right) = T_b \quad \text{for } |f| < 1/2T_b \tag{5.4}$$

then

$$p_r(nT_b) = \begin{cases} 1 & \text{for } n = 0 \\ 0 & \text{for } n \neq 0 \end{cases} \tag{5.5}$$

Proof

$p_r(t)$ is related to $P_r(f)$ by

$$p_r(t) = \int_{-\infty}^{\infty} P_r(f) \exp(j2\pi ft)\, df$$

The range of integration in the preceding equation can be divided into segments of length $1/T_b$ as

$$p_r(t) = \sum_{k=-\infty}^{\infty} \int_{(2k-1)/2T_b}^{(2k+1)/2T_b} P_r(f) \exp(j2\pi ft)\, df$$

and we can write $p_r(nT_b)$ as

$$p_r(nT_b) = \sum_k \int_{(2k-1)/2T_b}^{(2k+1)/2T_b} P_r(f) \exp(j2\pi fnT_b)\, df$$

Making a change of variable, $f' = f - k/T_b$, we can write the above equation as

$$p_r(nT_b) = \sum_k \int_{-1/2T_b}^{1/2T_b} P_r\left(f' + \frac{k}{T_b}\right) \exp(j2\pi f'nT_b)\, df'$$

Further, if we assume that the integration and summation can be interchanged, then the preceding equation can be rewritten as

$$p_r(nT_b) = \int_{-1/2T_b}^{1/2T_b} \left(\sum_k P_r\left(f + \frac{k}{T_b}\right)\right) \exp(j2\pi fnT_b)\, df$$

Finally, if (5.4) is satisfied, then

$$p_r(nT_b) = \int_{-1/2T_b}^{1/2T_b} T_b \exp(j2\pi fnT_b)\, df$$
$$= \frac{\sin(n\pi)}{n\pi}$$

which verifies that the $p_r(t)$ with a transform $P_r(f)$ satisfying (5.4) produces zero ISI.

The condition for the removal of ISI given in Equation (5.4) is called the Nyquist (pulse shaping) criterion.

Theorem 5.1 gives the condition for the removal of ISI using a $P_r(f)$ with a bandwidth larger than $r_b/2$. Proceeding along similar lines, it can be shown that ISI cannot be removed if the bandwidth of $P_r(f)$ is less than $r_b/2$.

The condition stated in Equation (5.4) does not uniquely specify $P_r(f)$. The particular choice of $P_r(f)$ for a given application is guided by two important considerations: the rate of decay of $p_r(t)$ and the ease with which shaping filters can be built. A pulse with a fast rate of decay and smaller values near $\pm T_b, \pm 2T_b, \ldots$ is desirable since these properties will yield a system in which modest timing errors will not cause large intersymbol interference. The shape of $P_r(f)$ determines the ease with which shaping filters can be realized. A $P_r(f)$ with a smooth roll-off characteristic is preferable over one with arbitrarily sharp cut-off characteristics, since the latter choice might lead to filters that will be hard to realize.

In practical systems where the bandwidth available for transmitting data at a rate of r_b bits/sec is between $r_b/2$ to r_b Hz, a class of $P_r(f)$ with a *raised cosine frequency characteristic* is most commonly used. A raised cosine frequency spectrum consists of a flat amplitude portion and a roll-off portion that has a sinusoidal form. The pulse spectrum $P_r(f)$ is specified in terms of a parameter β as

$$P_r(f) = \begin{cases} T_b, & |f| \le r_b/2 - \beta \\ T_b \cos^2 \dfrac{\pi}{4\beta}\left(|f| - \dfrac{r_b}{2} + \beta\right), & \dfrac{r_b}{2} - \beta < |f| \le \dfrac{r_b}{2} + \beta \\ 0, & |f| > r_b/2 + \beta \end{cases} \tag{5.6}$$

where $0 < \beta < r_b/2$. The pulse shape $p_r(t)$ corresponding to the $P_r(f)$ given above is

$$p_r(t) = \frac{\cos 2\pi\beta t}{1 - (4\beta t)^2}\left(\frac{\sin \pi r_b t}{\pi r_b t}\right) \tag{5.7}$$

Plots of $P_r(f)$ and $p_r(t)$ for three values of the parameter β are shown in

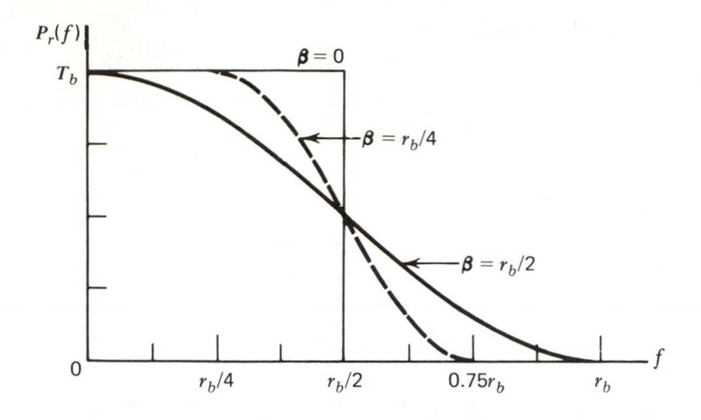

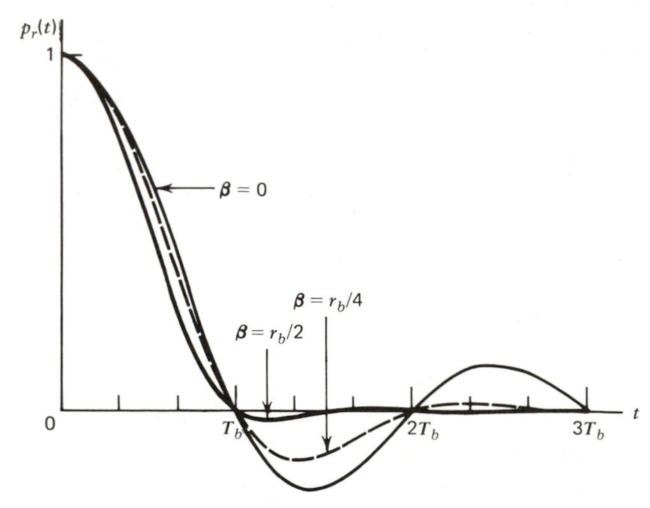

Figure 5.3 Pulses with raised cosine frequency characteristics. (*a*) $P_r(f)$ for three values of β. Note that $P_r(f) = P_r(-f)$. (*b*) $p_r(t)$ for three values of β

Figure 5.3. From Equations (5.6) and (5.7) and from Figure 5.3, the following observations can be made:

1. The bandwidth occupied by the pulse spectrum is $B = r_b/2 + \beta$. The minimum value of B is $r_b/2$ and the maximum value is r_b.
2. Larger values of β imply that more bandwidth is required for a given bit rate r_b. However, larger values of β lead to faster decaying pulses, which means that synchronization will be less critical and modest timing errors will not cause large amounts of ISI.

3. $\beta = r_b/2$ leads to a pulse shape with two convenient properties; the half amplitude pulse width is equal to T_b, and there are zero crossings at $t = \pm(\frac{3}{2})T_b, \pm(\frac{5}{2})T_b, \ldots$ in addition to zero crossings at $\pm T_b, \pm 2T_b, \ldots$. These properties aid in generating a timing signal for synchronization from the received signal.
4. $P_r(f)$ is real, nonnegative and $\int_{-\infty}^{\infty} P_r(f)\,df = 1$.

Summarizing the preceding discussion on the selection of a received pulse shape $p_r(t)$, we can say that a bandwidth of at least $r_b/2$ is required to generate a $p_r(t)$ producing zero ISI at a data rate r_b bits/sec. If additional bandwidth is available, then a $p_r(t)$ with an appropriate raised cosine spectrum given by (5.6) can be chosen. One must in general try to utilize all the available bandwidth up to r_b, and take advantage of faster decay of $p_r(t)$ with time. It must be pointed out that, strictly speaking, none of the raised cosine pulse spectra is physically realizable. A realizable frequency characteristic must have a time response that is zero prior to a time $t_0(t_0 > 0)$, which is not the case for the $P_r(f)$ given in Equation (5.6). However, a delayed version of $p_r(t)$, say $p_r(t - t_d)$, may be generated by causal filters if the delay t_d is chosen such that $p_r(t - t_d) \approx 0$ for $t < t_0$. A practical filter that generates such a waveform is given in Problem 5.11.

5.2.2 Optimum Transmitting and Receiving Filters

The transmitting and receiving filters are chosen to provide proper pulse shaping and noise immunity. One of the design constraints that we have for selecting the filters is the relationship between the Fourier transforms of $p_r(t)$ and $p_g(t)$,

$$P_g(f)H_T(f)H_c(f)H_R(f) = K_cP_r(f)\exp(-j2\pi f t_d) \tag{5.8}$$

where t_d is the time delay* in the system and K_c is a normalizing constant. In order to design optimum filters $H_T(f)$ and $H_R(f)$, we will assume that $P_r(f)$, $H_c(f)$, and $P_g(f)$ are known.

If we choose $p_r(t)$ to produce zero ISI, then the constraint in (5.8) specifies that the filters shape $p_g(t)$ to yield a delayed version of $p_r(t)$. Now we need only to be concerned with noise immunity, that is, we need to choose the transmit and receive filters to minimize the effect of noise. For a given data rate, transmitter power, noise power spectral density, $H_c(f)$, and $P_r(f)$, we

*If t_d is sufficiently large, then the response of the system $K_cp_r(t - t_d)$ may be assumed to be 0 for $t < t_0$, where t_0 is the time at which the input $p_g(t)$ is applied to the system. Hence the filters will be causal.

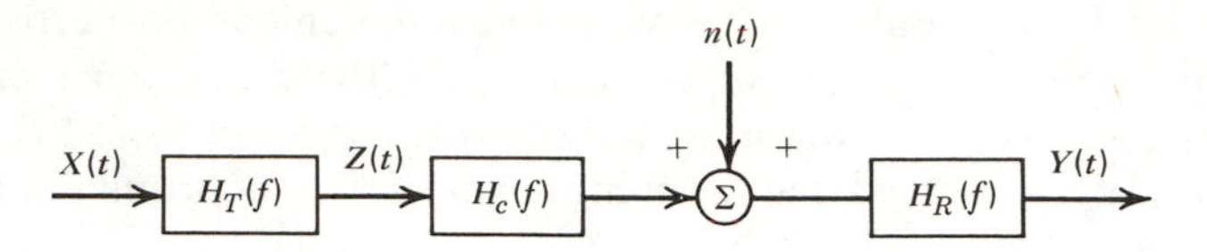

Figure 5.4 Portion of a baseband PAM system.

want to choose $H_T(f)$ and $H_R(f)$ such that the bit error probability is minimized (see Figure 5.4).

As a first step in deriving the optimum filters, let us derive an expression for the probability of a bit error. At the mth sampling time, the input to the A/D converter is

$$Y(t_m) = A_m + n_0(t_m), \quad t_m = mT_b + t_d$$

and the A/D converter output is 1 or 0 depending on whether $Y(t_m) > 0$ or < 0, respectively.* If we denote the mth input bit by d_m, then we can write an expression for the probability of incorrectly decoding the mth bit at the receiver as

$$\begin{aligned} P_e &= P[Y(t_m) > 0 | d_m = 0]P(d_m = 0) \\ &\quad + P[Y(t_m) < 0 | d_m = 1]P(d_m = 1) \end{aligned}$$

By virtue of our assumption of equiprobable bits, and due to the fact that $Y(t_m) = A + n_0(t_m)$ when $d_m = 1$, and $Y(t_m) = -A + n_0(t_m)$ when $d_m = 0$, where $A = K_c a$, we have

$$\begin{aligned} P_e &= \tfrac{1}{2}\{P[n_0(t_m) < -A] + P[n_0(t_m) > A]\} \\ &= \tfrac{1}{2}\{P[|n_0(t_m)| > A]\} \end{aligned}$$

The noise is assumed to be zero mean Gaussian at the input to $H_R(f)$, and hence the output noise $n_0(t)$ will also be zero mean Gaussian with a variance N_0 given by

$$N_0 = \int_{-\infty}^{\infty} G_n(f)|H_R(f)|^2 \, df \tag{5.9}$$

*In general, for minimizing the probability of error, the receiver threshold should be set at

$$\frac{N_0}{2A} \log_e \frac{P(d_m = 0)}{P(d_m = 1)}$$

where $P(d_m = 0)$ and $P(d_m = 1)$ denote the probability that the mth input bit is 0 and 1, respectively, and N_0 is the variance of the noise at the input to the A/D converter (see Problem 5.5).

Using the above property, we can write P_e as

$$P_e = \tfrac{1}{2}\int_{|x|>A}^{\infty} \frac{1}{\sqrt{2\pi N_0}} \exp(-x^2/2N_0)\,dx$$

$$= \int_A^{\infty} \frac{1}{\sqrt{2\pi N_0}} \exp(-x^2/2N_0)\,dx \tag{5.10}$$

A change of variable $z = x/\sqrt{N_0}$ yields

$$P_e = \int_{A/\sqrt{N_0}}^{\infty} \frac{1}{\sqrt{2\pi}} \exp(-z^2/2)\,dz = Q\left(\frac{A}{\sqrt{N_0}}\right) \tag{5.11}$$

where

$$Q(u) = \int_u^{\infty} \frac{1}{\sqrt{2\pi}} \exp(-z^2/2)\,dz$$

From Equation (5.11) we see that P_e decreases as $A/\sqrt{N_0}$ increases, and hence in order to minimize P_e we need to maximize the ratio $A/\sqrt{N_0}$. Thus, for maximum noise immunity the filter transfer functions $H_T(f)$ and $H_R(f)$ must be chosen to maximize the ratio $A/\sqrt{N_0}$. In order to do this maximizing, we need to express $A/\sqrt{N_0}$ or A^2/N_0 in terms of $H_T(f)$ and $H_R(f)$.

We start with the signal at the input to the transmitting filter

$$X(t) = \sum_{k=-\infty}^{\infty} a_k p_g(t - kT_b) \tag{5.12}$$

where $p_g(t)$ is a unit amplitude pulse having a duration less than or equal to T_b. Since the input bits are assumed to be independent and equiprobable, $X(t)$ is a random binary waveform with a psd (Chapter 3, Examples 3.8 and 3.9)

$$G_X(f) = \frac{|P_g(f)|^2}{T_b} E\{a_k^2\}$$

$$= \frac{a^2|P_g(f)|^2}{T_b} \tag{5.13}$$

Now, the psd of the transmitted signal is given by

$$G_Z(f) = |H_T(f)|^2 G_X(f)$$

and the average transmitted power S_T is

$$S_T = \frac{a^2}{T_b}\int_{-\infty}^{\infty} |P_g(f)|^2 |H_T(f)|^2\,df \tag{5.14}$$

Since $A_k = K_c a_k$ and $A = K_c a$, we can write

$$S_T = \frac{A^2}{K_c^2 T_b}\int_{-\infty}^{\infty} |P_g(f)|^2 |H_T(f)|^2\,df$$

or

$$A^2 = K_c^2 S_T T_b \left[\int_{-\infty}^{\infty} |P_g(f)|^2 |H_T(f)|^2 \, df \right]^{-1} \tag{5.15}$$

Now, the average output noise power or the variance of $n_0(t)$ is given by

$$N_0 = \int_{-\infty}^{\infty} G_n(f) |H_R(f)|^2 \, df \tag{5.16}$$

and hence the quantity we need to maximize, A^2/N_0, can be expressed as

$$\frac{A^2}{N_0} = S_T T_b \left[\int_{-\infty}^{\infty} |H_R(f)|^2 G_n(f) \, df \int_{-\infty}^{\infty} \frac{|P_r(f)|^2}{|H_c(f) H_R(f)|^2} \, df \right]^{-1} \tag{5.17}$$

Or, we need to minimize

$$\gamma^2 = \int_{-\infty}^{\infty} |H_R(f)|^2 G_n(f) \, df \int_{-\infty}^{\infty} \frac{|P_r(f)|^2}{|H_c(f) H_R(f)|^2} \, df \tag{5.18}$$

The minimization of the right-hand side of (5.18) with respect to $H_R(f)$ can be carried out by using Schwarz's inequality, which is stated as follows: If $V(f)$ and $W(f)$ are complex functions of f, then

$$\int_{-\infty}^{\infty} |V(f)|^2 \, df \int_{-\infty}^{\infty} |W^*(f)|^2 \, df \geq \left| \int_{-\infty}^{\infty} V(f) W^*(f) \, df \right|^2 \tag{5.19}$$

The minimum value of the left-hand side of the equality is reached when $V(f) =$ const times $W(f)$. Applying (5.19) to (5.18) with

$$|V(f)| = |H_R(f)| G_n^{1/2}(f)$$

$$|W(f)| = \frac{|P_r(f)|}{|H_c(f)||H_R(f)|}$$

we see that γ^2 is minimized when

$$|H_R(f)|^2 = \frac{K|P_r(f)|}{|H_c(f)| G_n^{1/2}(f)} \tag{5.20}$$

where K is an arbitrary positive constant. Substituting Equation (5.20) in (5.8), we can obtain the optimum transmitting filter transfer function as

$$|H_T(f)|^2 = \frac{K_c^2 |P_r(f)| G_n^{1/2}(f)}{K |P_g(f)|^2 |H_c(f)|} \tag{5.21}$$

These filters should have linear phase response resulting in a total time delay of t_d (Equation (5.8)).

Finally, we obtain the maximum value of A^2/N_0 as

$$\left(\frac{A^2}{N_0} \right)_{\max} = (S_T)(T_b) \left[\int_{-\infty}^{\infty} \frac{|P_r(f)| G_n^{1/2}(f)}{|H_c(f)|} \, df \right]^{-2} \tag{5.22}$$

by substituting (5.20) in (5.17). The bit error probability P_e is then equal to

$$P_e = Q(\sqrt{(A^2/N_0)_{\max}}) \tag{5.23}$$

A special case of significant practical interest occurs when the channel noise is white ($G_n(f) = \eta/2$), Gaussian, and when $P_g(f)$ is chosen such that it does not change much over the bandwidth of interest. The filter transfer functions now reduce to

$$|H_R(f)|^2 = K_1 \frac{|P_r(f)|}{|H_c(f)|} \tag{5.24a}$$

and

$$|H_T(f)|^2 = K_2 \frac{|P_r(f)|}{|H_c(f)|} \tag{5.24b}$$

where K_1 and K_2 are positive constants. From Equations (5.24a) and (5.24b) it follows that $|H_T(f)| = K_3|H_R(f)|$, where K_3 is a positive constant. With the exception of an arbitrary gain difference, the transmitting and receiving filters have the same frequency characteristics so that one design serves both filters. In a large data communication system, having identical transmitting and receiving filters makes production and maintenance easy. A simple pulse shape $p_g(t)$ that yields an approximately constant $P_g(f)$ over the bandwidth of interest is

$$p_g(t) = \begin{cases} 1 & \text{for } |t| < \tau/2; \quad \tau \ll T_b \\ 0 & \text{elsewhere} \end{cases} \tag{5.25}$$

That is, a rectangular pulse of width $\tau \ll T_b$ can be used at the input to the transmit filter.

5.2.3 Design Procedure and Example

We will now see how the relationships derived in the preceding sections can be used to design a binary baseband PAM system given the bit rate r_b, acceptable error probability P_e, channel transfer function $H_c(f)$, and the channel noise power spectral density $G_n(f)$. Unless otherwise specified we will assume that the input bits are independent and equiprobable, that the channel noise is zero mean Gaussian, and that the channel is lowpass with a bandwidth B ($r_b/2 \leq B \leq r_b$). If the channel bandwidth is less than $r_b/2$, we will have to use an M-ary signaling scheme that will be discussed later. If the channel bandwidth is much greater than r_b, then it would be wise to use some nonlinear modulation scheme to utilize the full bandwidth for reducing the effects of channel noise.

The design of the system consists of specifying the pulse shapes and spectra $P_r(f)$, $P_g(f)$, the transmitting and receiving filters $H_T(f)$, $H_R(f)$, and the transmitter power requirements S_T to meet the specified error probability. The steps involved in the design procedure are illustrated in the following example.

Example 5.1. Design a binary baseband PAM system to transmit data at a bit rate of 3600 bits/sec with a bit error probability less than 10^{-4}. The channel response is given by

$$H_c(f) = \begin{cases} 10^{-2} & \text{for } |f| < 2400 \\ 0 & \text{elsewhere} \end{cases}$$

The noise power spectral density is $G_n(f) = 10^{-14}$ watt/Hz.

Solution. We are given $r_b = 3600$ bits/sec, $P_e \leq 10^{-4}$, channel bandwidth $B = 2400$ Hz, and $G_n(f) = 10^{-14}$ watt/Hz.

If we choose a raised cosine pulse spectrum with $\beta = r_b/6 = 600$, then the channel bandwidth constraint is satisfied. Hence,

$$P_r(f) = \begin{cases} \dfrac{1}{3600}, & |f| < 1200 \\ \dfrac{1}{3600} \cos^2 \dfrac{\pi}{2400} (|f| - 1200), & 1200 \leq |f| < 2400 \\ 0, & |f| \geq 2400 \end{cases}$$

Let us choose a $p_g(t)$ to satisfy (5.24) as

$$p_g(t) = \begin{cases} 1, & |t| < \tau/2 \\ 0, \quad \text{elsewhere}; & \tau = T_b/10 = (0.28)(10^{-4}) \end{cases}$$

Then,

$$P_g(f) = \tau \left(\frac{\sin \pi f \tau}{\pi f \tau} \right)$$

$$P_g(0) = \tau, \qquad P_g(2400) = 0.973\tau \approx \tau$$

Hence, the variation of $P_g(f)$ over $0 < |f| < 2400$ is very small and we obtain $|H_T(f)|$ and $|H_R(f)|$ from Equation (5.20) and (5.21) as

$$|H_T(f)| = K_1 |P_r(f)|^{1/2}$$

$$|H_R(f)| = |P_r(f)|^{1/2}$$

We will choose $K_1 = (3600)(10^3)$ so that the overall response of H_T, H_c, and H_R to $P_g(f)$ produces $P_r(f)$, that is,

$$P_g(f)|H_T(f)||H_c(f)||H_R(f)| = P_r(f)$$

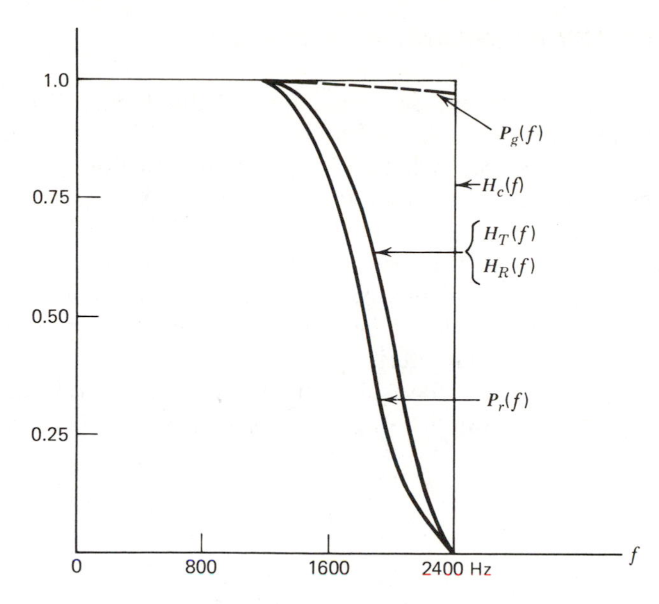

Figure 5.5 Normalized plots of P_g, H_c, H_T, H_R, and P_r. All functions are shown normalized with respect to their values at $f = 0$.

Plots of $P_g(f)$, $H_c(f)$, $H_T(f)$, $H_R(f)$, and $P_r(f)$ are shown in Figure 5.5.

Now, to maintain a $P_e \leqslant 10^{-4}$, we need $(A^2/N_0)_{\max}$ such that

$$Q(\sqrt{(A^2/N_0)_{\max}}) \leqslant 10^{-4}$$

Using the tabulated values of Q given in Appendix D we get

$$\sqrt{(A^2/N_0)_{\max}} \geqslant 3.75$$

or

$$(A^2/N_0)_{\max} \geqslant 14.06$$

From Equation (5.22) we obtain the transmitted power S_T as

$$\begin{aligned} S_T &= \frac{1}{T_b}\left(\frac{A^2}{N_0}\right)_{\max}\left[\int_{-\infty}^{\infty} \frac{|P_r(f)|G_n^{1/2}(f)}{|H_c(f)|}\,df\right]^2 \\ &= (3600)(14.06)\left(\frac{10^{-14}}{10^{-4}}\right)\left[\int_{-\infty}^{\infty} |P_r(f)|\,df\right]^2 \end{aligned}$$

For $P_r(f)$ with raised cosine shape $\int_{-\infty}^{\infty} |P_r(f)|\,df = 1$ and hence

$$S_T = (14.06)(3600)(10^{-10}) \approx -23\ \text{dBm}$$

which completes the design.

5.3 DUOBINARY BASEBAND PAM SYSTEM

In the preceding section we saw that a baseband binary PAM data transmission system requires a bandwidth of at least $r_b/2$ Hz in order to transmit data at a rate of r_b bits/sec, with zero ISI. If the bandwidth available is exactly $r_b/2$, then the only possible way in which binary PAM data transmission at a rate of r_b bits/sec can be accomplished without ISI would be to use ideal (rectangular) lowpass filters at the transmitter and receiver. Of course, such filters are physically unrealizable. Furthermore, any system that would closely approximate these filters would be extremely sensitive to perturbations in rate, timing, or channel characteristics.

In the past few years a class of signaling schemes known as *duobinary*, *polybinary*, or *partial response signaling* schemes has been developed to

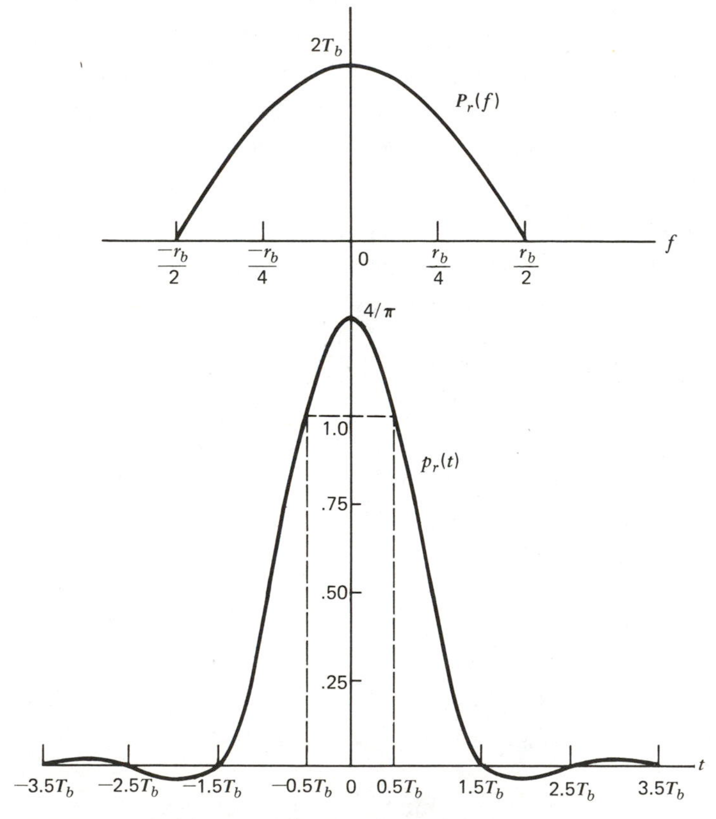

Figure 5.6 $P_r(f)$ and $p_r(t)$ for duobinary signaling scheme. Observe that the sampling is done at $t = (m \pm 0.5)T_b$.

overcome some of the difficulties mentioned in the preceding paragraph. The duobinary scheme utilizes *controlled amounts of ISI* for transmitting data at a rate of r_b bits/sec over a channel with a bandwidth of $r_b/2$ Hz. The shaping filters for the duobinary system are easier to realize than the ideal rectangular filters needed to accomplish the data transmission at the maximum rate with zero ISI. The main disadvantage of the duobinary scheme is that it requires more power than an ideal binary PAM data transmission scheme.

The duobinary signaling schemes use pulse spectra $P_r(f)$ that yield $Y(t_m) = A_m + A_{m-1}$, where A_m and A_{m-1} are amplitudes related to the input bits d_m and d_{m-1}. One such $P_r(f)$ is

$$P_r(f) = \begin{cases} 2T_b \cos(\pi f T_b), & |f| \leq 1/2T_b \\ 0, & |f| > 1/2T_b \end{cases} \tag{5.26}$$

The pulse response $p_r(t)$ corresponding to the above $P_r(f)$ is

$$p_r(t) = \frac{4\cos(\pi t/T_b)}{\pi(1 - 4t^2/T_b^2)} \tag{5.27}$$

Plots of $P_r(f)$ and $p_r(t)$ are shown in Figure 5.6. (See Problem 5.12 for other examples of $P_r(f)$.)

5.3.1 Use of Controlled ISI in Duobinary Signaling Scheme

The output $Y(t)$ of the receive filter can be written as

$$Y(t) = \sum_k A_k p_r(t - t_d - kT_b) + n_0(t) \tag{5.28}$$

where $p_r(t)$ is defined in (5.27).

If the output is sampled at $t_m = mT_b - T_b/2 + t_d$, then it is obvious that in the absence of noise

$$Y(t_m) = A_m + A_{m-1} \tag{5.29}$$

Equation (5.29) shows that the duobinary signaling scheme using a $P_r(f)$ given in Equation (5.26) introduces ISI. However, the intersymbol interference is controlled in that the interference comes only from the preceding symbol.

The A_m's in Equation (5.29) can assume one of two values, $\pm A$, depending on whether the mth input bit is 1 or 0. Since $Y(t_m)$ depends on A_m and A_{m-1}, $Y(t_m)$ can have one of the following *three* values (assuming no noise):

$$Y(t_m) = \begin{cases} +2A & \text{if the } m\text{th and } (m-1\text{st}) \text{ bits are both 1's} \\ 0 & \text{if the } m\text{th and } (m-1\text{st}) \text{ bits are different} \\ -2A & \text{if the } m\text{th and } (m-1\text{st}) \text{ bits are both zero} \end{cases}$$

That is, the receiving filter output is a *three-level* waveform. The decoding of the mth bit from the sampled value of $Y(t_m)$ is done by checking to see if the input to the A/D converter at the sampling time is at the top, bottom, or middle level.

One apparent drawback of the system is that errors tend to propagate. Since the mth bit is decoded based on the decoded value of the $(m-1\text{st})$ bit, any error in the $(m-1\text{st})$ bit is likely to introduce an error in the decoding of the mth bit. A method of avoiding error propagation was proposed by Lender. In his scheme, error propagation is eliminated by *precoding* the input bit stream at the transmitter. The input bit stream (coming from the source) $b_1, b_2, b_3, \ldots$ is converted to another binary stream $d_1, d_2, d_3, \ldots$ before transmission according to the rule

$$d_m = b_m \oplus d_{m-1} \tag{5.30}$$

(The symbol $\oplus$ stands for modulo-2 addition.) The binary sequence d_k is transmitted using two levels $+a$ and $-a$ for 1 and 0, respectively.

If the mth input bit b_m is 0, then $d_m = d_{m-1}$ and according to Equation (5.29) $Y(t_m)$ will be $2A$ or $-2A$. On the other hand, if $b_m = 1$, then d_m will be the complement of d_{m-1} and $Y(t_m)$ will be zero. Hence the mth input bit b_m can be decoded according to the rule

$$\begin{aligned} b_m &= 0 \quad \text{if } Y(t_m) = \pm 2A \\ b_m &= 1 \quad \text{if } Y(t_m) = 0 \end{aligned} \tag{5.31}$$

In the preceding rule the mth bit is decoded from the value of $Y(t_m)$ only, and error propagation does not occur. Also, the implementation of the decoding algorithm is simple; $Y(t)$ is rectified and a simple threshold binary decision with a threshold at level A yields the output.

5.3.2 Transmitting and Receiving Filters for Optimum Performance

The procedure used for deriving optimum transmitting and receiving filters for the duobinary signaling scheme is the same as the one used in Section 5.2.2. The received levels at the input to the A/D converter are $2A$, 0, and $-2A$ with probabilities $\frac{1}{4}$, $\frac{1}{2}$, and $\frac{1}{4}$, respectively. The probability of a bit error P_e is given by (assuming that the threshold is set at $\pm A$)

$$\begin{aligned} P_e &= \tfrac{1}{4}P\{n_0 < -A\} + \tfrac{1}{2}P\{|n_0| > A\} + \tfrac{1}{4}P\{n_0 > A\} \\ &= \tfrac{3}{2}P\{n_0 > A\} \end{aligned} \tag{5.32}$$

Since n_0 is a zero mean Gaussian random variable with a variance N_0', we can write P_e as

$$P_e = \tfrac{3}{2}Q(A/\sqrt{N_0'}) \tag{5.33}$$

The transmitting and receiving filters are chosen to maximize A^2/N_0' (in order to minimize P_e) subjected to the constraint that the overall pulse response $p_r(t)$ and its transform $P_r(f)$ satisfy Equations (5.26) and (5.27). It can be easily shown that the expressions for $H_T(f)$ and $H_R(f)$ will be the same as Equations (5.20) and (5.21).

The error probability P_e for the duobinary scheme will be higher than the error probability for the binary PAM system discussed in Section 5.1. For comparison purposes, let us assume that both schemes operate at the same bit rate r_b over an ideal channel ($H_c(f) = 1$) with additive Gaussian white noise. For the direct binary PAM case it can be verified from Equation (5.22) that

$$\left(\frac{A^2}{N_0}\right)_{\max} = S_T T_b \left(\frac{2}{\eta}\right)$$

where $\eta/2 = G_n(f)$ is the noise power spectral density. Hence, the probability of error is

$$(P_e)_{\text{binary}} = Q\left(\sqrt{2\frac{S_T T_b}{\eta}}\right) \tag{5.34}$$

For the duobinary case from Equation (5.22), we have

$$\left(\frac{A^2}{N_0'}\right)_{\max} = S_T T_b \frac{2}{\eta}\left[\int_{-\infty}^{\infty} |P_r(f)|\, df\right]^{-2}$$

where $P_r(f)$ is given in Equation (5.20). The integral in the preceding equation can be evaluated as

$$\int_{-\infty}^{\infty} |P_r(f)|\, df = \int_{-1/2T_b}^{1/2T_b} (2T_b)\cos(\pi f T_b)\, df$$
$$= \frac{2}{\pi}\int_{-\pi/2}^{\pi/2} \cos\theta\, d\theta = \frac{4}{\pi}$$

Hence,

$$\left(\frac{A^2}{N_0'}\right)_{\max} = 2\left(\frac{S_T T_b}{\eta}\right)\left(\frac{\pi}{4}\right)^2$$

and the probability of bit error for the duobinary scheme is given by Equation (5.33) as

$$(P_e)_{\substack{\text{duobinary}\\ \text{PAM}}} = \tfrac{3}{2}Q\left(\frac{\pi}{4}\sqrt{2\frac{S_T T_b}{\eta}}\right) \tag{5.35}$$

A comparison of Equations (5.34) and (5.35) reveals that the bit error probability for the duobinary scheme is always higher than the bit error probability for an ideal binary PAM scheme using pulses with raised cosine frequency characteristics. However, the duobinary scheme uses less bandwidth than binary PAM schemes.

Example 5.2. Compare a binary PAM system with the duobinary system for transmitting data at a rate of 4000 bits/sec over an ideal channel ($H_c(f) = 1$) with Gaussian white noise. Assume $G_n(f) = \eta/2 = 10^{-12}$ watt/Hz and an error probability of 10^{-3}.

Solution. For comparison let us consider binary PAM using a pulse with raised cosine frequency characteristics and $\beta = r_b/2$.

The bandwidth used is $r_b = 4000$ Hz.

$$P_e = Q\left(\sqrt{2\frac{S_T T_b}{\eta}}\right) < 10^{-3}$$

or

$$\sqrt{\frac{2S_T T_b}{\eta}} > 3.1$$

$$S_T > (3.1)^2\left(\frac{\eta}{2}\right)\frac{1}{T_b} \approx -44.2 \text{ dBm}$$

For the duobinary case, the bandwidth used is $r_b/2 = 2000$ Hz.

$$P_e = \tfrac{3}{2}Q\left(\frac{\pi}{4}\sqrt{\frac{2S_T T_b}{\eta}}\right) < 10^{-3}$$

or

$$\sqrt{\frac{2S_T T_b}{\eta}} > \left(\frac{4}{\pi}\right)(3.25) \quad \text{or} \quad S_T = -41.7 \text{ dBm}$$

The duobinary scheme uses $\frac{1}{2}$ the bandwidth of the binary PAM, but it requires about 2.5 dB more power than the binary PAM.

The discussion in the preceding section and the above example illustrate that the duobinary scheme can be used to transmit binary data at a rate of r_b bits/sec over a channel with a bandwidth of $r_b/2$ Hz. The shaping filters required by the scheme are easier to realize, and the pulse shape is such that small timing errors do not affect the performance of the system. The only comparable binary PAM system that can transmit r_b bits/sec over a channel with a bandwidth of $r_b/2$ Hz must use infinite cut-off ideal filters that are not physically realizable. Even if an approximation to the ideal filters can be realized, it would be extremely sensitive to any perturbations in timing or channel characteristics.

5.4 *M*-ARY SIGNALING SCHEMES

The baseband binary PAM systems discussed in the preceding sections use binary pulses, that is, pulses with one of two possible amplitude levels. In

M-ary baseband PAM systems, the output of the pulse generator is allowed to take on one of M possible levels ($M > 2$). Each level corresponds to a distinct input symbol, and there are M distinct input symbols. If symbols in the input sequence are equiprobable and statistically independent, then the information rate coming out of the pulse generator (Figure 5.7) is $r_s \log_2 M$ bits/sec, where r_s is the symbol rate.* Each pulse contains $\log_2 M$ bits of information. In the absence of noise and ISI, the receiver decodes input symbols correctly by observing the sampled values of the received pulse. As in the binary signaling scheme, noise and ISI introduce errors, and the major objectives of the design are then to eliminate ISI and reduce the effects of noise.

Before we discuss the design and analysis of M-ary signaling schemes, let us look at an example of such a scheme. Figure 5.7*a* shows the functional block diagram of an M-ary signaling scheme and Figure 5.7*b* shows the waveforms at various points in a *quarternary* ($M = 4$) system.

A comparison of Figures 5.7*a* and 5.7*b* with Figures 5.1 and 5.2 reveals that the binary and M-ary schemes are very similar, with the exception of the number of symbols in the input sequence and the corresponding number of amplitude levels of the pulses. In the quarternary scheme shown in Figure 5.7b, the source emits a sequence of symbols at a rate of r_s symbols/sec from a source alphabet consisting of four symbols A, B, C, and D. During each signaling interval of duration T_s ($= 1/r_s$), one of the four symbols is emitted by the source and the amplitude of the pulse generator output takes on one of four distinct levels. Thus, the sequence of symbols emitted by the source is converted to a four-level PAM pulse train by the pulse generator. The pulse train is shaped and transmitted over the channel, which corrupts the signal waveform with noise and distortion. The signal plus noise passes through the receiving filter and is sampled by the A/D converter at an appropriate rate and phase. The sampled value is compared against preset threshold values (also called *slicing levels*) and a decision is reached as to which symbol was transmitted, based on the sampled value of the received signal. Intersymbol interference, noise, and poor synchronization cause errors and the transmitting and receiving filters are designed to minimize the errors.

Procedures used for the design and analysis of multi-level PAM systems are similar to the ones we used for binary systems. We will assume that the input to the system is a sequence of statistically independent, equiprobable symbols produced by an ergodic source and that the channel noise is a zero-mean Gaussian random process with a psd of $G_n(f)$. We will see that an M-ary PAM scheme operating with the preceding constraints can transmit data at a bit rate of $r_s \log_2 M$ bits/sec and require a minimum bandwidth of $r_s/2$ Hz. In comparison, a binary scheme transmitting data at the same rate,

*The signaling speed is often given in the units of bauds.

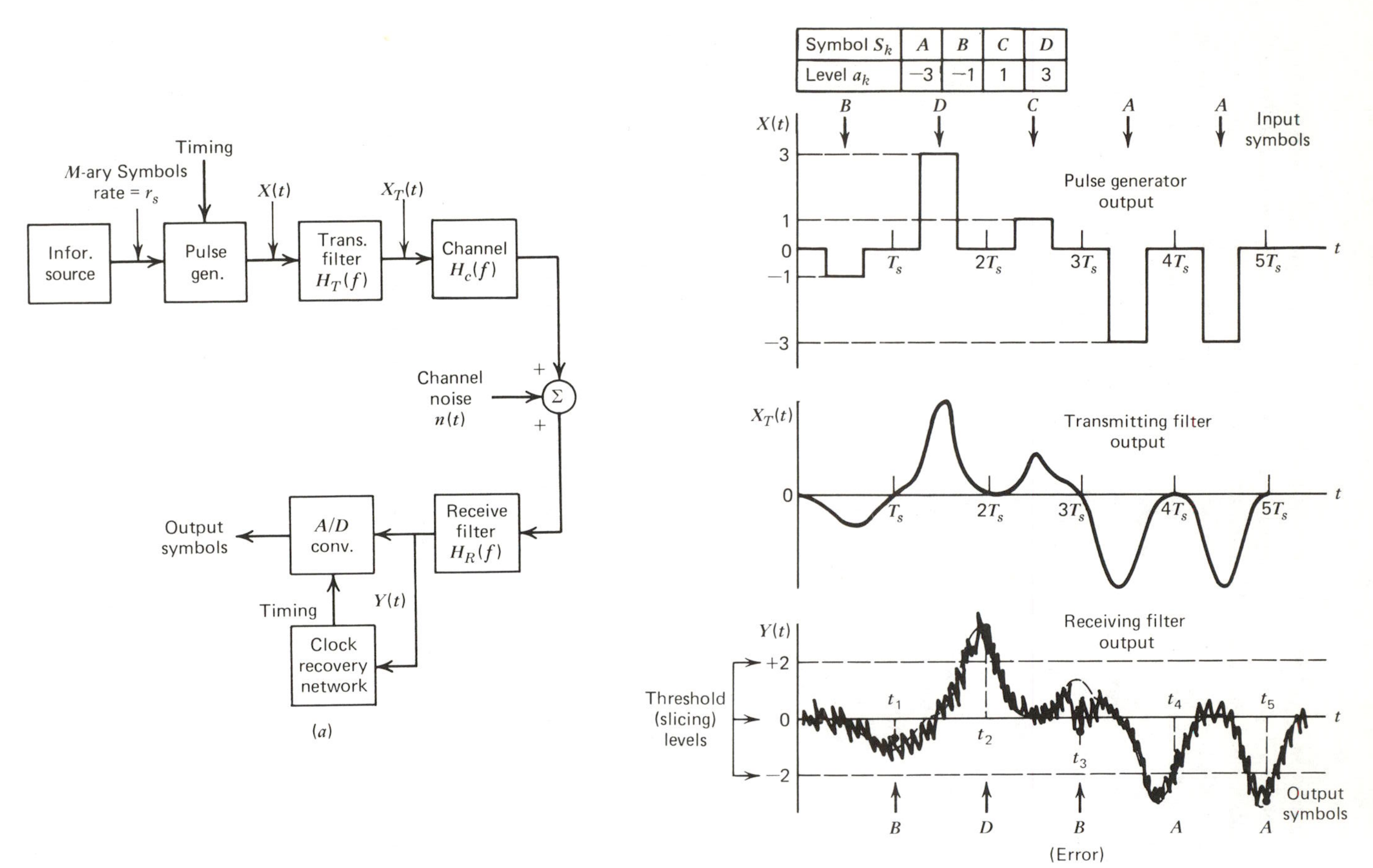

Figure 5.7 (a) Block diagram of an M-ary signaling scheme. (b) Signaling waveforms in a quarternary scheme.

$r_s \log_2 M$, will require a bandwidth of $(R_s \log_2 M)/2$ Hz. We will show that the price we pay for the bandwidth reduction that results from the use of M-ary schemes is the requirement for more power and more complex equipment.

5.4.1 Analysis and Design of *M*-ary Signaling Schemes

We start with the output of the pulse generator $X(t)$, which is given by

$$X(t) = \sum_{k=-\infty}^{\infty} a_k p_g(t - kT_s)$$

where $p_g(t)$ is the basic pulse whose amplitude a_k depends on the kth input symbol. If the source alphabet contains M letters, then a_k will take on one of M levels. We will assume that the spacing of amplitude levels is to be uniform and that a_k can take positive and negative values. It is easy to verify that for minimum power requirements, that is, for a minimum value of $E\{a_k - E\{a_k\}\}^2$, the M amplitude levels have to be centered at zero. For a separation of $2a$ between adjacent levels, the levels are given by

$$a_k = \begin{cases} 0, \pm 2a, \pm 4a, \ldots, \pm (M-1)a, & M \text{ odd} \\ \pm a, \pm 3a, \pm 5a, \ldots, \pm (M-1)a, & M \text{ even} \end{cases} \tag{5.36}$$

The signal power for the level spacings given in Equation (5.36) can be obtained from

$$E\{a_k^2\} = \frac{1}{M} \sum_{k=1}^{M} [(2k - M - 1)a]^2$$

When M is even, we have

$$\begin{aligned} E\{a_k^2\} &= \frac{2}{M} \{a^2 + (3a)^2 + (5a)^2 + \cdots + [(M-1)a]^2\} \\ &= \frac{(M^2 - 1)a^2}{3} \end{aligned} \tag{5.37}$$

A similar expression can be derived for the case when M is odd.

Following the procedure used for the binary case (Section 5.2) we can write the input to the A/D converter at the sampling time as (from Equations (5.1) and (5.2))

$$Y(t_m) = A_m + \sum_{k \neq m} A_k p_r[(m-k)T_s] + n_0(t_m) \tag{5.38}$$

where $t_m = mT_s + t_d$ and t_d is the total time delay in the system. A_m takes on one of M values given by

$$A_m = \begin{cases} 0, \pm 2A, \pm 4A, \ldots, \pm (M-1)A, & M \text{ odd}, A = K_c a \\ \pm A, \pm 3A, \pm 5A, \ldots, \pm (M-1)A, & M \text{ even} \end{cases} \tag{5.39}$$

The second term in Equation (5.38) represents ISI, which can be eliminated by choosing a $p_r(t)$ with zero crossings at $\pm T_s, \pm 2T_s, \ldots$. We can use pulses with raised cosine frequency characteristics with the choice governed by rate of decay and bandwidth requirements. Pulses with raised cosine frequency characteristics have spectra $P_r(f)$ and waveform shapes $p_r(t)$ given by (Equation (5.6) with T_b replaced by T_s)

$$P_r(f) = \begin{cases} T_s, & |f| \leq r_s/2 - \beta \\ T_s \cos^2 \dfrac{\pi}{4\beta}\left(|f| - \dfrac{r_s}{2} + \beta\right), & \dfrac{r_s}{2} - \beta < |f| \leq \dfrac{r_s}{2} + \beta \\ 0, & |f| > r_s/2 + \beta \end{cases} \tag{5.40}$$

where $0 < \beta < r_s/2$, and

$$p_r(t) = \frac{\cos 2\pi\beta t}{1 - (4\beta t)^2}\left(\frac{\sin \pi r_s t}{\pi r_s t}\right) \tag{5.41}$$

with

$$\int_{-\infty}^{\infty} |P_r(f)|\, df = 1 \tag{5.42}$$

Since $P_r(f) = 0$ for $|f| > r_s/2 + \beta$, the channel bandwidth requirements for an M-ary signaling scheme is $r_s/2 + \beta$. With $0 \leq \beta \leq r_s/2$, the maximum BW requirement for an M-ary scheme using pulses with raised cosine frequency characteristics is r_s Hz. The data rate of such a scheme is $r_s \log_2 M$ bits/sec. For a binary scheme with the same data rate, the bandwidth requirement will be $r_s \log_2 M$.

The transmitting and receiving filters are chosen to produce zero ISI and minimize the probability of error for a given transmitted power. The zero ISI condition is met if $P_r(f)$ has the form given in Equation (5.40), and

$$P_g(f)H_T(f)H_c(f)H_R(f) = K_c \exp(-j2\pi f t_d)P_r(f) \tag{5.43}$$

where t_d is the total time delay, $P_g(f)$ is the transform of the basic pulse output of the pulse generator, and K_c is a normalizing constant. To minimize the probability of error, let us begin with the following expression for the probability density function of $Y(t_m)$. For convenience, let us use Y to denote $Y(t_m)$, and let us use the case where $M = 4$, shown in Figure 5.8. The probability density function of Y is given by

$$\begin{aligned} f_Y(y) = {} & P(A_m = -3A) f_{Y|A_m=-3A}(y) + P(A_m = -A) f_{Y|A_m=-A}(y) \\ & + P(A_m = A) f_{Y|A_m=A}(y) + P(A_m = 3A) f_{Y|A_m=3A}(y) \end{aligned} \tag{5.44}$$

where $f_{Y|A_m=kA}(y)$ is the conditional pdf of Y given that $A_m = kA$.

From Equation (5.38), with zero ISI,

$$Y_m = A_m + n_0$$

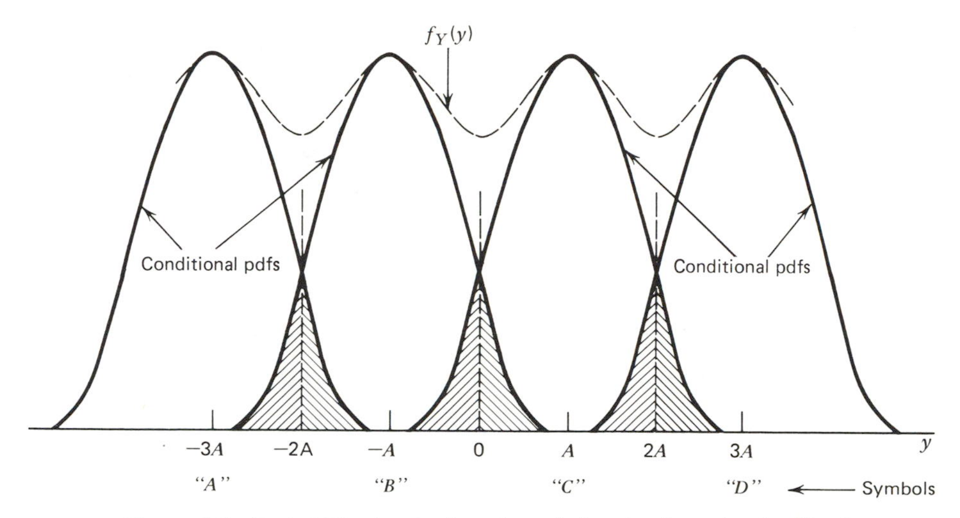

Figure 5.8 Probability density function of *signal pulse noise* for M = 4.

where n_0 is the noise term that has a zero mean Gaussian pdf

$$f_{n_0}(z) = \frac{1}{\sqrt{2\pi N_0}} \exp(-z^2/2N_0), \quad -\infty < z < \infty \tag{5.45}$$

and

$$N_0 = \int_{-\infty}^{\infty} G_n(f)|H_R(f)|^2\, df$$

Hence, for a given value of A_m, say $A_m = 3A$,

$$Y = 3A + n_0$$

and

$$f_{Y|A_m=3A}(y) = \frac{1}{\sqrt{2\pi N_0}} \exp(-(y-3A)^2/2N_0), \quad -\infty < y < \infty$$

The conditional probability density functions corresponding to the four input symbols (hence, the four levels $-3A$, $-A$, A, $3A$) are shown in Figure 5.8. By virtue of our assumption that the symbols are equiprobable, the optimum threshold levels are midway between the values of A_m, that is, the optimum decoding algorithm at the receiver is given by

$$\begin{aligned}
\text{if} \quad & Y(t_m) > 2A && \text{output symbol} = D \\
& 0 < Y(t_m) \leq 2A && \text{output symbol} = C \\
& -2A < Y(t_m) \leq 0 && \text{output symbol} = B \\
& Y(t_m) \leq -2A && \text{output symbol} = A
\end{aligned}$$

The probability of error can now be written as

$$\begin{aligned} P(\text{error}) = {} & P(\text{error}|D \text{ was sent})P(D \text{ sent}) \\ & + P(\text{error}|C \text{ was sent})P(C \text{ sent}) \\ & + P(\text{error}|B \text{ was sent})P(B \text{ sent}) \\ & + P(\text{error}|A \text{ was sent})P(A \text{ sent}) \end{aligned}$$

or

$$\begin{aligned} P_e = {} & P(Y(t_m) \leq 2A | A_m = 3A)P(D \text{ sent}) \\ & + P(Y(t_m) > 2A \quad \text{or} \quad \leq 0 | A_m = A)P(C \text{ sent}) \\ & + P(Y(t_m) > 0 \quad \text{or} \quad \leq -2A | A_m = -A)P(B \text{ sent}) \\ & + P(Y(t_m) > -2A | A_m = -3A)P(A \text{ sent}) \end{aligned} \tag{5.46}$$

Let us look at the second term in Equation (5.46) in detail. We have

$$P(C \text{ sent}) = \tfrac{1}{4}$$

and

$$\begin{aligned} & P(Y(t_m) > 2A \quad \text{or} \quad \leq 0 | A_m = A) \\ & \quad = P(Y(t_m) > 2A | A_m = A) + P(Y(t_m) \leq 0 | A_m = A) \\ & \quad = \int_{y>2A} f_{Y|A_m=A}(y)\, dy + \int_{y\leq 0} f_{Y|A_m=A}(y)\, dy \\ & \quad = \int_{y\geq 2A} \frac{1}{\sqrt{2\pi N_0}} \exp(-(z-A)^2/2N_0)\, dz + \int_{y\leq 0} \frac{1}{\sqrt{2\pi N_0}} \exp(-(z-A)^2/2N_0)\, dz \end{aligned}$$

A change of variable $u = (z - A)/\sqrt{N_0}$ yields

$$\begin{aligned} P(Y(t_m) > 2A \text{ or } \leq 0 | A_m = A) & = 2 \int_{u > A/\sqrt{N_0}} \frac{1}{\sqrt{2\pi}} \exp(-u^2/2)\, du \\ & = 2Q\left(\frac{A}{\sqrt{N_0}}\right) \end{aligned}$$

Using a similar expression for the remaining terms in Equation (5.46), we arrive at an expression for P_e as

$$P_e = \tfrac{1}{4}(6)Q(A/\sqrt{N_0})$$

In the preceding expression, the factor $\frac{1}{4}$ represents the probability of occurrence of each of the four symbols; the factor 6 represents the six areas shown marked in Figure 5.8, and $Q(A/\sqrt{N_0})$ represents the numerical value of each one of the areas. Extending the derivation to the M-ary case and using similar arguments, we can obtain an expression for the probability of error as

$$P_e = \frac{2(M-1)}{M} Q\left(\frac{A}{\sqrt{N_0}}\right) \tag{5.47}$$

A comparison of Equation (5.47) with the probability of error for the binary case (given in Equation (5.11)) reveals that the two expressions are identical with the exception of the factor $2(M-1)/M$. The probability of error in the M-ary case is minimized when A^2/N_0 is maximized and the maximizing is carried out using a procedure similar to the one for the binary case (described in Section 5.2). Some of the intermediate results in the derivation are given below:

1. Transmitted power $= S_T$

$$= \frac{A^2}{K_c^2 T_s}\left(\frac{M^2-1}{3}\right)\int_{-\infty}^{\infty} |H_T(f)P_g(f)|^2\, df \tag{5.48}$$

2.

$$\frac{A^2}{N_0} = \left(\frac{3S_T T_s}{M^2-1}\right)\left[\int_{-\infty}^{\infty} |H_R(f)|^2 G_n(f)\, df \int_{-\infty}^{\infty} \frac{|P_r(f)|^2}{|H_c(f)H_R(f)|^2}\, df\right]^{-1} \tag{5.49}$$

3. The optimum transmitting and receiving filters that maximize A^2/N_0 given above are

$$|H_R(f)|^2 = \frac{K|P_r(f)|}{|H_c(f)|G_n^{1/2}(f)} \tag{5.50a}$$

$$|H_T(f)|^2 = \frac{K_c^2|P_r(f)|G_n^{1/2}(f)}{K|P_g(f)|^2|H_c(f)|} \tag{5.50b}$$

where K is a positive constant. The filter phase responses are arbitrary as long as Equation (5.43) is satisfied.

4.

$$\left(\frac{A^2}{N_0}\right)_{\max} = \frac{3S_T T_s}{M^2-1}\left[\int_{-\infty}^{\infty} \frac{|P_r(f)|G_n^{1/2}(f)}{|H_c(f)|}\, df\right]^{-2} \tag{5.51}$$

and

$$P_e = 2\left(\frac{M-1}{M}\right) Q\left(\sqrt{\left(\frac{A^2}{N_0}\right)_{\max}}\right) \tag{5.52}$$

5. In the special case when

$$G_n(f) = \eta/2$$

and $p_g(t)$ is chosen as

$$p_g(t) = \begin{cases} 1, & |t| < \tau/2;\ (\tau \ll T_s) \\ 0, & \text{elsewhere} \end{cases}$$

the filter transfer functions reduce to

$$|H_R(f)|^2 = K_1 \frac{|P_r(f)|}{|H_c(f)|} \tag{5.53}$$

$$|H_T(f)|^2 = K_2 \frac{|P_r(f)|}{|H_c(f)|} \tag{5.54}$$

where K_1 and K_2 are positive constants. As in the binary case, we have identical frequency response characteristics for the transmitting and receiving filters. Now if we also have

$$|H_c(f)| = \begin{cases} 1, & |f| < r_s/2 + \beta \\ 0, & \text{elsewhere} \end{cases}$$

(i.e., an ideal lowpass channel) then,

$$|H_R(f)| = |H_T(f)| = K_3|P_r(f)|^{1/2}$$

where K_3 is a positive constant. In this case we also have

$$\left(\frac{A^2}{N_0}\right)_{\max} = \left(\frac{3S_T T_s}{M^2 - 1}\right)\left(\frac{2}{\eta}\right) \tag{5.55}$$

and

$$P_e = \left(\frac{2(M-1)}{M}\right) Q\left(\sqrt{\left(\frac{3S_T T_s}{M^2 - 1}\right)\left(\frac{2}{\eta}\right)}\right) \tag{5.56}$$

The design and analysis of M-ary systems are carried out using equations (5.48)–(5.56).

Example 5.3. Design a quarternary signaling scheme for transmitting the output of an ergodic source emitting an independent sequence of symbols from a source alphabet consisting of four equiprobable symbols A, B, C, and D. The symbol rate of the source is 5000 symbols per sec and the overall error probability has to be less than 10^{-4}. The channel available for transmitting the data has the following characteristics:

$$H_c(f) = \frac{1}{1 + j(f/5000)}$$

and

$$G_n(f) = 10^{-12} \text{ watt/Hz}$$

Solution. We are given $M = 4$, $r_s = 5000$, and $T_s = (2)(10^{-4})$. Also we are given

$$|H_c(f)| = [1 + (f/5000)^2]^{-1/2}$$

Since there are no strict channel bandwidth limitations, we can use a received pulse having raised cosine frequency spectrum with $\beta = r_s/2$:

$$P_r(f) = \begin{cases} T_s \cos^2(\pi f/2r_s), & |f| < r_s \\ 0, & \text{elsewhere} \end{cases}$$

In order to arrive at $H_T(f)$ and $H_R(f)$ with similar functional forms, we choose $p_g(t)$ as

$$p_g(t) = \begin{cases} 1, & |t| < \tau/2 \\ 0, & \text{elsewhere} \end{cases} \quad \tau = T_s/10 = (2)(10^{-5})$$

$P_g(f)$ for the above $p_g(t)$ can be computed as

$$P_g(f) \approx \tau \quad \text{for } |f| < r_s$$

From Equations (5.50a) and (5.50b), we obtain $H_T(f)$ and $H_R(f)$ as

$$|H_R(f)| = \begin{cases} K_1\,[1 + (f/5000)^2]^{1/4} \cos(\pi f/2r_s), & |f| < 5000 \text{ Hz} \\ 0, & \text{elsewhere} \end{cases}$$

$$|H_T(f)| = \begin{cases} [1 + (f/5000)^2]^{1/4} \cos(\pi f/2r_s), & |f| < 5000 \text{ Hz} \\ 0, & \text{elsewhere} \end{cases}$$

The constant K_1 and the phase shifts of the filters are chosen to yield

$$P_g(f)H_T(f)H_c(f)H_R(f) = P_r(f)\exp(-j2\pi f t_d)$$

Plots of $P_g(f)$, $H_c(f)$, $H_T(f)$, $H_R(f)$, and $P_r(f)$ are shown in Figure 5.9.

Now we need to find the transmitter power requirements to maintain $P_e < 10^{-4}$. From Equation (5.52), we have

$$P_e = 2 \cdot \tfrac{3}{4} Q\left(\sqrt{(A^2/N_0)_{\max}}\right) < 10^{-4}$$

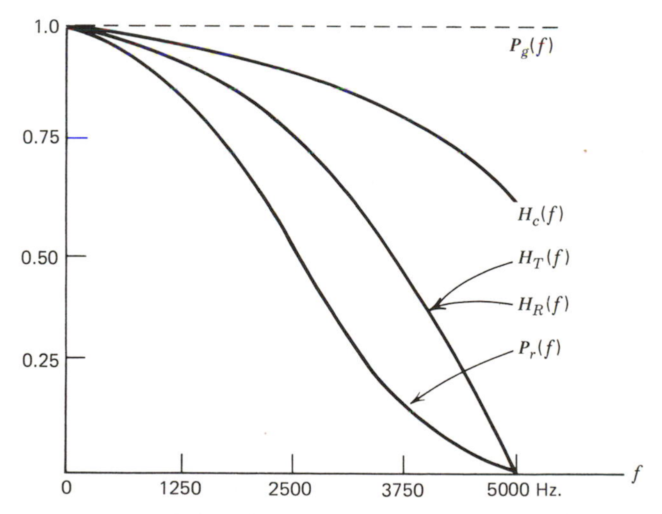

Figure 5.9 Normalized plots of P_g, H_c, H_T, H_R, and $P_r(f)$. All functions are shown normalized with respect to their values at $f = 0$.

which requires

$$Q(\sqrt{(A^2/N_0)_{\max}}) < (\tfrac{2}{3})(10^{-4})$$

or

$$(A^2/N_0)_{\max} > (3.8)^2 = 14.44$$

From Equation (5.51), we obtain

$$\left(\frac{A^2}{N_0}\right)_{\max} = \frac{3S_T T_s}{15}\left[\int_{-\infty}^{\infty} \frac{|P_r(f)|G_n^{1/2}(f)}{|H_c(f)|}\,df\right]^{-2} > 14.44$$

The integral in the preceding equation can be evaluated either in closed form (when possible) or by a numerical approximation:

$$\begin{aligned}\int_{-\infty}^{\infty} \frac{|P_r(f)|G_n^{1/2}(f)}{|H_c(f)|}\,df &= 10^{-6}\int_{-5000}^{5000} T_s\left(\cos\frac{\pi f}{2r_s}\right)^2\left[1+\left(\frac{f}{5000}\right)^2\right]^{1/2} df \\ &= (2)(10^{-6})\int_0^1 \left(\cos\frac{\pi}{2}x\right)^2 (1+x^2)^{1/2}\,dx \\ &\approx (1.20)10^{-6} \quad \text{(by numerical integration)}\end{aligned}$$

Hence

$$\left(\frac{S_T T_s}{5}\right)[(1.20)10^{-6}]^{-2} > 14.44$$

or

$$\begin{aligned}S_T &> (5)(5000)(1.20)^2(10^{-12})(14.44) \\ &> (-32.5)\ \text{dBm}\end{aligned}$$

5.4.2 Binary versus *M*-ary Signaling Schemes

We are now ready to compare binary and M-ary signaling schemes and determine which scheme should be used in a particular situation. Let us assume that:

1. The input to both systems comes from an ergodic information source that emits an independent sequence of equally likely binary digits at a rate of r_b bits/sec (no loss in generality results from this assumption since the output symbol sequence of any source may be coded into a binary sequence).
2. The channel is ideal lowpass, and the channel noise is zero mean Gaussian with a psd of $\eta/2$.
3. Both signaling schemes use pulses having appropriate raised cosine frequency characteristics with the maximum value of β.
4. Both signaling schemes are required to yield the same error probability P_e.

Table 5.1. Comparison of binary and M-ary signaling schemes

	Binary scheme	M-ary scheme
Bandwidth	r_b Hz	$r_s = (r_b/k)$ Hz
Probability of Error P_e	$Q\left(\sqrt{\frac{S_T}{r_s \log_2 M}\left(\frac{2}{\eta}\right)}\right)$ (Eqs. (5.22); (5.23))	$\frac{2(M-1)}{M} Q\left(\sqrt{\frac{3S_T}{(M^2-1)r_s}\left(\frac{2}{\eta}\right)}\right)$ (Eq. (5.56))
Transmitter power for a given P_e	Less	More
Equipment complexity	Less	More

Bit rate $= r_b$ bits/sec.
M-ary symbol rate $= r_s = r_b/k$ symbols/sec.

The M-ary signaling scheme is assumed to operate on blocks of k binary digits at a time ($M = 2^k$). Each block of k binary digits is translated to one of M levels at the transmitter, and each received level is decoded as a block of k binary digits at the receiver. The bandwidth and power requirements of the binary and M-ary schemes are shown in Table 5.1.

Comparison of binary and M-ary signaling schemes indicate that *binary transmission has lower power requirements*, and that *M-ary signaling schemes require lesser bandwidth*. For $M \gg 2$ and $P_e \ll 1$, the transmitter power must increase by a factor of $M^2/\log_2 M$, whereas the bandwidth is reduced by $1/\log_2 M$. M-ary schemes are more complex since the receiver has to decide on one of M-levels using $M-1$ comparators or level slicers. In the binary case, the decoding requires only one comparator.

Example 5.4. Compare the power-bandwidth requirements of binary and quarternary ($M = 4$) signaling schemes for transmitting the output of an ergodic source emitting an independent sequence of symbols from an alphabet consisting of four equiprobable letters A, B, C, and D. The symbol rate is 5000/sec and the signaling schemes are required to maintain $P_e \leq 10^{-4}$. Assume an ideal lowpass channel with additive Gaussian noise with a psd $\eta/2 = 10^{-12}$ watt/Hz.

Solution. The data rate for the problem is $5000 \log_2 4 = 10{,}000$ bits/sec. To design a binary signaling scheme, we need to convert the symbols into bit strings. This can be done by assigning 2-bit binary words to each letter in the source alphabet. For example, we can assign 00, 01, 10, and 11 to A, B, C, and

D, respectively. A symbol sequence such as *DBCAADCB* will be translated into a bit stream 1101100000111001 and transmitted as binary data. The receiver will first decode the individual binary digits, and then decode the letters by looking at two bit groups.

Hence, for the binary scheme we have

$$r_b = 10{,}000 \text{ bits/sec}$$

$$H_c(f) = 1, \ |f| < 10{,}000 \text{ Hz}$$

$$G_n(f) = \eta/2 = 10^{-12}$$

If we use a received pulse having raised cosine frequency characteristics with $\beta = r_b/2$, then the bandwidth required is r_b Hz, that is,

$$(\text{Bandwidth})_{\text{binary}} = 10{,}000 \text{ Hz}$$

Power requirement may be computed using

$$P_e = Q(\sqrt{(A^2/N_0)_{\max}}) < 10^{-4}$$

which requires

$$(A^2/N_0)_{\max} \geqslant (3.75)^2 = 14.06$$

From Equation (5.22) with $G_n(f) = 10^{-12}$, $H_c(f) = 1$, and $\int_{-\infty}^{\infty} P_r(f)\, df = 1$, we get

$$\left(\frac{A^2}{N_0}\right)_{\max} = \frac{S_T T_b}{(\eta/2)} = S_T(10^8)$$

Hence

$$(S_T)_{\text{binary}} \geqslant (14.06)(10^{-8}) = -38.52 \text{ dBm}$$

For the quarternary ($M = 4$) scheme the bandwidth required is (using $P_r(f)$ with $\beta = r_s/2$)

$$(\text{Bandwidth})_{\text{quarternary}} = 5000 \text{ Hz}$$

and

$$P_e = \tfrac{3}{2} Q(\sqrt{(A^2/N_0)_{\max}}) < 10^{-4}$$

or

$$(A^2/N_0)_{\max} \geqslant 14.44$$

From Equation (5.51), we have

$$\left(\frac{A^2}{N_0}\right)_{\max} = \left(\frac{S_T T_s}{5}\right)\left(\frac{2}{\eta}\right)$$

$$(S_T)_{\text{quarternary}} \geqslant (14.44)(5\eta/2T_s)$$
$$\geqslant -34.42 \text{ dBm}$$

The reader can, for comparison purposes, verify that a scheme with $M = 16$ can be designed for this problem. The requirements for $M = 16$ are

$$(\text{Bandwidth})_{M=16} = 2500 \text{ Hz}$$

$$(S_T)_{M=16} \geqslant -24.9 \text{ dBm}$$

M	Bandwidth (Hz)	Power (dBm)
2	10,000	−38.52
4	5000	−34.42
16	2500	−24.90

The results in the table show that as the bandwidth is reduced the power requirement increases sharply.

In the preceding comparison of binary and M-ary PAM systems, the symbol error probability was used as the basis for comparison. If the output of the information source is assumed to be binary digits, it is more meaningful to use the bit error probability for comparing M-ary and binary schemes. While there are no unique relationships between the bit error probability and M-ary symbol error probability, we can derive such relationships for two special cases.

In the first case we assume that whenever an M-ary symbol is in error, the receiver output is equally likely to be any one of the $2^k - 1$ erroneous k-bit sequences. Now, considering any arbitrary bit position in the M-fold set of k-bit sequences, $M/2$ sequences contain a binary 1 in that position and the remaining sequences contain a binary 0. Then considering an arbitrary bit in the input sequence, it is apparent that the same binary digit occurs in $2^{k-1} - 1$ of the remaining $2^k - 1$ possible sequences and the other binary digit occurs in that position in the remaining 2^{k-1} sequences. Under these assumptions, the average probability of bit error P_{eb} in M-ary transmission with a symbol error probability P_e is given by

$$P_{eb} = \frac{2^{k-1}}{2^k - 1} P_e \leqslant \tfrac{1}{2} P_e$$

Thus in comparing binary and M-ary schemes, we should use

$$P_e = \begin{cases} P_{eb} & \text{for binary transmission} \\ P_{eb}\left(\dfrac{2^k - 1}{2^{k-1}}\right) & \text{for } M\text{-ary transmission} \end{cases} \tag{5.57a}$$

where P_{eb} is the bit error probability of both schemes.

In practical M-ary PAM schemes, one often uses a binary-to-M-ary coding such that binary sequences corresponding to adjacent amplitude levels differ in only one bit position (one such coding is the so-called Gray code). Now, when an M-ary symbol error occurs, it is more likely that only one of the k input bits will be in error, especially when the signal-to-noise ratio at the receiver input is high. In such a case, we have

$$P_{eb} = P_e/k$$

Thus, for a given bit error probability, we should use

$$P_e = \begin{cases} P_{eb} & \text{for binary transmission} \\ kP_{eb} & \text{for } M\text{-ary transmission} \end{cases} \tag{5.57b}$$

for comparison purposes.

5.5 SHAPING THE TRANSMITTED SIGNAL SPECTRUM

In many applications, the spectrum of the PAM signal should be carefully shaped to match the channel characteristics in order to arrive at physically realizable transmitting and receiving filters. As an example, consider a channel that has a poor amplitude response (high attenuation) in the low-frequency range. Using a signaling scheme that has high power content in the low-frequency range will result in transmitting and receiving filters with unreasonably high gain in the low-frequency range (see Equations (5.20) and (5.21)). This problem could be avoided if the spectrum of the transmitted signal is altered so that it has small power content at low frequencies.

The spectrum of the transmitted signal in a PAM system will depend on the signaling pulse waveform and on the statistical properties of sequences of transmitted digits. The pulse waveform in a PAM system is specified by the ISI requirements. While it might be possible to shape the transmitted signal spectrum by changing the transmitted pulse shape, such changes might lead to increased ISI. An easier way to shape the transmitted signal spectrum is to alter the statistical properties of the transmitted bit or symbol sequence.

In the following section we will look at methods of changing the spectrum of the transmitted signal by changing the statistical properties of the amplitude sequence of the pulse train. We will also look at digital methods of generating accurately shaped signaling waveforms.

5.5.1 Effect of Precoding on the Signal Spectrum

The output of the pulse generator in a PAM system is a pulse train $X(t)$, which can be written as

$$X(t) = \sum_{k=-\infty}^{\infty} a_k p_g(t - kT)$$

where T is the symbol duration ($T = T_b$ for binary schemes and $T = T_s$ for M-ary schemes). The output of the transmitting filter $Z(t)$ is also a pulse train,

$$Z(t) = \sum_{k=-\infty}^{\infty} a_k p_t(t - kT)$$

where $p_t(t)$ is the response of the transmitting filter to $p_g(t)$. The power spectral density $G_Z(f)$ of $Z(t)$ is given by (Chapter 3, Examples 3.8 and 3.9)

$$G_Z(f) = \frac{|P_t(f)|^2}{T} G(f) \tag{5.58a}$$

where $G(f)$ has the form

$$G(f) = R(0) - m^2 + 2 \sum_{k=1}^{\infty} [R(k) - m^2] \cos 2\pi kfT \tag{5.58b}$$

In the preceding equations $P_t(f)$ is the Fourier transform of $p_t(t)$, $R(k)$ represents $E\{a_j a_{j+k}\}$, and $m = E\{a_k\}$. The factor $G(f)$ represents the way in which the spectrum is affected by the statistics of the amplitude level sequence $\{a_k\}$. We will consider three commonly used methods of precoding which alter the statistical properties of the transmitted sequence and hence the spectrum of the transmitted signal. We will illustrate the principles of precoding using a binary PAM system with an input bit stream $\{b_k\}$ that is assumed to be an equiprobable, independent sequence from an ergodic source. The bit sequence $\{b_k\}$ is converted into an amplitude sequence $\{a_k\}$ which amplitude modulates $p_t(t - kT)$ (see Figure 5.10). We will look at methods of mapping $\{b_k\}$ into $\{a_k\}$ and their effects on the shape of the transmitted spectrum. We will concentrate our attention on $G(f)$ in Equation (5.58).

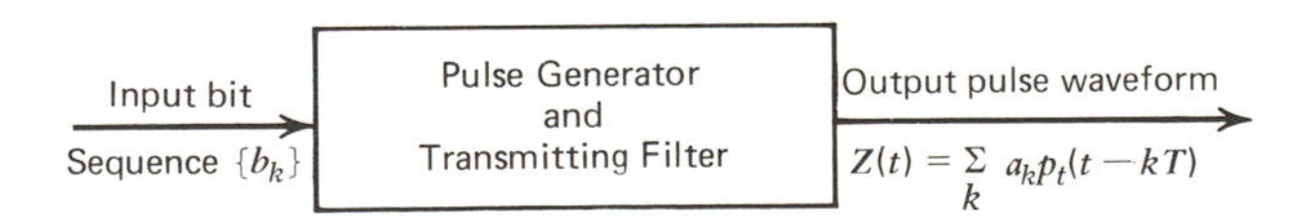

Figure 5.10 Portion of a binary PAM system.

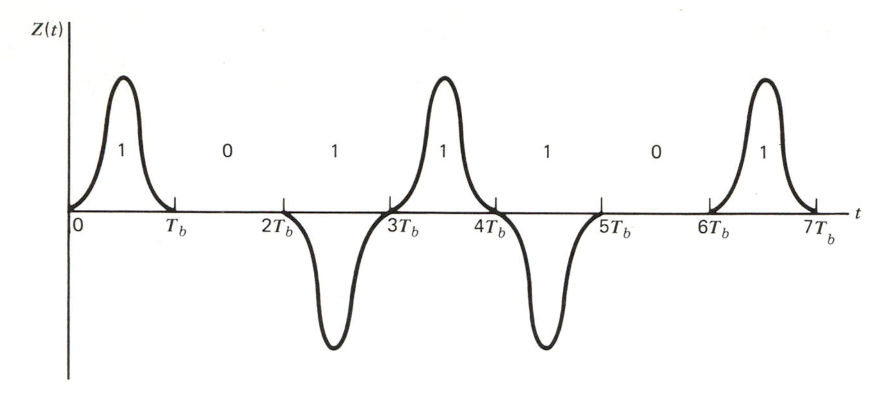

Figure 5.11 An example of bipolar coding.

Bipolar coding. The most widely used method of coding for baseband PAM data transmission is the *bipolar* coding. The bipolar coding uses three amplitude levels: zero and equal magnitude positive and negative levels (for convenience we will take these levels to be +1, 0, and −1). Binary 0 in the input is represented by $a_k = 0$, and binary 1 is represented by either +1 or −1. Each successive 1 that occurs is coded with opposite polarity as shown in Figure 5.11. The autocovariance $R(j)$ for this sequence can be calculated as follows:

$$R(0) = E\{a_k^2\}$$
$$= (1)^2 P(a_k = 1) + (-1)^2 P(a_k = -1) + (0)^2 P(a_k = 0)$$

It can be easily verified that $P(a_k = 1) = P(a_k = -1) = \frac{1}{4}$ and $P(a_k = 0) = \frac{1}{2}$. Hence

$$R(0) = \tfrac{1}{2}$$

$R(1)$ and $R(2)$ can be calculated using the joint probabilities given in Table 5.2.

The reader can verify that

$$R(1) = E\{a_k a_{k+1}\}$$
$$= \sum_{i=-1}^{1} \sum_{j=-1}^{1} ij P(a_k = i,\ a_{k+1} = j)$$
$$= -\tfrac{1}{4}$$

and

$$R(2) = E\{a_k a_{k+2}\}$$
$$= \sum_{i=-1}^{1} \sum_{j=-1}^{1} ij P(a_k = i,\ a_{k+2} = j)$$
$$= -\tfrac{1}{8} + \tfrac{1}{8} = 0$$

Table 5.2. Joint probabilities for $a_k a_{k+j}$

d_k	d_{k+1}	d_{k+2}	a_k	a_{k+1}	a_{k+2}	Prob.
0	0	0	0	0	0	$\frac{1}{8}$
0	0	1	0	0	1	$\frac{1}{8}$
0	1	0	0	1	0	$\frac{1}{8}$
0	1	1	0	1	-1	$\frac{1}{8}$
1	0	0	1	0	0	$\frac{1}{8}$
1	0	1	1	0	-1	$\frac{1}{8}$
1	1	0	1	-1	0	$\frac{1}{8}$
1	1	1	1	-1	1	$\frac{1}{8}$

(Assume the last 1 was coded as -1.)

a_{k+1} \ a_k	-1	0	1
-1	0	0	$\frac{1}{4}$
0	0	$\frac{1}{4}$	$\frac{1}{4}$
1	0	$\frac{1}{4}$	0

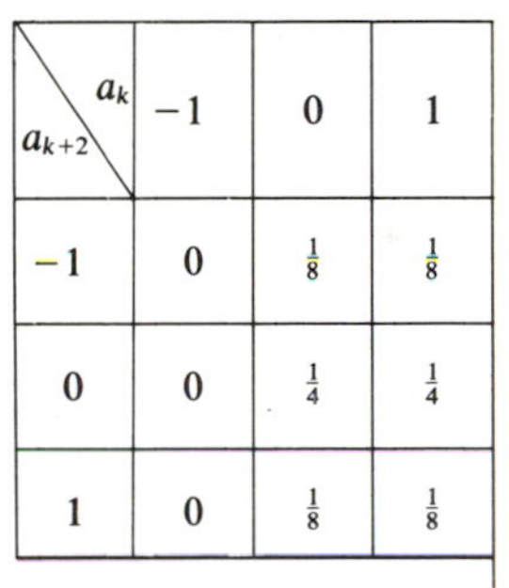

a_{k+2} \ a_k	-1	0	1
-1	0	$\frac{1}{8}$	$\frac{1}{8}$
0	0	$\frac{1}{4}$	$\frac{1}{4}$
1	0	$\frac{1}{8}$	$\frac{1}{8}$

(Entries inside these tables are joint probabilities.)

Following a similar procedure it can be shown that

$$R(j) = 0 \quad \text{for } j > 2$$

Hence the spectral component $G(f)$ for the bipolar coding becomes

$$G(f) = \tfrac{1}{2}(1 - \cos 2\pi f T_b) \tag{5.59}$$

A plot of $G(f)$ shown in Figure 5.12 reveals that this scheme can be used for baseband signaling without DC or near DC components (which makes it ideal for channels which have poor low-frequency response). Also, the average power is somewhat less since one half of the pulses have zero amplitude. However, the missing pulses will make clock recovery difficult.

Twinned Binary Coding. Another commonly used method of coding fo binary PAM data transmission is the twinned binary coding which is also called *split phase* or *Manchester coding*. In this method, each bit is represented by two successive pulses of opposite polarity, the two binary values

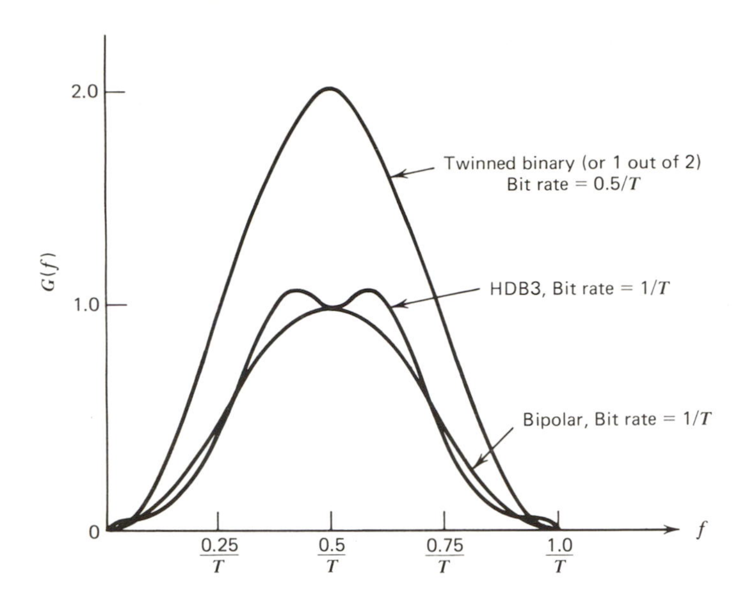

Figure 5.12 Power spectral density of the transmitted signal.

having the representations $+-$ and $-+$ (Figure 5.13). The bit rate is half that of unrestricted binary pulses. Hence, this method uses twice the bandwidth. For the twinned binary scheme, $R(0) = 1$, $R(1) = -\frac{1}{2}$, $T = T_b/2$, and

$$G(f) = 1 - \cos \pi f T_b \tag{5.60}$$

A plot of $G(f)$ for the twinned binary coding is shown in Figure 5.12. Like the bipolar coding scheme, twinned binary coding also has no near DC components. The main advantage of the twinned binary coding scheme is that every pulse position is occupied and hence clock recovery is easier. The disadvantage is that the twinned binary scheme requires larger bandwidth.

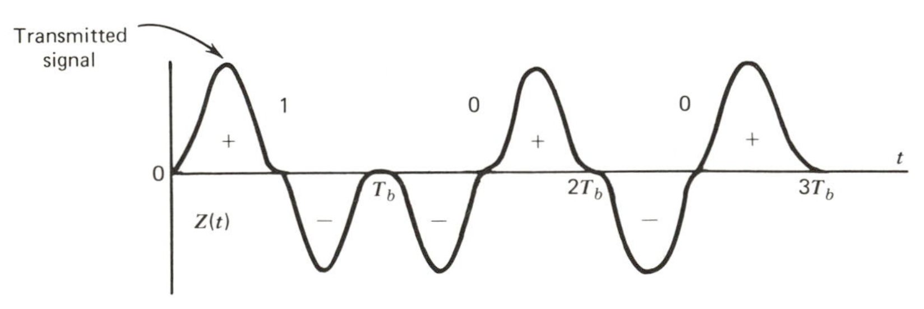

Figure 5.13 Twinned binary coding.

The twinned binary coding is also called a one out of two coding scheme since one out of every two pulses is positive and the other one is negative. An extension of this method leads to 2 out of 4, 3 out of 6, and 4 out of 8 schemes that produce spectra with sharper roll-offs. These methods are very useful for tailoring the transmitted signal spectrum to a finite frequency band.

High Density Bipolar Coding. A series of high density bipolar coding schemes known as HDB1, HDB2, HDB3, . . . are used to eliminate long strings of zero amplitude pulses that occur in regular bipolar coding schemes. Absence of signaling pulses make clock recovery difficult. In the regular bipolar coding scheme, successive pulses have opposite sign. In the HDB coding schemes, "bipolar" rule violations are used to carry extra information needed to replace strings of zeros. The HDBn scheme is designed to avoid the occurrence of more than n pulses of zero amplitude. The most important of the HDB codes is the HDB3.

The HDB3 code uses a bipolar coding scheme whenever possible. But if the string 0000 occurs in the input binary stream, special sequences other than 0000 are transmitted. The special sequences contain bipolar violations and hence can be easily distinguished. The special sequences used in HDB3 coding are 000D and 100D. In the special sequence, "1" is represented by amplitude level $a_k = +1$ or -1 *following the bipolar rule*, and "0" by level 0. The "D" is replaced by level $+1$ or -1 *violating the bipolar rule*. The choice of the special sequence 000D or 100D is made in such a way that the pulses violating the bipolar rule take on levels $+1$ and -1 alternately. 100D is used when there has been an even number of ones since the last special sequence. Special sequences can follow each other if the string of zeros continues. Two special sequences are necessary to assure that the special sequences can be distinguished from data sequences, and to guarantee that there will be zero crossings when special sequences follow each other. An example is shown below.

Input bit stream	1	0	1	1	0	0	0	0	0	1	0	0	0	0	0	0	0	0
Coded bit stream	1	0	1	1	[1	0	0	D]	0	1	[0	0	0	D]	[1	0	0	D]
Amplitude levels a_k	−	0	+	−	[+	0	0	+]	0	−	[0	0	0	−]	[+	0	0	+]

(Special sequences are shown enclosed. The choice of the first special sequence is arbitrary.)

The HDB3 waveform corresponding to the above example is shown in Figure 5.14.

The calculation of the autocovariance and $G(f)$ for HDB codes is rather lengthy and hence is not included here. However, a plot of $G(f)$ for the HDB3

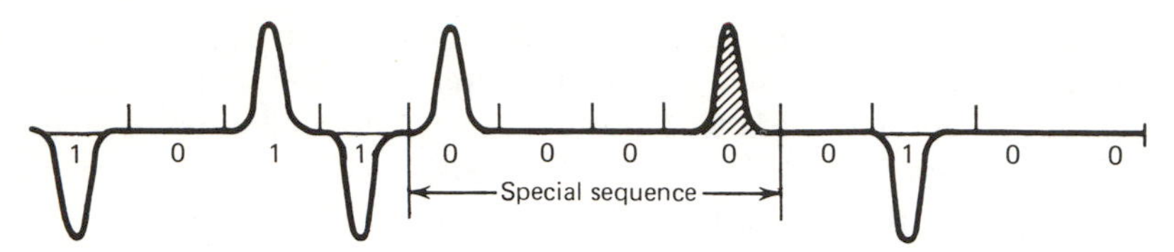

Figure 5.14 Example of HDB3 waveform. Shaded pulse represents a bipolar violation.

coding is shown in Figure 5.12 for comparison purposes. As may be expected, $G(f)$ for HDB3 lies close to the bipolar curve. The mean power for the HDB3 is 10% higher than the bipolar case since some of the 0 amplitude pulses in the bipolar case are replaced by nonzero amplitudes in HDB3 coding schemes.

5.5.2 Pulse Shaping by Digital Methods

Digital signaling schemes require accurately shaped pulse waveforms. The shape of a pulse can be specified either by its amplitude as a function of time or by its Fourier transform. If a pulse shape is specified by its transform in the frequency domain, then, in principle at least, we can design filters to shape the pulse. This design becomes difficult in a practical sense because of the need for linear phase characteristics for the filters. An alternate method is to generate the pulse waveform directly. Such a method should be capable of generating a signal composed of many overlapping pulses that are superposed. One of the most commonly used methods of direct waveform generation makes use of a binary transversal filter shown in Figure 5.15. The pulse waveform $p_t(t)$ which is to be amplitude modulated is sampled at intervals of Δ and the amplitude values at sampling times are $b_{-8}, b_{-7}, \ldots, b_{-1}, b_0, b_1, \ldots, b_8$. For purposes of simplicity we will assume that sample values outside the time interval -8Δ to 8Δ are small enough to be ignored. The transversal filter for this waveform consists of a 17-bit shift register with 17 outputs, each of which can be two possible levels 0 or 1. These outputs are attenuated by b_{-8} to b_8, respectively, and summed by an analog summing network.

The actual waveform we are trying to generate has the form

$$Z(t) = \sum_k d_k p_t(t - kT_b)$$

where d_k is the kth input bit having a value 0 or 1. In the example shown in Figure 5.14, we are approximating $Z(t)$ by the staircase waveform

$$\tilde{Z}(t) = \sum_k d_k \tilde{p}_t(t - kT_b)$$

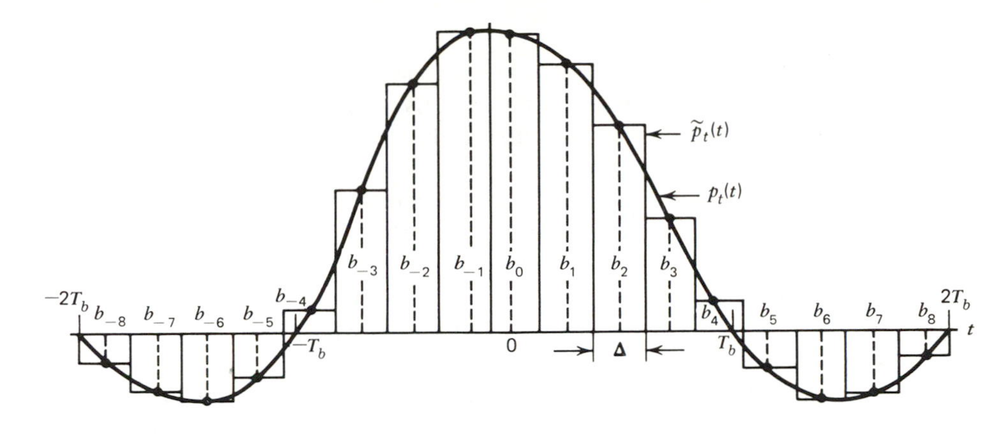

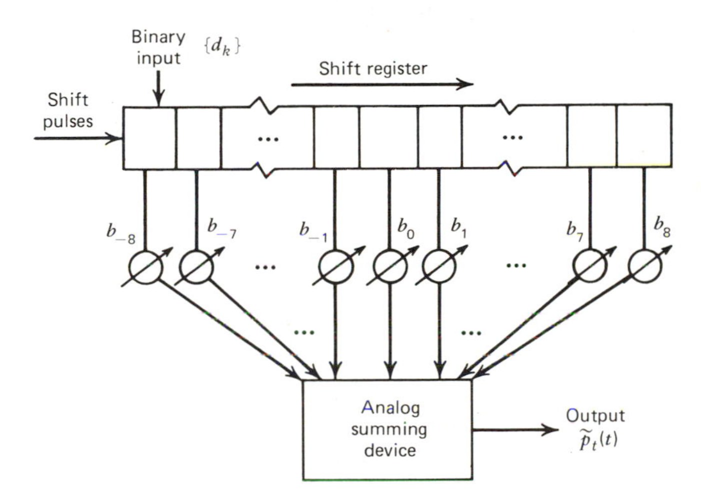

Figure 5.15 Waveform generation using a binary transversal filter.

that is, $p_t(t)$ is approximated by

$$\tilde{p}_t(t) = \sum_{j=-8}^{8} b_j p_s(t - j\Delta)$$

where

$$p_s(t) = \begin{cases} 1, & -\Delta/2 \leq t < \Delta/2 \\ 0, & \text{elsewhere} \end{cases}$$

The kth input bit d_k stays in the shift register for a duration of 17Δ seconds. It is shifted through the shift register at a rate equal to 17 times the bit rate.

Successive bits d_k are inserted into the shift register once every T_b seconds and the output due to successive input bits overlap and the summing device performs the superposition of individual pulses to form the composite waveform.

The digitally generated waveform has zero errors at sampling times $0, \pm T_b, \pm 2T_b, \ldots$. The spectrum of the staircase waveform is centered around the sampling frequency $(1/\Delta)$ and its harmonics. By choosing a suitably large sampling frequency the "noise" due to sampling can be separated from the spectrum of the desired pulses. Then, a simple lowpass filter can be used to smooth out the waveform.

For signaling methods with more than one amplitude level, we need additional shift registers attenuators, and summing devices. For example, if the signaling scheme demands three levels 1, 0, and -1, then two shift registers will be needed.

5.6 EQUALIZATION

In the design of baseband PAM systems we assumed that the frequency response $H_c(f)$ of the channel is completely known. Based on the knowledge of $H_c(f)$ we designed PAM systems to yield zero intersymbol interference (ISI). In almost all real systems some amount of residual ISI inevitably occurs due to imperfect filter design, incomplete knowledge of channel characteristics, changes in channel characteristics, and so forth. The only recourse to mitigate the residual distortion is to include within the system an adjustable filter or filters that can be "trimmed" to compensate for the distortion. The process of correcting channel induced distortion is called *equalization.* Equalizing filters are most often inserted between the receiving filter and the A/D converter, especially in systems using switched telephone lines where the specific line characteristics are not known in advance.

5.6.1 Transversal Equalizer

It is obvious that an equalizing filter should have a frequency response $H_{eq}(f)$ such that the actual channel response multiplied by $H_{eq}(f)$ yields the assumed channel response $H_c(f)$ that was used in the system design. However, since we are interested in the output waveform at a few predefined sampling times only, the design of an equalizing filter is greatly simplified. The most commonly used form of easily adjustable equalizer has been the transversal filter shown in Figure 5.16.

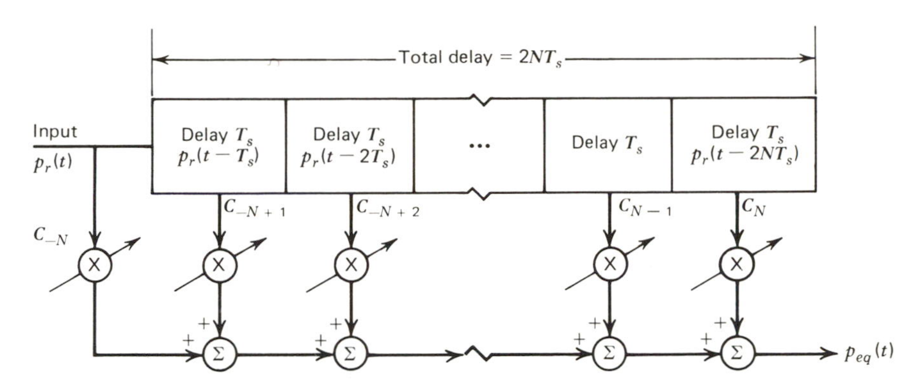

Figure 5.16 Transversal equalizer.

The transversal equalizer consists of a delay line tapped at T_s second intervals. Each tap is connected through a variable gain device to a summing amplifier. For convenience we will assume that the filter has $(2N+1)$ taps with gains C_{-N}, $C_{-N+1}, \ldots, C_0, C_1, \ldots, C_N$. The input to the equalizer is $p_r(t)$, which is known, and the output is $p_{eq}(t)$. We can write the output $p_{eq}(t)$ in terms of $p_r(t)$ and the tap gains as

$$p_{eq}(t) = \sum_{n=-N}^{N} C_n p_r[t-(n+N)T_s] \tag{5.61}$$

If $p_r(t)$ has its peak at $t=0$ and ISI on both sides, the output should be sampled at $t_k = (k+N)T_s$ and

$$p_{eq}(t_k) = \sum_{n=-N}^{N} C_n p_r[(k-n)T_s] \tag{5.62}$$

If we denote $p_r(nT_s)$ by $p_r(n)$ and t_k by k, we have

$$p_{eq}(k) = \sum_{n=-N}^{N} C_n p_r(k-n) \tag{5.63}$$

Ideally, we would like to have

$$p_{eq}(k) = \begin{cases} 1 & \text{for } k=0 \\ 0 & \text{elsewhere} \end{cases}$$

This condition cannot always be realized since we have only $(2N+1)$ variables (namely, the $2N+1$ tap gains) at our disposal. However, we can specify the value of $p_{eq}(t)$ at $2N+1$ points as

$$p_{eq}(k) = \begin{cases} 1 & \text{for } k=0 \\ 0, & k=\pm 1, \pm 2. \ldots, \pm N \end{cases} \tag{5.64}$$

Combining Equations (5.63) and (5.64), we have

$$\begin{matrix} N \\ \text{zeros} \\ \\ \\ N \\ \text{zeros} \end{matrix}\begin{bmatrix} 0 \\ 0 \\ \vdots \\ 0 \\ 1 \\ 0 \\ \vdots \\ 0 \end{bmatrix}_{(2N+1)} = \begin{bmatrix} p_r(0) & p_r(-1) & \cdots & p_r(-2N) \\ p_r(1) & p_r(0) & \cdots & p_r(-2N+1) \\ & & & \vdots \\ p_r(2) & p_r(1) & \cdots & \vdots \\ \vdots & & & \vdots \\ p_r(2N) & \cdots & \cdots & p_r(0) \end{bmatrix} \begin{bmatrix} C_{-N} \\ C_{-N+1} \\ \vdots \\ C_0 \\ \cdot \\ C_{N-1} \\ C_N \end{bmatrix} \tag{5.65}$$

Equation (5.65) represents a set of $(2N + 1)$ simultaneous equations that can be solved for the C_n's. The equalizer described in Equation (5.64) is called a zero forcing equalizer since $p_{eq}(k)$ has N zero values on either side. This equalizer is optimum in that it minimizes the peak intersymbol interference. The main disadvantage of a zero forcing equalizer is that it increases the noise power at the input to the A/D converter (Problem 5.21). But this effect is normally more than compensated for by the reduction in the ISI.

Example 5.5. Design a three tap equalizer to reduce the ISI due to the $p_r(t)$ shown in Figure 5.17*a*.

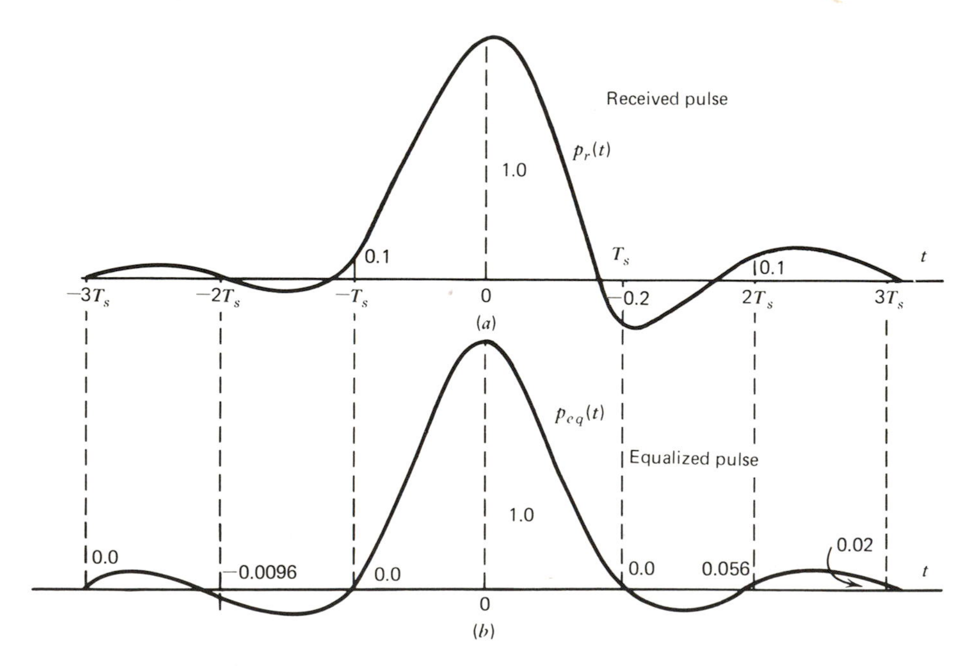

Figure 5.17 Three tap equalizer discussed in Example 5.5.

Solution. With a three tap equalizer we can produce one zero crossing on either side of $t = 0$ in the equalized pulse. The tap gains for this equalizer are given by

$$\begin{pmatrix} 0 \\ 1 \\ 0 \end{pmatrix} = \begin{pmatrix} 1.0 & 0.1 & 0 \\ -0.2 & 1.0 & 0.1 \\ 0.1 & -0.2 & 1.0 \end{pmatrix} \begin{pmatrix} C_{-1} \\ C_0 \\ C_1 \end{pmatrix}$$

or

$$\begin{pmatrix} C_{-1} \\ C_0 \\ C_1 \end{pmatrix} = \begin{pmatrix} -0.09606 \\ 0.9606 \\ 0.2017 \end{pmatrix}$$

The values of the equalized pulse can be computed from Equation (5.63) as

$$\begin{aligned} p_{eq}(-3) &= 0.0 & p_r(-3) &= 0.0 \\ p_{eq}(-2) &= -0.0096 & p_r(-2) &= 0.0 \\ p_{eq}(-1) &= 0.0 & p_r(-1) &= 0.1 \\ p_{eq}(0) &= 1.0 & p_r(0) &= 1.0 \\ p_{eq}(1) &= 0.0 & p_r(1) &= -0.2 \\ p_{eq}(2) &= 0.0557 & p_r(2) &= 0.1 \\ p_{eq}(3) &= 0.02016 & p_r(3) &= 0.0 \end{aligned}$$

The equalized pulse is shown in Figure 5.17*b*. While the equalized pulse has one zero crossing on either side of $t = 0$, it has small ISI at points further out from $t = 0$ where the unequalized pulse had zero ISI.

5.6.2 Automatic Equalizers

The design and adjustment of the tap gains of the zero forcing equalizer described in the preceding section involves the solution of a set of simultaneous equations. In the manual mode, the "trimming" of the equalizer involves the following steps:

1. Send a test pulse through the system.
2. Measure the output of the receiving filter $p_r(t)$ at appropriate sampling times.
3. Solve for the tap gains using Equation (5.65).
4. Set the gains on the taps.

Highly accurate and simply instrumented *automatic systems* for setting up the gains have been developed in recent years. These systems are usually divided into two groups: the *preset type* that uses a special sequence of pulses prior to or during breaks in actual data transmission, and the *adaptive type* that adjusts itself continuously during data transmission by operating on the data signal itself. Automatic equalizers use iterative techniques to arrive at optimum tap gains. Before we take a detailed look at the operation of automatic equalizers, let us briefly review an iterative technique commonly used for solving a set of simultaneous equations.

The equations we have to solve, given in (5.65), can be written as

$$I = XC \tag{5.66}$$

where I is a $(2N+1)$ column vector whose components are all zero except the zeroth, X is a $(2N+1)$ square matrix whose (i, j)th element is $p_r(i-j)$, and C is a column vector whose jth element C_j represents the gain of the jth tap; the indices i, j run from $-N$ to N. The iterative method assumes that at the end of the kth iteration we have a solution vector C^k that yields an error in the solution

$$\epsilon^k = XC^k - I \tag{5.67}$$

The components of the error vector are denoted by ϵ_j^k ($j = -N$ to N). The new, adjusted value of the solution vector C^{k+1} is obtained by

$$C^{k+1} = C^k - \Delta\,\mathrm{sgn}(\epsilon^k) \tag{5.68}$$

where

$$\mathrm{sgn}(y) = \begin{cases} +1, & y > 0 \\ 0, & y = 0 \\ -1, & y < 0 \end{cases}$$

and Δ represents a positive increment. The iterations are continued until C^k and C^{k+1} differ by less than some arbitrarily small value.

Under some mild assumptions, it has been shown that the system distortion per tap can be reduced to near the minimum possible value of $\Delta/2$. In this method (called the fixed increment method) Δ must be made as small as possible. Two examples of automatic equalizers are presented next.

Preset Equalizer. A simple preset equalizer is shown in Figure 5.18. In this system, components of the error vector ϵ^k $(k=1, 2, \ldots)$ are measured by transmitting a sequence of widely separated pulses through the system and observing the output of the equalizer ($= XC^k$) at the sampling times. Fixed increment iteration, with stepsize Δ, is used to adjust the tap gains. The sampling of the filter output is done using a timing circuit triggered by a peak detector. The center sample is sliced at $+1$ (or compared with $+1$) and the polarity of the error component ϵ_0^k is obtained. The polarities of the other

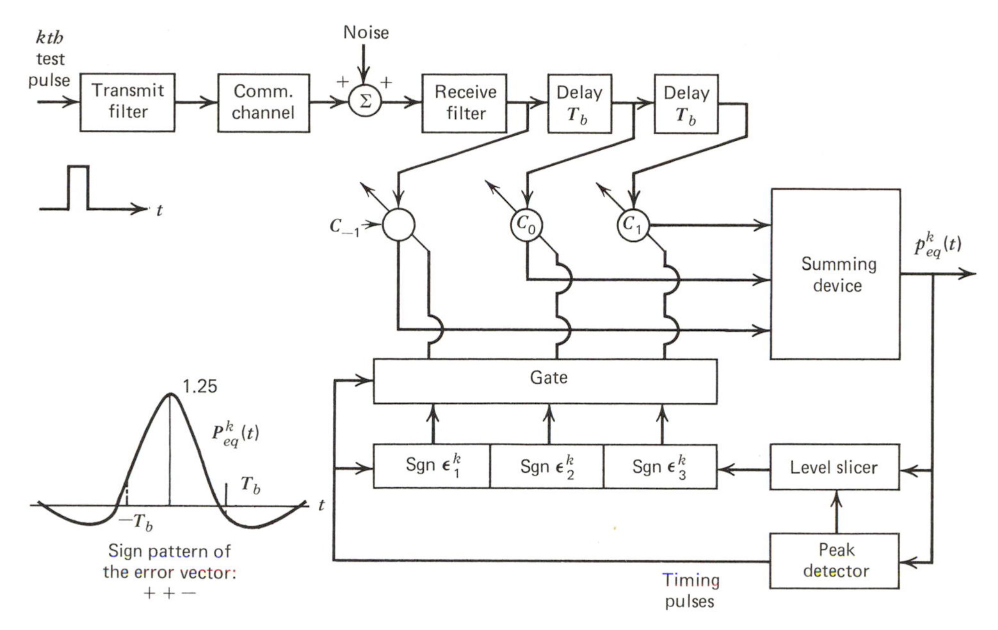

Figure 5.18 A three tap preset equalizer.

error components ϵ_j^k ($j=-N$ to N) are obtained from the value of the filter output at $t=\pm jT_s$. (If we denote the values of the equalized pulse at the end of the kth iteration by $p_{eq}^k(t)$, then $\epsilon_j^k=p_{eq}^k(jT_s)$ for $j\neq 0$, and $\epsilon_0^k=p_{eq}^k(0)-1$.)

At the end of the kth test pulse, the gate is opened, and depending on the polarity of the components of ϵ^k, the tap gains are moved up or down by Δ according to (5.68). This iterative "training session" is carried on until the procedure converges. The training procedure might involve hundreds of pulses.

A major problem in "training" a preset equalizer is the presence of noise in the observed values of $p_{eq}(t)$. The effects of noise can be somewhat minimized by averaging the measured values of $p_{eq}(t)$ over a number of test pulses before the tap gains are changed. One of the difficulties with averaging is that the rate of convergence is slowed down somewhat. Other methods for reducing the effects of noise may be found in Reference 1 listed at the end of this chapter.

Adaptive Equalizer. In adaptive equalizers, the error vector ϵ^k is continually estimated during the normal course of data transmission. Such schemes have the ability to adapt to changes during data transmission and eliminate the need for long training sessions. Adaptive equalizers are quite

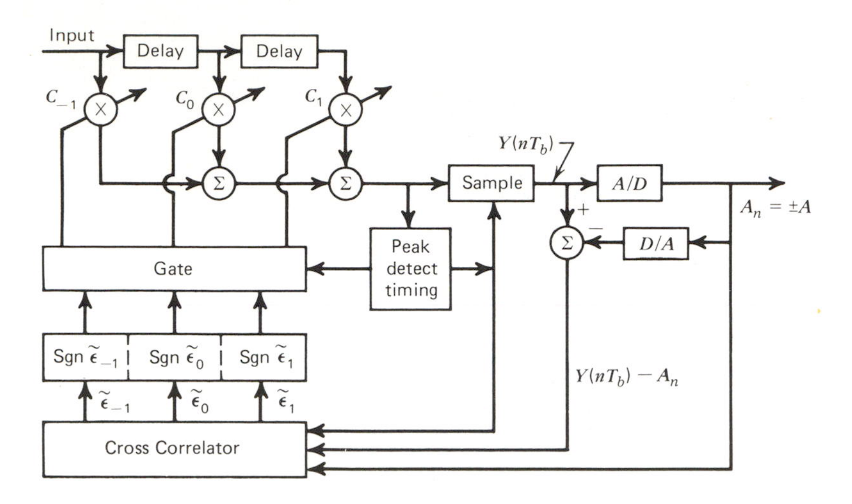

Figure 5.19 A three tap adaptive equalizer.

common in practice and are more accurate, versatile, and cheaper than preset equalizers.

A simple adaptive equalizer for a binary data transmission scheme is shown in Figure 5.19. The output of the equalizer $Y(t)$ at sampling times should be $\pm A$, $+A$ if the actual input bit corresponding to the sampling time is 1 and $-A$ if the input bit is 0. In an actual system, due to ISI, the values $Y(jT_b)$ will vary about $\pm A$ depending on the input sequence. For a random data sequence these variations will also be random. If the ISI is not very large, we can still decode the data and generate a sequence of ideal or desired levels A_j, where $A_j = \pm A$. From the ideal sequence A_j and the actual measured values $Y(jT_b)$, we can generate an estimate of the error sequence needed for adjusting the tap gains. The estimate* most commonly used is given by [Lucky, page 160]

$$\hat{\epsilon}_j = \frac{1}{A^2 m} \sum_{n=1}^{m} (A_{n-j})[Y(nT_b) - A_n] \tag{5.69}$$

where m is the sequence length used for estimation. The jth tap gain is adjusted according to $C_j^{k+1} = C_j^k - \Delta \operatorname{sgn}\{\hat{\epsilon}_j\}$; $j = -N$ to N, where k denotes the number of the iteration cycle.

In order for the adaptive equalizer to work effectively the input bit sequence and the sequence of received samples $Y(nT_b)$ must be random. Further, adaptive equalizers have a difficult time establishing initial equaliza-

*$\hat{\epsilon}_j$ is the maximum likelihood estimator of ϵ_j, the jth component of the error vector. The right-hand side of Equation (5.69) represents the cross correlation operation.

tion. Once correct equalization is acquired, the error estimates are accurate and the equalization loop tracks changes in channel characteristics easily. A procedure used often to circumvent this difficulty is to use a hybrid system in which data transmission is delayed during a brief reset period in which a quasi-random sequence is transmitted and regenerated at the receiver. When the equalization is reasonably good, the equalizer is switched to an adaptive mode and data transmission is started.

5.7 MISCELLANEOUS TOPICS

5.7.1 Eye Diagram

The performance of baseband PAM systems depends on the amount of ISI and channel noise. The distribution of ISI and channel noise in the system can be readily observed by displaying the received waveform $Y(t)$ on an oscilloscope using a sweep rate that is a fraction of the symbol rate r_s. The resulting display shape resembles a human eye and is widely known as the *eye pattern* of the system. To understand and interpret eye patterns let us consider a binary PAM system. The received waveform with no noise and no distortion is shown in Figure 5.20*a*. When segments of this waveform are superimposed on each other, the "open" eye pattern results. A vertical line drawn through the center of the eye pattern reveals that if the sampling time is correct, then all sampled values are $\pm A$.

Figure 5.20*b* shows a distorted version of the waveform and the corresponding eye pattern. The eye pattern appears "closed" and the sampled values are now distributed about $\pm A$. Decoding of the received waveform is somewhat difficult now. Finally, 5.20*c* shows a noisy distorted version of the received waveform and the corresponding eye pattern. Plots shown in these figures reveal that the eye pattern displays useful information about the performance of the system. For comparison purposes typical eye patterns of a duobinary signaling scheme are shown in Figure 5.21.

Eye patterns are often used for monitoring the performance of baseband PAM systems. If the signal-to-noise ratio at the receiver is high, then the following observations can be made from the eye pattern shown simplified in Figure 5.22:

1. The best time to sample the received waveform is when the eye opening is largest.
2. The maximum distortion is indicated by the vertical width of the two branches at sampling time.

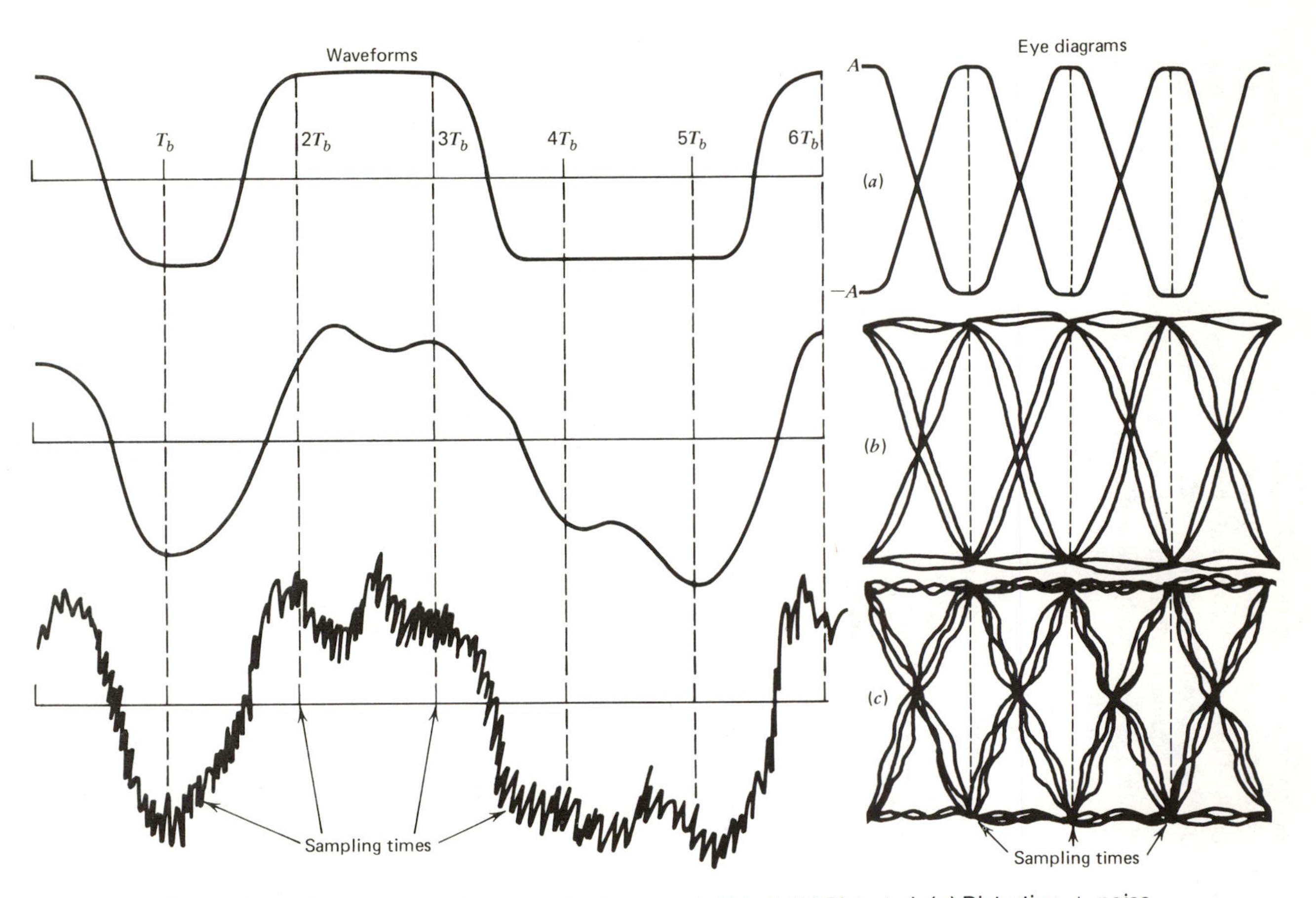

Figure 5.20 Eye diagrams of a binary PAM system. (*a*) Ideal. (*b*) Distorted. (*c*) Distortion + noise.

(*a*) Noise free

(*b*) Noisy

Figure 5.21 Eye pattern of a duobinary system.

3. The noise margin or immunity to noise is proportional to the width of the eye opening.
4. The sensitivity of the system to timing errors is revealed by the rate of closing of the eye as sampling time is varied.
5. The sampling time is midway between zero crossings; if the clock information is derived from zero crossings, then the amount of distortion of zero crossings indicates the amount of "*jitter*" or variations in clock rate and phase.

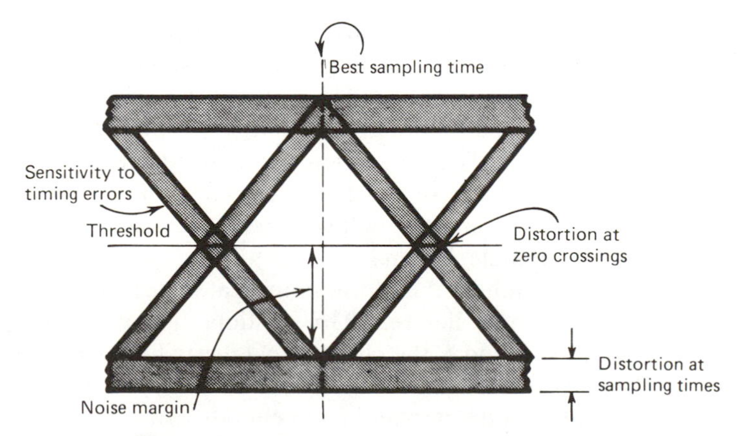

Figure 5.22 Characteristics of an eye pattern.

6. Asymmetries in the eye pattern indicate nonlinearities in the channel since in a strictly linear system with truly random data all the eye openings will be identical.

5.7.2 Synchronization

In baseband PAM systems the output of the receiving filter $Y(t)$ must be sampled at precise sampling instants $t_m = mT_s + t_d$. To do this sampling we need a clock signal at the receiver that is synchronized with the clock signal at the transmitter. Three general methods in which this synchronization can be obtained are:

1. derivation of clock information from a primary or secondary standard; for example, the transmitter and receiver can be slaved to a master clock;
2. transmitting a synchronizing clock signal;
3. derivation of the clock signal from the received waveform itself.

The first method is often employed in large data communication networks. On point-to-point data transmission at low rates this method is not necessary, and the high cost of this method does not justify its use. The second method involves the transmission of a clock signal along with data which means that a small part of the channel's information capacity needs to be given over to the clock signal. If the available capacity is large compared to the data rate requirements, then this method is most reliable and least expensive.

The third method, *self-synchronization*, is a very efficient method of synchronizing the receiver to the transmitter. An example of a system used to derive a clock signal from the received waveform is shown in Figure 5.23.

The clock recovery network consists of a voltage controlled oscillator (VCO) and a phase comparator consisting of the phase comparison logic and transistor controlled current switches. The phase comparison logic circuit is triggered by the one shot multivibrator that outputs a pulse of duration $T_b/2$ when the input $Y(t)$ is ≤ 0. The correction or error signal comes out of the phase comparator in the form of I_c. The charging and discharging of the capacitor is controlled by I_c and the voltage across the capacitor controls the VCO, which generates the clock signal.

To illustrate the operation of the phase comparator network, let us look at the timing diagram shown in Figure 5.23*b*. At time t_1, the clock signal is in phase and between time t_1 and t_2 the clock signal drifts by a small amount. As $Y(t)$ goes negative at t_1, the one shot is triggered and it puts out a pulse of duration $0.5T_b$. The phase comparison logic generates two equal width pulses QC and $Q\bar{C}$, and the current I_c has a waveform with equal width, equal

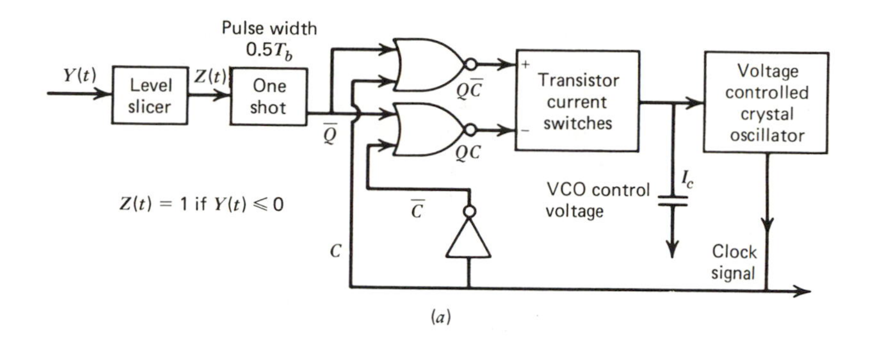

(a)

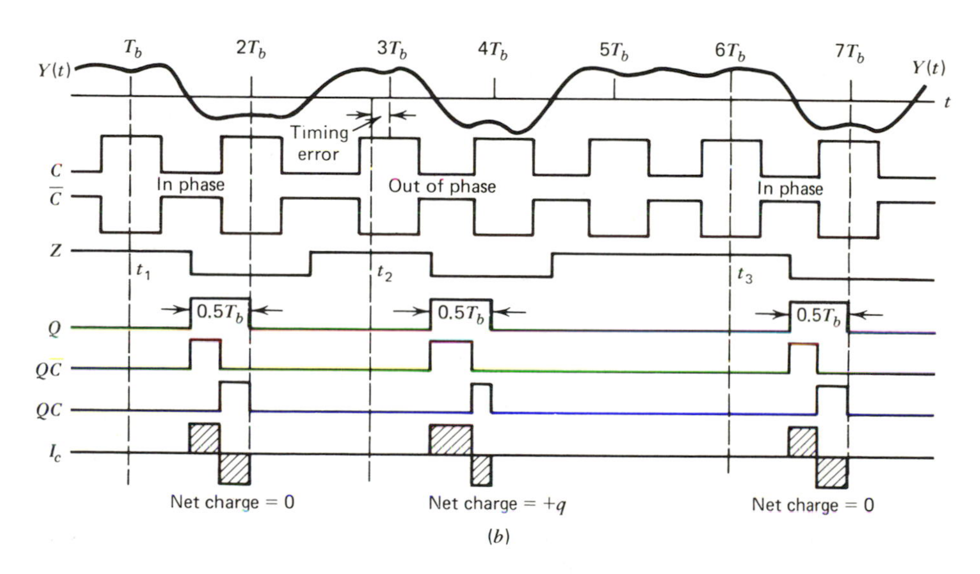

(b)

Figure 5.23 Clock recovery network.

amplitude positive and negative portions. The net change in the charge across the capacitor is zero, the VCO control voltage remains constant, and no adjustment is made on the rate and phase of the clock signal.

As the clock signal drifts out of phase, the phase comparison operation at time t_2 results in a current pulse I_c with a more positive component. Now, there is a change q in the capacitor charge and hence a change in the VCO control voltage. This change in the VCO control voltage results in a correction of the clock phase. In the example shown in Figure 5.23*b*, the clock phase is shown corrected before the next phase comparison operation is initiated at t_3.

Several versions of the clock recovery network shown in Figure 5.23 are used in practice. Almost all self-synchronizing networks depend on level changes or zero crossings in the received waveform $Y(t)$. The performance of these networks will degrade considerably if the signal stays at a constant level for long periods of time and if the zero crossings in $Y(t)$ are obliterated by noise. The lack of level changes in the data can be corrected by using one of the coding methods suggested in the preceding section. The effect of noise on the zero crossing can be minimized by setting an error threshold in the phase comparator output below which no clock phase correction is attempted. Another way of minimizing the effect of noise is to use an averaged estimate of the zero crossing times in the phase comparison network.

5.7.3 Scrambler and Unscrambler

Binary communication systems are designed to convey a sequence of bits from the source to the receiver. While the system is expected to convey all possible sequences, there may be some sequences that are not conveyed correctly. For example, the clock recovery in the system might be affected if a long series of 1's is sent and hence the system might start losing data due to timing errors. A binary communication system is said to have *bit sequence transparency* if it can convey *any* given sequence of bits. Several methods are used to preserve bit transparency and most of the methods involve an encoding procedure to restrict the occurrence of periodic sequences and sequences containing long strings of ones or zeros.

Scrambler. Many subsystems in data communication systems work best with random bit sequences. Examples of such subsystems are adaptive equalizers and self-synchronization networks. While strings of ones or zeros, or periodic sequences might appear in the output of an information source, such sequences have to be recoded for transmission if the data transmission system has difficulty in conveying these sequences. A device commonly used for recoding undesirable bit strings is called a scrambler. While it may not be possible to prevent the occurrence of all undesirable sequences with absolute certainty, at least most of the common repetitions in the input data can be removed by the use of a scrambler.

The scrambler shown in Figure 5.24*a* consists of a "feedback" shift register and the matching unscrambler has a "feed forward" shift register structure. In both the scrambler and unscrambler the outputs of several stages of shift register are added together, modulo-2, and then added to the data stream again in modulo-2 arithmetic. The contents of the registers are shifted at the bit rate of the system.

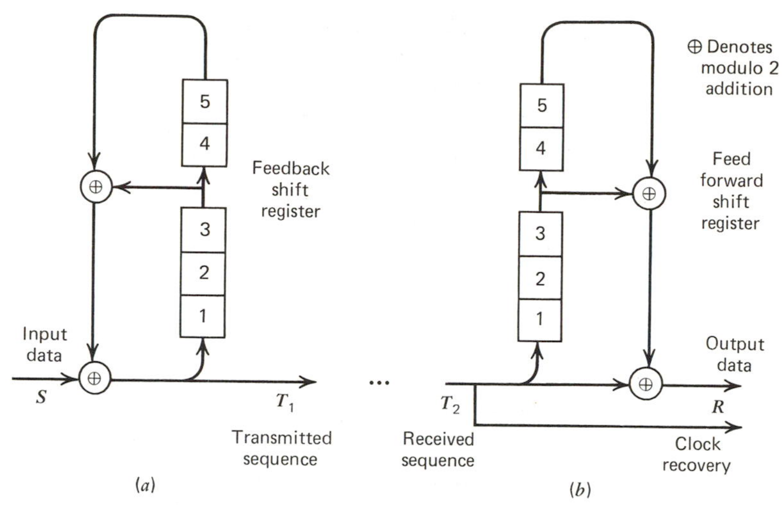

Figure 5.24 (*a*) Scrambler. (*b*) Unscrambler.

In order to analyze the operation of the scrambler and unscrambler, let us introduce an operator "D" to denote the effect of delaying a bit sequence by one bit. Thus DS represents a sequence S delayed by one bit and D^kS represents the sequence S delayed by k bits. Using the delay operator, we can establish the following relationships:

Starting with the unscrambler, we have $R = T_2 \oplus D^3T_2 \oplus D^5T_2 = (1 \oplus F)T_2$, where F stands for the operator $D^3 \oplus D^5$. In the scrambler, T_1 is operated on by $F = D^3 \oplus D^5$ and added to S. That is, $T_1 = S \oplus FT_1$ or

$$T_1 = \frac{S}{1 \oplus F} \tag{5.70}$$

where $F = D^3 \oplus D^5$ and division stands for inverse operator. In the absence of errors, we have $T_2 = T_1$ and hence the unscrambled output

$$\begin{aligned} R &= (1 \oplus F)T_2 \\ &= (1 \oplus F)T_1 = \frac{1 \oplus F}{1 \oplus F}S \\ &= S \end{aligned} \tag{5.71}$$

Thus the input sequence is exactly duplicated at the output of the unscrambler.

To illustrate the effect of the scrambler on periodic sequences and on long strings of ones or zeros, let us consider an input sequence shown in Table 5.3 and assume that the initial content of the register is zeros. This table

Table 5.3. Input and output bit streams of the scrambler shown in Figure 5.24

Input S	1 0 1 0 1 0 1 0 0 0 0 0 1 1
D^3T_1	0 0 0 1 0 1 1 1 0 0 0 1 1 0
D^5T_1	0 0 0 0 0 1 0 1 1 1 0 0 0 1
Output T_1	1 0 1 1 1 0 0 0 1 1 0 1 0 0

illustrates that the scrambler can effectively remove periodic sequences and long strings of zeros by scrambling the data input. The scrambler in general produces a pseudo-random sequence given zeros as a data input, assuming it starts from a nonzero state. With an appropriate choice of taps, an n-bit shift register scrambler can be made to produce a sequence of $2^n - 1$ bits before it repeats itself. The design of the feedback and feed forward registers used in scramblers and unscramblers is rather involved. The interested reader can find a good treatment on the analysis and synthesis of these devices in texts dealing with algebraic coding schemes (see, for example, Peterson's book on coding).

Introduction of scramblers affects the error performance of the communication system in that a single channel error may cause multiple errors at the output of the unscrambler. This is due to the propagation of the error bit in the shift register at the unscrambler. Fortunately, the error propagation effect lasts over only a finite, often small, number of bits. In the scrambler/unscrambler shown in Figure 5.24, each isolated error bit causes three errors in the final output. It must also be pointed out that some random bit patterns might be scrambled to all zeros or all ones.

5.8 SUMMARY

In this chapter we developed procedures for designing and analyzing baseband PAM data transmission systems. The main objectives of the design were to eliminate intersymbol interference and minimize the effects of noise. Several methods of data transmission using PAM techniques were considered. The performance of baseband PAM systems was compared in terms of power-bandwidth requirements for a given data rate and error rate. Methods of shaping the transmitted signal and its spectrum were discussed. The problems of equalization and clock recovery were considered and methods of automatic equalization and synchronization were presented.

REFERENCES

A detailed and thorough treatment of several aspects of baseband system design may be found in the book by Lucky et al. (1968). This book is written for advanced graduate students and the average reader may find it hard to read. Practical aspects of baseband data transmission are dealt with, rather nicely, in the book by Davies and Barber (1975). Introductory level treatment of baseband PAM systems may be found in many undergraduate texts [Bennet and Davey (1965), Carlson (1975), and Ziemer and Tranter (1976)]. Carlson's book contains an easily readable treatment of PAM systems with several examples.

1. R. W. Lucky et al. *Principles of Data Communication.* McGraw-Hill, New York (1968), Chapter 4.
2. D. W. Davies and D. L. A. Barber. *Communication Networks for Computers.* Wiley, New York (1975), Chapter 5.
3. W. R. Bennet and J. R. Davey. *Data Transmission.* McGraw-Hill, New York (1965).
4. A. B. Carlson. *Communication Systems.* McGraw-Hill, New York (1975).
5. R. E. Ziemer and W. H. Tranter. *Principles of Communications.* Houghton Mifflin, Boston (1976).
6. W. W. Peterson. *Error Correcting Codes.* MIT Press, Cambridge, MA (1961).

PROBLEMS

Section 5.1

5.1. A baseband binary PAM system uses a signaling waveform

$$p_g(t) = \frac{\sin \pi r_b t}{\pi r_b t}$$

to transmit binary data at a bit rate of r_b bits/sec. The amplitude levels at the pulse generator output are $+1$ volt or -1 volt, $+1$ if the input bit is 1 and -1 if the input bit is 0. Sketch the waveform of the pulse generator output when the input bit string is 0 0 1 0 1 1 0.

5.2. Suppose that the received pulse in a baseband binary PAM system is given by

$$p_r(t) = \frac{\sin \pi r_b t}{\pi r_b t}$$

with amplitude levels ± 1 mvolt. The received waveform is sampled at $t_k = kT_b + (T_b/10)$ $(k = 0, \pm 1, \ldots, \pm M)$, that is, there is a timing error of one tenth of a bit duration. Assuming that the input to the system is a sequence of $2M+1$ bits of alternating 0's and 1's, find the value of the ISI term at t_0.

5.3. Suppose that the received pulse in a baseband binary system has the shape shown in Figure 5.25. Consider the input to the A/D converter

$$Y(t) = \sum_k A_k p_r(t - t_d - kT_b), \quad A_k = \pm 1$$

Assuming $t_d = T_b/2$ and $\tau = 2T_b$, find the value of the ISI term when the input bit sequence is a long string of alternating zeros and ones.

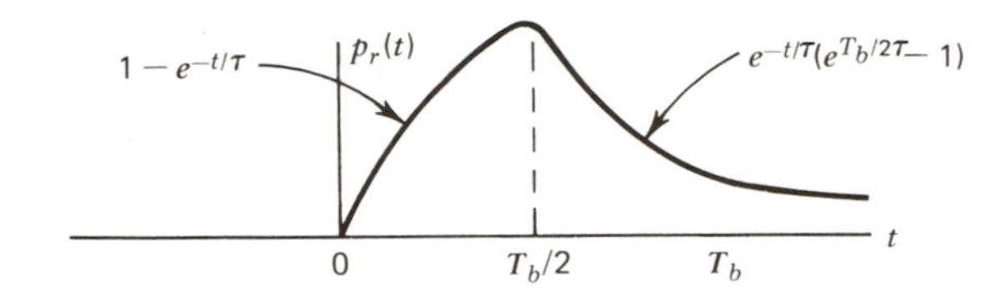

Figure 5.25 $p_r(t)$ for Problem 5.3.

Sections 5.2 *and* 5.3

5.4. We want to select a $P_r(f)$ for transmitting binary data at a rate of r_b bits/sec. Which one of the three shown in Figure 5.26 would you choose? Give the reasons for your choice.

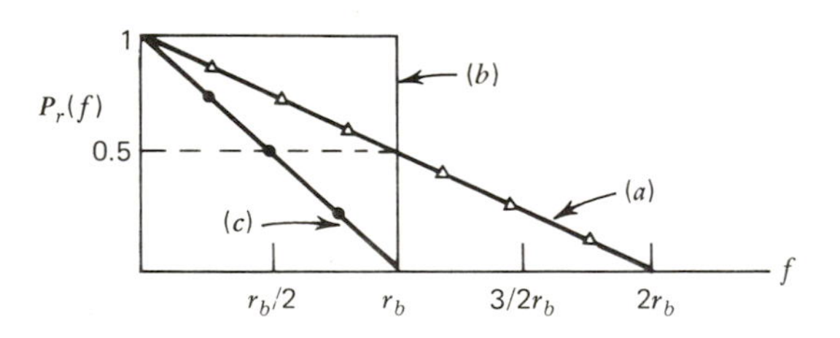

Figure 5.26 $P_r(f)$ for Problem 5.4.

5.5. Derive the result stated in the footnote on page 198.

5.6. In a binary PAM system, the sampled value of the received waveform Y has the following probability density functions depending on the input bit:

$$f_{Y|1\text{ sent}}(y) = \frac{1}{\sqrt{2\pi}} \exp(-(y-1)^2/2), \quad -\infty < y < \infty$$

$$f_{y|0\text{ sent}}(y) = \frac{1}{\sqrt{2\pi}} \exp(-(y+1)^2/2), \quad -\infty < y < \infty$$

$$P(1\text{ sent}) = p, \quad P(0\text{ sent}) = 1 - p$$

The receiver compares Y against a threshold value T and outputs a 1 if $Y > T$ and a 0 if $Y \leq T$.

(a) Derive an expression for the threshold T that minimizes the probability of incorrectly decoding a bit. Find T for $p = 0.2, 0.4, 0.5, 0.6,$ and 0.8.

(b) Calculate the probability of error for each of the above values of p.

5.7. Calculate P_e for the system of Problem 5.6 with $T = 0$ for $p = 0.2, 0.4, 0.5, 0.6,$ and 0.8 and compare with the results obtained in that problem.

5.8. The sampled values of the received waveform in a binary PAM system suffer from ISI such that

$$Y(t_m) = \begin{cases} A_m + n(t_m) + Q & \text{when the input bit} = 1 \\ A_m + n(t_m) - Q & \text{when the input bit} = 0 \end{cases}$$

where Q is the ISI term. The ISI term has one of three values with the following probabilities:

$$P(Q = +q) = \tfrac{1}{4}$$
$$P(Q = 0) = \tfrac{1}{2}$$
$$P(Q = -q) = \tfrac{1}{4}$$

(a) Assume that $n(t_m)$ is a Gaussian random variable with a variance of σ^2 and that $A_m = +A$ or $-A$ depending on whether the transmitted bit is 1 or 0. Derive an expression for the probability of error in terms of A, σ, and q.

(b) Find P_e for $A/\sigma = 3.0$, and $q/A = 0.1$ and 0.25. How much does the ISI affect the probability of error in each case?

5.9. Design a binary baseband PAM system to transmit data at a rate of 9600 bits/sec with a bit error probability $P_e < 10^{-5}$. The channel available is an ideal lowpass channel with a bandwidth of 9600 Hz. The noise can be assumed to be white, Gaussian with a two-sided psd $\eta/2 = 10^{-13}$ watt/Hz. Sketch the shape of $|H_T(f)|$, $|H_R(f)|$, $|P_g(f)|$, and find the transmitter power requirements.

5.10. Repeat Problem 5.9 with

$$H_c(f) = \frac{1}{1 + j(f/f_c)}, \quad f_c = 4800\text{ Hz}$$

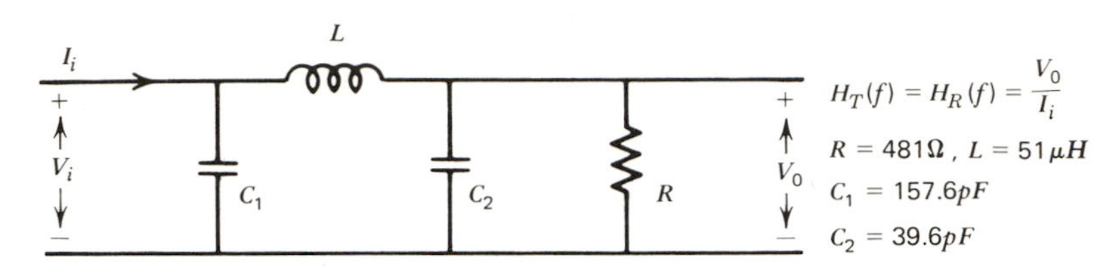

Figure 5.27 Filter network for Problem 5.11.

5.11. The filter shown in Figure 5.27 is used for both the transmitting and receiving filter in a binary PAM system. The channel is ideal lowpass with additive Gaussian noise.

(a) Assume that the bit rate at the input is $(6.28)(10^6)$ bits/sec and $p_g(t)$ is a rectangular pulse with a width $= T_b$. Find $p_r(t)$ and sketch it.

(b) Is there any ISI in the received waveform? Can you use this filter for binary data transmission at all?

5.12. The following $P_r(f)$ are used in binary data transmission with controlled ISI:

(a) $4T_b \cos^2 \pi f T_b \begin{cases} & \text{for } |f| \leq \frac{1}{2T_b} \\ 0 & \text{elsewhere} \end{cases}$

(b) $2T_b \sin 2\pi f T_b \begin{cases} & \text{for } |f| \leq \frac{1}{2T_b} \\ 0 & \text{elsewhere} \end{cases}$

(c) $4T_b \sin^2(2\pi f T_b) \begin{cases} & \text{for } |f| \leq \frac{1}{2T_b} \\ 0 & \text{elsewhere} \end{cases}$

For each of the above cases, find $p_r(t)$, and the number of received levels.

5.13. A source emits one of three equiprobable symbols in an independent sequence at a symbol rate of 1000/sec. Design a three-level PAM system to transmit the output of this source over an ideal lowpass channel with additive Gaussian noise having a psd of $\eta/2 = 10^{-14}$ watt/Hz. The symbol error probability has to be maintained at or below 10^{-6}. Specify the power, bandwidth requirements, and $H_T(f)$, $H_R(f)$, and $p_g(t)$.

5.14. The received waveform in a three-level system has the values -1 volt, 0 and $+1$ volt in the absence of noise. The probabilities of these levels are $\frac{1}{4}$, $\frac{1}{2}$, and $\frac{1}{4}$, respectively. The additive noise in the system is Gaussian with a standard deviation of $\frac{1}{4}$ volt.

(a) Find the optimum threshold settings for decoding the levels.

(b) Find the probability of error P_e for the optimum decoding scheme.

5.15. Design a PAM system to transmit the output of a source emitting an equiprobable, independent bit stream at a rate of 10,000 bits/sec over an ideal lowpass channel of width 5000 Hz and additive Gaussian noise with a psd $= 10^{-12}$ watt/Hz. P_e has to be maintained at or below 10^{-4}.

5.16. Calculate the capacity of the discrete channel discussed in Problem 5.15 if the bit transition rate is limited to 10,000/sec.

Section 5.5

5.17. Verify the spectral component $G(f)$ for the twinned binary coding scheme given in Equation (5.60).

5.18. A baseband binary communication system uses a received pulse with a spectrum $P_r(f)$ given by

$$P_r(f) = \begin{cases} T_b \cos^2(\pi f/2r_b), & |f| < r_b \\ 0, & \text{elsewhere} \end{cases}$$

The channel and noise characteristics dictate a $H_T(f) = \sqrt{[P_r(f)]}$. The system uses a bipolar coding scheme for the pulse amplitudes with

$$p_g(t) = \begin{cases} 1 & \text{for } |t| < T_b/20 \\ 0 & \text{elsewhere} \end{cases}$$

Assuming an independent stream of equiprobable bits at the input to the pulse generator, compute the power spectral density of the transmitting filter output.

Section 5.6

5.19. The unequalized pulse in a PAM system has the following values at sampling times:

$$p_r(kT_b) = p_r(k) = \begin{cases} 0.2, & k = 1 \\ 0.8, & k = 0 \\ 0.2, & k = -1 \end{cases}$$

$$p_r(k) = 0 \quad \text{for } |k| > 1$$

(a) Design a three-tap zero forcing equalizer so that the equalizer output is 1 at $k = 0$ and 0 at $k = \pm 1$.
(b) Calculate $p_{eq}(k)$ for $k = \pm 2, \pm 3$.

5.20. Would a five-tap equalizer for Problem 5.19 yield $p_{eq}(k) = 1$ for $k = 0$ and $p_{eq}(k) = 0$ for $k \neq 0$?

5.21. A baseband binary PAM system was designed with the assumption that the channel behaved like an ideal lowpass filter with a bandwidth $B = r_b$ Hz. The channel noise was assumed to be white and the pulse spectrum was chosen to be

$$P_r(f) = \begin{cases} T_b \cos^2(\pi f/2r_b), & \text{for } |f| < r_b \\ 0, & \text{elsewhere} \end{cases}$$

(a) Calculate the design value of $(A^2/N_0)_{max}$.

(b) Suppose that the channel response turned out to be

$$H_c(f) = 1/(1 + jfT_b)$$

and an equalizing filter with $H_{eq}(f) = 1/H_c(f)$ was used at the receiver. Calculate the (A^2/N_0) assuming that the transmitting and receiving filters are the same as before. (Note: The signal power in both (a) and (b) will be the same but the equalizer will change the value of the noise power.)

(c) By what factor should the transmitter power be increased to maintain the same error probability?

5.22. A four-level PAM signaling scheme is used to transmit data over an ideal lowpass channel having a bandwidth B. The additive channel noise is zero mean, Gaussian with a power spectral density of $\eta/2$, and the signal-to-noise ratio at the output of the channel is S/N.

(a) Plot C/B versus S/N (in dB) for this channel (C is the channel capacity).

(b) Develop the discrete channel model for the four-level PAM scheme. Find the rate of information transmission D_t over the discrete channel and sketch D_t/B. Compare your results with the plot shown in Figure 4.14 (Chapter 4). (Assume that the input symbols are equiprobable and occur in independent sequences.)

5.23. Repeat Problem 5.22(b) for $M = 8$, 16, and 32.

6

ANALOG SIGNAL TRANSMISSION

6.1 INTRODUCTION

In Chapter 5 we were primarily concerned with the transmission of messages that consisted of sequences of symbols. Each symbol was chosen from a source alphabet consisting of a finite number of symbols. Corresponding to each symbol a particular electrical waveform was transmitted over the channel. Thus messages were represented by sequences of waveforms, each of which was selected from a finite set of known waveforms. The receiver had the simple task of detecting which one of the finite number of known waveforms was transmitted during each symbol interval.

In contrast, we will now be concerned with the transmission of messages that are continuous (or analog) signals. Each message waveform is chosen from an uncountably infinite number of possible waveforms. For example, in radio or television broadcasting we have an uncountably infinite number of possible messages and the corresponding waveforms are not all known. Such a collection of messages and waveforms can be conveniently modeled by continuous random processes wherein each member function of the random process corresponds to a message waveform. We will use the random signal model in Chapter 7 when we discuss the effects of random noise in analog communication systems. For purposes of analysis, let us define analog signal transmission as the transmission of an arbitrary finite energy lowpass signal $x(t)$ over a given channel. In some cases we will let $x(t)$ to be a single (sinusoidal) tone or a power signal.

If the channel is lowpass in nature, the lowpass signal waveform may be transmitted over the channel without modulation. Such signal transmission is referred to as *baseband communication.* The majority of practical channels have bandpass characteristics and modulation is necessary for translating the lowpass signal spectrum to match the bandpass channel characteristics. While we will take a look at both baseband systems and analog modulation systems in this chapter, our emphasis will be on analog modulation systems.

6.1.1 Need for Modulation

Before we start a quantitative discussion and analysis of modulation systems, let us look at the advantages of using modulated signals for information transmission. We have already mentioned that modulation is required to match the signal to the channel. However, this match involves several important aspects that deserve further amplification.

Modulation for Ease of Radiation. If the communication channel consists of free space, then antennas are needed to radiate and receive the signal. Efficient electromagnetic radiation requires antennas whose dimensions are the same order of magnitude as the wavelength of the signal being radiated. Many signals, including audio signals, have frequency components down to 100 Hz or lower. For these signals, antennas some 300 km long will be necessary if the signal is radiated directly. If modulation is used to impress the message signal on a high-frequency carrier, say at 100 MHz, then antennas need be no more than a meter or so across.

Modulation for Multiplexing. If more than one signal utilizes a single channel, modulation may be used to translate different signals to different spectral locations thus enabling the receiver to select the desired signal. Applications of multiplexing include data telemetry, FM stereophonic broadcasting, and long distance telephone.

Modulation to Overcome Equipment Limitations. The performance of signal processing devices such as filters and amplifiers, and the ease with which these devices can be built depend on the signal location in the frequency domain and on the ratio of the highest to lowest signal frequencies. Modulation can be used for translating the signal to a location in the frequency domain where design requirements are easily met. Modulation can also be used to convert a "wideband signal" (a signal for which the ratio of highest to lowest signal frequencies is large) to a "narrow band" signal.

Occasionally in signal processing applications the frequency range of the signal to be processed and the frequency range of the processing apparatus may not match. If the processing apparatus is elaborate and complex, it may be wise to leave the processing equipment to operate in some fixed frequency range and, instead, translate the frequency range of the signal to correspond to this fixed frequency range of the equipment. Modulation can be used to accomplish this frequency translation.

Modulation for Frequency Assignment. Modulation allows several radio or television stations to broadcast simultaneously at different carrier frequencies and allows different receivers to be "tuned" to select different stations.

Modulation to Reduce Noise and Interference. The effect of noise and interference cannot be completely eliminated in a communication system. However, it is possible to minimize their effects by using certain types of modulation schemes. These schemes generally require a transmission bandwidth much larger than the bandwidth of the message signal. Thus bandwidth is traded for noise reduction—an important aspect of communication system design.

6.1.2 Types of Analog CW Modulation

In Chapter 5 we treated discrete pulse modulation. In this chapter, we will consider continuous carrier wave (CW) modulation in which a parameter (such as the amplitude or phase) of a high-frequency carrier is varied in proportion to the low-frequency message signal. The carrier is usually sinusoidal in shape and the modulated carrier has the mathematical form

$$x_c(t) = A(t)\cos[\omega_c t + \phi(t)]\,, \omega_c = 2\pi f_c \tag{6.1}$$

where $A(t)$ is the instantaneous *amplitude* of the carrier, f_c is the *carrier frequency*, and $\phi(t)$ is instantaneous *phase deviation* of the carrier.

If $A(t)$ is linearly related to the message signal $x(t)$, then we have *linear modulation.* If $\phi(t)$ or the time derivative of $\phi(t)$ is linearly related to $x(t)$, then we have *phase* and *frequency modulation*, respectively. The common name *angle modulation* is used to denote both phase and frequency modulation. A substantial portion of this chapter is devoted to the analysis of linear and angle modulation schemes.

While modulation is the process of transferring information to a carrier, the reverse operation of extracting the information bearing signal from the modulated carrier is referred to as demodulation. For various types of modulation schemes we will consider different methods of demodulation and we will assume that demodulation is done in the absence of noise. The effect of noise on the output signal quality of various modulated transmission methods will be the subject of discussion in the following chapter.

In our analysis of CW modulation schemes, we will pay particular attention to three important parameters. These are the transmitted power, the transmission bandwidth, and complexity of the modulating and demodulating equipment. These parameters, along with the output signal quality in the presence of noise, will provide the basis for comparing various modulation schemes.

Finally, we will discuss how modulation can be used for simultaneously transmitting several signals over a single channel using examples from stereo FM broadcasting, simultaneous sound and picture transmission in TV broadcasting, and multiplexed long distance telephone systems.

6.2 ANALOG BASEBAND SIGNAL TRANSMISSION

Communication systems in which signal transmission takes place without modulation are called *baseband systems*. The functional elements of a baseband communication system are shown in Figure 6.1. The transmitter and receiver amplify signal power and perform appropriate filtering operations. No modulation and demodulation operations are performed in the system. Noise, and signal distortion due to non-ideal characteristics of the channel cause the output signal $y(t)$ to be different from the input signal $x(t)$. We will now identify different types of distortion, their causes, and possible cures. The effects of noise on signal quality, and optimum transmitter and receiver design that minimizes the effects of noise will be discussed in the next chapter.

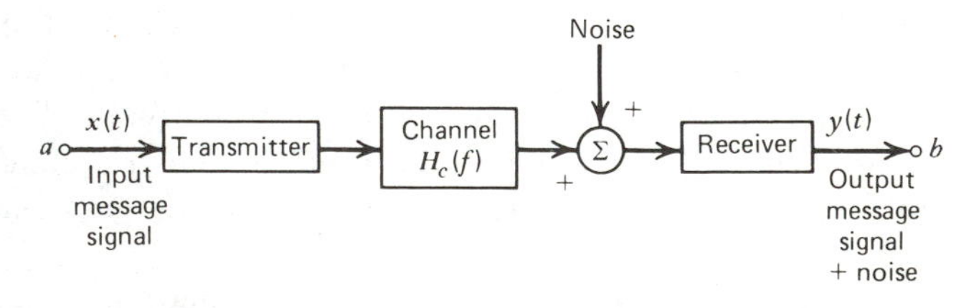

Figure 6.1 A baseband communication system.

6.2.1 Signal Distortion in Baseband Transmission

The output signal $y(t)$ is said to be undistorted if it "looks like" the input signal $x(t)$. More specifically, if $y(t)$ differs from $x(t)$ by a constant of proportionality and a finite time delay, then the transmission is said to be *distortionless*. That is,

$$y(t) = Kx(t - t_d) \tag{6.2}$$

for distortionless transmission. The constant K is the attenuation and t_d is the time delay. The power loss in transmission is $20 \log_{10} K$ and typical values of transmission loss for several transmission media are given in Table 6.1.

The requirement for distortionless transmission stated in Equation (6.2) can be met if the overall transfer function of the system between points a and b in Figure 6.1 is

$$H(f) = K \exp(-j2\pi f t_d) \quad \text{for } |f| < f_x \tag{6.3}$$

where f_x is the bandwidth of the baseband signal. If we assume the transmitter and receiver cause no signal distortion, then the channel response has to satisfy

$$H_c(f) = K \exp(-j2\pi f t_d) \quad \text{for } |f| < f_x \tag{6.4}$$

for distortionless transmission.

The condition stated in Equation (6.4) is very stringent and, at best, real channels can only approximately satisfy this condition. Thus some distortion will always occur in signal transmission even though it can be minimized by proper design. A convenient approach for minimizing signal distortion is to identify various types of distortion and attempt to minimize their bad effects separately.

Three common types of distortion encountered in a channel are:

1. Amplitude distortion due to $|H_c(f)| \neq K$.

Table 6.1. Typical values of transmission loss.

Transmission medium	Typical frequencies	Power loss (dB/km)
Twisted pair wire	1 kHz	0.05
(16 gauge)	100 kHz	3
Coaxial cable	100 kHz	1
(1 cm diameter)	3 MHz	4
Helical wave guide	100 GHz	1.5
Low loss optical fiber	$10^{14} - 10^{16}$ MHz	2–3

2. Phase (or delay) distortion due to

$$\text{angle}\{H_c(f)\} \neq -2\pi f t_d \pm m\pi \quad (m \text{ is an integer} > 0)$$

3. Nonlinear distortion due to nonlinear elements present in the channel.

The first two categories are called *linear distortion* and the third one is called *nonlinear distortion.* We now examine these individually.

6.2.2 Linear Distortion

If the amplitude response of the channel is not flat over the range of frequencies for which the input spectrum is nonzero, then different spectral components of the input signal are modified differently. The result is *amplitude distortion.* The most common forms of amplitude distortion are excess attenuation, or enhancement of high or low frequencies in the signal spectrum. Experimental results indicate that if $|H_c(f)|$ is constant to within ± 1 dB in the message band, then the amplitude distortion will be negligible. Beyond such qualitative observations, little more can be said about amplitude distortion without detailed analysis.

If the phase shift is arbitrary, various components of the input signal suffer different amounts of time delay resulting in *phase or delay distortion.* A spectral component of the input at frequency f suffers a delay $t_d(f)$,

$$t_d(f) = -\frac{\text{angle of }\{H(f)\}}{2\pi f} \tag{6.5}$$

The reader can verify that angle of $\{H(f)\} = -2\pi t_d f \pm m\pi$ will result in $y(t) = \pm x(t - t_d)$, that is, no distortion. Any other phase response including a constant phase shift θ, $\theta \neq \pm m\pi$, will result in distortion.

Delay distortion is a critical problem in pulse (data) transmission. However, the human ear is surprisingly insensitive to delay distortion and hence delay distortion is seldom of concern in audio transmission.

6.2.3 Equalization

The theoretical cure for linear distortion is equalization shown in Figure 6.2. If the transfer function of the equalizer satisfies

$$H_{eq}(f) = \frac{K \exp(-j2\pi f t_d)}{H_c(f)} \quad \text{for } |f| < f_x \tag{6.6}$$

we then have $H_c(f)H_{eq}(f) = K \exp(-j2\pi f t_d)$ and no distortion will be encoun-

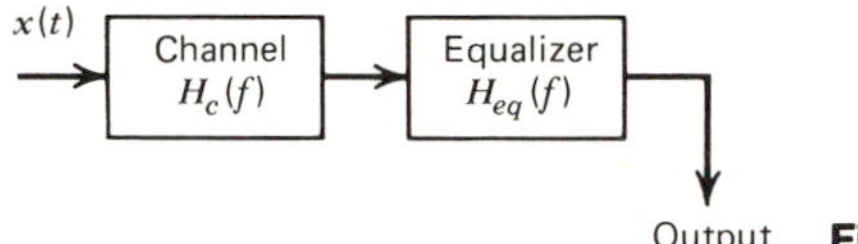

Figure 6.2 Channel equalizer.

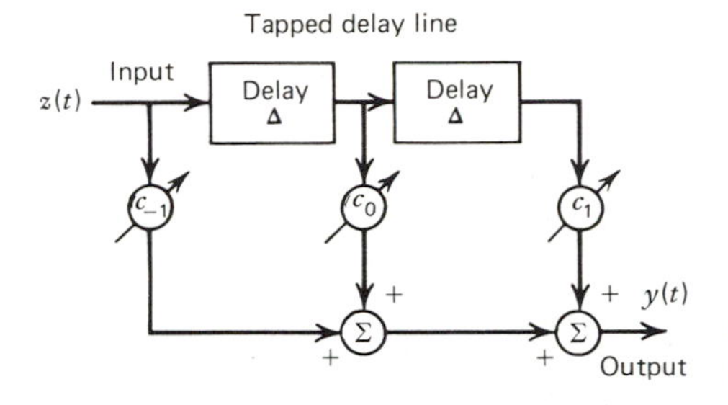

Figure 6.3 A three tap delay line equalizer (transversal filter).

tered. However, it is very rare that an equalizer can be designed to satisfy Equation (6.6) exactly. But excellent approximations are possible, especially with a tapped-delay line equalizer shown in Figure 6.3.

The output of the equalizer shown in Figure 6.3 can be written as

$$y(t) = c_{-1}z(t) + c_0 z(t-\Delta) + c_1 z(t-2\Delta)$$

from which we obtain the transfer function of the filter as

$$H_{eq}(f) = c_{-1} + c_0 \exp(-j\omega\Delta) + c_1 \exp(-j\omega 2\Delta), \quad \omega = 2\pi f$$

Generalizing this to an equalizer with $2M + 1$ taps, we have

$$H_{eq}(f) = \exp(-j\omega M\Delta)\left(\sum_{m=-M}^{M} c_m \exp(-j\omega m\Delta)\right) \tag{6.7}$$

which is in the form of an exponential Fourier series with periodicity $1/\Delta$. Hence, if the channel is to be equalized over the message band f_x, we can approximate the right-hand side of Equation (6.6) by a Fourier series (in the frequency domain) with periodicity $1/\Delta \geq 2f_x$. If the Fourier series approximation has $2M + 1$ terms, then an equalizer with $2M + 1$ taps is needed. The tap gains of the equalizer are equal to the Fourier series coefficients!

6.2.4 Nonlinear Distortion and Companding

Practical channels and electronic devices such as amplifiers often exhibit nonlinear transfer characteristics that result in nonlinear signal distortion. An example of the transfer characteristic of a memoryless nonlinear element is

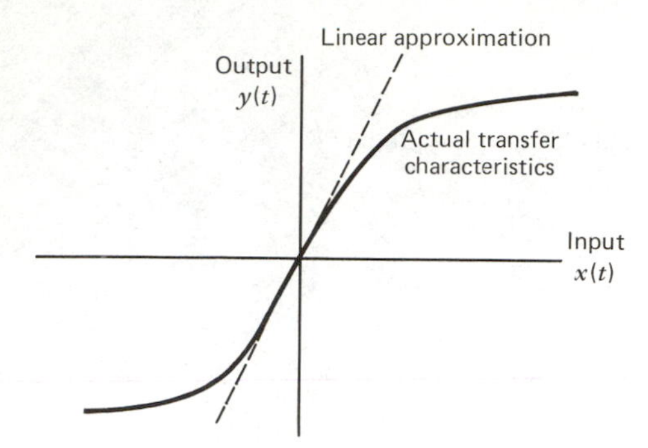

Figure 6.4 Transfer characteristic of a nonlinear device.

shown in Figure 6.4. Such devices, in general, act linearly when the input $x(t)$ is small, but distort the signal when the input amplitude is large.

To investigate the nature of nonlinear signal distortion let us assume that the transfer characteristic of the nonlinear device can be modeled by

$$y(t) = a_1x(t) + a_2x^2(t) + a_3x^3(t) + \cdots \tag{6.8}$$

Now, if the input is the sum of two cosine waves, say $\cos 2\pi f_1 t + \cos 2\pi f_2 t$, then the output will contain *harmonic distortion* terms at frequencies $2f_1$, $2f_2$ and *intermodulation distortion* terms at frequencies $f_1 \pm f_2$, $2f_2 \pm f_1$, $2f_1 \pm f_2$, and so forth. In a general case, if $x(t) = x_1(t) + x_2(t)$, then $y(t)$ will contain the terms $x_1^2(t)$, $x_2^2(t)$, $x_1(t)x_2(t)$, and so forth. It is easy to see in the frequency domain (see Problem 6.8) that even though $X_1(f)$ and $X_2(f)$ may be separated in frequency, the spectrum of $x_1(t)x_2(t)$ (obtained from $X_1(f) * X_2(f)$) may overlap $X_1(f)$ or $X_2(f)$ or both. This form of intermodulation distortion (or cross talk) is of important concern in systems where a number of different signals are multiplexed and transmitted over the same channel.

The transfer characteristic shown in Figure 6.4 suggests that one cure for minimizing nonlinear distortion is to keep the signal amplitude within the linear operating range of the transfer characteristic. This is usually accomplished by using two nonlinear devices, a compressor and expander as shown in Figure 6.5.

A compressor essentially reduces the amplitude range of an input signal so that it falls within the linear range of the channel. For a positive-valued signal

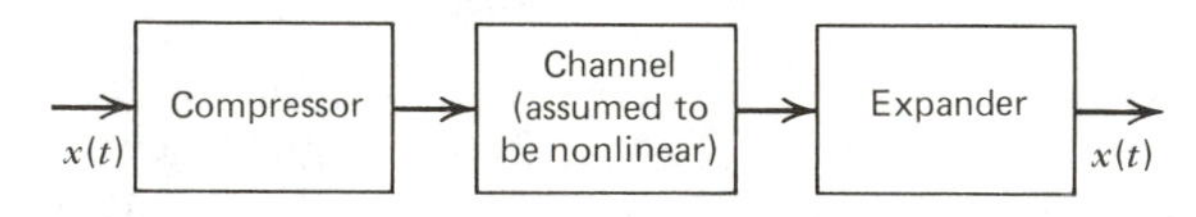

Figure 6.5 Companding.

$x(t)$, for example, we can use a compressor with a transfer characteristic $g_{comp}[x(t)] = \log_e[x(t)]$. Since a compressor reduces the range of the input signal, it also reduces the range of the output signal. The output signal is expanded to the appropriate level by the expander that operates on the output of the channel. Ideally, an expander has a transfer characteristic g_{exp} that yields $g_{exp}\{g_{comp}[x(t)]\} = x(t)$. For example, if $g_{comp}[x(t)] = \log_e[x(t)]$, then $g_{exp}[y(t)] = \exp[y(t)]$ will yield $g_{exp}\{g_{comp}[x(t)]\} = x(t)$. The combined operation of compressing and expanding is called *companding*. Companding is widely used in telephone systems to reduce nonlinear distortion and also to compensate for signal level difference between loud and soft talkers.

6.3 LINEAR CW MODULATION SCHEMES

Linear modulation refers to the direct frequency translation of the message spectrum using a sinusoidal carrier. The modulated carrier is represented by

$$x_c(t) = A(t) \cos \omega_c t \tag{6.9}$$

in which the carrier amplitude $A(t)$ is linearly related to the message signal $x(t)$. Depending on the nature of the spectral (frequency domain) relationship between $x(t)$ and $A(t)$, we have the following types of linear modulation schemes: double-sideband modulation (DSB), amplitude modulation (AM), single-sideband modulation (SSB), and vestigial-sideband modulation (VSB). Each of these schemes has its own distinct advantages, disadvantages, and practical applications. We will now look at these different types of linear modulation schemes emphasizing such matters as signal spectra, power and bandwidth, demodulation methods, and the complexity of transmitters and receivers.

In our discussion of linear modulation schemes, we will use one of three different models for the message signal $x(t)$: a single tone of frequency $\leq f_x$, a combination of tones restricted in frequency to $\leq f_x$, or an arbitrary finite energy lowpass signal with a Fourier transform $X(f)$ which is identically equal to zero for $|f| > f_x$.

6.3.1 Double-Sideband Modulation (DSB)

An easy way to translate the spectrum of a lowpass message signal $x(t)$ is to multiply it with a carrier waveform $A_c \cos \omega_c t$, as shown in Figure 6.6*a*. The modulated waveform $x_c(t)$

$$x_c(t) = A_c x(t) \cos \omega_c t = A(t) \cos \omega_c t, \quad \omega_c = 2\pi f_c \tag{6.10}$$

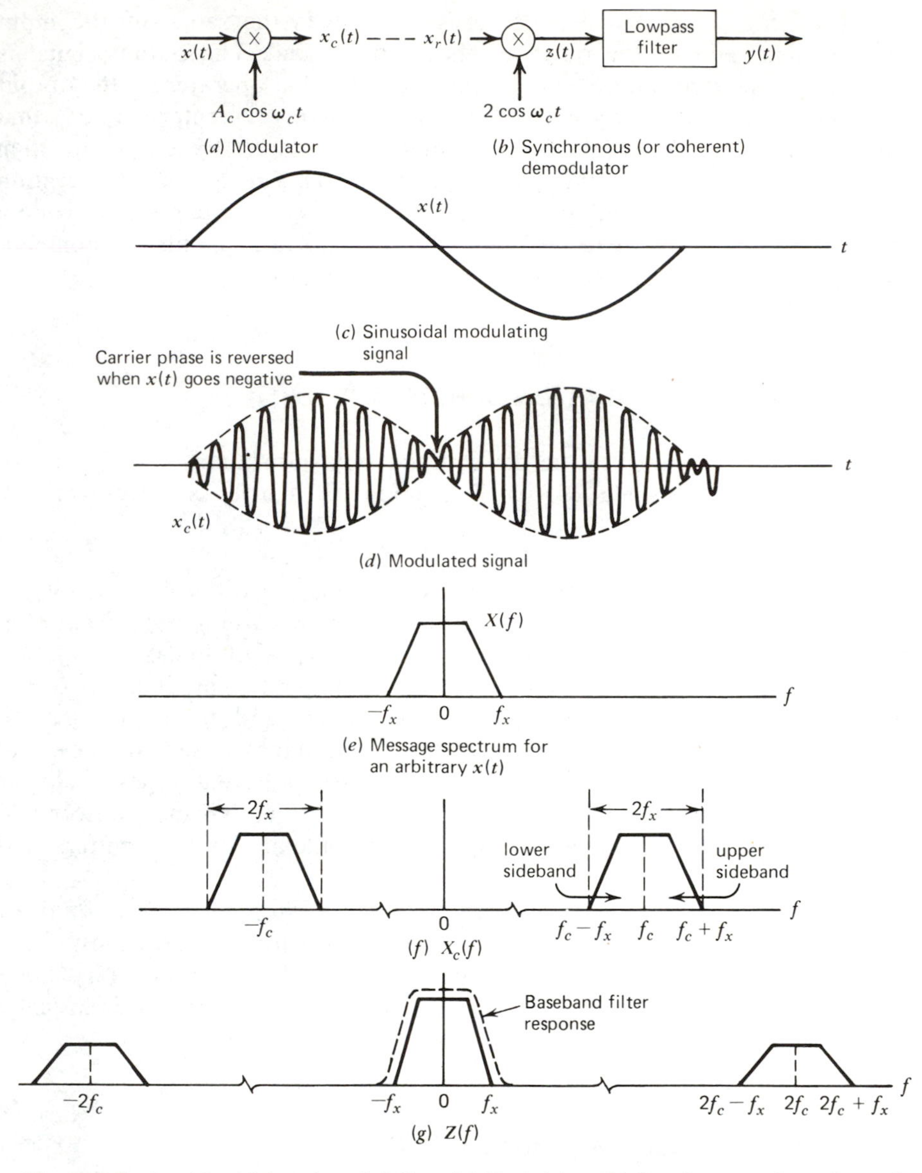

Figure 6.6 Double sideband modulation. (*a*) Modulator. (*b*) Synchronous (or coherent) demodulator. (*c*) Sinusoidal modulating signal. (*d*) Modulated signal. (*e*) Message spectrum for an arbitrary $x(t)$. (*f*) $X_c(f)$. (*g*) $Z(f)$.

is called the *double-sideband modulated signal.* Equation (6.10) reveals that the instantaneous amplitude of the carrier $A(t)$ is proportional to the message signal $x(t)$. A time domain example of the modulated waveform $x_c(t)$ is shown in Figure 6.6*d* for a sinusoidal message signal.

It follows from the modulation theorem (see section 2.3.3) that the spectrum of the DSB signal given in Equation (6.10) is

$$X_c(f) = \tfrac{1}{2}A_c[X(f+f_c) + X(f-f_c)] \tag{6.11}$$

where $f_c = \omega_c/2\pi$. The frequency domain representation of $X(f)$ and $X_c(f)$ are shown in Figures 6.6*e* and 6.6*f* for a lowpass message signal.

The spectral range occupied by the message signal is called the *baseband frequency range* and the message signal is usually referred to as the baseband signal. The operation of multiplying signals is called *mixing* or *heterodyning.* In the translated signal, spectral components of the baseband signal that correspond to positive values of f appear in the range f_c to $f_c + f_x$. This part of the modulated signal is called the *upper sideband signal.* The part of the modulated signal that lies between $f_c - f_x$ to f_c is called the *lower sideband signal.* The carrier signal of frequency f_c is also referred to as the *local oscillator signal,* the *mixing signal,* or the *heterodyne signal.* The carrier frequency is normally much higher than the bandwidth of the baseband signal. That is,

$$f_c \gg f_x \tag{6.12}$$

Transmitted Signal Power and Bandwidth. From Figure 6.6*f* we see that the bandwidth B_T required for transmitting a message signal of bandwidth f_x via double-sideband modulation is $2f_x$ Hz:

$$B_T = 2f_x \tag{6.13}$$

To compute the average transmitted power S_T of the modulated signal, let us assume $x(t)$ to be a power signal. Then,

$$\begin{aligned} S_T &= \lim_{T\to\infty} \frac{1}{T}\int_{-T/2}^{T/2} A_c^2 x^2(t)\cos^2(\omega_c t)\,dt \\ &= \lim_{T\to\infty}\frac{1}{T}\left[\int_{-T/2}^{T/2}\frac{A_c^2}{2}x^2(t)\,dt + \int_{-T/2}^{T/2}\frac{A_c^2}{2}x^2(t)\cos 2\omega_c t\,dt\right] \end{aligned}$$

The value of the second integral is zero (see Problem 6.13), and if we define the average signal power S_x to be

$$S_x = \lim_{T\to\infty}\frac{1}{T}\int_{-T/2}^{T/2} x^2(t)\,dt$$

then

$$S_T = S_c S_x \tag{6.14}$$

where $S_c = A_c^2/2$ is the average carrier power.

Recovery of the Baseband Signal (Demodulation). If we assume the channel to be ideal, then the received signal $x_r(t)$ will have the same form as $x_c(t)$. That is,

$$x_r(t) = a_c x(t) \cos \omega_c t$$

where a_c/A_c is the channel attenuation. The baseband message signal $x(t)$ can be recovered from the received signal $x_r(t)$ by multiplying $x_r(t)$ with a local carrier and lowpass filtering the product signal (Figure 6.6*b*). The output of the multiplier is

$$\begin{aligned} z(t) &= [a_c x(t) \cos \omega_c t] 2 \cos \omega_c t \\ &= a_c x(t) + a_c x(t) \cos 2\omega_c t \end{aligned}$$

and the spectrum of $Z(f)$ is given by

$$Z(f) = a_c X(f) + \tfrac{1}{2} a_c [X(f - 2f_c) + X(f + 2f_c)]$$

The spectrum of $Z(f)$ is shown in Figure 6.6*g* from which it is obvious that if

$$f_x < 2f_c - f_x \quad \text{or} \quad f_c > f_x$$

then $X(f)$ does not overlap with $X(f - 2f_c)$ and $X(f + 2f_c)$. Hence filtering $Z(f)$ with a lowpass filter having a cut-off frequency B, $f_x < B < 2f_c - f_x$ will yield an output signal $y(t)$,

$$y(t) = a_c x(t)$$

which is a replica of the transmitted message signal $x(t)$.

While the bandwidth of the lowpass filter can be between f_x and $2f_c - f_x$, it should be as small as possible to reduce the effects of any noise that may accompany the received signal. If noise is present at the receiver input, then a bandpass filter with a center frequency f_c and bandwidth $2f_x$ should be inserted before the multiplier in Figure 6.6*b* to limit the noise power going into the demodulator.

The signal recovery scheme shown in Figure 6.6*b* is called a *synchronous* or *coherent* demodulation scheme. This scheme requires that there be available at the receiver a local oscillator signal that is precisely synchronous with the carrier signal used in generating the modulated signal. This is a very stringent requirement and cannot be met easily in practical systems. Lack of synchronism will result in signal distortion. Suppose that the local oscillator signal has a frequency offset $\Delta\omega$ and phase offset θ. Then the product signal $z(t)$ will have the form

$$z(t) = a_c x(t) \cos(\Delta\omega t + \theta) + \text{double frequency terms}$$

and the output signal $y(t)$ will be

$$y(t) = a_c x(t) \cos(\Delta\omega t + \theta) \tag{6.15}$$

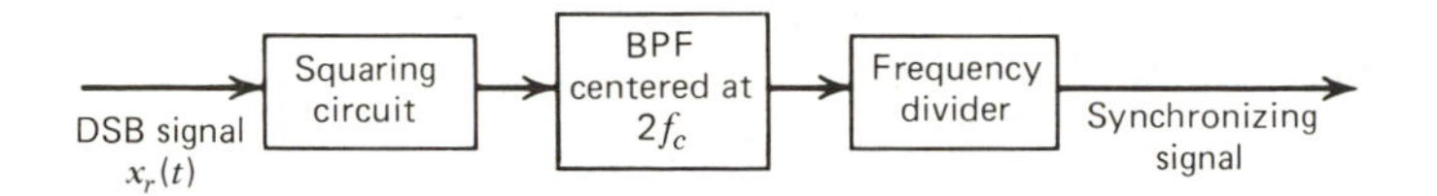

Figure 6.7 A squaring synchronizer.

The reader can verify that when $\Delta\omega = 0$ and $\theta = \pi/2$, the signal is lost entirely. When $\theta = 0$, then $y(t) = a_c x(t) \cos \Delta\omega t$ will warble resulting in a serious signal distortion. This is especially a serious problem since usually $f_c \gg f_x$ so that even a small percentage error in f_c will cause a deviation Δf that may be comparable to or larger than f_x! Experimental evidence indicates that for audio signals, $\Delta f > 30$ Hz becomes unacceptable. For audio signals it may be possible to manually adjust the frequency and phase of the local carrier until the output "sounds" right. Unfortunately, the carrier phase and frequency offsets are often time-varying quantities thus requiring almost continuous adjustments!

There are several techniques that are used to generate a coherent carrier for demodulation. In the method shown in Figure 6.7, a carrier component is extracted from the DSB signal using a squaring circuit and a bandpass filter. If $x(t)$ has a zero DC value, then $x_c(t)$ does not have any spectral component at f_c. However, $x^2(t)$ will have a nonzero DC component and hence a discrete frequency component at $2f_c$ can be extracted from the spectrum of $x_r^2(t)$ using a narrow bandpass filter. The frequency of this component can be halved to yield the desired carrier for demodulation.

In the second method shown in Figure 6.8, a small (pilot) carrier signal is transmitted along with the DSB signal; at the receiver the pilot carrier can be extracted, amplified, and used as a synchronous local carrier for demodulation (Figure 6.8*b*).

If the amplitude of the inserted carrier is large enough, then the received signal can be demodulated without having to regenerate the carrier at the

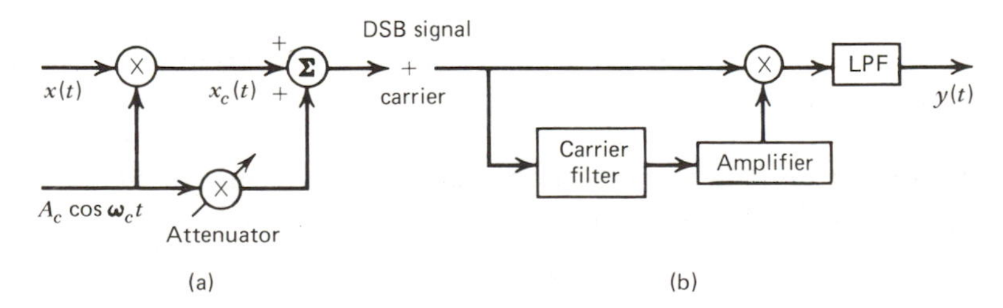

Figure 6.8 Pilot carrier DSB system. (*a*) Transmitter. (*b*) Receiver.

receiver. A DSB signal with a large discrete carrier component is called an amplitude modulated (AM) signal.

6.3.2 Amplitude Modulation (AM)

An amplitude modulated signal is generated by adding a large carrier component to the DSB signal. The AM signal has the form

$$x_c(t) = A_c[1 + x(t)] \cos \omega_c t \tag{6.16}$$

$$= A(t) \cos \omega_c t \tag{6.17}$$

where $A(t)$ is the *envelope* of the modulated carrier. For easy signal recovery using simple demodulation schemes, the signal amplitude has to be small, and the DC component of the signal has to be zero, that is,

$$|x(t)| < 1 \quad \text{and} \quad \lim_{T\to\infty} \frac{1}{T} \int_{-T/2}^{T/2} x(t)\, dt = 0$$

We will explain the need for these restrictions shortly.

In the frequency domain, the spectrum of the AM signal is given by

$$X_c(f) = \tfrac{1}{2}A_c[X(f - f_c) + X(f + f_c)] + \tfrac{1}{2}A_c[\delta(f - f_c) + \delta(f + f_c)] \tag{6.18}$$

Examples of AM signals, in both time domain and frequency domain, are shown in Figure 6.9.

Two unique features of the AM signal are that a carrier frequency component is present, and that the envelope $A(t)$ of the modulated carrier has the same shape of $x(t)$ as long as $f_c \gg f_x$, and as long as $A(t) = A_c[1 + x(t)]$ does not go negative. Our assumption that $|x(t)| < 1$ guarantees that $A(t)$ will not go negative. If $x(t)$ is less than -1, then $A(t)$ goes negative and envelope distortion results, as shown in Figure 6.10. An important parameter of an AM signal is its *modulation index m* which is defined as

$$m = \frac{[A(t)]_{\max} - [A(t)]_{\min}}{[A(t)]_{\max} + [A(t)]_{\min}} \tag{6.19}$$

When m exceeds 1 the carrier is said to be *overmodulated*, resulting in envelope distortion.

Transmitted Signal Power and Bandwidth. From Figure 6.9*d* we see that the bandwidth of the AM signal is

$$B_T = 2f_x$$

Assuming $x(t)$ to be a power signal we can compute the average transmitted

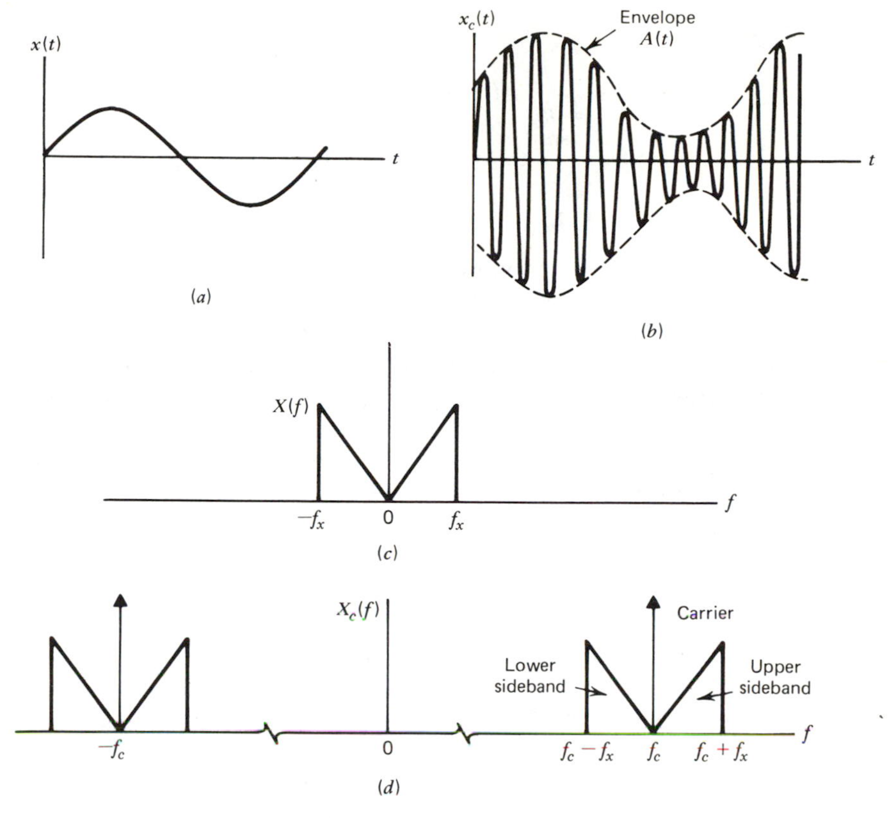

Figure 6.9 Amplitude modulation. (*a*) Sinusoidal message signal. (*b*) AM signal. (*c*) Message spectrum for an arbitrary $x(t)$. (*d*) Modulated signal spectrum.

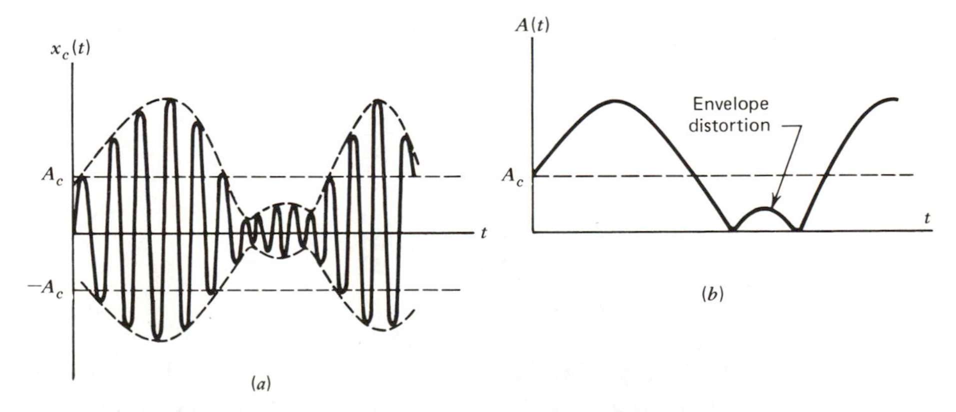

Figure 6.10 Envelope distortion of an AM signal. (*a*) Modulated signal. (*b*) Envelope $A(t)$.

signal power as

$$S_T = \lim_{T\to\infty} \frac{1}{T} \int_{-T/2}^{-T/2} A_c^2[1 + x(t)]^2 \cos^2 \omega_c t \, dt$$

$$= \lim_{T\to\infty} \frac{1}{T} \int_{-T/2}^{T/2} \frac{A_c^2}{2} [1 + x^2(t) + 2x(t)][1 + \cos 2\omega_c t] \, dt$$

$$= S_c + S_c S_x \tag{6.20}$$

where $S_c = A_c^2/2$ and S_x is the normalized average signal power.

The carrier waveform by itself without modulation does not convey any information to the receiver. Hence we may conclude that a portion of the transmitted power S_T is "wasted" in the carrier. We will see shortly that the simplicity of AM demodulators depends on this power, and hence the carrier is not altogether a waste.

For AM signals, the percentage of total power that conveys information is used as a measure of *power efficiency*. We denote the power efficiency by E and define it as

$$E = \frac{S_c S_x}{S_c + S_c S_x} \tag{6.21}$$

It is left as an exercise for the reader to show that the maximum efficiency for an arbitrary signal $x(t)$ is 50% and that the maximum efficiency for a sine wave message signal is 33.3% (remember that $|x(t)| < 1$ and hence $S_x \leq 1$).

Example 6.1. A commercial AM station is broadcasting with an average transmitted power of 10 kW. The modulation index is set at 0.707 for a sinusoidal message signal. Find the transmission power efficiency and the average power in the carrier component of the transmitted signal.

Solution. For a sinusoidal message signal with a modulation index of 0.707, the modulated signal is given by

$$x_c(t) = A_c(1 + 0.707 \cos \omega_x t) \cos \omega_c t$$

Hence

$$S_x = \tfrac{1}{2}(0.707)^2 = 0.25$$

$$E = \frac{0.25 S_c}{S_c + 0.25 S_c} = 20\%$$

Now, $S_c + 0.25 S_c = 10$ kW, hence $S_c = 8$ kW.

Demodulation of AM Signals. The baseband message signal $x(t)$ can be recovered from the AM signal $x_r(t)$ using the simple circuit shown in Figure 6.11*a*. As long as $|x(t)| < 1$, the envelope of the received signal never goes

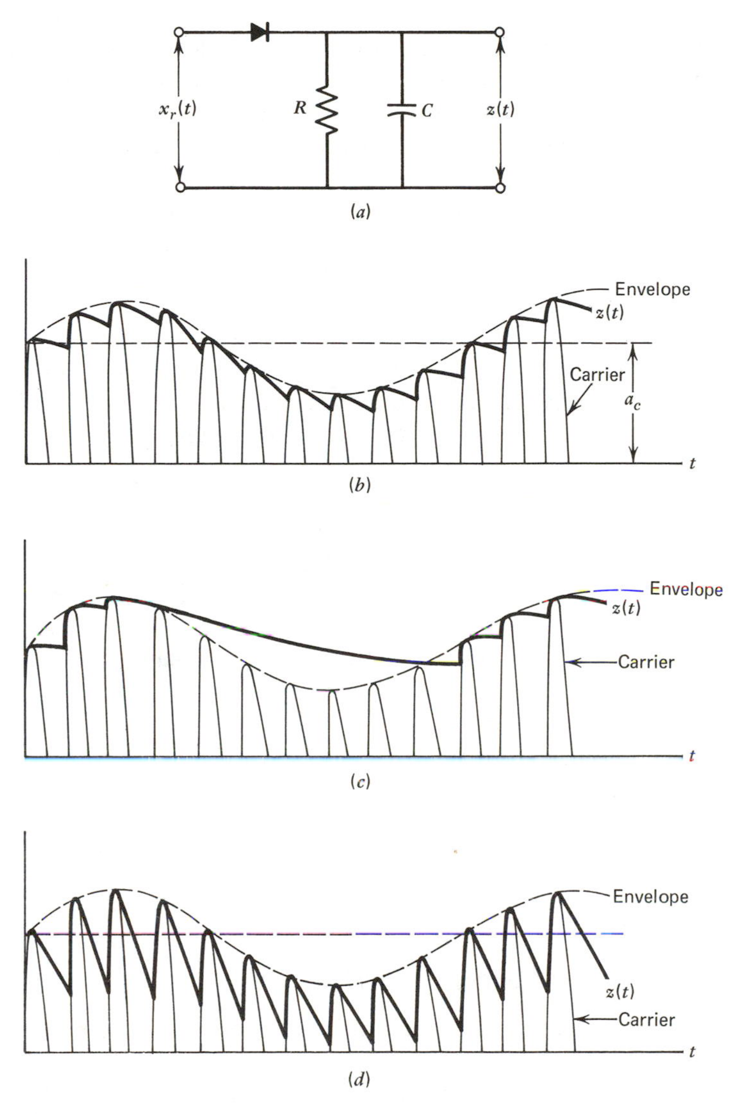

Figure 6.11 Envelope demodulation of AM signals. (*a*) Envelope detector. (*b*) Correct demodulation. (*c*) *RC* too large. (*d*) *RC* too small.

through zero and the positive portion of the envelope approximates the message signal $x(t)$. The positive portion of the envelope is recovered by rectifying $x_r(t)$ and smoothing the rectified waveform using an RC network.

For best operation, the carrier frequency should be much higher than f_x, and the discharge time constant RC should be adjusted so that the maximum negative rate of the envelope will never exceed the exponential discharge rate. If the time constant is too large, then the envelope detector fails to follow the envelope (Figure 6.11*c*). Too small a time constant will generate a ragged waveform (Figure 6.11*d*) and demodulation becomes inefficient (see Problem 6.19).

Under ideal operating conditions, the output of the demodulator is

$$z(t) = k_1 + k_2 x(t)$$

where k_1 is a DC offset due to the carrier and k_2 is the gain of the demodulator circuit. A coupling capacitor or a transformer can be used to remove the DC offset; however, any DC term in the message signal $x(t)$ will be removed as well (one of the reasons for our assumption that the DC value of $x(t) = 0$). In addition to removing DC components, the DC removal filter will attenuate the low-frequency components of the message signal. Hence AM is not suitable for transmitting message signals having significant low-frequency content.

The AM signal can also be demodulated by passing $x_r(t)$ through a square law device (or any other type of nonlinearity not having odd function symmetry) and filtering the squared output. Details of square law demodulation are left as an exercise for the reader (see Problem 6.20).

The envelope and square law demodulators for AM signals do not require a synchronized (or coherent) local oscillator signal. These detectors are simple, efficient, and cheap to build. The savings in the cost of building receivers alone is sufficient for using AM in many applications such as in commercial radio where millions of receivers are involved. Factors offsetting the advantage of equipment simplicity are the power wasted in the carrier and the poor low-frequency response of AM systems using envelope or square law demodulators.

6.3.3 Suppressed Sideband Modulation

Transmitter power requirement and transmission bandwidth are two important parameters of a communication system. Savings in power requirement and bandwidth are highly desirable. The AM scheme is wasteful of both transmitted power and transmission bandwidth. The DSB modulation scheme has lesser power requirements than the AM, but uses the same bandwidth as AM. Both DSB modulation and AM retain the upper and lower sidebands of the message signal resulting in a transmission bandwidth that is twice the bandwidth of the message signal.

The spectrum of any real-valued signal $x(t)$ must exhibit the symmetry condition (Chapter 2, Problem 2.15)

$$X(f) = X^*(-f)$$

and hence the sidebands of AM and DSB are uniquely related to each other by symmetry. Given the amplitude and phase of one, we can always reconstruct the other. Hence transmission bandwidth can be cut in half if one sideband is entirely suppressed. This leads to *single-sideband modulation* (*SSB*). In SSB modulation, bandwidth saving is accompanied by a considerable increase in equipment complexity.

In addition to equipment complexity, practical SSB systems have poor low-frequency response. A reduction in equipment complexity and an improvement in low-frequency response is possible if a sideband is only partially suppressed rather than being entirely suppressed. Modulation schemes in which one sideband plus a trace or vestige of the second sideband are transmitted are called *vestigial-sideband* (*VSB*) *modulation schemes.* VSB modulation is widely used for transmitting message signals that have very large bandwidths and significant low-frequency contents (such as in high speed data transmission, and in television broadcasting).

SSB Modulation. In SSB modulation, only one of the two sidebands that result from the multiplication of the message signal $x(t)$ with a carrier is transmitted. The generation of an upper sideband SSB signal by filtering a DSB signal is shown in Figure 6.12*a*. The recovery of the baseband signal by synchronous demodulation is shown in Figure 6.12*b*. Frequency domain representation of the important operations in an SSB modulation scheme are shown in Figures 6.12*c* through 6.12*e*. Time domain description of SSB signals is somewhat more difficult, except for the case of tone modulation (see Problem 6.22).

From Figure 6.12, we can verify that the bandwidth of the SSB signal is

$$B_T = f_x \tag{6.22}$$

and the average transmitted power is

$$S_T = \tfrac{1}{2} S_c S_x \tag{6.23}$$

The modulation and demodulation operations for the SSB signal as shown in Figure 6.12 look very simple. However, practical implementation is exceedingly difficult for two reasons. First, the modulator calls for an ideal bandpass sideband filter; second, the demodulator requires a synchronous carrier.

The sharp cutoff characteristics required of the sideband filter $H_{\text{SSB}}(f)$ cannot be synthesized exactly. So, one must either attenuate a portion of the desired sideband or pass a portion of the unwanted sideband. Fortunately, many (not all) message signals have little or no low-frequency content. Such signals (for example, voice and music) have "holes" at zero frequency and these holes appear as a vacant space centered at the carrier frequency. The

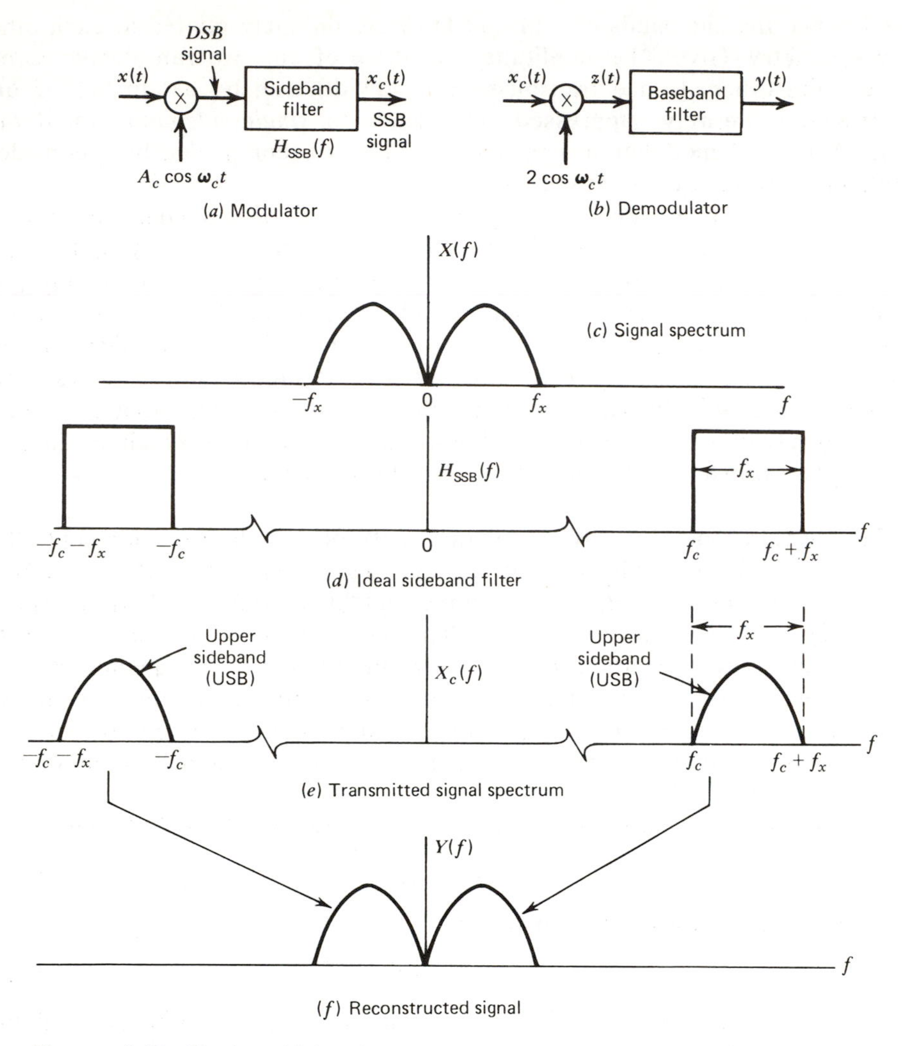

Figure 6.12 Single sideband modulation. (*a*) Modulator. (*b*) Demodulator, (*c*) Signal spectrum. (*d*) Ideal sideband filter. (*e*) Transmitted signal spectrum. (*f*) Reconstructed signal.

transition region of a practical sideband filter can be fitted into this region as shown in Figure 6.13. As a rule of thumb the ratio $2\alpha/f_c$ cannot be smaller than 0.01 if the cutoff rate for the filter is to be reasonable. The width of the transition region 2α is limited to the width of the "hole" in the spectrum and for a given f_c it may not be possible to obtain a reasonable value for the ratio $2\alpha/f_c$. For such cases, the modulation process can be carried out in two or

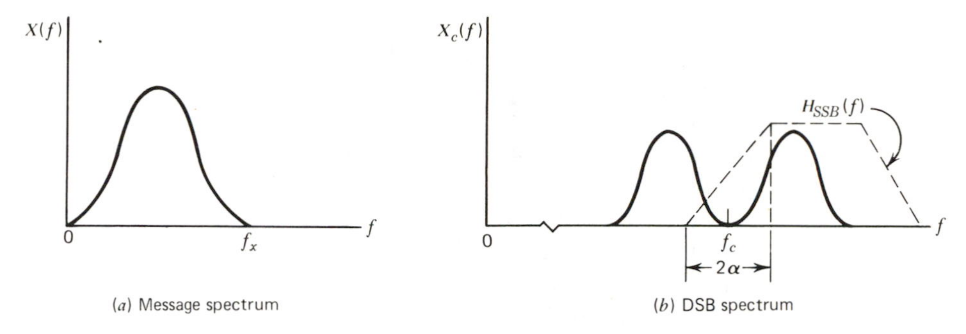

Figure 6.13 Sideband filter characteristics. (*a*) Message spectrum. (*b*) DSB spectrum.

more steps using one or more intermediate carrier frequencies (see Problem 6.23).

The SSB signal can be generated by another method, called the *phase-shift method*, which does not require a sideband filter. To illustrate how this method works, let us assume that the message signal has the form

$$x(t) = \sum_{i=1}^{n} X_i \cos(2\pi f_i t + \theta_i), \quad f_n \leq f_x \tag{6.24}$$

Then the SSB (upper sideband) signal corresponding to $x(t)$ is given by

$$x_c(t) = \frac{A_c}{2} \sum_{i=1}^{n} X_i \cos[2\pi(f_c + f_i)t + \theta_i]$$

We can rewrite $x_c(t)$ as

$$\begin{aligned} x_c(t) &= \frac{A_c}{2} \left\{ \left[\sum_{i=1}^{n} X_i \cos(2\pi f_i t + \theta_i) \right] \cos 2\pi f_c t \right. \\ &\quad \left. - \left[\sum_{i=1}^{n} X_i \sin(2\pi f_i t + \theta_i) \right] \sin 2\pi f_c t \right\} \\ &= \frac{A_c}{2} [x(t) \cos 2\pi f_c t] - \frac{A_c}{2} \hat{x}(t) \sin 2\pi f_c t \end{aligned} \tag{6.25}$$

where $\hat{x}(t)$ is defined as

$$\hat{x}(t) = \sum_{i=1}^{n} X_i \sin(2\pi f_i t + \theta_i) \tag{6.26}$$

Equations (6.24), (6.25), and (6.26) suggest that an SSB signal can be generated from two DSB signals having quadrature carriers $\frac{1}{2}A_c \cos \omega_c t$ and $\frac{1}{2}A_c \sin \omega_c t$ modulated by $x(t)$ and $\hat{x}(t)$. The *quadrature signal component**

*$\hat{x}(t)$ is referred to as the Hilbert transform of $x(t)$.

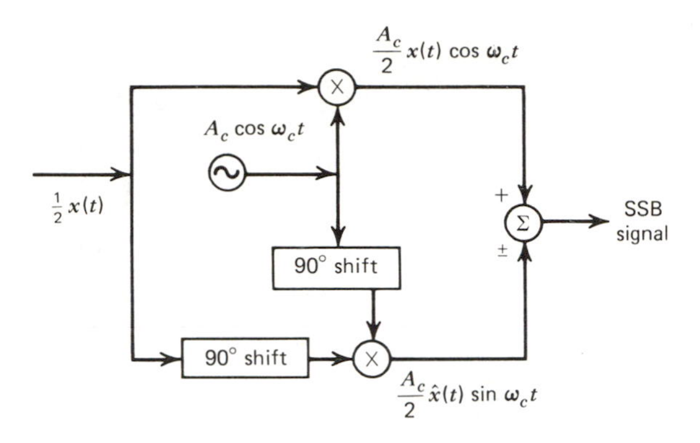

Figure 6.14 Phase-shift SSB modulator.

$\hat{x}(t)$ is obtained from $x(t)$ by shifting the phase of every spectral component of $x(t)$ by 90°. A phase-shift SSB modulator consisting of two DSB (product) modulators and appropriate phase-shift networks is shown in Figure 6.14. The design of the phase-shift circuitry is not trivial and imperfect design generally results in distortion of the low-frequency components. An alternate method of generating SSB signals is discussed in Problem 6.24.

Rather than using a synchronous demodulator, we may add a carrier component to the SSB signal (preferably at the transmitter) and attempt to demodulate the SSB signal using an envelope demodulator. However, this procedure will lead to some signal distortion and waste of transmitted power as discussed in the following section.

Vestigial-Sideband Modulation (VSB). Many message signals such as television video, facsimile, and high speed data signals have very large bandwidth and significant low-frequency content. SSB modulation may be used to conserve bandwidth, but practical SSB modulation systems have poor low-frequency response. While DSB works well for messages with significant low-frequency content, DSB transmission bandwidth is twice that of SSB. A modulation scheme that offers the best compromise between bandwidth conservation, improved low-frequency response, and improved power efficiency is vestigial sideband (VSB) modulation.

VSB modulation is derived by filtering DSB or AM signals in such a fashion that one sideband is passed almost completely while only a trace of the other sideband is included. A typical VSB filter transfer function is shown in Figure 6.15. An important and essential requirement of the VSB filter $H_{\text{VSB}}(f)$ is that it must have odd symmetry about f_c and a relative response of $\frac{1}{2}$ at f_c.

The VSB sideband filter has a transition interval of width 2α Hz, and the

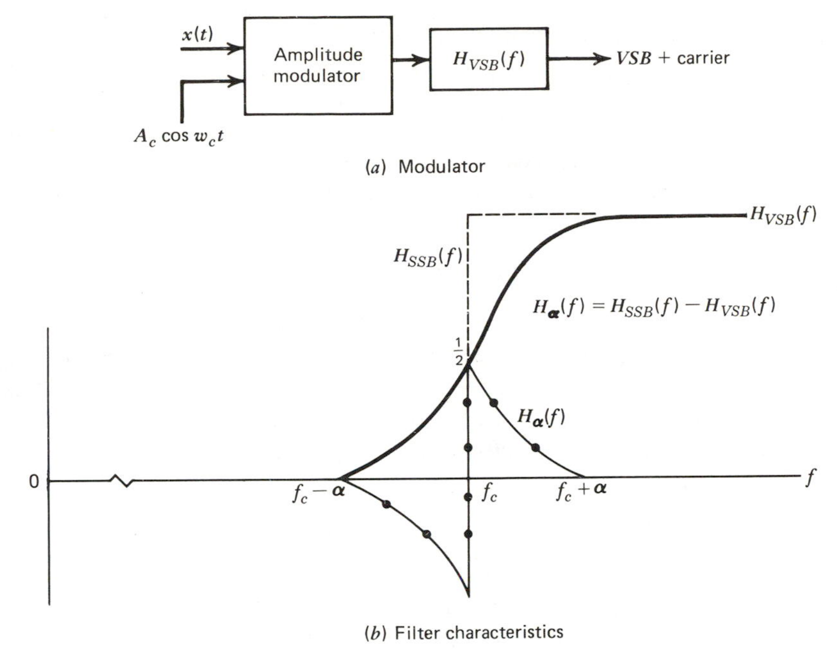

Figure 6.15 VSB modulation. (*a*) Modulator. (*b*) Filter characteristics.

transmission bandwidth of the VSB signal is

$$B_T = f_x + \alpha, \quad \alpha < f_x \tag{6.27}$$

To derive a time domain expression for the VSB signal, let us express $H_{\text{VSB}}(f)$ as

$$H_{\text{VSB}}(f) = H_{\text{SSB}}(f) - [H_\alpha(f)] \tag{6.28}$$

where $H_\alpha(f)$ represents the difference between the response of the SSB and VSB filters. $H_\alpha(f)$ is required to have odd symmetry about f_c (the reason for this requirement will be clear when Problem 6.27 is worked). The input to the VSB filter is $A_c[1 + x(t)] \cos \omega_c t$, and the output signal can be expressed in the form

$$\underbrace{x_c(t)}_{\substack{\textit{VSB +}\\ \textit{carrier}}} = \underbrace{\tfrac{1}{2}A_c \cos \omega_c t}_{\textit{carrier}} + \underbrace{\underbrace{\tfrac{1}{2}A_c[x(t) \cos \omega_c t - \hat{x}(t) \sin \omega_c t]}_{\textit{SSB signal}} - \tfrac{1}{2}A_c x_\alpha(t) \sin \omega_c t}_{\textit{VSB signal}} \tag{6.29}$$

In Equation (6.29), $\frac{1}{2}A_c x_\alpha(t) \sin \omega_c t$ is the response of $H_\alpha(f)$ to the input

$A_c x(t) \cos \omega_c t$. We can also write Equation (6.29) as

$$x_c(t) = \tfrac{1}{2}A_c[1 + x(t)] \cos \omega_c t - \tfrac{1}{2}A_c\gamma(t) \sin \omega_c t \tag{6.30}$$

where $\gamma(t) = \hat{x}(t) + x_\alpha(t)$. If $\gamma(t) = 0$, then Equation (6.30) reduces to an AM signal, and when $\gamma(t) = \hat{x}(t)$, we have an SSB + carrier signal.

While it is not easy to derive an exact expression for average transmitted power for VSB modulation, we can obtain bounds for S_T as

$$S_c + \tfrac{1}{2}S_c S_x \leq S_T \leq S_c S_x + S_c \tag{6.31}$$

where S_c is the carrier power and S_x is the signal power.

The reader can verify that the VSB signal can be demodulated using a synchronous demodulator. However, it turns out that we can demodulate a VSB signal with a small amount of signal distortion using envelope demodulation if a sufficiently large carrier component had been added to the VSB signal at the transmitter.

Envelope Demodulation of Suppressed Sideband Signals. It is often desirable to combine the envelope demodulation of AM with bandwidth conservation of suppressed sideband signals. Perfect distortion-free envelope demodulation requires both sidebands and a large carrier signal. Adding a carrier to a VSB signal, we have

$$x_c(t) = A_c\{[1 + x(t)] \cos \omega_c t - \gamma(t) \sin \omega_c t\} \tag{6.32}$$

For AM, $\gamma(t) = 0$; and $\gamma(t) = \hat{x}(t)$ for SSB + carrier. For VSB + carrier, $\gamma(t)$ takes an intermediate value.

The envelope of $x_c(t)$ is found by writing

$$x_c(t) = R(t) \cos[\omega_c t + \phi(t)]$$

where $R(t)$ is the envelope given by

$$\begin{aligned} R(t) &= A_c\{[1 + x(t)]^2 + [\gamma(t)]^2\}^{1/2} \\ &= A_c\{[1 + x(t)]\}\left\{1 + \left[\frac{\gamma(t)}{1 + x(t)}\right]^2\right\}^{1/2} \end{aligned} \tag{6.33}$$

Equation (6.33) shows that the envelope is distorted. (The distortion-free envelope, as in the AM case, is $A_c[1 + x(t)]$.) However, if $|\gamma(t)| \ll 1$, the distortion is negligible and $R(t) \approx A_c[1 + x(t)]$, as in the AM case. Thus the key to the success of envelope demodulation of suppressed sideband signals is to keep the quadrature component $\gamma(t)$ small.

For the SSB + carrier signal, $\gamma(t) = \hat{x}(t)$, and hence $\gamma(t)$ cannot be ignored. Furthermore, a substantial amount of power is wasted in the carrier, far more than AM. For a VSB signal with not too small a vestigial sideband, $|\gamma(t)|$ is small compared to $|x(t)|$ most of the time. Thus envelope demodulation can be used without excessive distortion. Also, for a VSB + carrier signal, the average transmitted power can be shown to be

$$S_T \approx S_c + S_c S_x$$

which is essentially the same as AM.

The permissible width of the vestigial sideband will depend on the spectral characteristics of $x(t)$ and the amount of distortion that can be tolerated. Commercial TV broadcasts use VSB + carrier with a 30% vestigial sideband. While the distortion may be quite sizable, experimental evidence indicates that the picture quality is not detracted much. We will discuss several interesting aspects of commercial TV signals later on.

6.3.4 Frequency Conversion (Mixing)

Frequency translation, also referred to as frequency conversion or *mixing*, is the single most important operation in linear modulation systems. Modulation translates the message spectrum upwards in frequency and demodulation is basically a downward frequency translation operation. Frequency translation is also used often to translate a bandpass signal at one carrier frequency to a new center frequency. This can be accomplished by multiplying the bandpass signal by a periodic signal as shown in Figure 6.16.

The important mathematical operation that takes place in all frequency translation operations is the multiplication of a signal by a periodic waveform. We now take a look at ways of performing this operation using practical devices.

Balanced Modulator. A balanced modulator that generates a DSB (product) signal is shown in Figure 6.17. The balanced modulator consists of summing devices (operational amplifiers) and two matched nonlinear elements (such as appropriately biased diodes). If we assume that the nonlinearity can be represented by a power series transfer characteristic with two terms, then

$$\begin{aligned} y(t) &= a_1[A_c \cos \omega_c t + x(t)] + a_2[A_c \cos \omega_c t + x(t)]^2 \\ &\quad - a_1[A_c \cos \omega_c t - x(t)] - a_2[A_c \cos \omega_c t - x(t)]^2 \\ &= 2a_1 x(t) + 4a_2 x(t) A_c \cos \omega_c t \end{aligned}$$

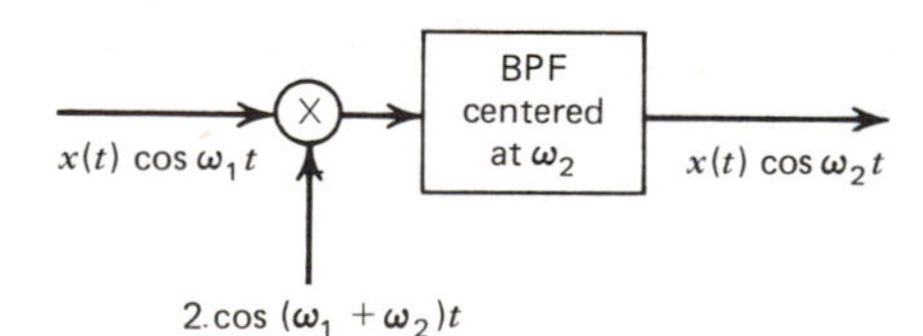

Figure 6.16 Frequency conversion or mixing.

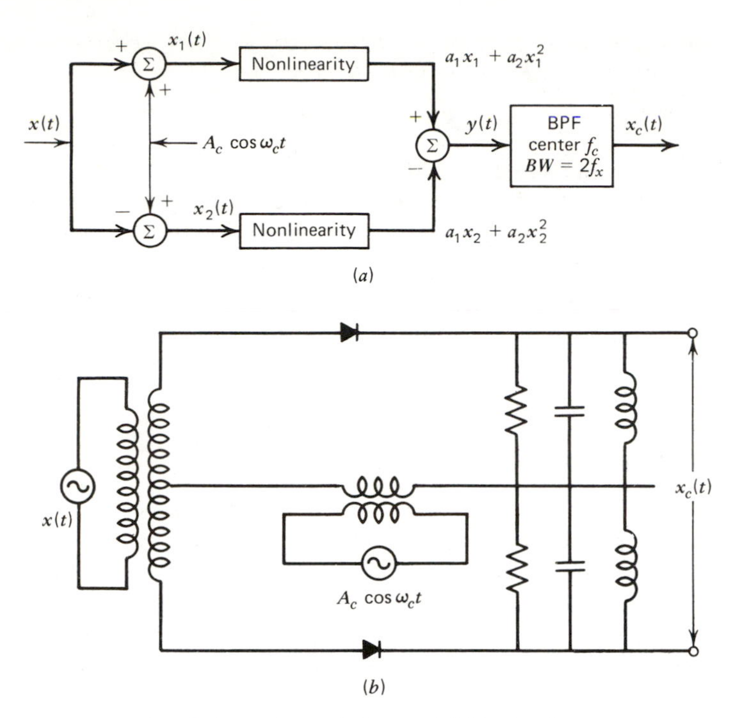

Figure 6.17 Balanced Modulator. (*a*) Block diagram of a balanced modulator. (*b*) Circuit diagram of a balanced modulator.

If $x(t)$ is bandlimited to f_x and if $f_c > 2f_x$, then the bandpass filter output will be

$$z(t) = (4a_2A_c)x(t) \cos \omega_c t$$

which is the desired product signal. Semiconductor diodes are often used for nonlinear devices in balanced modulators. The performance of this type of modulator is dependent on how close the characteristics of the diodes can be matched.

Switching Modulator. Another circuit that is used for mixing two signals is shown in Figure 6.18. When the carrier voltage is positive the output voltage $v(t)$ is present, and when the carrier is negative the output voltage is zero. Thus the diodes operate as switches operating at a rate of f_c and we can write the output $v(t)$ as

$$v(t) = x(t)s(t)$$

where $s(t)$ is a switching function with a period f_c. Assuming zero DC value for

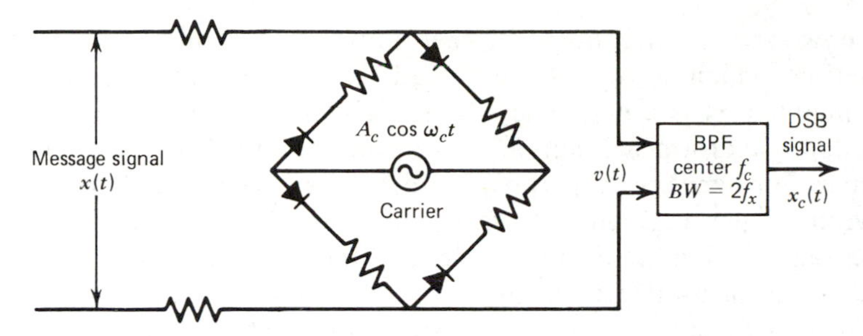

Figure 6.18 Switching modulator.

the message signal $x(t)$, and using the Fourier series expansion for $s(t)$, we can write $v(t)$ as

$$\begin{aligned} v(t) &= k_0 x(t) + k_1 x(t)\cos(\omega_c t) + k_3 x(t)\cos(3\omega_c t) + \cdots \\ &= \text{baseband term} + \text{DSB term} + \text{harmonics} \end{aligned} \tag{6.34}$$

In deriving Equation (6.34) we have assumed $s(t)$ to be a symmetric square wave, and hence the Fourier series coefficient $k_2, k_4, \ldots$ are zero. By bandpass filtering $v(t)$ we get the output as

$$x_c(t) = k_1 x(t) \cos \omega_c t$$

which is the desired DSB signal.

In practical systems, the oscillators, summing devices, and filters are constructed using discrete RLC components and active devices such as transistors and integrated circuit operational amplifiers. At microwave frequencies these devices become distributed parameter systems.

6.4 ANGLE MODULATION

All linear modulation schemes discussed in the preceding sections have the following features in common:

1. The spectrum of the modulated signal is basically the translated message spectrum.
2. All operations performed on the signal are linear operations so that superposition applies.
3. The transmission bandwidth never exceeds twice the message bandwidth.
4. The signal-to-noise ratio at the receiver output can be increased only by increasing the transmitted signal power (we will establish this property in Chapter 7).

We now turn our attention to a new type of modulation, namely, the angle modulation, which is not characterized by the features mentioned above. Angle modulation is a nonlinear process and the spectral components of the modulated waveform are not related in any simple fashion to the message spectrum. Furthermore, superposition does not apply (Problem 6.36) and the bandwidth of the angle modulated signal is usually much greater than twice the message bandwidth. The increase in bandwidth and system complexity is compensated for by improved signal-to-noise ratio at the receiver output. We will show in Chapter 7 that, with angle modulated signals, we can trade bandwidth for improved signal-to-noise ratio without having to increase transmitted signal power.

Angle modulated signals have the general form

$$x_c(t) = A_c \cos[\omega_c t + \phi(t)]$$

where A_c and ω_c are constants and the phase angle $\phi(t)$ is a function of the baseband signal $x(t)$. While there are many forms of angle modulation, only two—phase modulation (PM) and frequency modulation (FM)—have proved to be practical. We will now consider PM and FM systems, paying particular attention to signal spectra, power, bandwidth, and methods of generating and demodulating angle modulated signals.

6.4.1 Angle Modulated Signals

An angle modulated signal, also referred to as an *exponentially modulated* signal, has the form

$$x_c(t) = A_c \cos[\omega_c t + \phi(t)] = \text{Re}(A_c \exp(j\omega_c t + j\phi(t))) \tag{6.35}$$

where "Re" denotes the "real part of" a complex number. The *instantaneous phase* of $x_c(t)$ is defined as

$$\theta_i(t) = \omega_c t + \phi(t) \tag{6.36}$$

and the *instantaneous frequency* of the modulated signal is defined as

$$\omega_i(t) = \frac{d\theta_i}{dt} = \omega_c + \frac{d\phi}{dt} \tag{6.37}$$

The functions $\phi(t)$ and $d\phi/dt$ are referred to as the (instantaneous) *phase* and *frequency deviations*, respectively.

The phase deviation of the carrier $\phi(t)$ is related to the baseband message signal $x(t)$. Depending on the nature of the relationship between $\phi(t)$ and $x(t)$ we have different forms of angle modulation.

In *phase modulation, the instantaneous phase deviation* of the carrier is proportional to the message signal,* that is,

$$\phi(t) = k_p x(t) \tag{6.38}$$

where k_p is the *phase deviation constant* (expressed in radian/volt). For *frequency modulated signals, the frequency deviation of the carrier is proportional to the message signal,* that is,

$$\frac{d\phi}{dt} = k_f x(t) \tag{6.39}$$

or

$$\phi(t) = k_f \int_{t_0}^{t} x(\lambda)\, d\lambda + \phi(t_0) \tag{6.40}$$

where k_f is the *frequency deviation constant* (expressed in (radian/sec)/volt) and $\phi(t_0)$ is the initial angle at $t = t_0$. It is usually assumed that $t_0 = -\infty$ and

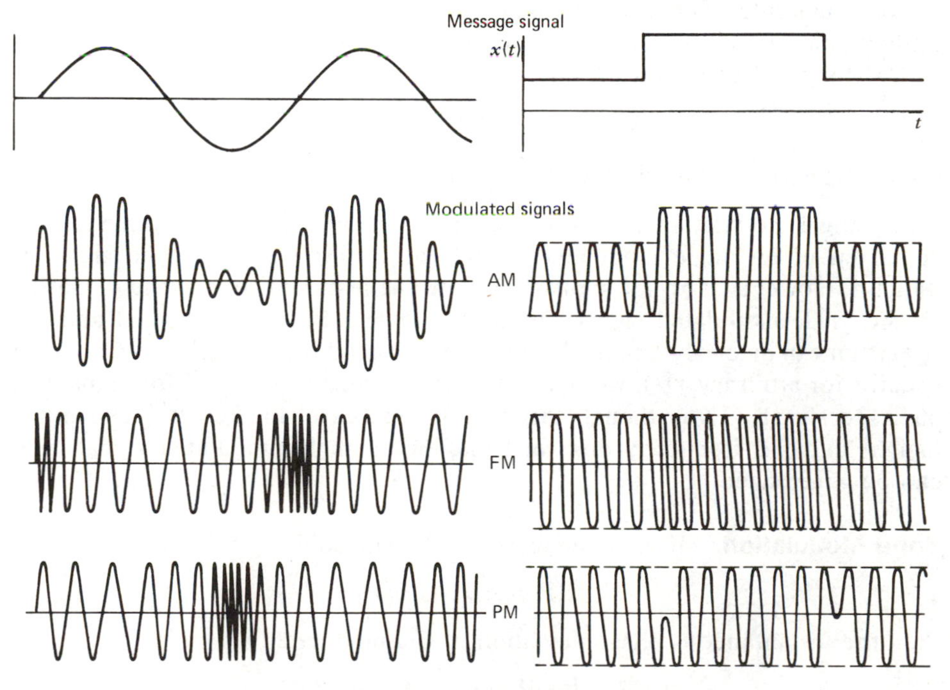

Figure 6.19 AM, FM, and PM waveforms.

*For unique demodulation of a PM signal, $\phi(t)$ must not exceed $\pm 180°$ since there is no physical distinction between the angles $+270°$ and $-90°$, for instance.

$\phi(-\infty) = 0$. Combining Equations (6.39) and (6.40) with Equation (6.35), we can express the angle modulated signal as

$$x_c(t) = \begin{cases} A_c \cos[\omega_c t + k_p x(t)] & \text{for PM} \\ A_c \cos\left[\omega_c t + k_f \int_{-\infty}^{t} x(\tau)\, d\tau\right] & \text{for FM} \end{cases} \tag{6.41}$$

Equation (6.41) reveals that PM and FM signals are similar in functional form with the exception of the integration of the message signal in FM.

Figure 6.19 shows typical AM, FM, and PM waveforms for two different message waveforms. These figures reveal an important feature of angle modulation, namely, that the amplitude of an FM or PM waveform is always constant. Because of this property we can conclude that the message resides in the zero crossings of the angle modulated signal when the carrier frequency is large. Figure 6.19 also shows that in some cases, such as with tone modulation, it is impossible to visually distinguish between FM and PM waveforms.

The similarities between FM and PM waveforms allow us to present a unified analysis for both systems. We will emphasize FM leaving the details of PM for the reader to work out.

6.4.2 Spectra of Angle Modulated Signals

Since angle modulation is a nonlinear process, an exact description of the spectrum of an angle modulated signal for an arbitrary message signal is difficult. However if $x(t)$ is sinusoidal, then the instantaneous phase deviation of the angle modulated signal (for both FM and PM) is sinusoidal and the spectrum can be easily obtained. Therefore, instead of examining FM and PM spectra for arbitrary $x(t)$, we will examine the spectra in detail for sinusoidal message signals. Even though this is a special case, the results yield much insight into the frequency domain behavior of angle modulated signals and can be generalized.

Tone Modulation. If we assume $x(t)$ to be sinusoidal (or a tone)

$$x(t) = A_m \cos \omega_m t$$

then the instantaneous phase deviation of the modulated signal is

$$\phi(t) = \begin{cases} k_p A_m \cos \omega_m t & \text{for PM} \\ \dfrac{k_f A_m}{\omega_m} \sin \omega_m t & \text{for FM} \end{cases} \tag{6.42}$$

(we assume that $\phi(-\infty) = 0$ for the FM case). The modulated signal, for the

FM case, is given by

$$x_c(t) = A_c \cos(\omega_c t + \beta \sin \omega_m t) \tag{6.43}$$

where the parameter β is called the *modulation index* defined as

$$\beta = \frac{k_f A_m}{\omega_m} \quad \text{for FM} \qquad \text{and} \qquad \beta = k_p A_m \quad \text{for PM} \tag{6.44}$$

The parameter β is defined only for tone modulation and it represents the maximum phase deviation produced by the modulating tone.

Returning to the problem of computing the spectrum of $x_c(t)$ given in Equation (6.43), we can express $x_c(t)$ as

$$x_c(t) = A_c \operatorname{Re}\{\exp(j\omega_c t) \exp(j\beta \sin \omega_m t)\} \tag{6.45}$$

In the preceding expression, $\exp(j\beta \sin \omega_m t)$ is periodic with a period $T_m = 2\pi/\omega_m$. Hence, we can represent it in a Fourier series of the form

$$\exp(j\beta \sin \omega_m t) = \sum_{n=-\infty}^{\infty} C_{x_c}(nf_m) \exp(j2\pi n f_m t) \tag{6.46}$$

$$\begin{aligned} C_{x_c}(nf_m) &= \frac{\omega_m}{2\pi} \int_{-\pi/\omega_m}^{\pi/\omega_m} \exp(j\beta \sin \omega_m t) \exp(-jn\omega_m t)\, dt \\ &= \frac{1}{2\pi} \int_{-\pi}^{\pi} \exp[j(\beta \sin \theta - n\theta)]\, d\theta = J_n(\beta) \end{aligned} \tag{6.47}$$

where $J_n(\beta)$ are *Bessel functions of the first kind* whose values have been well tabulated (see, for example, the handbook for radio engineers). A short listing of Bessel functions is given in Table 6.2.

Combining Equations (6.47), (6.46), and (6.45), we can obtain the following expression for the FM signal with tone modulation:

$$x_c(t) = A_c \sum_{n=-\infty}^{\infty} J_n(\beta) \cos[(\omega_c + n\omega_m)t] \tag{6.48}$$

The spectrum of $x_c(t)$ is easily obtained from the preceding equation. An example is shown in Figure 6.20.

The spectrum of an FM signal has several important properties:

1. The FM spectrum consists of a carrier component plus an infinite number of sideband components at frequencies $f_c \pm nf_m$ $(n = 1, 2, 3, \ldots)$. In comparison, the spectrum of an AM signal with tone modulation has only three spectral components (at frequencies f_c, $f_c + f_m$ and $f_c - f_m$).
2. The relative amplitude of the spectral components of an FM signal depend on the values of $J_n(\beta)$. The relative amplitude of the carrier depends on $J_0(\beta)$ and its value depends on the modulating signal (unlike in linear

Table 6.2. Bessel functions $J_n(\beta)$:
$J_{-n}(\beta) = (-1)^n J_n(\beta)$; $\frac{1}{2}[J_0(\beta)]^2 + [J_1(\beta)]^2 + [J_2(\beta)]^2 + \cdots = \frac{1}{2}$

n \ β	0.2	0.5	1	2	5	8
0	0.990	0.938	0.765	0.224	−0.178	0.172
1	0.100	0.242	0.440	0.577	−0.328	0.235
2	0.005	0.031	0.115	0.353	0.047	−0.113
3			0.020	0.129	0.365	−0.291
4			0.002	0.034	0.391	−0.105
5				0.007	0.261	0.186
6					0.131	0.338
7					0.053	0.321
8					0.018	0.223
9						0.126
10						0.061

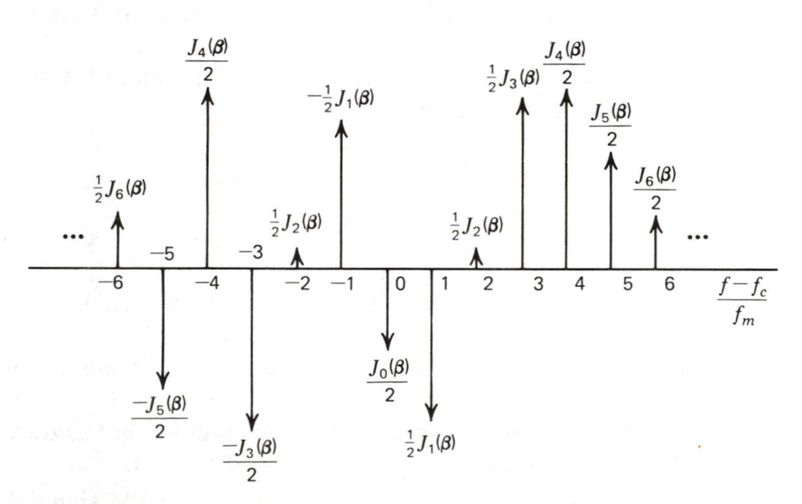

Figure 6.20 Spectrum of an FM signal, $\beta = 5$, $A_c = 1$, and $f_c \gg f_m$. Negative values are shown plotted downward.

modulation where the residual carrier amplitude does not depend on the value of the modulating signal).

3. The phase relationship between the sideband components is such that the odd-order lower sidebands are reversed in phase.
4. The number of significant spectral components is a function of β (see Table 6.2). When $\beta \ll 1$, only J_0 and J_1 are significant so that the spectrum will consist of carrier plus two sideband components, just like AM spectrum with the exception of the phase reversal of the lower sideband component.
5. A large value of β implies a large bandwidth since there will be many significant sideband components.

The above observations apply to phase modulated signals as well. Indeed, for tone modulation we can easily obtain the spectra of a PM signal using the analysis given above by defining the PM modulation index β to be

$$\beta = k_p A_m$$

where A_m is the amplitude of the modulating signal.

Despite the complexity of the spectra of angle modulated signals, there are a few modulating signals for which we can calculate the spectra without too much difficulty. One such signal is the rectangular waveform. This important and informative example is left as an exercise for the reader (Problems 6.39 and 6.40).

6.4.3 Power and Bandwidth of FM Signals

In the preceding section we saw that a tone modulated FM signal has an infinite number of sideband components and hence the FM spectrum seems to have infinite extent. Fortunately, it turns out that for any β a large portion of the total power is confined to some finite bandwidth. No serious baseband signal distortion results if the sidebands outside this bandwidth are left out. Hence the determination of FM transmission bandwidth boils down to the question of how many significant sidebands need to be included for transmission if the distortion is to be within certain tolerable limits. The answer to this question is based on experimental evidence which indicates that baseband signal distortion is negligible if 98% or more of the FM signal power is contained within the transmission band. This rule of thumb, based on experimental studies, leads to useful approximate relationships between transmission bandwidth, message signal bandwidth, and modulation index.

To determine FM transmission bandwidth, let us define a power ratio S_n as the fraction of the total power contained in the carrier plus n sidebands on each side of the carrier. That is, define S_n to be

$$S_n = \frac{\frac{1}{2}A_c^2 \sum_{k=-n}^{n} J_k^2(\beta)}{\frac{1}{2}A_c^2 \sum_{k=-\infty}^{\infty} J_k^2(\beta)} \tag{6.49}$$

The denominator of the preceding equation represents the average transmitted power S_T. Now the amplitude of an angle modulated waveform is always constant. Therefore, regardless of the message $x(t)$, the average transmitted power is

$$\begin{aligned} S_T &= \tfrac{1}{2}A_c^2 \\ &= \tfrac{1}{2}A_c^2 \sum_{k=-\infty}^{\infty} J_k^2(\beta) \end{aligned} \tag{6.50}$$

Substituting Equation (6.50) into (6.49), we have

$$S_n = \sum_{k=-n}^{n} J_k^2(\beta)$$

To find the transmission bandwidth of the FM signal for a given modulation index β, we have to find the smallest value of n that yields $S_n \geq 0.98$. We note that the double lines in Table 6.2, which indicate the value of n for which $S_n \geq 0.98$, always occur just after $n = \beta + 1$. Thus, for tone modulation, the bandwidth of the FM signal is given by

$$B_T \approx 2(\beta + 1)f_m \tag{6.51}$$

For an arbitrary message $x(t)$, we cannot use the preceding expression to determine B_T since β is defined only for tone modulation. For arbitrary message signals bandlimited to f_x, we can define a *deviation ratio D* (which is analogous to the modulation index β) as

$$\begin{aligned} D &= \frac{\text{peak frequency deviation}}{\text{bandwidth of } x(t)} \\ &= \frac{k_f \max|x(t)|}{2\pi f_x} = \frac{f_\Delta}{f_x} \end{aligned} \tag{6.52}$$

Using D in place of β in Equation (6.51) results in the generally accepted expression for bandwidth:

$$B_T = 2(D + 1)f_x \tag{6.53a}$$

$$= 2(f_\Delta + f_x) \tag{6.53b}$$

where $f_\Delta = Df_x$ is the *maximum frequency deviation.* The preceding expression for bandwidth is referred to as *Carson's rule,* which indicates that the FM bandwidth is twice the sum of the maximum frequency deviation and the bandwidth of the message signal.

FM signals are classified into two categories based on the value of D (or β). If D (or β) $\ll 1$, the FM signal is called a narrowband FM (NBFM) signal and the bandwidth of the NBFM signal is equal to $2f_x$, which is the same as the bandwidth of the DSB signal. When D (or β) $\gg 1$, the FM signal is called a wideband FM (WBFM) signal and its bandwidth is approximately $2f_\Delta$.

Example 6.2. A 20 MHz carrier is frequency modulated by a sinusoidal signal such that the peak frequency deviation is 100 kHz. Determine the modulation index and the approximate bandwidth of the FM signal if the frequency of the modulating signal is: (a) 1 kHz; (b) 50 kHz; (c) 500 kHz.

Solution. We are given $f_\Delta = 100$ kHz and $f_c = 20$ MHz $\gg f_m$. For sinusoidal modulation, $f_\Delta = \beta f_m$, that is, $\beta = D = f_\Delta/f_m$.

(a) With $f_m = 1000$ Hz, $\beta = f_\Delta/f_m = 10^5/1000 = 100$. This is WBFM and $B_T \approx 2(\beta + 1)f_m = 202$ kHz.
(b) With $f_m = 50$ kHz, $\beta = 2$ and $B_T \approx 300$ kHz.
(c) With $f_m = 500$ kHz, $\beta = 0.2$. This is NBFM and $B_T \approx 2f_m = 1$ MHz.

Narrowband FM is in many ways similar to DSB or AM signals. By way of illustration let us consider the NBFM signal

$$x_c(t) = A_c \cos[\omega_c t + \phi(t)]$$

where

$$\phi(t) = k_f \int_{-\infty}^{t} x(\tau)\, d\tau$$

For NBFM, the maximum value of $|\phi(t)|$ is much less than one (another definition for NBFM) and hence we can write $x_c(t)$ as

$$\begin{aligned} x_c(t) &= A_c[\cos \phi(t) \cos \omega_c t - \sin \phi(t) \sin \omega_c t] \\ &\approx A_c \cos \omega_c t - A_c\phi(t) \sin \omega_c t \end{aligned} \tag{6.54}$$

using the approximations $\cos \phi \approx 1$ and $\sin \phi \approx \phi$, when ϕ is very small. Equation (6.54) shows that a NBFM signal contains a carrier component and a quadrature carrier linearly modulated by (a function of) the baseband signal. Since $x(t)$ is assumed to be bandlimited to f_x, the reader can easily verify that $\phi(t)$ is also bandlimited to f_x. Hence, the bandwidth of NBFM is $2f_x$ and the NBFM signal has the same bandwidth as an AM signal.

It turns out that NBFM has no inherent advantages over linear modulation and is seldom used for transmission purposes except at UHF frequencies where NBFM signals are easier to generate than AM. NBFM is also used as an intermediate step in the generation of WBFM signals. We will discuss this application of NBFM in the following section.

6.4.4 Generation of FM Signals

There are two basic methods for generating FM signals known as *direct* and *indirect* methods. The direct method makes use of a device called *voltage controlled oscillator* (VCO) whose oscillation frequency depends linearly on the modulation voltage. In the indirect method, a narrowband FM signal is produced first using an amplitude modulator. The NBFM signal is then converted to a WBFM signal by frequency multiplication. We shall now examine these two methods.

Direct FM. A system that can be used for generating a PM or FM signal is shown in Figure 6.21. The reader can verify that the combination of message differentiation followed by FM produces a PM signal. The physical device that generates the FM signal is the VCO whose output frequency depends on the applied control voltage. VCO's are easily implemented at microwave frequencies using the reflex klystron. Reactance tube modulators are used at lower frequencies. The reactance tube modulator consists of a pentode that is operated in such a manner as to produce a capacitance which is proportional to the grid voltage. Integrated circuit VCO's are also used at lower frequencies.

At low carrier frequencies it may be possible to generate an FM signal by varying the capacitance of a parallel resonant circuit. Suppose that the capacitance in the tuned circuit shown in Figure 6.22 has a time dependence of the form

$$C(t) = C_0 - kx(t)$$

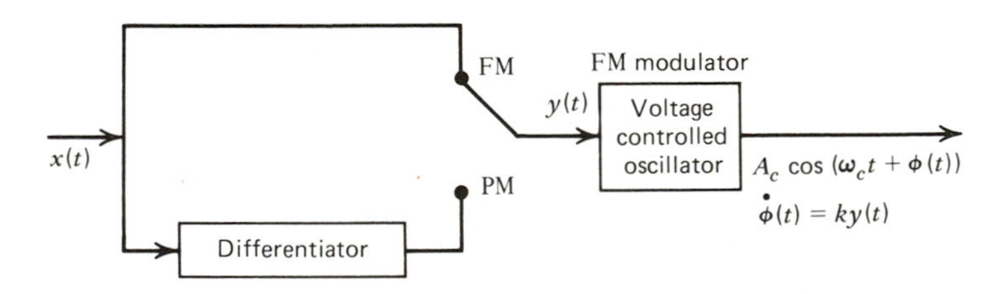

Figure 6.21 Direct method of generating FM or PM signal.

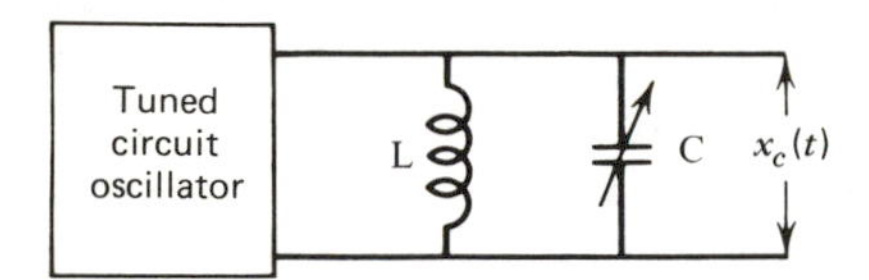

Figure 6.22 Variable reactance FM modulator.

Now, if we assume $kx(t)$ is small and slowly varying, then the output frequency of the oscillator is given by

$$f_i = \frac{1}{2\pi\sqrt{LC(t)}} = \frac{1}{2\pi\sqrt{L[C_0 - kx(t)]}}$$
$$= \frac{1}{2\pi}\frac{1}{\sqrt{LC_0}}\left(1 - \frac{k}{C_0}x(t)\right)^{-1/2}$$

Letting $f_c = [2\pi\sqrt{LC_0}]^{-1}$ and assuming $|(k/C_0)x(t)| \ll 1$, we can use the approximation $(1-z)^{-1/2} \approx 1 + \frac{1}{2}z$ and obtain

$$f_i = f_c\left(1 + \frac{1}{2}\frac{k}{C_0}x(t)\right) = f_c + \frac{k_f}{2\pi}x(t) \tag{6.55}$$

Thus the instantaneous frequency deviation of the output waveform is proportional to $x(t)$ and hence $x_c(t)$ is an FM signal.

The approximation used in obtaining Equation (6.55) is valid (good within 1%) when

$$|(k/C_0)x(t)| < 0.01 \quad \text{or} \quad |k/C_0| < 0.01$$

assuming $|x(t)| \leq 1$. The restriction $|k/C_0| < 0.01$ limits the maximum attainable frequency deviation to

$$f_\Delta = \frac{k_f}{2\pi} = \frac{k}{2C_0}f_c \leq 0.005 f_c$$

The variable reactance required for generating FM signals can be obtained using reactance tubes, saturable reactor elements, or reverse baised (varactor) diodes. The main advantage of direct FM is that large frequency deviations are possible. The major disadvantage is that the carrier frequency tends to drift and additional circuitry is required for frequency stabilization.

Indirect FM. Equation (6.54) suggests that it is possible to generate NBFM using a system such as the one shown in Figure 6.23. The signal is integrated prior to modulation and a DSB modulator is used to generate the quadrature

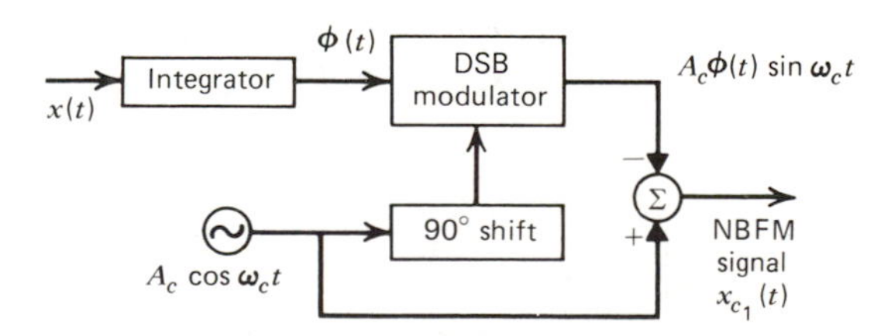

Figure 6.23 NBFM signal generation.

component of the NBFM signal. The carrier is added to the quadrature component to generate an approximation to a true NBFM signal. The output of the modulator can be approximated by

$$x_{c_1}(t) \approx A_c \cos[\omega_{c_1}t + \phi(t)]$$

The approximation is good as long as the deviation ratio $D_1 = f_{\Delta_1}/f_x$ is very small.

A WBFM signal can be obtained from a NBFM signal using a device called a *frequency multiplier* shown in Figure 6.24. The frequency multiplier is a nonlinear device designed to multiply the frequencies of an input signal by a given factor. For example, suppose that the nonlinearity is an ideal square law device with input-output characteristics $e_0(t) = ae_i^2(t)$. If the input is a NBFM signal,

$$e_i(t) = A_c \cos(\omega_{c_1}t + \beta \sin \omega_x t)$$

the output is

$$e_0(t) = \tfrac{1}{2}aA_c^2 + \tfrac{1}{2}aA_c^2 \cos(2\omega_{c_1} + 2\beta \sin \omega_x t)$$

The filter rejects the DC term and we get an output

$$x_c(t) = A_c' \cos(2\omega_{c_1}t + 2\beta \sin \omega_x t), \quad A_c' = \tfrac{1}{2}aA_c^2 \tag{6.56}$$

Equation (6.56) shows that both the carrier frequency and the modulation index have been doubled by the frequency multiplier. By using an nth law device followed by a filter, we could increase the modulation index by a factor of n.

WBFM signal generation using frequency multiplication was first suggested by Armstrong in 1936 and is still used in commercial FM broadcasting. Limitations of this method include the fact that losses incurred in the harmonic generation require additional amplification, and small phase instabilities in the multiplication process accumulate and appear as noise in the output. Furthermore, use of frequency multiplication often increases the carrier frequency to too high a value for a given application. To avoid this, a frequency down conversion (using a mixer or DSB modulator) is necessary.

By way of illustrating the use of indirect method of WBFM signal generation, let us look at early transmitters for commercial FM. These transmitters had $f_{c_1} = 200$ kHz, $f_{\Delta_1} = 25$ Hz, and $f_x = 15$ kHz so that the deviation ratio of

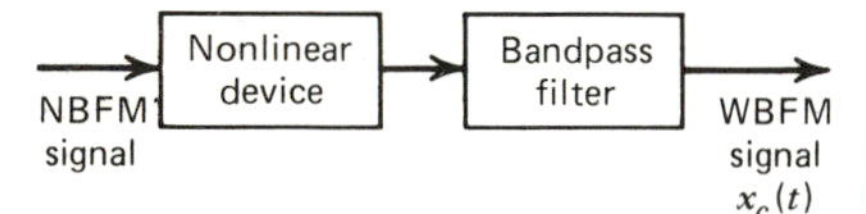

Figure 6.24 Frequency multiplier for narrowband to wideband conversion.

the NBFM signal before frequency multiplication is 0.00167 (indeed a very NBFM signal). Since the final peak frequency deviation for commercial FM is 75 kHz, the frequency multiplication factor required is $(75)(10^3)/25 = 3000$. After multiplication, the new carrier frequency becomes $(200)(10^3)(3000) = 600$ MHz, which is outside the FM band of 88 to 108 MHz. A frequency down converter with a local oscillator frequency in the range of 600 ± 100 MHz is used to translate the spectrum from 600 MHz down to the FM band.

6.4.5 Demodulation of FM Signals

An FM demodulator is required to produce an output voltage that is linearly proportional to the input frequency. Circuits that produce such response are called *discriminators*. If the input to a discriminator is an FM signal,

$$x_c(t) = A_c \cos\left[\omega_c t + k_f \int_{-\infty}^{t} x(\tau)\, d\tau\right]$$

the discriminator output will be

$$y_d(t) = k_d k_f x(t)$$

where k_d is the *discriminator sensitivity*. The characteristics of an ideal discriminator are shown in Figure 6.25*a*.

An approximation to the ideal discriminator characteristics can be obtained by the use of a differentiator followed by an envelope detector (Figure 6.25*b*). If the input to the differentiator is $x_c(t)$, then the differentiator output is

$$z_c(t) = -A_c[\omega_c + k_f x(t)] \sin[\omega_c t + \phi(t)] \tag{6.57}$$

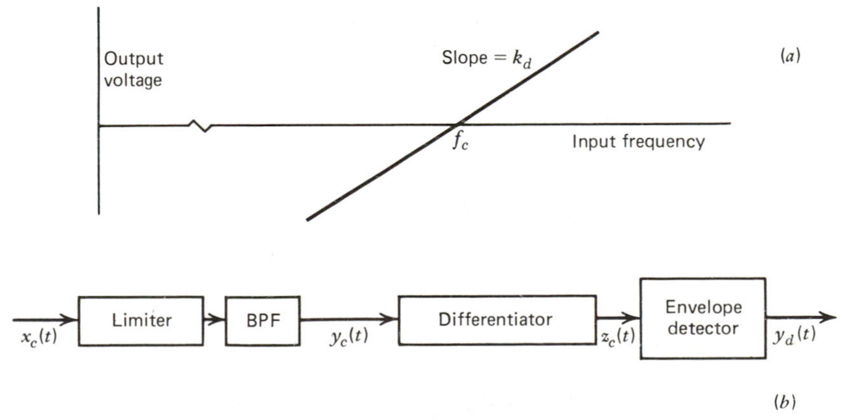

Figure 6.25 (*a*) Ideal discriminator. (*b*) FM discriminator with limiter.

With the exception of the phase deviation $\phi(t)$, the output of the differentiator has the form of an AM signal. Hence envelope detection can be used to recover the message signal. The baseband signal is recovered without any distortion if $\max\{k_f|x(t)|\} = 2\pi f_\Delta < \omega_c$, which is true in most practical systems.

The method described in the preceding section is also referred to as *slope detection*, and differentiating an FM signal to convert it to an AM/FM signal is called *FM to AM conversion.* One of the main problems with this method is that the detector responds to spurious amplitude variations of the FM signal due to noise and due to unintended FM to AM conversion that takes place prior to the detector. In order to ensure that the amplitude at the input to the differentiator is constant, a limiter is placed before the differentiator. The limiter clips the input signal and converts it to a frequency modulated square wave whose zero crossing contains the message signal. The rectangular waveform is filtered to remove harmonic contents before differentiation (Problem 6.41).

A second problem with slope detection is that the linear range of the frequency-to-voltage characteristic is often quite small. A *balanced discriminator* such as the one shown in Figure 6.26 can be used to extend linearity. A balanced discriminator has two resonant circuits, one tuned above f_c and the other below f_c. The overall frequency-to-voltage characteristics has the S shape shown in Figure 6.26. The DC component is automatically blocked and the response to low modulating frequencies is good. Balanced discriminators are widely used in the microwave range of frequencies with resonant cavities as tuned circuits and crystal diodes as envelope detectors.

Before we discuss other methods of demodulating FM signals, let us take a qualitative look at how WBFM can be used to provide increased signal-to-noise ratios without increased transmitted power. We have just seen that the output of the FM demodulator is proportional to the frequency deviation and one can increase output signal power by increasing the deviation. Since the carrier amplitude remains constant in FM, the increase in output power that results from increasing deviation is realized without increasing transmitted power. In AM systems, for example, the output signal power can be increased only by increasing the transmitted power.

The use of large frequency deviation in WBFM requires a greater transmission bandwidth. Thus in WBFM we are trading bandwidth for signal power, an important concept in the design of communication systems.

6.4.6 Phase-Lock Loops

A different approach for demodulating angle modulated signals is to use a feedback system. *Feedback demodulators* perform better than discriminators

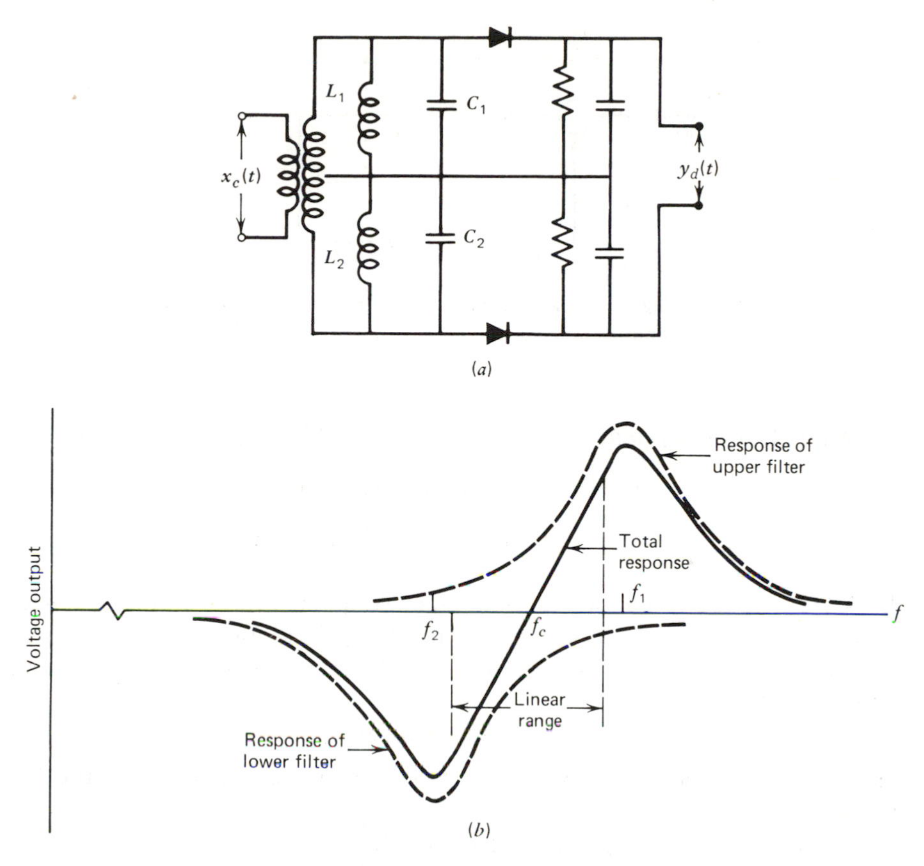

Figure 6.26 Balanced discriminator. (*a*) Circuit. (*b*) Frequency to voltage characteristics: $f_1 = \dfrac{1}{2\pi\sqrt{L_1C_1}}$, $f_2 = \dfrac{1}{2\pi\sqrt{L_2C_2}}$

in the presence of noise (a fact we will discuss in Chapter 7). Among the demodulators in this group are the phase-lock loop (PLL) demodulator and the FM demodulator with feedback (FMFB). PLL demodulators are widely used in today's communication systems because of their superior performance, ease of alignment, and ease of implementation using inexpensive integrated circuits. We will center our discussion on the PLL and refer the reader to advanced texts on communication theory for details of FMFB demodulators.

A block diagram of a PLL demodulator is shown in Figure 6.27. The system consists of two main functional blocks, a *phase comparator*, and a voltage controlled oscillator (VCO).

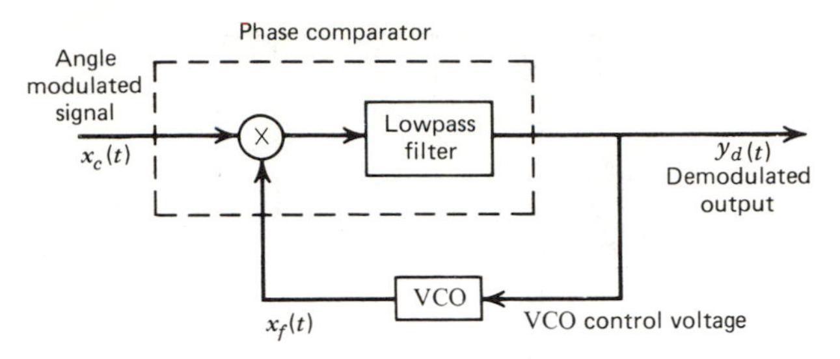

Figure 6.27 Block diagram of a phase-lock loop for demodulating FM signals.

The input to the system is assumed to be an FM signal of the form

$$x_c(t) = A_c \sin[\omega_c t + \theta_i(t)]$$
$$= A_c \sin\left[\omega_c t + k_f \int x(\tau)\, d\tau\right]$$

The VCO is tuned to operate at an angular frequency of ω_c in the absence of control voltage. When a control voltage $y_d(t)$ is present, then the instantaneous frequency deviation of the VCO is proportional to the control voltage. That is,

$$\omega_f(t) = \omega_c + g_v y_d(t)$$

where g_v is the sensitivity of the VCO in (radian/sec)/volt. Thus the VCO output has the form*

$$x_f(t) = A_f \cos\left(\int \omega_f(\tau) d\tau\right)$$
$$= A_f \cos\left(\omega_c t + g_v \int_{-\infty}^{t} y_d(\tau)\, d\tau\right)$$
$$= A_f \cos[\omega_c t + \theta_f(t)] \tag{6.58}$$

The reader can easily verify that the phase comparator output is given by

$$y_d(t) = \tfrac{1}{2} A_f A_c \sin[\theta_i(t) - \theta_f(t)] = g_p \sin[\theta_i(t) - \theta_f(t)] \tag{6.59}$$

When the loop is operating in lock, the phase error $\theta_i(t) - \theta_f(t)$ will be small and

$$\sin[\theta_i(t) - \theta_f(t)] \approx \theta_i(t) - \theta_f(t)$$

This condition allows the sinusoidal nonlinearity to be neglected and the PLL becomes a linear feedback control system shown in Figure 6.28.

*The assumptions of $A_c \sin \omega_c t$ as the unmodulated carrier and $A_f \cos \omega_c t$ as the VCO output are required so that the comparator output is zero when $\theta_i = \theta_f$.

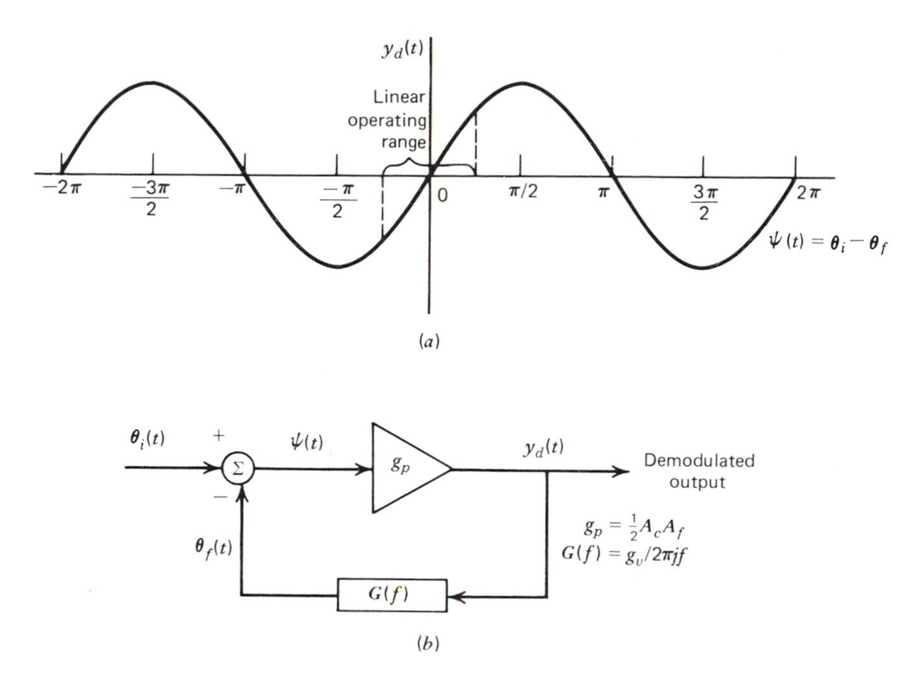

Figure 6.28 (*a*) Linear approximation of PLL characteristics. (*b*) Linearized model of PLL.

Principle of Operation. To illustrate how the PLL operates, assume that the loop is in lock at $t = 0$ and that at $t = 0^+$ the frequency of the input FM signal changes by ω (corresponding to a step change in the message signal). Then, at $t = 0^+$,

$$\Psi(t) = [\theta_i(t) - \theta_f(t)] = \omega t \quad \text{since } \frac{d}{dt}[\theta_i(t) - \theta_f(t)] = \omega$$

As $\Psi(t)$ begins to increase, the phase comparator output will be positive, which in turn increases the frequency of the VCO. A new equilibrium point will be reached when the frequency of the VCO is equal to the frequency of the input signal. Since the VCO frequency deviation is proportional to $y_d(t)$, $y_d(t)$ then represents the frequency deviation of the input FM signal. Thus $y_d(t)$ is the demodulated output.

When equilibrium is established, the frequency of the input signal and the frequency of the VCO output will be identical; but, the phase difference will no longer be 0°. For, if the VCO is to operate at a frequency other than ω_c, there must be a control voltage $y_d(t)$ that is generated by a phase difference other than 0°. (A phase difference of 0° will yield $y_d(t) = 0$.) The reader can easily verify that, at equilibrium after a step change,

$$\theta_i(t) = \omega t, \quad \theta_f(t) = \omega t - \frac{\omega}{g_p g_v}, \quad \frac{d\theta_i(t)}{dt} = \frac{d\theta_f(t)}{dt}$$

$$y_d(t) = \omega / g_v \quad \text{and} \quad \Psi(t) = \Psi_e = \omega / g_p g_v$$

where Ψ_e is the new equilibrium point.

Now, if the frequency of the input FM signal changes continuously at a rate slower than the time required for the PLL to establish a new equilibrium, the PLL output will be continuously proportional to the frequency deviation of the carrier.

Initially, with an unmodulated carrier, the operating point of the phase comparator is adjusted to be at the origin of the plot shown in Figure 6.28*a*. Under stable operating conditions, the equilibrium point moves up and down along the Ψ versus $y_d(t)$ curve, shown in Figure 6.28*a*, between $-\pi/2$ to $\pi/2$.

First-Order PLL and Its Operating Range. The operation of the PLL shown in Figure 6.27 and 6.28*b* can be described by a first-order differential equation, and hence it is called a first-order PLL. The differential equation that describes the first-order PLL, operating in the linear range of the phase comparator, is given by

$$\frac{d}{dt}[\theta_f(t)] = g_v y_d(t) = g_v g_p \Psi(t),$$

or

$$\frac{d\Psi(t)}{dt} + \frac{\Psi(t)}{\tau} = \frac{d}{dt}[\theta_i(t)] \tag{6.60}$$

where $\tau = 1/g_p g_v$. The linear approximation for the PLL holds and the PLL is in normal operating range when

$$|\Psi| \ll \pi/2 \tag{6.61}$$

Now, suppose that the input signal changes in frequency by ω, that is,

$$\frac{d}{dt}\theta_i(t) = \omega$$

Then, subject to initial conditions, the solution to Equation (6.60) is

$$\Psi(t) = \omega\tau(1 - \exp(-t/\tau)), \quad |\Psi| \ll \pi/2$$

The new equilibrium value of Ψ, denoted by Ψ_e, is

$$\Psi_e = \omega\tau$$

The operating range of the PLL is the range of angular frequency changes of the input carrier that will keep the steady-state operating point in the linear

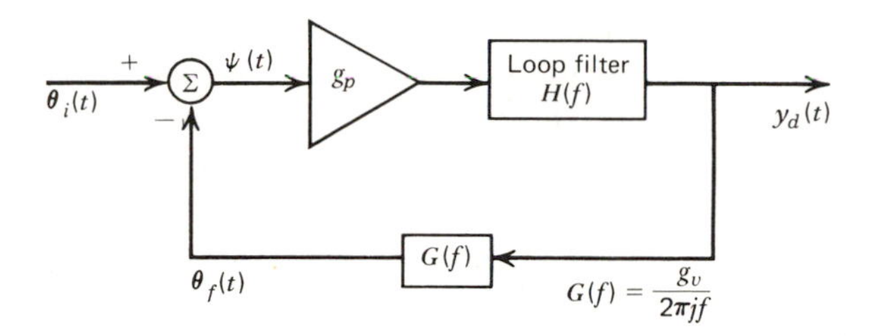

Figure 6.29 Higher order PLL.

range. This range is defined by

$$\Psi_e = \omega_{\max}\tau \ll \pi/2$$

or

$$\omega_{\max} \ll \pi/2\tau \tag{6.62}$$

where $\omega_{\max}$ is the maximum value of step type changes in the frequency of the input signal. Equation (6.62) reveals that by having a small value for τ, the PLL can be made to stay in the linear range and track step type changes in the input frequency. As τ becomes smaller, the speed with which the PLL responds to frequency changes increases. Smaller values of τ also imply a larger bandwidth, which is undesirable when noise is present.

Second-Order PLL.* The performance of the PLL can be improved considerably if a loop filter is included in the system as shown in Figure 6.29. The loop filter transfer function often used in the second-order PLL is

$$H(f) = \frac{1 + 2\pi jf\tau_2}{2\pi jf\tau_1} \tag{6.63}$$

The overall transfer function of the second-order PLL is

$$\frac{Y_d(f)}{\theta_i(f)} = \frac{g_p(2\pi jf)(1 + 2\pi jf\tau_2)}{\tau_1[(2\pi jf)^2 + 2\xi\omega_n(2\pi jf) + \omega_n^2]} \tag{6.64}$$

where $\omega_n = \sqrt{g_p g_v/\tau_1}$ is the natural frequency of the loop and $\xi = \omega_n\tau_2/2$ is the damping factor. The 3-dB bandwidth of the PLL is given by

$$B_{\text{PLL}} \approx (1 + \sqrt{2}\xi)\omega_n/2\pi, \quad \tfrac{1}{2} < \xi < 1$$

A criterion used for demodulating a signal having a bandwidth f_x is to choose

$$2\omega_n\xi > 2\pi f_x$$

*The reader is referred to textbooks on control systems for detailed analysis of second-order systems (see reference [11] for example).

Other parameters of the PLL are chosen such that the value of the phase error is in the linear range of the phase comparator, and the damping factor ξ is in the range of $\frac{1}{2} < \xi < 1$. The peak value of phase error for a step-type change in the frequency of the input signal is $\leq \Delta\omega/(2\xi\omega_n)$ (Problem 6.44), where $\Delta\omega$ is the magnitude of the step.

The second-order PLL provides good performance and is widely used in a variety of applications. The reader can easily verify that by integrating the output $y_d(t)$, we can use the PLL described in this section for demodulating PM signals.

Sweep Acquisition in PLL. In our analysis, we have assumed that initially the VCO frequency is the same as the frequency of the unmodulated FM carrier. If this is not true, then the PLL must go through a mode of acquisition to gain frequency lock. If the frequency difference is less than the loop bandwidth, the PLL will lock quickly. If the range of frequency differences is much larger than the loop bandwidth, the VCO frequency must be swept manually (as in the tuning of an FM receiver) or automatically to search for the signal. This sweep is similar to the sweep in a scanning spectrum analyzer and is subjected to the restriction that the maximum sweep rate be less than the square of the loop bandwidth.

The frequency sweep can be accomplished by applying a ramp voltage as the VCO control voltage (Figure 6.30). As the loop locks on to the input frequency, the ramp voltage is disconnected and the loop is closed in a normal fashion. A lock indicator is used to control this switching.

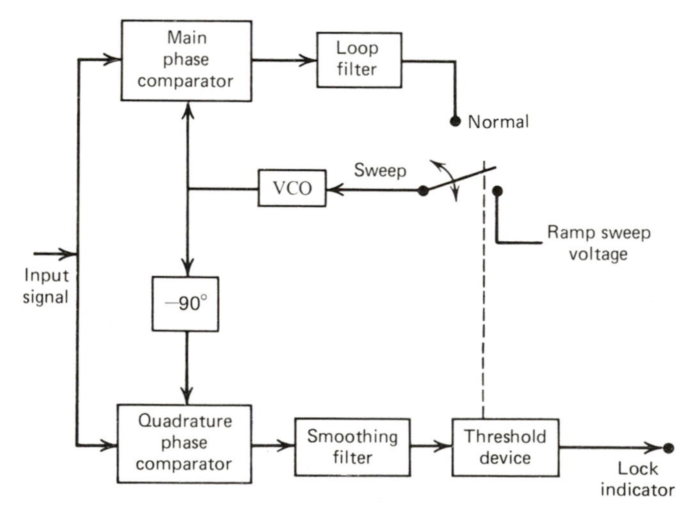

Figure 6.30 PLL sweep aquisition.

A *quadrature phase comparator* is commonly used as a lock indicator. As the loop pulls into lock, the main phase comparator output $\sin(\theta_i - \theta_f) \rightarrow 0$ whereas the quadrature phase comparator output $\cos(\theta_i - \theta_f) \rightarrow 1$. Thus the locked condition is sensed when the output of the quadrature phase comparator exceeds a threshold value. The lock indicator turns off the sweep acquisition circuitry and turns on an indicator to signal the lock condition (as in the case of the stereo indicator in an FM receiver, or the signal indicator in a data modem using phase modulation).

6.5 FREQUENCY DIVISION MULTIPLEXING

Simultaneous transmission of several message signals over a single channel is called multiplexing. There are two basic multiplexing techniques: *frequency division multiplexing* (FDM) and *time division multiplexing* (discussed in Chapter 10). In FDM, the available channel bandwidth is divided into a number of nonoverlapping "slots" and each message signal is assigned a slot of frequencies within the passband of the channel. Individual signals can be extracted from the FDM signal by appropriate filtering. FDM multiplexing is used in long distance telephone, FM stereo and TV broadcasting, space probe telemetry, and other applications.

The principle of FDM is illustrated in Figure 6.31 for three message signals that are assumed to be bandlimited. In general, if the message signals are not strictly bandlimited, then lowpass filtering will be necessary. The bandlimited signals individually modulate the *subcarriers* at frequencies f_{c_1}, f_{c_2}, and f_{c_3}. The subcarrier modulation shown in the example is SSB, but any CW modulation technique can be employed. The modulated signals are summed to produce a composite multiplexed signal $x(t)$ whose spectrum is shown in Figure 6.31*c*.

If the subcarrier frequencies are properly chosen, then each message signal occupies a frequency slot without any overlap. While the individual messages are clearly identified in the frequency domain, the multiplexed signal will bear no resemblance to the message signals in the time domain! The multiplexed signal $x(t)$ may be transmitted directly, or used to modulate another carrier at frequency f_c before transmission.

Recovery of individual message signals is shown in Figure 6.32. The first step in the recovery is the demodulation to extract $x(t)$ from $x_c(t)$. Bandpass filtering of $x_c(t)$ separates $x_{c_1}(t)$, $x_{c_2}(t)$, and $x_{c_3}(t)$. Finally, the messages are recovered by individually demodulating $x_{c_1}(t)$, $x_{c_2}(t)$, and $x_{c_3}(t)$. The multiplexing and demultiplexing equipment are often referred to by the abbreviation "*MUX*."

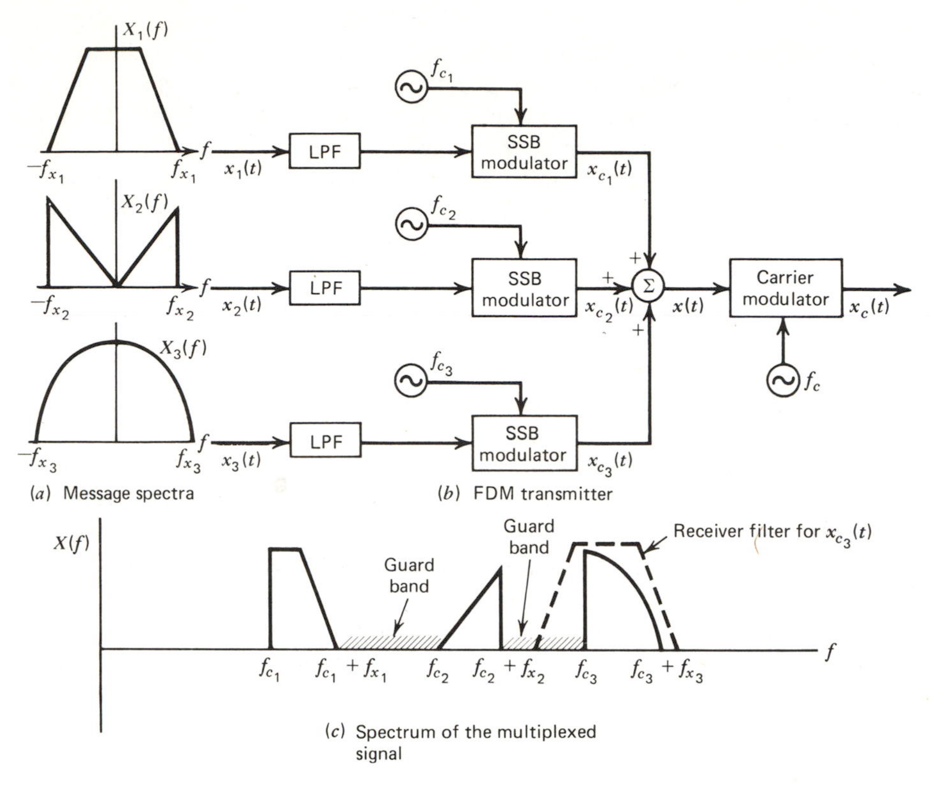

Figure 6.31 Frequency division multiplexing (FDM). (*a*) Message spectra. (*b*) FDM transmitter. (*c*) Spectrum of the multiplexed signal.

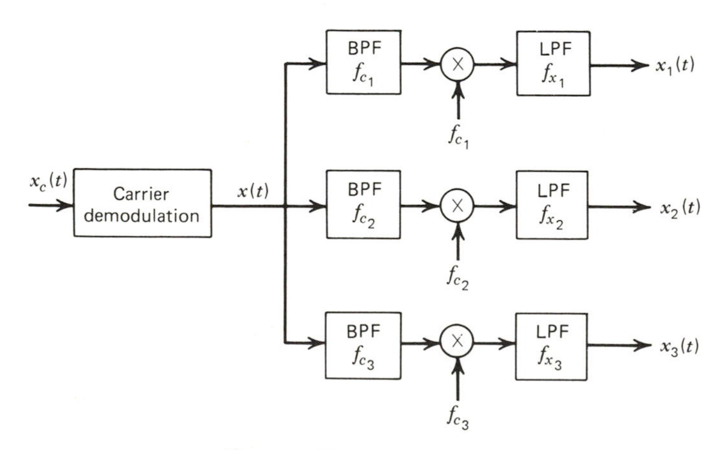

Figure 6.32 FDM receiver.

Table 6.3. Bell system FDM hierarchy.

Designation	Composition	Frequency range (kHz)	Bandwidth (kHz)	Number of voice channels
Group	12 voice channels	60–108	48	12
Super group	5 groups	312–552	240	60
Master group	10 super groups	564–3084	2520	600
Super master group	6 master groups	500–17,500	17,000	3600

One of the major problems with FDM is *cross talk*, the unwanted cross coupling of one message to another. Cross talk (intermodulation) arises mainly because of nonlinearities in the system and considerable care must be taken to reduce nonlinearities in devices that process FDM signals. A second source of cross talk is imperfect spectral separation of signals due to imperfect filtering and due to subcarrier frequency drifts. To reduce the possibility of spectral overlap, the modulated spectra are spaced out in frequency by *guard bands* into which the filter transition regions can be fitted.

The minimum bandwidth of an FDM signal is equal to the sum of the bandwidths of all the message signals. If a modulation scheme other than SSB is used for multiplexing, the FDM signal bandwidth will be higher. The presence of guard bands increases the bandwidth further.

FDM is widely used in telephone systems for transmitting a large number of voice signals over a single channel. The Bell system uses a modular FDM structure where up to 3600 voice channels are multiplexed for transmission via coaxial cable. The multiplexing is done using SSB modulation with subcarriers that are derived from a common oscillator and a 512 kHz pilot carrier is transmitted along with the multiplexed signal for synchronization. Grouping of channels as shown in Table 6.3 makes it possible to avoid excessive guardband requirements. Furthermore, grouping facilitates easy switching and routing of the signal and allows modular design for the equipment (see Problem 6.49).

Other applications of FDM such as in FM stereo and TV broadcasting will be discussed in the following sections.

6.6 COMMERCIAL BROADCASTING

Commercial broadcasting using electromagnetic radiation over the airways began in the early part of the 20th century with the invention of the AM radio.

The first regular radio broadcast was started by radio station WWJ of Detroit and KDKA of Pittsburgh in 1920. The CBS network started commercial radio broadcasting in 1927 with a 16-station network. Today, the broadcast industry consists of thousands of AM, FM, and TV stations. The frequency allocation and the licensing of airway transmission in the United States is regulated by the Federal Communication Commission (FCC).

In this section, we take a brief look at the important aspects of commercial broadcasting with emphasis on AM, FM, and TV broadcasting. We will discuss the spectra of these signals in detail and describe the receivers used for detecting these signals.

6.6.1 AM Radio Broadcasting and Reception

The carrier frequency allocation for commercial AM radio broadcasting in the U.S. ranges from 540 to 1600 kHz with 10 kHz spacing. Stations in local proximity are usually assigned carrier frequencies that are separated by 30 kHz or more. Transmission bandwidth is restricted to 10 kHz, resulting in a message bandwidth of 4 to 5 kHz. Interference between transmissions is controlled by a combination of frequency allocation, transmitter power, transmitting antenna pattern, and other restrictions such as carrier stability (± 20 Hz). The permissible average power, for an unmodulated carrier, is from 1 kW for local stations to 50 kW for regional stations. All stations are required to maintain an 85 to 95% modulation.

The block diagram of an AM receiver of the *superheterodyne* type is shown in Figure 6.33. It consists of an RF amplifier that is tuned to the desired carrier frequency. A local oscillator-mixer arrangement translates the radio

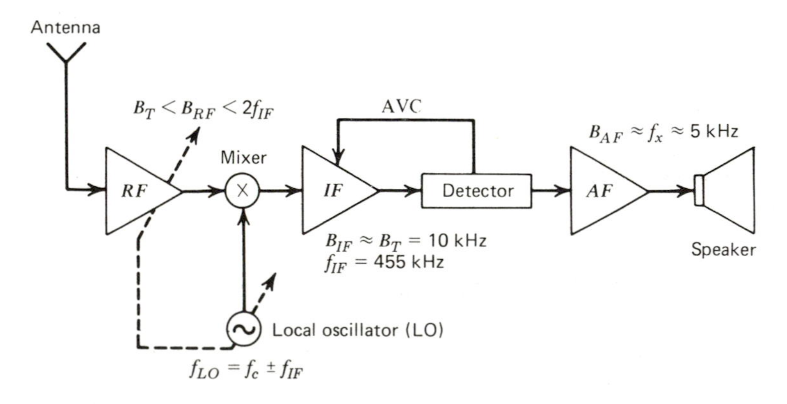

Figure 6.33 Superheterodyne receiver.

frequency (RF) signal to an intermediate frequency (IF) signal at a carrier frequency $f_{IF} = 455$ kHz. The IF amplifier is tuned to the fixed IF frequency of 455 kHz, and provides most of the gain and frequency selectivity in the system. The audio amplifier follows the detector and it brings up the power level to that required for the loudspeaker. The power requirements of the speaker range from a fraction of a watt to several watts (at a speaker impedance of 8 or 16 ohms).

The RF tuning range is of course 540 to 1600 kHz, while the local oscillator must tune over 995 to 2055 kHz if $f_{LO} = f_c + f_{IF}$, and over 85 to 1145 kHz if $f_{LO} = f_c - f_{IF}$. The tuning range is approximately 2:1 for the first case and 13:1 for the second case, which is harder to implement. Hence $f_{LO} = f_c + f_{IF}$ is taken to be the local oscillator frequency.

The bandwidth of the RF amplifier must be equal to the transmission bandwidth B_T. If B_{RF} is much larger, signals from two stations might reach the IF amplifier resulting in interference. Suppose that there are two RF signals at $f_c = f_{LO} - f_{IF}$ and $f_c' = f_{LO} + f_{IF}$. Now, if we are trying to receive the station at $f_c = f_{LO} - f_{IF}$, the local oscillator signal will have to be at f_{LO}. When the local oscillator signal is mixed with the incoming signals, we will also pick up the signal at f_c', which is called the *image frequency*. The reader can verify that the image frequency signal will be rejected by the RF filter if the RF filter bandwidth is less than $2f_{IF}$.

Receivers sometimes respond to spurious frequencies other than image frequencies. For example, when a strong signal of frequency $\frac{1}{2}f_{IF}$ gets into the IF input, any second harmonic produced by nonlinearities in the IF amplifier will be at the IF frequency. This second harmonic will be amplified and appear at the detector output as interference. Elaborate design using multiple mixers and IF amplifiers are required to permit sharper discrimination against such interference.

An important feature of AM receivers is the AVC or the automatic volume control. The RF signal at the receiver input often changes slowly with respect to time due to fading. This results in changes in the output audio signal level that could be annoying. AVC takes care of these variations by using the DC-offset of the detector output to control the gain of the IF amplifier. Since the DC offset is proportional to the RF signal level (see the section on envelope demodulation), the feedback arrangement compensates for RF signal level variations by controlling the gain of the IF amplifiers by appropriate biasing.

6.6.2 FM Radio Broadcasting and Reception

In 1936, Armstrong made the case for FM radio in his classical paper entitled "A Method of Reducing Disturbance in Radio Signalling by a System of

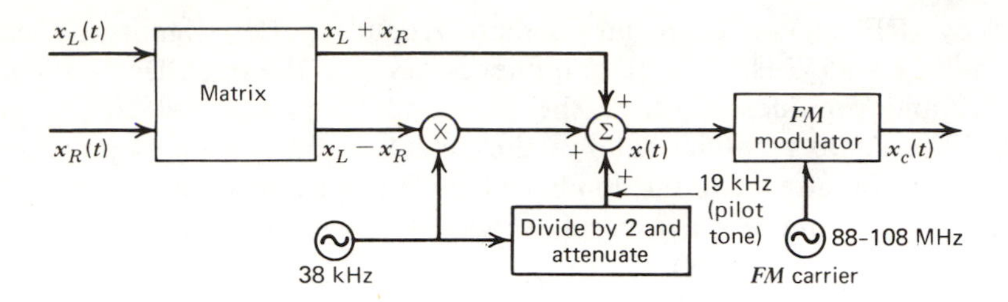

Figure 6.34 Stereo FM transmitter.

Frequency Modulation." FM broadcasting began shortly thereafter and today there are several thousand AM-FM and FM broadcasting stations. Significant changes in FM broadcasting since its beginnings in the 1930s include two-channel stereo broadcasting (authorized by the FCC in 1961) and quadraphonic FM broadcasting (1975).

FCC assigns carrier frequencies for commercial FM broadcasting in the United States at 200 kHz intervals in the range 88 to 108 MHz. The peak frequency deviation is fixed at 75 kHz, and power outputs are specified at 0.25, 1, 3, 5, 10, 25, 50, and 100 kW. The 200 kHz available to each station allows the transmission of one or more high quality audio signals with room to spare.

A block diagram of a commercial stereo FM transmitter is shown in Figure 6.34. Two audio signals labelled left and right (representing different orchestra sections, different speakers, etc.) are first processed by a "matrix" to produce sum and difference signals. The sum signal is heard with a "monophonic" receiver and is inserted directly into the baseband input of the FM transmitter. The difference signal DSB modulates a 38 kHz carrier and is added to the sum signal along with a 19 kHz pilot tone. The pilot tone is used for synchronization at the receiver. The composite spectrum of the baseband stereo signal is shown in Figure 6.35. (Observe that FDM is used for

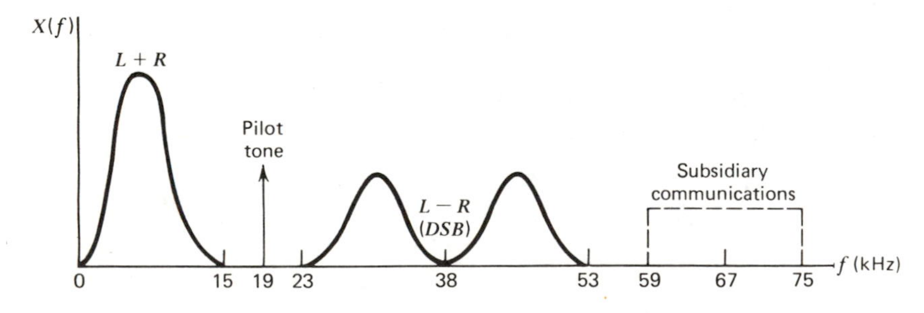

Figure 6.35 Baseband spectrum of multiplexed stereo FM signal.

Table 6.4. Maximum deviation allowed for various signals. Entries in the table denote percent deviation allowed for each signal acting alone. Peak deviation of the composite signal is limited to 75 kHz.

	Broadcast Mode		
Signals	Mono + SCA	Stereo	Stereo + SCA
L + R	70%	90%	80%
L − R	0	90%	80%
Pilot	0	10%	10%
SCA	30%	0	10%

multiplexing stereo signals.) In addition to the two audio signals, FCC authorizes FM stations to carry another signal referred to as SCA (subsidiary communication authorization). The SCA transmissions carry no commercial messages and are intended for private subscribers for background music in stores, offices, and so forth.

The various signals shown in Figure 6.35 modulate the FM carrier (88 to 108 MHz) to varying extents as shown in Table 6.4. The block diagram of an FM stereo multiplex receiver is shown in Figure 6.36. Here we show how the pilot tone is used for synchronously demodulating the difference signal. It is possible to use envelope demodulation if the pilot carrier is added to the difference signal before envelope detection.

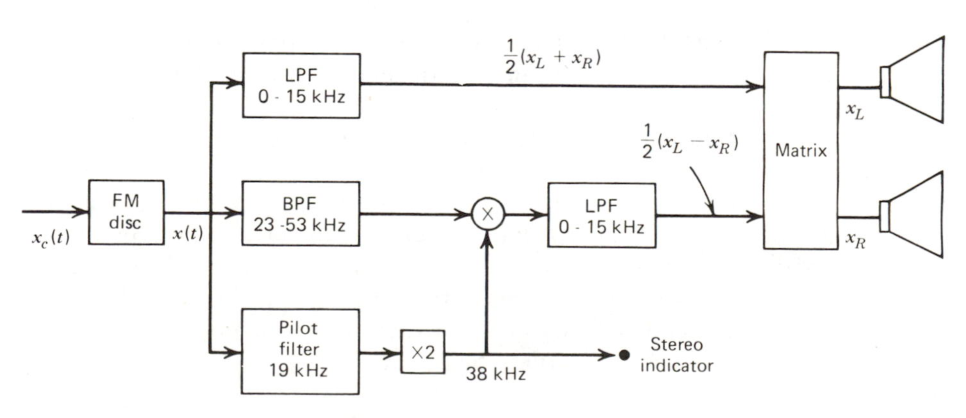

Figure 6.36 FM stereo multiplex receiver.

Not shown in Figures 6.34 and 6.36 are the preemphasis filter at the transmitter and the deemphasis filter at the receiver which are used for minimizing the effects of noise. We will discuss how preemphasis/deemphasis filters can be used for noise reduction in the next chapter. We have also not discussed the details of quadraphonic (four-channel) audio transmission. The quadraphonic system is not widely used at present because of the higher cost associated with complex receivers and additional speakers.

6.6.3 Commercial TV Broadcasting and Reception

The feasibility of black and white television, with voice, was demonstrated in 1930 and black and white TV broadcasting began in the middle 1930s. Early TV sets used very small cathode ray tubes and larger picture tubes were not developed until the late forties (the first 16″ tube came out in 1948). The feasibility of color television was demonstrated in 1940 and FCC authorized color broadcasting in 1941. However, large scale commercial network color broadcasting started only in 1965. Today, the TV broadcast industry consists of three major commercial networks (CBS, NBC, and ABC) and a large number of smaller independent networks. In addition to commercial networks, nonprofit organizations broadcast a variety of educational and other public service programs. TV broadcasting stations number in the several hundreds today. Television standards vary from country to country and we will only discuss the NTSC (National Television Systems Committee) system presently used in the United States. Other systems in use are the PAL (Phase Alternating Line) system in England and the SECAM (Sequential Couleur a' M'emoire) system in France. For details of these systems the reader is referred to Patchett's book listed at the end of this chapter.

Format of Black and White TV Signals. The scanning technique used for converting picture information to an electrical waveform is shown in Figure 6.37. The scanning spot starts at the upper left-hand corner A and moves at a constant rate to B during which time the scanning device produces a voltage or current proportional to the image intensity along the path of the scanning spot. Upon reaching B, the scanning spot returns rapidly to C (horizontal retrace) and begins another scan line. After reaching D, the scanning spot returns vertically (vertical retrace) to point E to start the *interlaced pattern* that ends at F. The two scan patterns, shown by solid and broken lines, are called the first and second fields and together they constitute a *frame.* To eliminate discernible flicker, alternate fields are sent at a rate of 60 fields/sec (or 30 frames/sec).

Timing information is added to the picture (video) signal at the transmitter

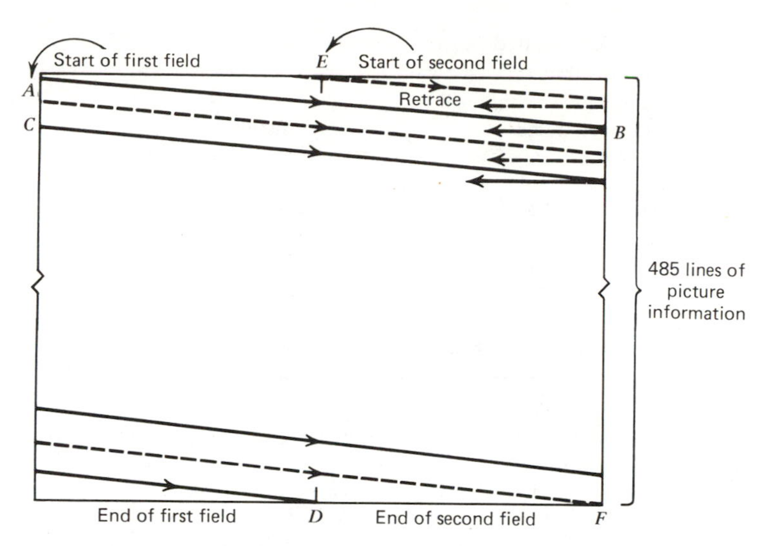

Figure 6.37 Interlaced scanning raster. Separation between lines is shown exaggerated.

so that the receiver can maintain proper line and field synchronization. The composite video signal is shown in Figure 6.38. Blanking pulses are inserted during retrace intervals so that the retrace lines are blanked at the receiver picture tube. Line synchronizing pulses are added on top of the blanking pulse to synchronize the receiver's horizontal sweep circuit. A special sequence of vertical synchronizing pulses is transmitted at the end of each field to synchronize the receiver's vertical sweep circuit. The important parameters of United States television standards are shown in Table 6.5.

The bandwidth of the video signal can be calculated as follows. The picture

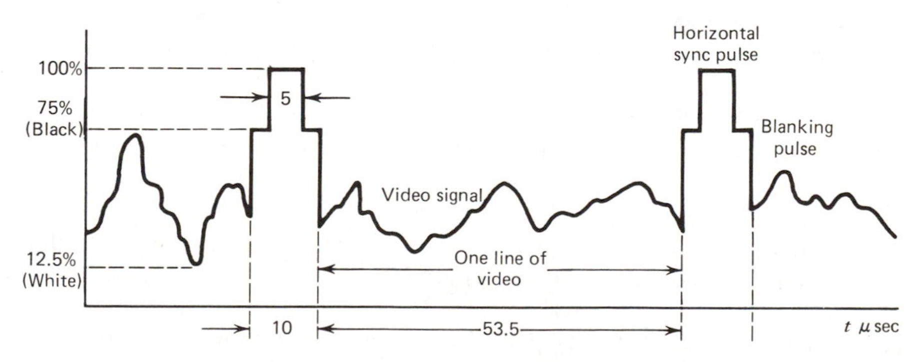

Figure 6.38 Composite video signal.

Table 6.5. United States TV standards.

Aspect ratio (width/height)	4/3
Total lines/frame	525/frame
Line rate	15,750 lines/sec
Line time	63.5 μsec
Horizontal retrace	10 μsec
Vertical retrace	1.27 msec (20 lines/field)
Picture information/frame	485 lines/frame

Table 6.6. VHF and UFH television allocations.

Channel number	Frequency band
VHF 2, 3, 4	54–60, 60–66, 66–72 MHz
VHF 5, 6	76–82, 82–88 MHz
VHF 7–13	174–216 MHz; 6 HMz intervals
UHF 14–83	470–890 MHz; 6 MHz intervals

may be viewed as an array of 525 lines to (525)(4/3) columns resulting in a total of 367,500 picture elements/frame. Thus during each frame time we have to transmit 367,500 intensity samples/frame or 11.0 million samples/sec. If we assume these samples to be independent, then by virtue of sampling theorem a minimum bandwidth of 5.5 MHz is required for transmitting the video signal. In actual pictures, the intensity of adjacent picture elements are highly dependent (redundant) and hence the actual bandwidth of the video signal is smaller. The video bandwidth in commercial TV broadcasting is set at 4.2 MHz. The audio signal is frequency division multiplexed with the video signal resulting in a composite message signal that has a bandwidth of approximately 4.5 MHz.

The transmission bandwidth allowed by FCC for TV broadcasting is 6 MHz. The frequency bands are shown below in Table 6.6. The 6 MHz bandwidth allocated for each station is not wide enough to transmit the TV signal using DSB. Hence, suppressed sideband modulation is used in TV broadcasting.

TV Transmitters and Receivers. The video signal has large bandwidth and significant low-frequency content. Bandwidth conservation suggests the use of SSB modulation, but SSB systems have poor low-frequency response. While DSB modulation yields good low-frequency response, the bandwidth becomes excessively large. Furthermore, it is important to keep the TV

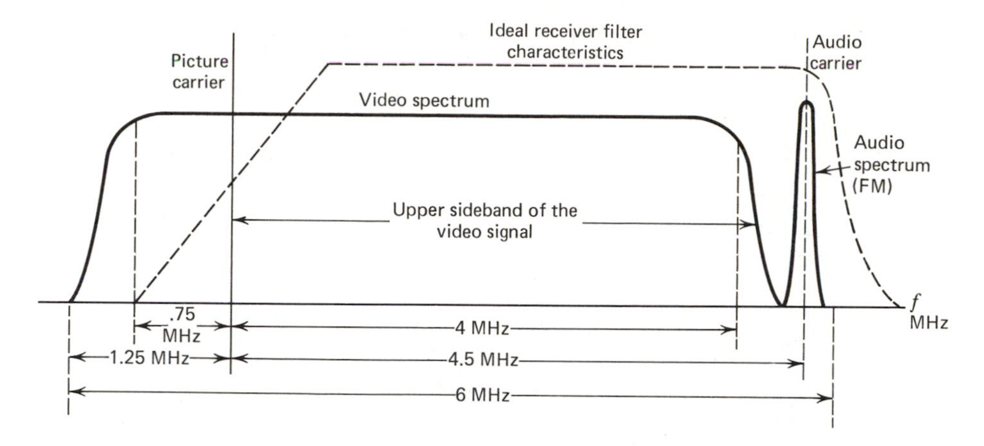

Figure 6.39 Spectrum of black and white TV signal (not to scale).

receiver simple since millions of receivers are involved. This last requirement suggests a signaling scheme suitable for envelope demodulation. The modulation scheme that best meets the above requirements is VSB + carrier.

The details of the TV signal spectrum, as transmitted, are shown in Figure 6.39. The upper sideband of the video signal is transmitted without attenuation up to 4 MHz. Thereafter it is attenuated so that it does not interfere with the lower sideband of the sound carrier located 4.5 MHz away from the carrier. The lower sideband of the picture carrier is transmitted without attenuation over the range 0 to 0.75 MHz and is entirely attenuated at 1.25 MHz. The reader should note that the modulated video signal, as transmitted, is not shaped like a VSB signal which is required to have some symmetries (Section 6.3.3). Precise VSB signal shaping takes place at the receiver before demodulation. The VSB shaping is more easily carried at the IF amplifier in the receiver where the power level is low.

The audio signal is frequency modulated on a separate carrier with a frequency deviation $f_\Delta = 25$ kHz. With an audio bandwidth of 10 kHz, we have a deviation ratio of 2.5 and an FM bandwidth of approximately 70 kHz. The total bandwidth of the modulated TV signal (audio plus video) is about 5.75 MHz leaving a 250 kHz guard band between adjacent TV channels. The details of a TV transmitter are shown in Figure 6.40.

The details of a TV receiver are shown in Figure 6.41. TV receivers are of the superheterodyne type with the main IF amplifier operating in the 41 to 46 MHz range and providing VSB shaping. The audio signal is also amplified by the IF amplifier.

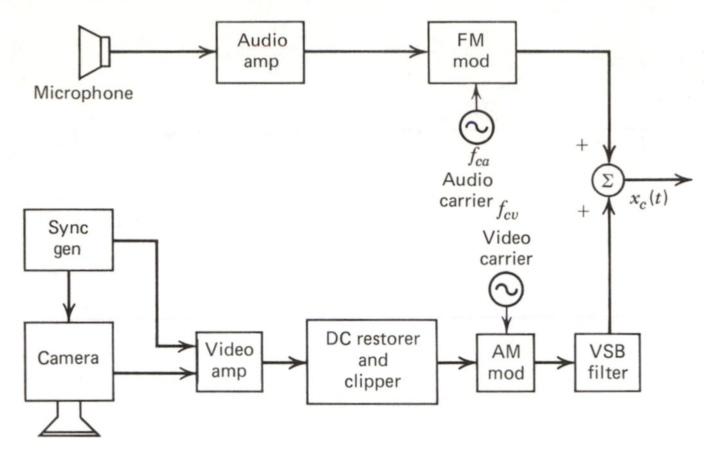

Figure 6.40 Black and white TV transmitter.

The input of the envelope detector has the form

$$y(t) = \underbrace{A_1[1 + x(t)] \cos \omega_{cv}t + A_1\gamma(t) \sin \omega_{cv}t}_{\text{VSB} + \text{C (video)}}$$

$$\underbrace{+ A_2 \cos[(\omega_{cv} + \omega_a)t + \phi(t)]}_{\text{FM (audio)}}$$

Since $|\gamma(t)| \ll 1$ for a VSB signal (see Equations (6.32) and (6.33)), we can write the envelope of $y(t)$ as

$$R(t) \approx A_1[1 + x(t)] + A_2 \cos[\omega_a t + \phi(t)]$$

Thus the envelope detector output has the video and audio signals. The video amplifier has a lowpass filter that removes the audio component. It also has a DC restorer that electronically clamps the blanking pulses and restores the correct DC level to the video signal. The audio component is demodulated using an FM discriminator.

The timing information in the form of horizontal and vertical synchronizing pulses are extracted from the composite video signal by the sync separator. These pulses synchronize the sweep generators. The brightness control allows for manual adjustment of the DC level and the contrast control is used to adjust the gain of the IF amplifier. The interested reader is referred to electronics textbooks for detailed circuits used in TV receivers (see the bibliography at the end of this chapter).

Color Television. All colors found in nature can be approximated by suitable mixing of three additive primary colors: red, green, and blue. Accordingly, early attempts at color TV broadcasting involved sequential

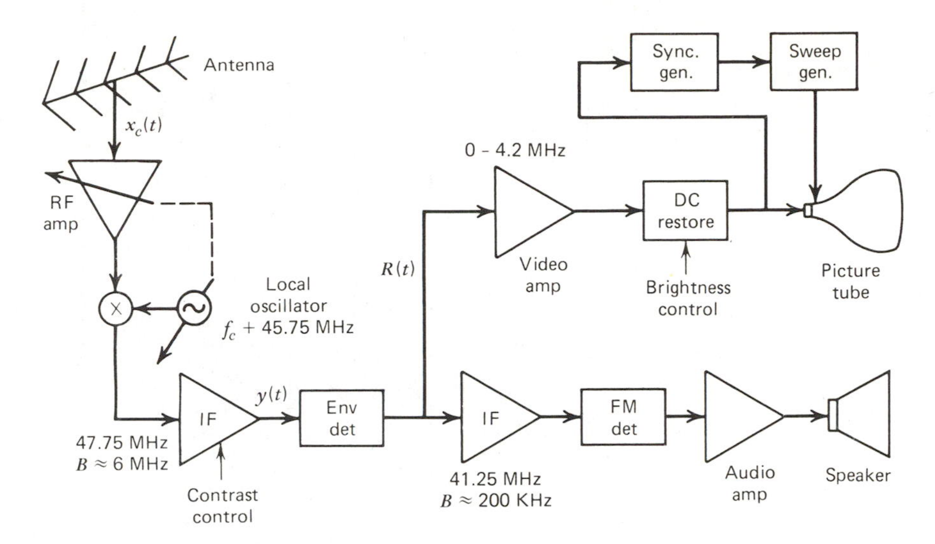

Figure 6.41 Black and white TV receiver.

transmission of three video signals, say $x_R(t)$, $x_G(t)$, and $x_B(t)$—one for each primary color. Unfortunately, this method is not compatible with existing monochrome systems. Furthermore, the bandwidth required is much larger. In the early 1950s a committee, the National Television Systems Committee (NTSC), was formed to set the standards for a fully compatible color TV signal that will fit into the monochrome channel using certain characteristics of human color perception. The resulting NTSC system design was authorized by FCC in 1953 and is the system presently in use in the United States. We briefly describe the salient features of the NTSC color system which are shown in Figures 6.42*a*–6.42*d*. For additional details the reader is referred to the *Proceedings of the IRE*, January 1954, which is devoted entirely to the subject of color television.

To begin with, the three primary color signals can be uniquely represented by any three other signals that are independent linear combinations of $x_R(t)$, $x_G(t)$, and $x_B(t)$. One of these three linear combinations may be made to yield an intensity or luminance signal which is similar to the conventional black and white TV (monochrome) signal. The remaining two signals, called the chrominance signals, contain additional information that can be used to reconstruct $x_G(t)$, $x_B(t)$, and $x_R(t)$ from which the original color can be reconstructed. The three signals that are used in the NTSC system have the form

$$\begin{aligned} x_L(t) &= 0.3x_R(t) + 0.59x_G(t) + 0.11x_B(t) \\ x_I(t) &= 0.60x_R(t) - 0.28x_G(t) - 0.32x_B(t) \\ x_Q(t) &= 0.21x_R(t) - 0.52x_G(t) + 0.31x_B(t) \end{aligned} \tag{6.65}$$

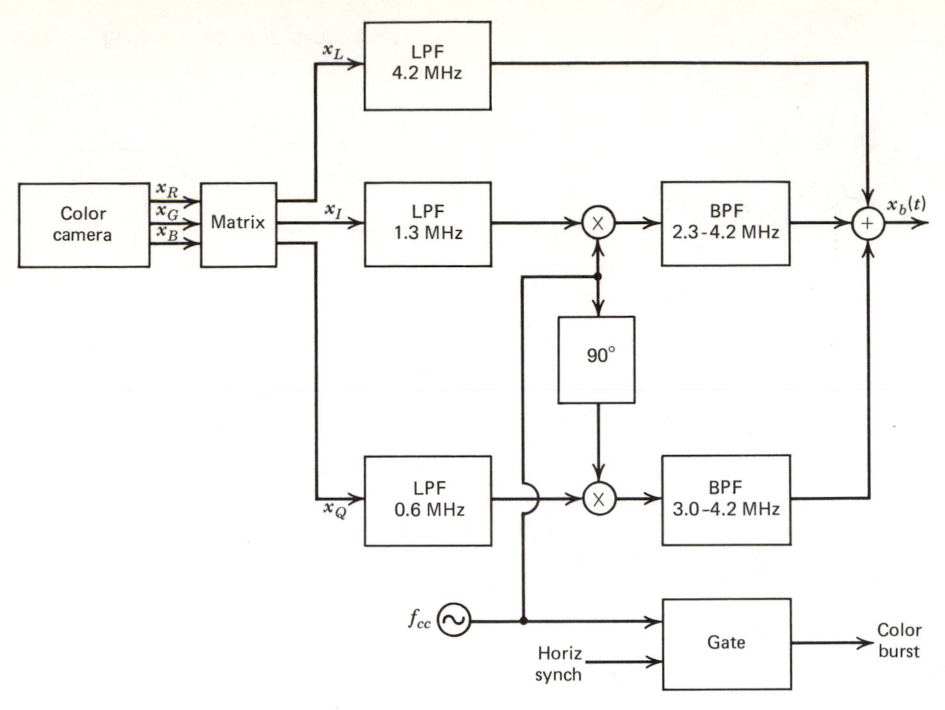

(a) Color TV transmitter

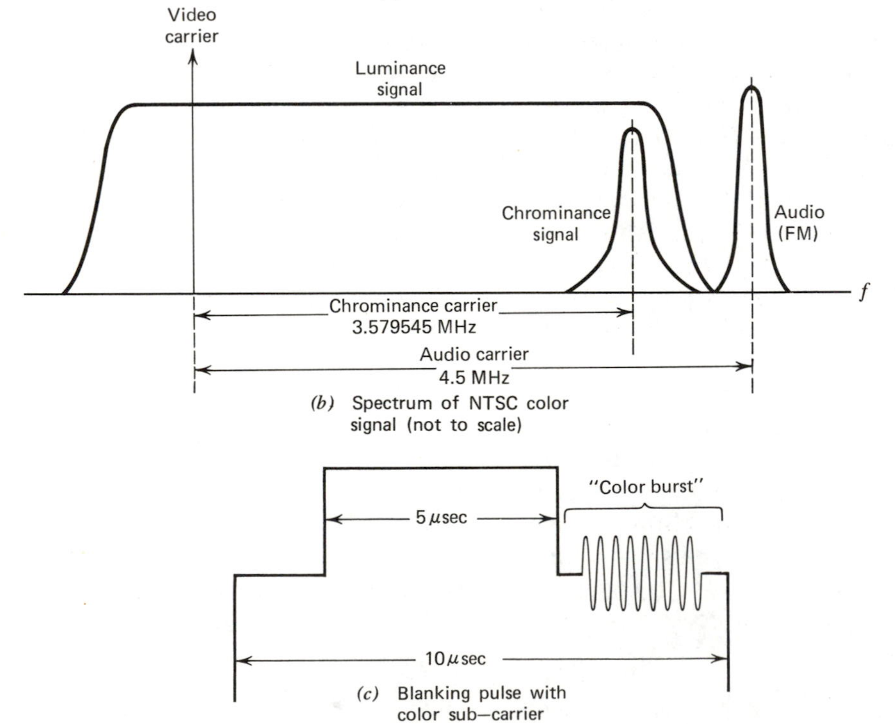

(b) Spectrum of NTSC color signal (not to scale)

(c) Blanking pulse with color sub-carrier

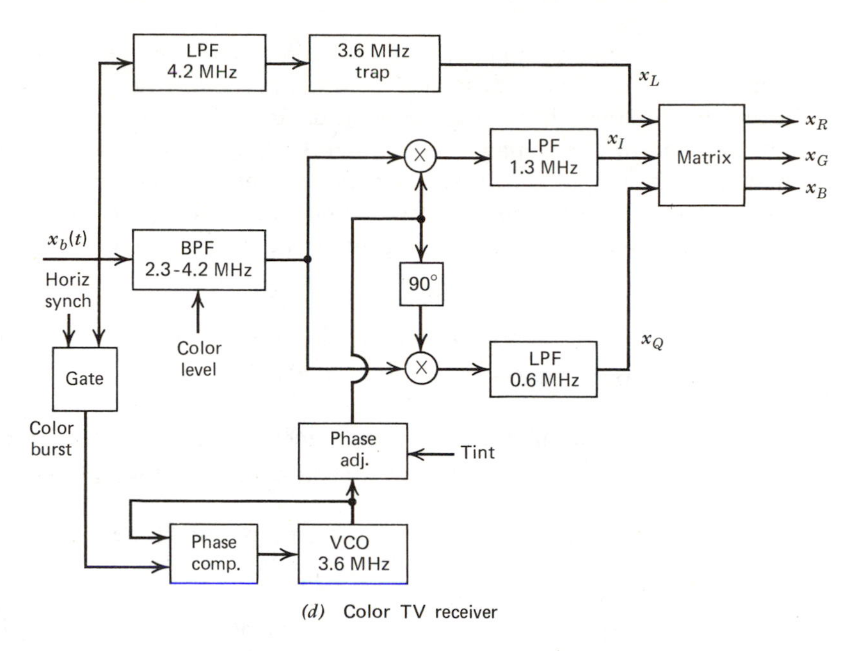

(*d*) Color TV receiver

Figure 6.42 (*a*) Color TV transmitter. (*b*) Spectrum of NTSC color TV signal (not to scale). (*c*) Blanking pulse with color sub-carrier. (*d*) Color TV receiver.

The color signals $x_R(t)$, $x_G(t)$, and $x_B(t)$ are normalized such that the luminance signal $x_L(t)$ is never negative.

The luminance signal $x_L(t)$ is allotted 4.2 MHz baseband bandwidth and transmitted as a VSB signal (same form as black and white TV signal). The chrominance signals are *quadrature multiplexed* (see Problem 6.48) on a color subcarrier at $f_c + 3.579545$ MHz, which corresponds to a gap in the spectrum of $x_L(t)$ (the spectrum of $x_L(t)$ has periodic gaps between the harmonics of the line frequency and 3.579545 MHz falls exactly between the 227th and 228th harmonic of the line frequency). The baseband color signal has the form

$$x_b(t) = \underbrace{x_L(t)}_{\substack{\text{Baseband}\\ \text{portion}\\ \text{of } x_L(t)}} + \underbrace{x_Q(t)\sin\omega_{cc}t}_{\substack{\text{DSB}\\ \text{modulation}\\ \text{for } x_Q(t)}} + \underbrace{x_I(t)\cos\omega_{cc}t + \hat{x}_I(t)\sin\omega_{cc}t}_{\substack{\textit{VSB modulation}\\ \textit{for } x_I(t)}}$$

where ω_{cc} is the color subcarrier frequency. Horizontal and vertical synchronizing pulses are added to $x_b(t)$ at the transmitter. Additionally, an eight-cycle piece of the color subcarrier known as the "colorburst" is put on the "back porch" of the blanking pulses for color subcarrier synchronization at the receiver.

At the receiver, the video signal $x_b(t)$ is recovered by envelope demodulation. The luminance signal is at baseband frequencies and hence it does not require further processing other than amplification and a 3.6 MHz trap filter to eliminate flicker due to chrominance components. A synchronous color carrier is generated from the "color burst" and the chrominance components are synchronously demodulated. The recovered luminance and chrominance signals are linearly combined (matrixed) to produce the primary color signals as

$$\begin{aligned} x_R(t) &= x_L(t) - 0.96x_I(t) + 0.62x_Q(t) \\ x_G(t) &= x_L(t) - 0.28x_I(t) - 0.64x_Q(t) \\ x_B(t) &= x_L(t) - 1.10x_I(t) + 1.70x_Q(t) \end{aligned} \tag{6.66}$$

Depending on the picture tube used in the receiver, these primary color signals are combined in different ways to produce natural color pictures on the receiver screen.

When a color signal is applied to a monochrome picture tube, the viewer does not see the sinusoidal variations produced by the color subcarrier and its sidebands. This is because all of these sinusoids are exactly an odd multiple of one-half the line frequency, and they reverse in phase from line to line. This produces flickering in small areas that average out over time and space!

If a black and white signal is applied to a color TV receiver, all three color signals will be equal and the reproduced signal will be black and white. This is termed reverse compatibility.

The reader should note that the color TV system represents a very complex communication system. Three different modulation schemes (DSB, VSB, and FM) as well as two multiplexing schemes (FDM for message signals and TDM for synchronizing signals) are used in the system. In spite of its complexity, a color TV receiver is relatively inexpensive due to advances in components technology and mass production.

6.7 SUMMARY

Modulation is the process of varying the parameter of a carrier in a one-to-one correspondence with a message signal. Primary uses of modulation are for frequency translation and multiplexing. Several methods of CW modulation, where the carrier is continuous, were discussed. Particular attention was given to the spectral characteristics of the modulated signal that determine the power and transmission bandwidth of the modulated signal.

CW modulation schemes are divided into two broad categories: linear modulation and angle modulation. In linear modulation, the amplitude of the carrier is varied in proportion to the message signal. In angle modulation, the phase or frequency of the carrier is varied according to the message signal.

The simplest form of linear modulation is double-sideband modulation (DSB), which results when the message waveform is multiplied with a carrier. Proper demodulation of DSB requires a coherent reference carrier at the receiver. Given carrier references, DSB modulation and demodulation can be implemented using product devices (or balanced mixers). The addition of a large carrier to the DSB signal results in amplitude modulation (AM). A simple envelope detector can correctly demodulate an AM signal. Both AM and DSB modulation requires a transmission bandwidth equal to twice the bandwidth of the message signal.

To conserve bandwidth, one sideband of a DSB signal is often fully or partially suppressed. This type of modulation is called single-sideband (SSB) or vestigial-sideband (VSB) modulation. SSB demands exact filter design and coherent demodulation. VSB + carrier offers a compromise between SSB and DSB in terms of bandwidth conservation and equipment simplicity.

The variation of the phase or frequency of the carrier in proportion to the amplitude of the message signal is called phase and frequency modulation, respectively. Angle modulation results in infinite bandwidth. Fortunately, most of the signal energy or power is contained in a finite bandwidth. The bandwidth of a frequency modulated signal is estimated using Carson's rule, which says that the transmission bandwidth of an angle modulated signal is approximately equal to the sum of twice the peak frequency deviation plus twice the message bandwidth.

A wideband FM signal can be generated either by generating a narrowband FM signal and increasing the modulation index by frequency multiplication, or by direct methods. Demodulation can be accomplished by the use of a discriminator or by a phase-lock loop. The discriminator is implemented as a differentiator followed by an envelope detector.

Wideband angle modulated systems offer an improvement in output signal-to-noise ratio over linear modulation schemes; we will establish this property in the next chapter and compare the performance of various modulation schemes in the presence of noise.

Finally, several message signals can be transmitted over a single channel using frequency division multiplexing where the channel bandwidth is divided into slots, and nonoverlapping slots are assigned to different messages. Individual messages are placed in their respective slots by frequency translation using any one of the modulation schemes discussed in this chapter. Different carrier frequencies are used for different channels. The signals are separated at the receiver by bandpass filtering. Examples of multiplexing include long distance telephone, stereo FM broadcasting, and TV broadcasting.

REFERENCES

CW modulation schemes are dealt with, in great detail, in a number of undergraduate and graduate level texts. Easily readable treatments of modulation theory, at about the same level as this text, may be found in the books by Carlson (1975), Gregg (1977), Stremler (1976), Taub and Schilling (1970), and Ziemer and Tranter (1976).

A thorough and exhaustive treatment of modulation theory and spectral analysis may be found in Panter's book (1965). This book is written at the beginning graduate level and contains an excellent treatment of FM systems.

Numerous aspects of communication electronics are dealt with in books by Alley and Atwood (1973), Clark and Hess (1971), and Hansen (1969).

The subject of phase-locked loops is dealt with in Blanchard (1977). Detailed analysis of second-order systems may be found in texts on control systems; see, for example, Kuo (1963). Digital implementation of phase-locked loops may be found in recent issues of communication systems journals.

1. A. B. Carlson. *Communication Systems*. McGraw-Hill, New York (1975).
2. W. David Gregg. *Analog and Digital Communication Systems*. Wiley, New York (1977).
3. H. Taub and D. L. Schilling. *Principles of Communication Systems*. McGraw-Hill, New York (1971).
4. F. G. Stremler. *Introduction to Communication Systems*. Addison-Wesley, Reading, Mass. (1977).
5. R. E. Ziemer and H. W. Tranter. *Principles of Communication Systems*. Houghton-Mifflin, Boston (1976).
6. P. F. Panter. *Modulation, Noise and Spectral Analysis*. McGraw-Hill, New York (1965).
7. K. K. Clark and D. T. Hess. *Communication Circuits: Analysis and Design*. Addison-Wesley, Reading, Mass. (1971).
8. C. L. Alley and K. W. Atwood. *Electronic Engineering*. Wiley, New York (1973).
9. G. L. Hansen. *Introduction to Solid State Television Systems*. Prentice-Hall, Englewood Cliffs, N.J. (1969).
10. A. Blanchard. *Phase-Locked Loops*. Wiley, New York (1977).
11. B. C. Kuo. *Automatic Control Systems*. Prentice-Hall, Englewood Cliffs, N.J. (1968).
12. G. N. Patchett. *Color TV Systems*. Pragman, London (1968).

PROBLEMS

Sections 6.1 and 6.2

6.1. Two signals $x_1(t)$ and $x_2(t)$ having Fourier transforms $X_1(f)$ and $X_2(f)$ shown in Figure 6.43 are combined to form

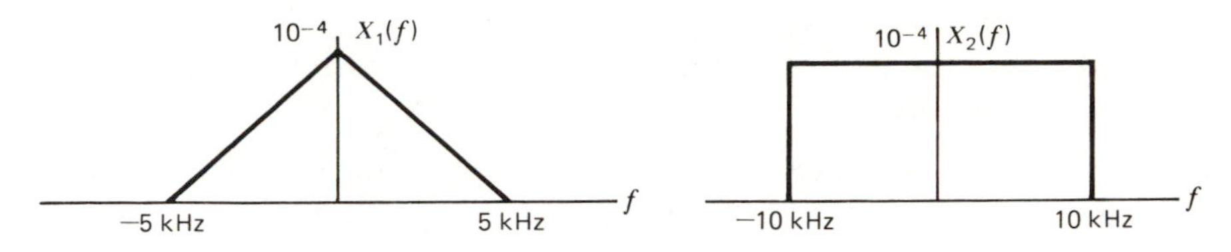

Figure 6.43 Spectra of $x_1(t)$, and $x_2(t)$ for Problem 6.1.

$$y(t) = x_1(t) + 2x_2(t) \cos 2\pi f_c t, \quad f_c = 20{,}000 \text{ Hz}$$

(a) Find the bandwidth of the signal $y(t)$.
(b) Given $y(t)$, how would you separate $x_1(t)$ and $2x_2(t) \cos 2\pi f_c t$?

6.2. A signal $m(t)$ has a Fourier transform

$$M(f) = \begin{cases} 1, & f_1 \leq |f| \leq f_2, \; f_1 = 1 \text{ kHz}; \; f_2 = 10 \text{ kHz} \\ 0, & \text{elsewhere} \end{cases}$$

Suppose we form a signal $y(t) = m(t) \cos 2\pi(10^6)t$. Find the range of frequencies for which $y(t)$ has nonzero spectral components. Also find the ratio of the highest to lowest frequencies [for which $|Y(f)| \neq 0$] of $y(t)$. Compare this ratio with f_2/f_1. Is $y(t)$ a narrowband signal? (A bandpass signal is said to be a narrowband signal if $f_{\text{high}}/f_{\text{low}} \ll 1$.)

Section 6.2

6.3. Consider a system with the amplitude and phase response shown in Figure 6.44 and the following three inputs:

$$x_1(t) = \cos 500\pi t + \cos 2000\pi t$$
$$x_2(t) = \cos 500\pi t + \cos 2500\pi t$$
$$x_3(t) = \cos 2500\pi t + \cos 3500\pi t$$

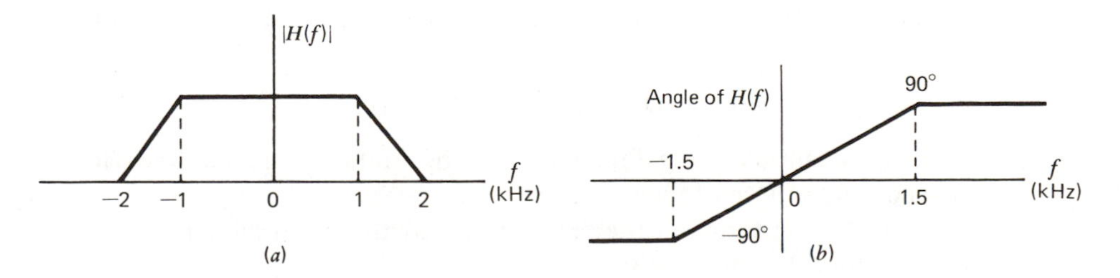

Figure 6.44 Response of the system for Problem 6.3. (*a*) Amplitude response. (*b*) Phase response

(a) Find the outputs $y_1(t)$, $y_2(t)$, and $y_3(t)$.
(b) Identify the type of distortion, if any, suffered by each of the three input signals.

6.4. Show that an RC lowpass filter gives nearly distortionless transmission if the input to the filter is bandlimited to $f_x \ll f_0 = 1/2\pi RC$.

6.5. Assume that a transfer function with "ripples" in the amplitude response can be approximated by

$$H(f) = \begin{cases} (1 + \alpha \cos \omega t_0) \exp(-j\omega t_d), & |\alpha| < 1, |f| < f_x \\ 0 & \text{elsewhere} \end{cases}$$

where f_x is the bandwidth of the input signal $x(t)$. Show that the output $y(t)$ is given by

$$y(t) = x(t - t_d) + \frac{\alpha}{2}[x(t - t_d + t_0) + x(t - t_d - t_0)]$$

(i.e., $y(t)$ has a pair of echoes.)

6.6. The transfer function of a channel is shown in Figure 6.45. The input to the channel is a lowpass signal $x(t)$ having a bandwidth of f_x Hz. Design a five-tap equalizer for this channel. (*Hint*: Expand $1/H_c(f)$ as a Fourier series in the interval $[-f_x, f_x]$ and use the Fourier series coefficients to set the tap gains of the equalizer.)

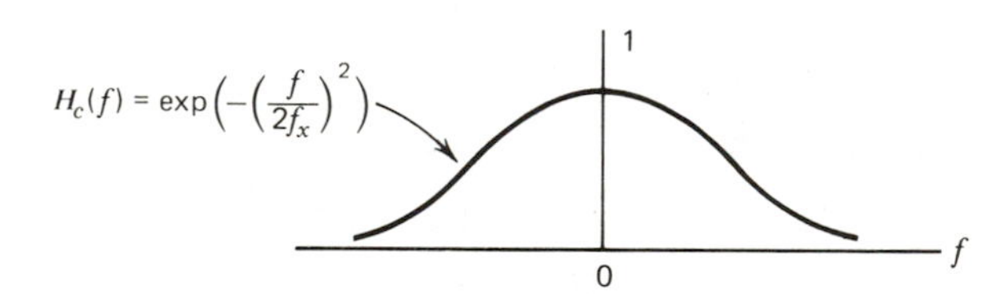

Figure 6.45 $H_c(f)$ for Problem 6.6.

6.7. A nonlinear element in a communication system has the transfer characteristic

$$y(t) = x(t) + 0.2x^2(t) + 0.02x^3(t)$$

The desired output is the first term. If the input to the nonlinearity is $x(t) = \cos 700\pi t + \cos 150\pi t$, find:
(a) the distortion terms at the frequencies of the input signal;
(b) the second harmonic distortion terms;
(c) the third harmonic distortion terms;
(d) the intermodulation distortion terms.

6.8. Consider a signal $x(t) = x_2(t) + x_1(t) \cos 2\pi f_c t$, where $x_1(t)$ and $x_2(t)$ have the spectra shown in Figure 6.43 and $f_c = 20{,}000$ Hz. Suppose $x(t)$ is applied to a nonlinearity with a transfer characteristic $y(t) = x(t) + 0.002x^2(t)$. Sketch the components that make up the spectrum of $y(t)$ and identify intermodulation (cross-product) terms.

Section 6.3

6.9. A lowpass signal $x(t)$ having a bandwidth of 10 kHz is multiplied by $\cos \omega_c t$ to produce $x_c(t)$. Find the value of f_c so that the bandwidth of $x_c(t)$ is 1% of f_c.

6.10. The modulating signal $x(t) = 2 \cos 2000\pi t + \sin 4000\pi t$ is applied to a DSB modulator operating with a carrier frequency of 100 kHz. Sketch the power spectral density of the modulator output.

6.11. DSB signals can be generated by multiplying the message signal with a nonsinusoidal carrier and filtering the product waveform as shown in Figure 6.46.

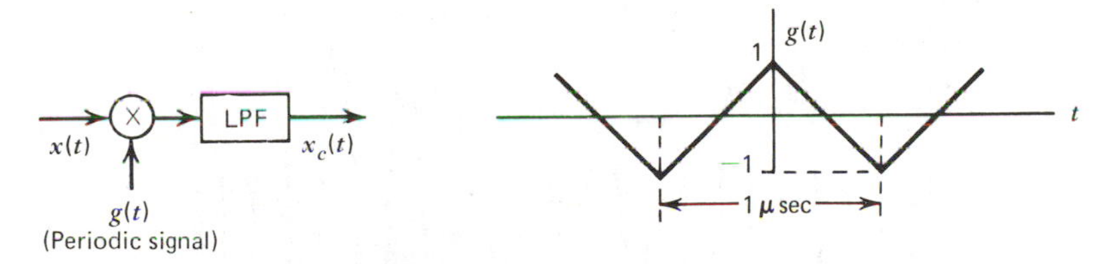

Figure 6.46 DSB modulator for Problem 6.11.

(a) Show that the scheme shown above will work if $g(t)$ has no DC component and the filter cut-off frequency is $f_c + f_x$, where f_c is the fundamental frequency of $g(t)$ and f_x is the bandwidth of $x(t)$.

(b) Assume that $x(t) = 2 \cos 1000\pi t$ and $g(t)$ is as shown in Figure 6.46. Find the filter bandwidth. Write an expression for the output $x_c(t)$.

(c) How would you modify the system if $g(t)$ has a DC component?

6.12. Show that it is possible to demodulate a DSB signal $x_c(t) = A_c x(t) \cos 2\pi f_c t$ by multiplying it with a rectangular waveform with a period $T = 1/f_c$ and a lowpass filtering the output. (Assume that the rectangular waveform is an even function of t.)

6.13. Prove that the average power of a DSB signal $x_c(t) = A_c x(t) \cos \omega_c t$ is

$S_c S_x$ (Equation 6.14) by showing that

$$\lim_{T\to\infty} \frac{1}{T} \int_{-T/2}^{T/2} x^2(t) \cos 2\omega_c t \, dt = 0$$

(Assume $x(t)$ to be bandlimited to f_x and $f_c \gg f_x$.)

6.14. An amplitude modulated waveform has the form

$$x_c(t) = 10(1 + 0.5 \cos 2000\pi t + 0.5 \cos 4000\pi t) \cos 20{,}000\pi t$$

(a) Sketch the amplitude spectrum of $x_c(t)$.
(b) Find the average power content of each spectral component including the carrier.
(c) Find the total power, the sideband power, and power efficiency.
(d) What is the modulation index?

6.15. An AM waveform is shown in Figure 6.47. Assume the message signal to be sinusoidal.
(a) Find the modulation index.
(b) Calculate S_c, S_x, and power efficiency.

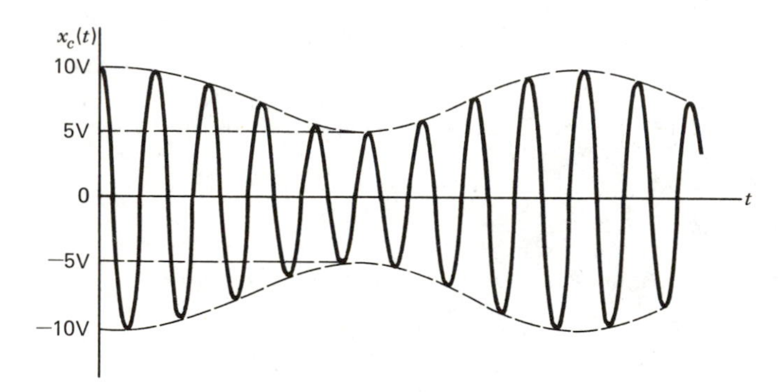

Figure 6.47 AM waveform for Problem 6.15.

6.16. Show that the maximum power efficiency of an AM modulator is 50%.

6.17. An AM transmitter develops an unmodulated power output of 400 watts across a 50 ohm resistive load. The carrier is modulated by a single tone with a modulation index of 0.8.
(a) Write the expression for the AM signal $x_c(t)$ assuming $f_x = 5$ kHz and $f_c = 1$ MHz.
(b) Find the total average power of the modulator output.
(c) Find the power efficiency of the modulator.

6.18. Practical modulators often have a *peak power limitation* in addition to average power limitation. Assume that a DSB modulator and an AM modulator are operating with

$$x(t) = 0.8 \cos 200\pi t$$

and a carrier waveform $10 \cos 2\pi f_c t$ ($f_c \gg 100$ Hz).

(a) Find the peak (instantaneous) power of the DSB and AM signals.

(b) Obtain the ratio of peak power to average sideband power for the DSB and AM signals and compare the ratios.

6.19. The input to an envelope detector is an AM signal of the form $x_c(t) = (1 + k \cos \omega_x t) \cos \omega_c t$, where k is a constant, $0 < k < 1$, and $\omega_c \gg \omega_x$.

(a) Show that if the demodulator output is to follow the envelope of $x_c(t)$, it is required that at any time t_0

$$\frac{1}{RC} \geq \frac{\omega_x k \sin \omega_x t_0}{1 + k \cos \omega_x t_0}$$

(*Hint*: use linear approximations for computing the change in the envelope value between successive carrier peaks and for the capacitor discharge.)

(b) If the demodulator is to follow the envelope at all times, show that

$$RC \leq \frac{1}{\omega_x} \frac{\sqrt{1 - k^2}}{k}$$

(*Hint*: find the maximum value of the right-hand side of the result obtained in part (a), with respect to t_0.)

6.20. Consider the switching modulator shown in Figure 6.48.

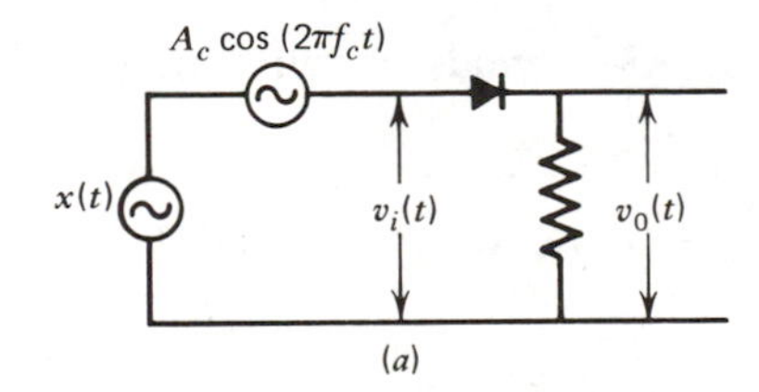

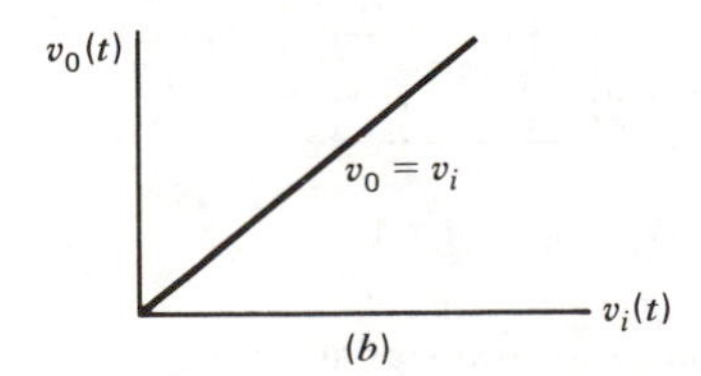

Figure 6.48 (*a*) Switching modulator. (*b*) Ideal diode characteristics.

(a) Assuming that $\max|x(t)| \ll A_c$ and that the diode acts as an ideal switch, show that

$$v_0(t) \approx A_c[\cos(2\pi f_c t) + mx(t)]g_p(t)$$

where $g_p(t)$ is a rectangular pulse train with a period $1/f_c$ and a duty cycle of $\frac{1}{2}$.

(b) By substituting the Fourier series for $g_p(t)$ in the preceding equation, show that $v_0(t)$ has a component of the form $A[1 + mx(t)] \cos(2\pi f_c t)$.

(c) Assuming $x(t)$ to be a lowpass signal bandlimited to f_x Hz ($f_x \ll f_c$), show that it is possible to generate an AM signal by bandpass filtering $v_0(t)$.

6.21. An AM signal of the form $R_x(t) \cos(2\pi f_c t)$ passes through a bandpass channel with a transfer function $H(f) = K \exp(j\theta(f))$. The phase response of the channel is such that it can be approximated by a Taylor series with two terms as

$$\theta(f + f_c) \approx \theta(f_c) + f \left.\frac{d\theta(f)}{df}\right|_{f=f_c}$$

Show that the signal at the output of the channel can be represented by

$$y(t) = KR_x(t - t_R) \cos[2\pi f_c(t - t_c)]$$

where the *carrier delay* t_c and the *envelope delay* t_R are given by

$$t_R = \frac{-1}{2\pi} \left.\frac{d\theta(f)}{df}\right|_{f=f_c}$$

$$t_c = -\frac{\theta(f_c)}{2\pi f_c}$$

6.22. Consider the square law demodulator for AM signals shown below in Figure 6.49.

(a) Sketch the spectrum of the output $\tilde{x}(t)$.

(b) Show that if $|x(t)| \ll 1$, then $\hat{x}(t) \approx a + kx(t)$, where a, and k are constants.

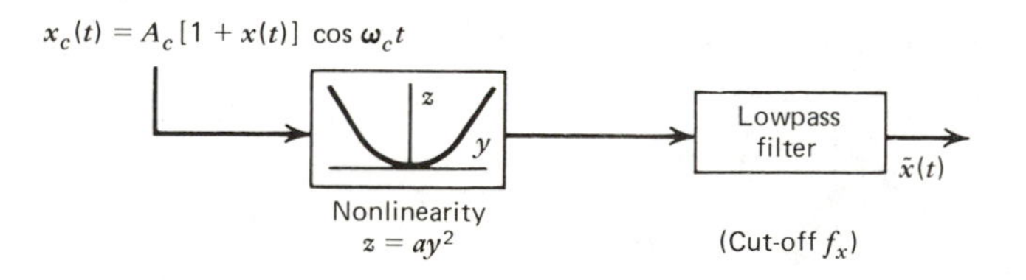

Figure 6.49 Square law demodulator for AM signal.

6.23. The signal $x_c(t) = 2(1 + 0.4 \cos 6000\pi t) \cos 10^6\pi t$ is applied to a square law device having a transfer characteristic $y = (x + 4)^2$. The output of the square law device is filtered by an ideal LPF with a cut-off frequency of 8000 Hz. Sketch the amplitude spectrum of the filter output.

6.24. A signal $x(t) = 2 \cos 1000\pi t + \cos 2000\pi t$ is multiplied by a carrier $10 \cos 10^5\pi t$. Write down the expression for the upper sideband terms of the product signal.

6.25. A multistage modulation scheme is often used to generate a SSB signal using filters with $2\alpha/f_c < 0.01$ (Figure 6.13). Assume that we want to use the scheme shown in Figure 6.50 to generate an SSB signal with a carrier frequency $f_c = 1$ MHz. The spectrum of the modulating signal is shown in Figure 6.51. Assume that bandpass filters are available that will provide 60 dB of attenuation in a frequency interval which is about 1% of the filter center frequency. Specify the carrier frequencies and filter characteristics for this application.

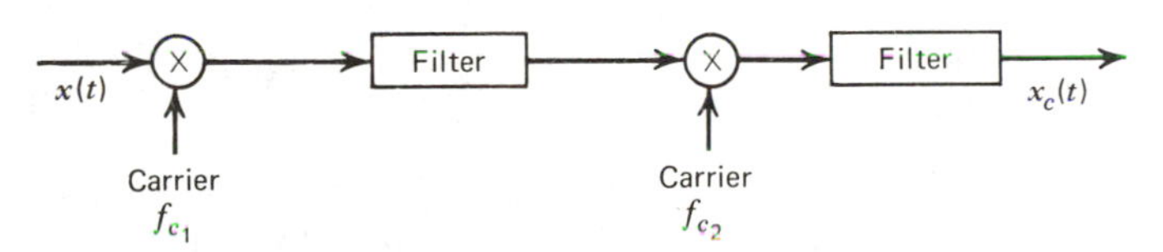

Figure 6.50 A two-stage SSB modulator.

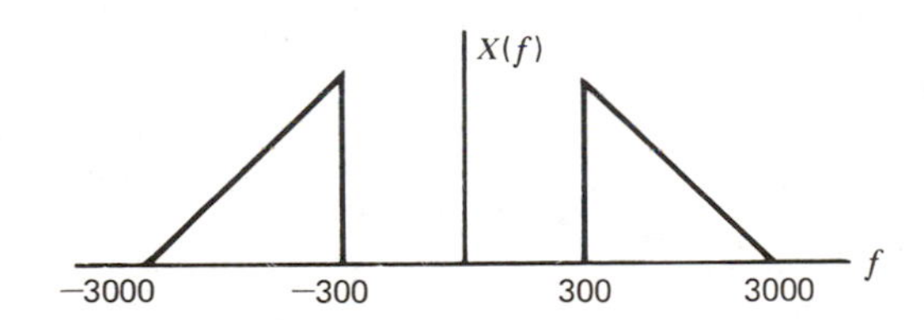

Figure 6.51 Signal spectrum for Problem 6.25.

6.26. Figure 6.52 shows *Weaver's SSB modulator.* Analyze its operation by taking $x(t) = \cos 2\pi f_x t$ $(f_x < 2B)$. Show that $x_c(t)$ is an SSB signal.

6.27. Draw the schematic diagram of a synchronous demodulator for a VSB signal. Show that for correct demodulation the VSB filter that is used for generating the VSB signal should have the symmetry shown in Figure 6.15.

6.28. Verify the statement following Equation (6.29).

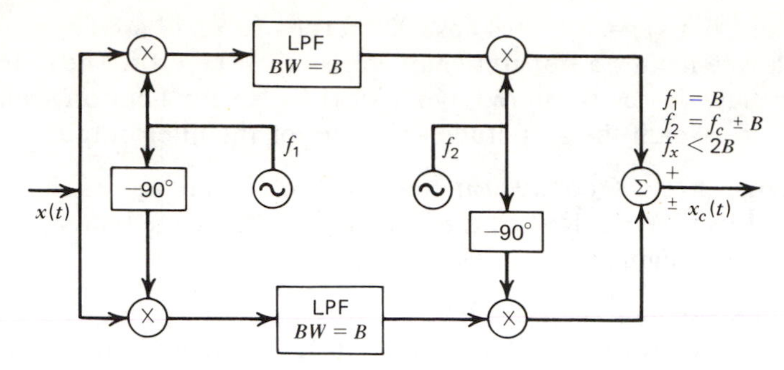

Figure 6.52 Weaver's SSB modulator. (Compare this modulator with the SSB modulator shown in Figure 6.14.)

6.29. Obtain an expression for a VSB signal generated with $x(t) = \cos 2\pi f_x t$ and $H_{VSB}(f_c + f_x) = 0.5 + a$, $H_{VSB}(f_c - f_x) = 0.5 - a$ $(0 < a < 0.5)$. Write the answer in envelope and phase form and in quadrature form. Take $a = 0.25$ and evaluate the distortion term in Equation (6.33).

6.30. Draw the block diagram for an AM modulator using a nonlinear device having the transfer characteristic $v_{out} = a_1 v_{in} + a_3 v_{in}^3$.

6.31. Suppose that the nonlinear elements used in a balanced modulator (Figure 6.17*a*) are not matched. That is, one nonlinear device has the transfer characteristic $v_{out} = a_{11}v_{in} + a_{12}v_{in}^2 + a_{13}v_{in}^3$, while the second nonlinear device has the transfer characteristic

$$v_{out} = a_{21}v_{in} + a_{22}v_{in}^2 + a_{23}v_{in}^3$$

Find the output signal.

Section 6.4

6.32. An angle modulated signal

$$x_c(t) = 10\cos[10^8\pi t + 3\sin(2\pi)(10^3)t]$$

is present across a 50 ohm resistive load. Find:
(a) the total average power;
(b) peak phase deviation;
(c) peak frequency deviation.
(d) Can you determine whether this is FM or PM?

6.33. A sinusoidal message signal of frequency 1000 Hz is used as a modulating signal in an AM and FM system. The unmodulated carrier amplitude is the same in both systems. The peak frequency deviation of the FM system is set to four times the bandwidth of the AM system. The magnitude of the spectral components at $f_c \pm 1000$ Hz are equal for both systems. Determine the modulation index for the AM and FM systems.

6.34. Consider an angle modulated signal

$$x_c(t) = 10\cos(\omega_c t + 3\cos\omega_m t), \quad f_m = 1000 \text{ Hz}$$

(a) Assume the modulation to be FM. Determine the modulation index and find the transmission bandwidth when (i) ω_m is increased by a factor of 4, and (ii) when ω_m is decreased by a factor of 4.

(b) Assume the modulation to be PM. Calculate the modulation index and the bandwidth when (i) ω_m is increased by a factor of 4 and (ii) decreased by a factor of 4.

6.35. An audio signal $x(t)$ has a bandwidth of 15 kHz. The maximum value of $|x(t)|$ is 1 volt. This signal is used to frequency modulate a carrier at 100 MHz. Estimate the bandwidth of the FM signal if the modulator has a frequency deviation constant (a) 1000 Hz/volt; (b) 5000 Hz/volt; (c) 50,000 Hz/volt.

6.36. Plot S_n (Equation (6.49)) versus n for $\beta = 1, 2, 5$, and 8 and for $n = 0, 1, 2, \ldots, \beta + 2$.

6.37. In addition to Carson's rule, the following formulas are also used to estimate the bandwidth of an FM signal: $B_T = (2D + 1)f_m$ and $B_T = 2(D + 2)f_m$. Compute and compare the bandwidths using these formulas for the FM signal with $f_\Delta = 75$ kHz and $f_m = 15$ kHz.

6.38. Let $x_1(t)$ and $x_2(t)$ be two message signals and let $x_{c_1}(t)$ and $x_{c_2}(t)$ be the modulated signals corresponding to $x_1(t)$ and $x_2(t)$, respectively.

(a) Show that, if the modulation is DSB (or, for that matter, SSB or VSB), then $x_1(t) + x_2(t)$ will produce a modulated signal equal to $x_{c_1}(t) + x_{c_2}(t)$.

(b) Show that if angle modulation is used, then the modulated signal produced by $x_1(t) + x_2(t)$ will not be $x_{c_1}(t) + x_{c_2}(t)$, that is, show that superposition does not apply to angle modulated signals.

6.39. An arbitrary periodic message signal $m(t)$ with period T_m is used to modulate the angle of a carrier. The modulated signal has the form

$$x_c(t) = A_c \cos[\omega_c t + \phi(t)]$$

Since $m(t)$ is periodic, $\phi(t)$ will also be periodic. Let $\{\phi_n\}$ be the Fourier series coefficient of $\exp(j\phi(t))$. Show that $x_c(t)$ can be expressed as

$$x_c(t) = \text{real part of} \sum_{n=-\infty}^{\infty} A_c\phi_n \exp(jn\omega_m t)\exp(j\omega_c t)$$

where $\omega_m = 2\pi/T_m$.

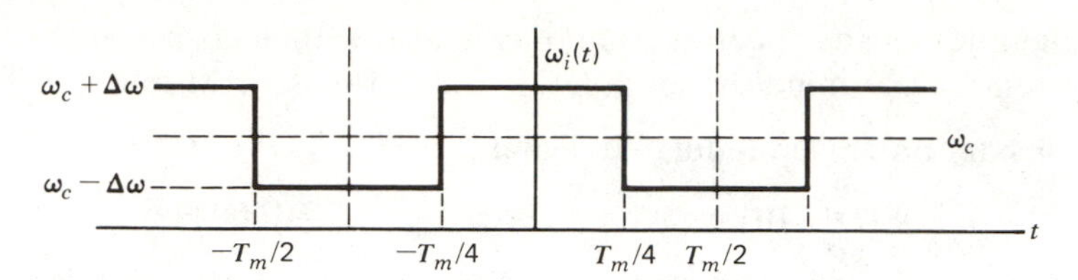

Figure 6.53 Frequency modulation by square wave.

6.40. Suppose that a square wave message signal is used to generate an FM signal whose instantaneous frequency is shown in Figure 6.53.
(a) Sketch the instantaneous phase deviation.
(b) Express the instantaneous phase deviation as a Fourier series.
(c) Show that $x_c(t)$ can be written as

$$x_c(t) = A_c \sum_{n=-\infty}^{\infty} \frac{2\beta \sin[(\beta - n)\pi/2]}{\pi(\beta^2 - n^2)} \cos(\omega_c + n\omega_m t)$$

where $\beta = \Delta\omega/\omega_m$, $\omega_m = 2\pi/T_m$.

6.41. The circuit shown in Figure 6.54 is used to generate an FM signal. The variable capacitor (voltage controlled) is a reverse biased diode (varactor) whose capacitance is related to the biasing voltage v by

$$C_v = (100/\sqrt{1+2v})\, pF$$

The capacitance $C_0 = 100pF$ and L is adjusted to yield a resonant frequency of 2 MHz when a fixed reverse voltage $v = 5$ is applied to the diode. Assume the modulation voltage to have the form

$$v(t) = 5 + 0.05 \sin 2\pi(10^3)t$$

If the oscillator output amplitude is 10 volt, write an expression for the angle modulated signal across the tuned circuit.

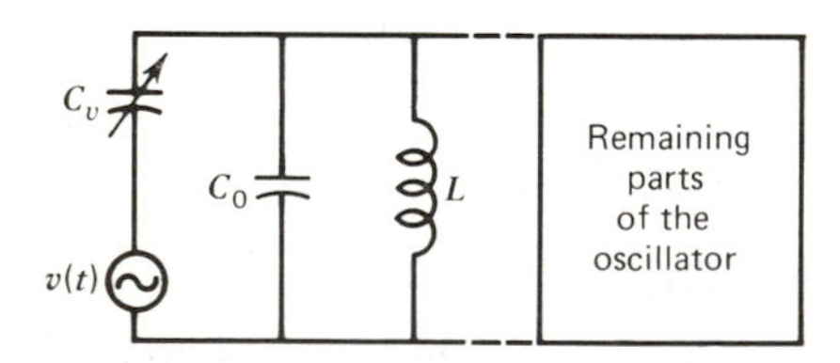

Figure 6.54 An oscillator circuit for generating FM signals.

6.42. An Armstrong-type FM modulator is shown in Figure 6.55. A 200 kHz crystal oscillator provides the carrier waveform for modulation and for the final frequency down conversion. In order to avoid distortion it is

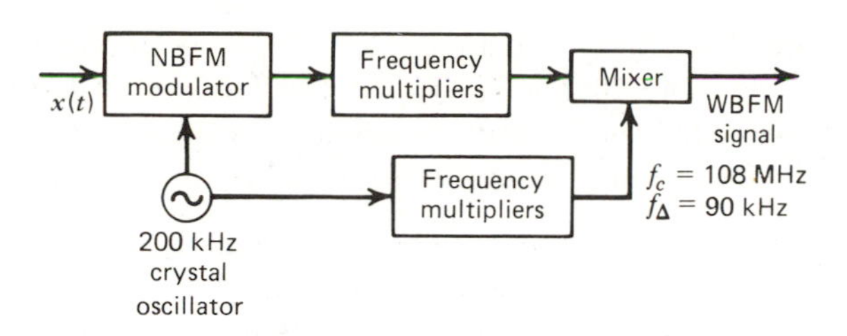

Figure 6.55 Armstrong FM modulator.

desired to keep the maximum angular deviation of the NBFM signal to 0.15 radians. The message signal has frequency components in the range of 30 to 15 kHz. The modulator output is to have a carrier frequency of 108 MHz and a frequency deviation of 90 kHz.

(a) Select appropriate multipliers and mixer oscillator frequencies to accomplish this WBFM modulation.

(b) If the 200 kHz oscillator drifts by 0.1 Hz, find the drift in the carrier of the WBFM signal.

(c) If the WBFM carrier frequency is to be within ±1 Hz of 108 MHz, what is the maximum permissible drift of the 200 kHz oscillator?

6.43. A bandpass limiter is shown in Figure 6.56. Suppose that the input is $v_{\text{in}}(t) = A_c(t)\cos\theta_c(t)$, $A_c(t) \gg A$. Although $v_{\text{in}}(t)$ is not necessarily periodic, $v_{\text{out}}(t)$ may be viewed as a periodic function of θ_c, namely, a square wave of amplitude A and period 2π.

(a) Verify the preceding statement by sketching θ_c versus v_{out}.

(b) Show that

$$v_{\text{out}} = \frac{4A}{\pi}(\cos\theta_c - \tfrac{1}{3}\cos 3\theta_c + \tfrac{1}{5}\cos 5\theta_c - \cdots)$$

(c) Show that an appropriate choice of the BPF characteristics will yield

$$z(t) = \frac{4A}{\pi}\cos[\theta_c(t)]$$

a pure angle modulated signal. (Assume that $\theta_c(t) = \omega_c + \phi(t)$, $\max|\dot{\phi}(t)| \ll \omega_c$.)

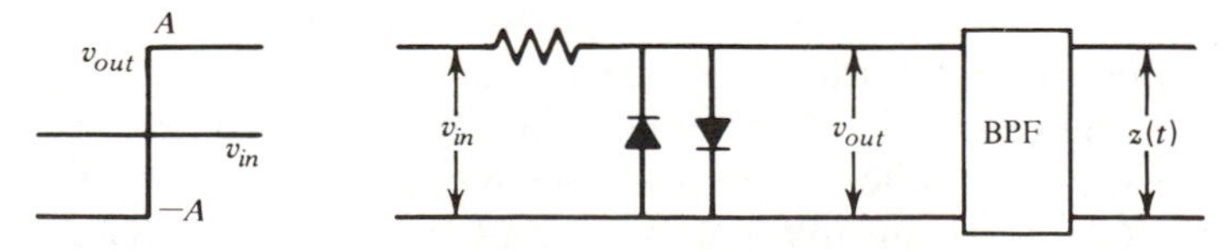

Figure 6.56 Bandpass limiter.

6.44. An FM signal $x_c(t) = A_c \cos(\omega_c t + \beta \sin \omega_m t)$ is applied to an RC high pass filter. Assume that $\omega RC \ll 1$ in the frequency band occupied by $x_c(t)$ and show that the output voltage across the resistor is an amplitude modulated signal. Find the modulation index of the AM signal.

6.45. Consider the FM discriminator shown in Figure 6.57.
(a) Plot the magnitude of the transfer function $V(f)/X_c(f)$. From this plot, determine a suitable carrier frequency and the discriminator constant k_d.
(b) What is the maximum allowable peak frequency deviation of the input?

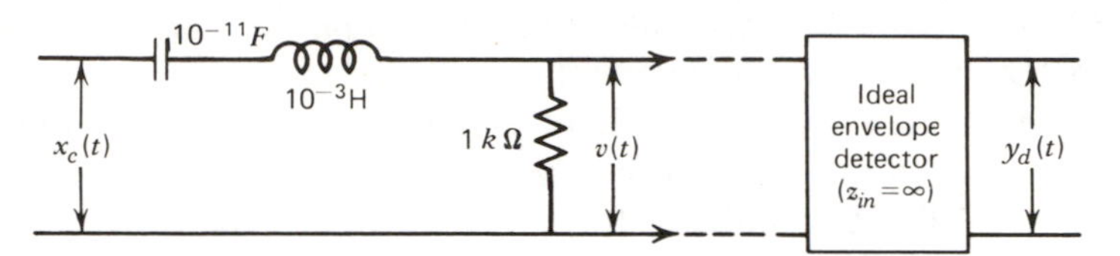

Figure 6.57 FM discriminator for Problem 6.45.

6.46. A second-order PLL uses a loop filter given in Equation (6.63). Show that if the frequency of the input signal changes by $\Delta\omega$ at $t = 0$, then

(a) $$\Psi(t) = \frac{\Delta\omega}{\omega_n\sqrt{1-\xi^2}} \exp(-\xi\omega_n t) \sin(\omega_n t\sqrt{1-\xi^2})u(t), \ \xi^2 < 1$$

where $u(t)$ is the unit step function.
(b) Show that the peak phase error is

$$\Psi_p \leq \frac{\Delta\omega}{\omega_n\sqrt{1-\xi^2}}$$

6.47. Determine the parameters of a minimum bandwidth PLL for demodulating commercial FM. Assume sinusoidal modulation with $\Delta f = \Delta\omega/2\pi = 75$ kHz, $0 \leq f_m \leq 75$ kHz, $\xi = 0.707$, and $\tau_1 = \tau_2$ (Equation (6.63)).

Section 6.5

6.48. Two message signals $x_1(t)$ and $x_2(t)$ can be modulated on the same carrier using the *quadrature multiplexing* scheme shown in Figure 6.58.
(a) Verify the operation of this quadrature multiplexing scheme.
(b) If the local oscillator at the receiver has a phase offset of $\Delta\theta$ with respect to the transmitter carrier, find the outputs $y_1(t)$ and $y_2(t)$. (Assume $\Delta\theta \ll 1$.)

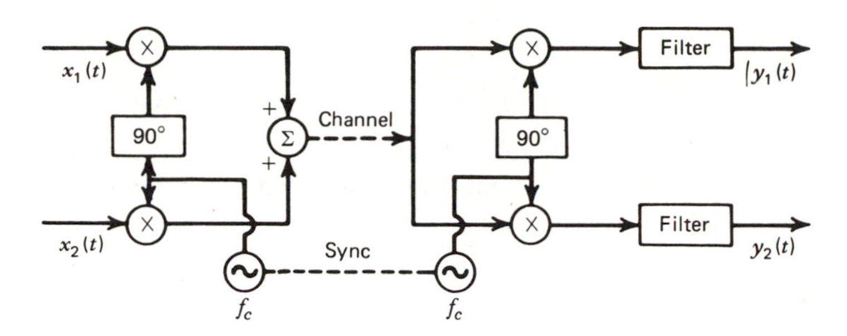

Figure 6.58 Quadrature multiplexing scheme.

6.49. Sixty telephone-grade voice signals are multiplexed using FDM. The bandwidth of the voice signal is 3 kHz and a 1 kHz guardband is required between adjacent voice channels. The subcarrier modulation is SSB (USB) and $f_{c_1} = 0$.
(a) Sketch the typical spectrum of the multiplexed signal.
(b) If all channels are multiplexed directly, calculate the required number of oscillators and SSB modulators.
(c) Suppose that multiplexing is done using five groups of 12 channels each to form a super group of 60 channels. Draw a block diagram of the multiplexer indicating all subcarrier frequencies. How many oscillators and modulators are needed to implement this multiplexing scheme?

6.50. The BPF's in the FDM receiver for the preceding problem have

$$|H(f - f_c')| = \left[1 + \left(\frac{f - f_c'}{B}\right)^{2n}\right]^{-1/2}$$

where f_c' is the subcarrier frequency + 1.5 kHz and B is the 3-dB bandwidth of the filter. The filter bandwidth is to be 3 kHz, and the filter is required to have an attenuation of at least 20 dB in the rejection region, that is,

$$|H(f - f_c')| < -20\text{ dB} \quad \text{for } |f - f_c'| > 1.5\text{ kHz}$$

Find a suitable value for n.

6.51. Two signals $x_1(t)$ and $x_2(t)$ are multiplexed to form

$$x(t) = x_1(t) + x_2(t)\cos 2\pi f_c t$$

$x_1(t)$ and $x_2(t)$ are lowpass signals bandlimited to 5 kHz with $X_1(f) = X_2(f) = 0.0001$ for $|f| < 5$ kHz and $f_c = 15$ kHz. The channel over which $x(t)$ is transmitted has nonlinear transfer characteristics and the channel output is

$$y(t) = x(t) + 0.2x^2(t)$$

(a) Sketch the spectrum of $x(t)$ and $y(t)$. Explain the difficulties associated with the demodulation of $x_1(t)$ and $x_2(t)$ from $y(t)$.
(b) Which of the demodulated signals suffers the worst distortion?

6.52. With reference to Problem 6.51, assume that the multiplexed signal is

$$x(t) = x_1(t) \cos 2\pi f_{c_1} t + x_2(t) \cos 2\pi f_{c_2} t,$$
$$f_{c_1} \gg 5\,\text{kHz} \quad \text{and} \quad f_2 = f_{c_1} + 20\,\text{kHz}$$

(a) Sketch the spectrum of $x(t)$ and $y(t)$.
(b) Can $x_1(t)$ and $x_2(t)$ be recovered from $y(t)$ without distortion?

6.53. The Picturephone® image is square, the active line time is 100 μsec and there are about 230 lines. Calculate the video bandwidth requirement. Referring to Table 6.2, how many (equivalent) voice channels will be required to transmit the Picturephone® signal? (Picturephone® signal is actually transmitted using a digital pulse modulation scheme—not CW modulation. See Chapter 10 for details.)

6.54. Select a suitable modulation scheme (explaining the reasons for your choice) for the following applications:
(a) Data transmission from a satellite over a noisy radio link; satellite has limited power capability.
(b) Multiplexed voice transmission over coaxial cable; primary objective is to transmit as many signals as possible over a single cable.
(c) Point to point voice communication (short distance, single channel) over a twisted pair of wires.
(d) Record high fidelity music on tape using a recorder having a usable bandwidth extending from 500 to 50 kHz.

® Registered service mark of AT & T.

7

NOISE IN ANALOG COMMUNICATION SYSTEMS

7.1 INTRODUCTION

Noise is present in varying degrees in all communication systems. In some parts of a communication system where the noise level is low compared to the signal the effects of noise can be ignored. However, there are a number of points in a communication system where the signal level is often low and the presence of even low level noise at these points can seriously degrade the performance of the system. Examples of this include the input stages of an AM or FM receiver tuned to a distant station, and the input stages of a long-range radar.

The effects of noise can be minimized by the use of modulation and appropriate filtering. In this chapter, we investigate the performance of baseband and CW analog communication systems in the presence of noise. The average signal-to-noise power ratio at the receiver output or the destination point, $(S/N)_d$, will be used as the measure of performance for comparing analog signal transmission schemes.

We begin our analysis of the effects of noise with a study of how the effects of noise can be minimized in baseband systems. We will show that this is accomplished by appropriate filtering of the baseband message signal at the transmitter and the receiver. We will derive expressions for the transfer functions of the transmitting (preemphasis) and receiving (deemphasis) filters that will yield the maximum value for the signal-to-noise power ratio at the output of the system. This value of $(S/N)_d$ will be used as the basis for comparing the performance of other analog signal transmission schemes.

In the second part of this chapter, we will analyze the performance of a number of CW modulation schemes to determine the effects of noise. We will establish the important principle that with the use of nonlinear modulation, such as FM, one can obtain *improved performance* at the expense of *increased transmission bandwidth.* This principle allows us the flexibility of trading bandwidth for signal power, and vice versa, while maintaining a specified $(S/N)_d$.

Finally, we will show that preemphasis and deemphasis filtering can be used in CW modulation schemes to further improve the output signal quality. The use of preemphasis/deemphasis in commercial FM broadcasting will be discussed in detail.

In the following analysis, we will be mainly concerned with the additive noise that accompanies the signal at the input to the receiver. The reason for concentrating on the noise at the receiver input is due to the fact that it is at this point in the system that the signal is weakest. Any attempt to increase signal power by amplification will increase the noise power also. Thus the additive noise at the receiver input will have considerable influence on the quality of the output signal. We will assume that this noise can be modeled by a zero mean stationary, Gaussian random process (Chapter 3, Section 3.7).

While comparing various communication systems, we will pay particular attention to the signal-to-noise ratio at the output, transmitter power and bandwidth requirements, and equipment complexity.

7.2 NOISE IN BASEBAND SYSTEMS

7.2.1 System Model and Parameters

Consider the baseband communication system shown in Figure 7.1. The baseband message signal $X(t)$ is to be transmitted over a baseband channel.

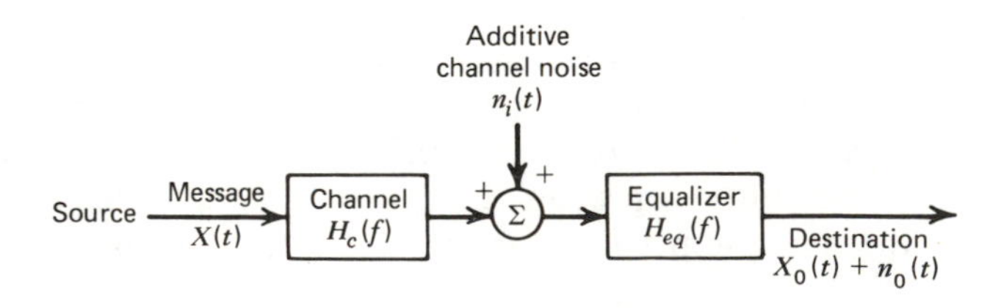

Figure 7.1 Model of a baseband communication system.

The channel has a transfer function $H_c(f)$, and the linear distortion introduced by the channel is removed by an equalizer with a transfer function

$$H_{eq}(f) = K[H_c(f)]^{-1} \exp(-j2\pi f t_d) \tag{7.1}$$

so that the signal component at the destination point is distortionless. That is,

$$X_0(t) = KX(t - t_d) \tag{7.2}$$

The channel also corrupts the signal with additive noise $n_i(t)$ that produces an additive noise component $n_0(t)$ at the destination point.

We will assume that the signal $X(t)$ and the front-end noise $n_i(t)$ are random processes with the following properties:

1. $X(t)$ is a stationary zero mean lowpass random process bandlimited to f_X with a power spectral density function $G_X(f)$.
2. $n_i(t)$ is a stationary, zero mean Gaussian random process (Chapter 3, Section 3.7) with a power spectral density function $G_{n_i}(f)$.
3. $X(t)$ and $n_i(t)$ are independent.

With these assumptions, we can now proceed to analyze the effects of noise in baseband communication systems.

7.2.2 Signal-to-Noise Ratio at the Output of a Baseband System

The signal quality at the output of analog communication systems is usually measured by the average signal power to noise power ratio defined as

$$\left(\frac{S}{N}\right)_d = \frac{E\{X_0^2(t)\}}{E\{n_0^2(t)\}} \tag{7.3}$$

In systems designed for transmitting audio signals, this ratio ranges from 10 dB for barely intelligible voice signals to 30 dB for telephone quality voice signals, and 60 dB for high-fidelity audio signals.

For the system shown in Figure 7.1, if we assume

$$H_{eq}(f)H_c(f) = \begin{cases} K \exp(-j2\pi f t_d) & \text{for } |f| < f_X \\ 0 & \text{elsewhere} \end{cases}$$

then we have

$$X_0(t) = KX(t - t_d)$$

and
$$E\{X_0^2(t)\} = K^2E\{X^2(t-t_d)\} = K^2\int_{-f_X}^{f_X} G_X(f)\,df \tag{7.4}$$

The average noise power at the output is computed as

$$\begin{aligned} E\{n_0^2(t)\} &= \int_{-\infty}^{\infty} G_{n_0}(f)\,df \\ &= \int_{-f_X}^{f_X} G_{n_i}(f)|H_{eq}(f)|^2\,df \\ &= \int_{-f_X}^{f_X} K^2\frac{G_{n_i}(f)\,df}{|H_c(f)|^2} \end{aligned} \tag{7.5}$$

The output signal-to-noise ratio is given by

$$\left(\frac{S}{N}\right)_d = \frac{\displaystyle\int_{-f_X}^{f_X} G_X(f)\,df}{\displaystyle\int_{-f_X}^{f_X} G_{n_i}(f)[|H_c(f)|^2]^{-1}\,df} \tag{7.6}$$

Special Case—Ideal Channel and Additive White Noise. In the special case when we have an ideal channel with additive white noise,

$$H_c(f) = \begin{cases} K\exp(-j2\pi f t_d) & \text{for } |f| < f_X \\ 0 & \text{elsewhere} \end{cases}$$

and

$$G_{n_i}(f) = \eta/2$$

the output signal-to-noise ratio is given by*

$$\left(\frac{S}{N}\right)_d = \frac{K^2\displaystyle\int_{-f_X}^{f_X} G_X(f)\,df}{\eta f_X} = \frac{S_r}{\eta f_X} \tag{7.7}$$

where S_r is the average received signal power. Now, the average transmitted signal power in this case is given by

$$S_T = E\{X^2(t)\} = \int_{-f_X}^{f_X} G_X(f)\,df$$

Substituting S_T in Equation (7.7), we obtain

$$\left(\frac{S}{N}\right)_d = K^2\left(\frac{S_T}{\eta f_X}\right) \tag{7.8}$$

*Instead of an equalizer, we need to have an ideal lowpass filter with a bandwidth f_X at the receiving end.

The denominator in Equation (7.8) is often called the *inband noise power* and it represents the noise power in the message bandwidth. The ratio given in Equation (7.8) is generally (but not always) an upper bound for analog baseband performance that may or may not be achieved in practical systems due to the non-ideal nature of the channel itself and other elements in the system such as nonideal filters. Nevertheless, we will use the ratio given in Equation (7.8) as a basis for comparing the performance of other systems.

7.2.3 Signal-to-Noise Ratio Improvement through Preemphasis/Deemphasis Filtering

When the channel noise is nonwhite and/or the channel frequency response changes considerably in the message bandwidth, then it is possible to improve the output signal-to-noise ratio by using specially designed filters at the transmitting and receiving ends of the channel (Figure 7.2). These filters are known as the *preemphasis/deemphasis filters or the transmitting and receiving filters*, respectively. The transfer functions for these filters are chosen such that $(S/N)_d$ is maximized.

The preemphasis/deemphasis filters serve two purposes, eliminating any linear distortion produced by the channel and maximizing the output signal-to-noise ratio. As far as distortionless transmission is concerned, $H_T(f)$ and $H_R(f)$ should be chosen to satisfy

$$H_T(f)H_R(f)H_c(f) = K \exp(-j2\pi f t_d), \quad \text{for } |f| < f_X \tag{7.9}$$

Now, if $H_R(f)$ is chosen to minimize the output noise and $H_T(f)$, $H_R(f)$ satisfy Equation (7.9), then we have maximized $(S/N)_d$ and the output signal is undistorted. That is, we have chosen optimum terminal filters. The optimization has an additional subtle constraint in that the transmitted power S_T must be kept at a specified (bounded) level. Accordingly, we seek to maximize $(S/N)_d/S_T$ or minimize $S_T/(S/N)_d$.

Starting with Equation (7.9) (with $K = 1$ to simplify the algebra) we have

$$X_0(t) = X(t - t_d) \tag{7.10a}$$

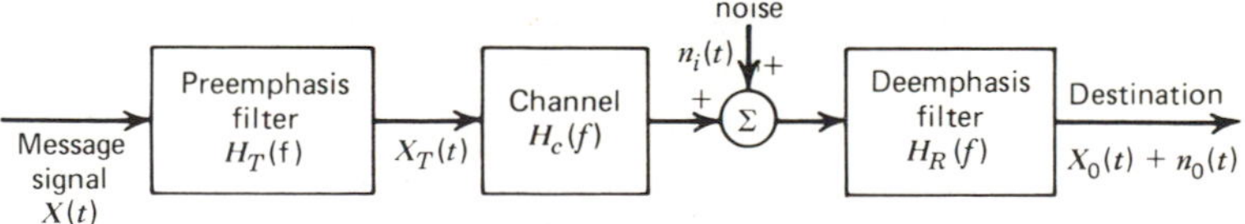

Figure 7.2 Preemphasis/deemphasis filtering.

$$E\{X_0^2(t)\} = E\{X^2(t)\} = S_0 \tag{7.10b}$$

$$E\{n_0^2(t)\} = \int_{-\infty}^{\infty} |H_R(f)|^2 G_{n_i}(f)\, df \tag{7.11}$$

and

$$S_T = E\{X_T^2(t)\} = \int_{-\infty}^{\infty} G_X(f)|H_T(f)|^2\, df \tag{7.12}$$

Substituting for $|H_T(f)|^2$ from Equation (7.9), we have the following expression*:

$$\frac{S_T E\{n_0^2(t)\}}{E\{X_0^2(t)\}} = \frac{\int_{-\infty}^{\infty} [G_X(f)/|H_c H_R(f)|^2]\, df \int_{-\infty}^{\infty} G_{n_i}(f)|H_R(f)|^2\, df}{\int_{-\infty}^{\infty} G_X(f)\, df} \tag{7.13}$$

which is to be minimized. Observe that the minimizing has to be done by appropriate selection of $H_R(f)$, and that the denominator does not involve $H_R(f)$.

The minimizing of the expression in the numerator of Equation (7.13) can be done using the Schwarz's inequality, which is stated as

$$\left|\int_{-\infty}^{\infty} V(f)W^*(f)\, df\right|^2 \leq \int_{-\infty}^{\infty} |V^2(f)|\, df \int_{-\infty}^{\infty} |W^2(f)|\, df \tag{7.14}$$

where $V(f)$ and $W(f)$ are complex functions of f. In (7.14) the equality holds when $V(f) = cW(f)$, where c is an arbitrary positive constant. Applying Schwarz's inequality to the numerator of Equation (7.13) with

$$V = |H_R(f)|G_{n_i}^{1/2}(f), \qquad W = G_X^{1/2}(f)/|H_c(f)H_R(f)|$$

we see that $S_T/(S/N)_d$ is minimized when

$$|H_R(f)|^2 = \frac{cG_X^{1/2}(f)}{|H_c(f)|G_{n_i}^{1/2}(f)}, \quad \text{for } |f| < f_X \tag{7.15a}$$

Substituting Equation (7.15a) in Equation (7.9) with $K = 1$, we have

$$|H_T(f)|^2 = \frac{G_{n_i}^{1/2}(f)}{c|H_c(f)|G_X^{1/2}(f)}, \quad \text{for } |f| < f_X \tag{7.15b}$$

*Note that H_c, H_R, H_T, G_X, and G_n are functions of f; the argument f may be left out in some of the lengthy expressions.

Finally, from Equation (7.13) we obtain the maximum value for the output signal-to-noise ratio as

$$\left(\frac{S}{N}\right)_{d_{\max}} = \frac{S_T \int_{-\infty}^{\infty} G_X(f)\, df}{\left| \int_{-\infty}^{\infty} [G_X^{1/2}(f) G_{n_i}^{1/2}(f)/|H_c(f)|]\, df \right|^2} \tag{7.15c}$$

The limits on the integrals may be changed to $\pm f_X$, since $G_X(f) = 0$ for $|f| > f_X$.

By way of interpreting Equations (7.15*a*) and (7.15*b*), we can say that $H_R(f)$ *deemphasizes* (or attenuates) those frequencies where the noise psd is large and the signal psd is small (a very reasonable thing to do) while $H_T(f)$ does the reverse. The phase angles of $H_T(f)$ and $H_R(f)$ are somewhat arbitrary as long as Equation (7.9) is satisfied.

The optimum filters will yield a significant improvement over the baseband scheme using a simple equalizer (or a lowpass filter) at the receiving end when $H_c(f)$ and/or $G_X(f)$, $G_{n_i}(f)$ change appreciably in the message bandwidth. Otherwise, the slight improvement that results from preemphasis/deemphasis filtering may not justify the expense of putting in these filters.

We will see in a later section of this chapter that in commercial FM broadcasting, preemphasis/deemphasis filtering significantly improves the output signal-to-noise ratio. For now, we will illustrate the advantage of using preemphasis/deemphasis filters with the following example.

Example 7.1. Consider a baseband system with the following characteristics:

$$G_X(f) = 10^{-8} \text{ watt/Hz}, \quad |f| \le 10 \text{ kHz}$$
$$G_{n_i}(f) = 10^{-14} \text{ watt/Hz}$$

(a) Assume that the channel is ideal and compute the signal-to-noise ratio at the destination point.
(b) Assume that

$$|H_c(f)| = \begin{cases} \dfrac{1}{1 + (f/f_c)^4}, & \text{for } |f| \le 10 \text{ kHz},\ f_c = 5 \text{ kHz} \\ 0 & \text{for } |f| > 10 \text{ kHz} \end{cases}$$

and compute the signal-to-noise ratio at the destination point for a system that uses an equalizing filter at the receiver with a transfer function

$$H_{eq}(f) = \begin{cases} \dfrac{1}{H_c(f)}, & \text{for } |f| \le 10 \text{ kHz} \\ 0 & \text{for } |f| > 10 \text{ kHz} \end{cases}$$

(c) Assume the channel characteristics specified in part (b) and compute the $(S/N)_{d_{\max}}$ for the system that uses optimum preemphasis/deemphasis filters.

Solution

(a) The only filter needed for this case is an ideal lowpass filter at the receiving end to limit the out of band noise. The filter cutoff frequency should be 10 kHz, and

$$E\{X_0^2(t)\} = \int_{-10^4}^{10^4} (10^{-8})\, df = 2(10^{-4}) \text{ watt}$$

$$E\{n_0^2(t)\} = \int_{-10^4}^{10^4} (10^{-14})\, df = (2)(10^{-10}) \text{ watt}$$

$$(S/N)_d = 10^6 = 60 \text{ dB}$$

$$S_T = E\{X_0^2(t)\} = -7 \text{ dBm}$$

(b) For this case $(S/N)_d$ is given by Equation (7.6). With $K = 1$,

$$E\{X_0^2(t)\} = E\{X^2(t)\} = (2)(10^{-4}) \text{ watt}$$

$$E\{n_0^2(t)\} = \int_{-10^4}^{10^4} [10^{-14}][1 + (f/f_c)^4]^2\, df$$

$$= (2)(10^{-14})\left(f + \frac{1}{9}\frac{f^9}{f_c^8} + \frac{2}{5}\frac{f^5}{f_c^4}\right)_0^{10^4}$$

$$= (71.7)(10^{-10})$$

Hence

$$(S/N)_d = 44.5 \text{ dB}$$

$$S_T = E\{X^2(t)\} = (2)(10^{-4}) \text{ watt} = -7 \text{ dBm}$$

(c) With optimum filters, $(S/N)_d$ is given by Equation (7.15c). The transfer function of the optimum terminal filters are (with $c = 1$)

$$|H_T(f)|^2 = G_n^{1/2}/|H_c|G_X^{1/2} \qquad |H_R(f)|^2 = G_X^{1/2}/|H_c|G_{n_i}^{1/2}$$

and

$$S_T = \int_{-f_X}^{f_X} G_X^{1/2}(f)G_{n_i}^{1/2}(f)|H_c(f)|^{-1}\, df$$

Substituting for S_T in Equation (7.15c), we have

$$\left(\frac{S}{N}\right)_{d_{\max}} = E\{X^2(t)\}\left[\int_{-f_X}^{f_X} G_X^{1/2} G_{n_i}^{1/2} |H_c|^{-1}\, df\right]^{-1}$$

and

$$\int_{-f_X}^{f_X} G_X^{1/2} G_{n_i}^{1/2} |H_c|^{-1}\, df = 10^{-11} \int_{-10^4}^{10^4} \left[1 + \left(\frac{f}{f_c}\right)^4\right] df$$
$$= (8.4)(10^{-7})$$

Hence

$$\left(\frac{S}{N}\right)_{d_{\max}} = \frac{(2)(10^{-4})}{(8.4)(10^{-7})} = 23.8 \text{ dB}$$

and

$$S_T = (8.4)(10^{-7}) = -30.8 \text{ dBm}$$

Now, if S_T is raised to $(2)(10^{-4})$, as in case (a) and (b), then $(S/N)_{d_{\max}}$ will be raised by a factor of $(2)(10^{-4})/(8.4)(10^{-7})$ or by a factor of 23.8 dB. Thus for the same transmitter power we have

(a) $(S/N)_d$ for an ideal channel = 60 dB

(b) $(S/N)_d$ for a nonideal channel with equalizer = 44.5 dB

(c) $(S/N)_d$ for a nonideal channel with optimum filters = 47.6 dB

Thus the optimum filter yields approximately a 3-dB power advantage over the scheme using a simple equalizer arrangement. The 3-dB advantage in power requirement is significant and hence preemphasis/deemphasis filters should be considered.

7.3 NOISE IN LINEAR CW MODULATION SYSTEMS

We are now ready to analyze the performance of linear modulation schemes (DSB, SSB, AM, and VSB) in the presence of additive noise. In linear as well as exponential modulation schemes, we have bandpass transmission as opposed to baseband transmission. Models of narrowband (bandpass) noise developed in Chapter 3 will prove to be quite useful in analyzing the noise performance of CW modulation schemes. Many of the concepts discussed in the preceding section on baseband systems are also applicable to the band-pass case we will be dealing with now.

Our analysis begins with a discussion of the system models and parameters. For the purpose of analyzing the noise performance, we will assume that all subsystems (such as modulators, filters, and demodulators) are ideal in nature and that the performance of the system is degraded by additive noise at the input to the receiver. With this idealized model, we will derive expressions for the signal-to-noise power ratio at the output of various modulation systems. To facilitate fair comparison, we will assume that all systems operate with the same average transmitted (or received) power level and that the noise environment (noise psd) is the same for all systems.

7.3.1 System Model and Parameters

An idealized model of a CW communication system is shown in Figure 7.3. The signal $X(t)$ is modeled as a zero mean stationary, lowpass random process with a psd $G_X(f)$ that is bandlimited to f_X. The peak value of $|X(t)|$ is assumed to be unity. The transmitter is assumed to be ideal, and the transmitter output is assumed to have the form

$$X_c(t) = A(t)\cos[\omega_c t + \phi(t)].$$

$X_c(t)$ is the modulated signal with a bandwidth B_T.

The channel is assumed to give distortionless transmission over the transmission bandwidth B_T. At the output of the channel, the signal is accompanied by additive noise that is modeled as a zero mean, stationary Gaussian random process with a power spectral density of $G_{n_i}(f)$.

The receiver front-end (RF/IF stages) is modeled as an ideal bandpass filter with a bandwidth B_T, which is the same as the bandwidth of $X_c(t)$. Thus the front-end filter, also known as a *predetection filter*, passes $X(t)$ without

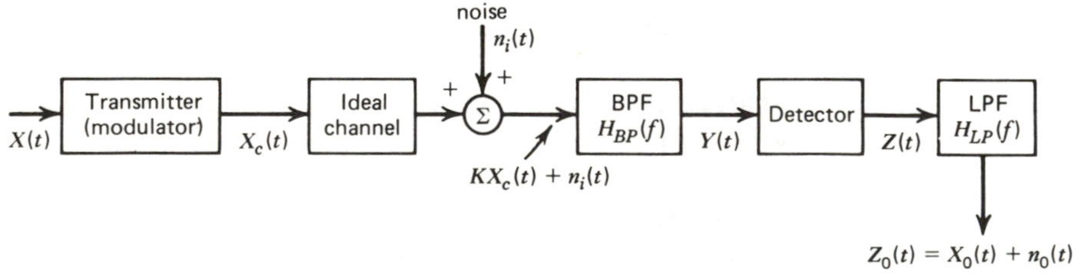

Figure 7.3 Model of a CW communication system.

distortion, but limits the amount of out of band noise that reaches the detector (demodulator).

The input to the detector consists of a signal component $KX_c(t)$ plus a noise component $n(t)$ that is the filtered version of $n_i(t)$. The bandpass noise $n(t)$ can be represented in the form (see section 3.7)

$$n(t) = n_c(t)\cos\omega_c t - n_s(t)\sin\omega_c t$$

and hence we can express the input to the detector as

$$Y(t) = KX_c(t) + n(t) \tag{7.16a}$$

$$= R_Y(t)\cos[\omega_c t + \theta_Y(t)] \tag{7.16b}$$

$$= Y_c(t)\cos\omega_c t - Y_s(t)\sin\omega_c t \tag{7.16c}$$

The detector (demodulator) responds to $Y(t)$ and its response is modeled by the following idealized characteristics:

$$Z(t) = \begin{cases} k_1 Y_c(t) & \text{synchronous detector} \\ k_2 R_Y(t) & \text{envelope detector} \\ k_3\theta_Y(t) & \text{phase detector} \\ k_4(d\theta_Y(t)/dt) & \text{frequency detector} \end{cases} \tag{7.17}$$

In Equation (7.17), k_1, k_2, k_3, and k_4 are the detector gains that are often assumed to be unity.

The detector output is lowpass filtered to remove out of band noise and harmonic signal terms. The lowpass filter, also referred to as a baseband or *post-detection filter*, has a bandwidth of f_X Hz.

The analysis we present below for computing the output signal-to-noise ratio is valid for any given form of $G_{n_i}(f)$. In most practical systems the front end noise can be assumed to be white with a psd of

$$G_{n_i}(f) = \eta/2 \text{ watt/Hz} \tag{7.18}$$

White noise assumption simplifies the calculations considerably and allows us to look at the conceptually important aspects of analysis rather than getting mired in algebraic details.

With additive noise at the input to the detector, it is reasonable to anticipate that the output of the system will consist of a signal component $X_0(t)$ and an additive noise component $n_0(t)$. While this is not the case in general, it is true in most cases and the output signal-to-noise ratio is obtained by computing the ratio of $E\{X_0^2(t)\}$ and $E\{n_0^2(t)\}$.

The output signal quality will depend, to a large extent, on the quality of the signal at the detector input. The input signal quality is measured by the signal-to-noise power ratio at the detector input, which is defined as

$$\left(\frac{S}{N}\right)_i = \frac{K^2E\{X_c^2(t)\}}{B_T\eta}$$

The numerator in the preceding equation is the average signal power at the receiver input and is denoted by S_r.

We are now ready to address the important problem: Given $X_c(t)$ and the type of detector, how do we calculate the output signal-to-noise ratio $(S/N)_d$ in terms S_r, f_X (or B_T), and $\eta/2$. We answer this question in the following sections.

7.3.2 Noise in DSB and SSB Systems

In a DSB system, the transmitted signal $X_c(t)$ has the form*

$$X_c(t) = A_cX(t)\cos\omega_ct \tag{7.19}$$

and the average value of the transmitted power is

$$S_T = E\{X_c^2(t)\} = \tfrac{1}{2}A_c^2S_X \tag{7.20}$$

where

$$S_X = E\{X^2(t)\} = \int_{-f_X}^{f_X} G_X(f)\,df \tag{7.21}$$

and the average value of the received power is $S_r = K^2E\{X_c^2(t)\}$.
The demodulator portion of the DSB system is shown in Figure 7.4. The predetection filter has a center frequency f_c and bandwidth $B_T = 2f_X$. The output of the filter is

$$Y(t) = KX_c(t) + n(t) = a_cX(t)\cos\omega_ct + n(t) \tag{7.22}$$

where $n(t)$ is narrowband Guassian noise with a psd $G_n(f)$ given by

$$G_n(f) = \begin{cases} \eta/2 & \text{for } |f - f_c| < f_X \\ 0 & \text{elsewhere} \end{cases} \tag{7.23}$$

and $a_c = KA_c$.

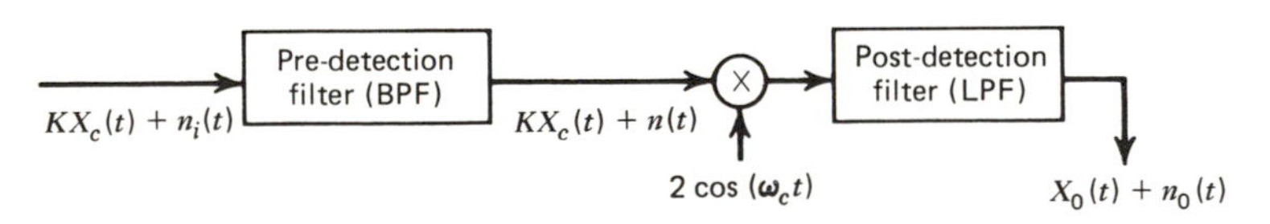

Figure 7.4 DSB modulator.

*Strictly speaking, we need a random phase term in the carrier to make $X_c(t)$ stationary.

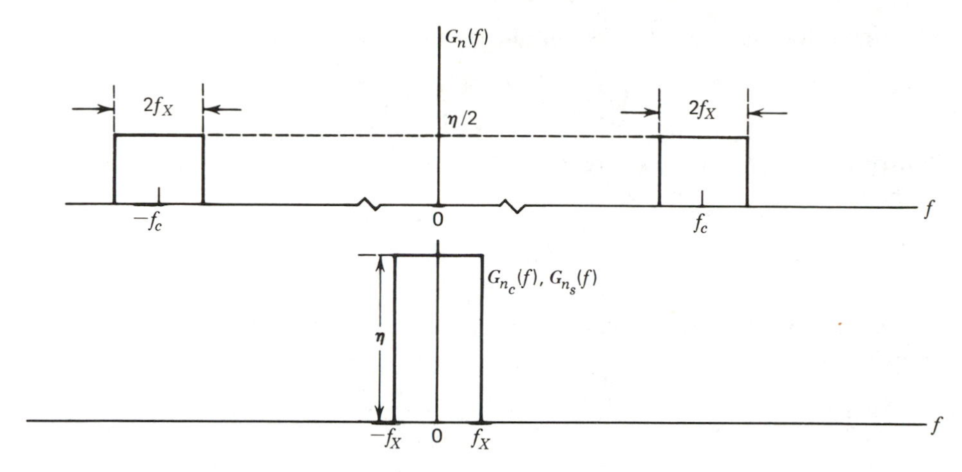

Figure 7.5 Spectral densities of $n(t)$, $n_c(t)$, and $n_s(t)$.

Using the results of Section 3.7.2., we can write $n(t)$ as

$$n(t) = n_c(t) \cos \omega_c t - n_s(t) \sin \omega_c t \tag{7.24}$$

where $n_c(t)$ and $n_s(t)$ are lowpass Gaussian random processes with power spectral densities (see Figure 7.5)

$$G_{n_c}(f) = G_{n_s}(f) = \begin{cases} \eta & \text{for } |f| < f_X \\ 0 & \text{elsewhere} \end{cases}$$

Substituting Equation (7.24) in (7.22), we have

$$Y(t) = [a_c X(t) + n_c(t)] \cos \omega_c t - n_s(t) \sin \omega_c t \tag{7.25}$$

and

$$2Y(t) \cos \omega_c t = [a_c X(t) + n_c(t)] + [a_c X(t) + n_c(t)] \cos 2\omega_c t - n_s(t) \sin 2\omega_c t$$

Since the cutoff frequency of the post-detection filter is f_X and $f_X \ll f_c$, the double frequency terms are removed by the filter and the output is given by

$$X_0(t) + n_0(t) = a_c X(t) + n_c(t) \tag{7.26}$$

The output signal component is

$$X_0(t) = a_c X(t) \quad \text{and} \quad E\{X_0^2(t)\} = a_c^2 S_X$$

The output noise component is

$$n_0(t) = n_c(t) \quad \text{and} \quad E\{n_c^2(t)\} = 2\eta f_X$$

The signal-to-noise ratio at the output is easily computed as

$$\left(\frac{S}{N}\right)_d = \frac{E\{X_0^2(t)\}}{E\{n_0^2(t)\}} = \frac{a_c^2 S_X}{2\eta f_X}$$

Substituting $\frac{1}{2}a_c^2 S_X = S_r$, where S_r is the average power at the receiver input, we have

$$\left(\frac{S}{N}\right)_d = \frac{S_r}{\eta f_X} \tag{7.27}$$

The reader can verify that the signal-to-noise ratio at the input of the detector is

$$\left(\frac{S}{N}\right)_i = \frac{S_r}{2\eta f_X}$$

and

$$\frac{(S/N)_d}{(S/N)_i} = \alpha_d = 2 \tag{7.28}$$

The ratio α_d is known as the *detection gain* and is often used as a figure of merit for the demodulator.

The detection gain of 2 for the coherent demodulator is somewhat questionable for it appears that we have gained a 3 dB improvement in the signal-to-noise ratio. This is true however, since the demodulator effectively suppresses a quadrature noise component (see Equations (7.25) and (7.26)). A comparison of the $(S/N)_d$ for the DSB scheme with the $(S/N)_d$ for the baseband system (Equation (7.7)) indicates that the destination signal to noise is the same for DSB with coherent demodulation and the baseband system. This result can be justified as follows. The transmission bandwidth B_T is $2f_X$ for DSB as compared to $B_T = f_X$ for the baseband system and thus the noise power at the input to the demodulator of the DSB system is twice the noise power in a baseband system. The 3-dB detection gain is just enough to overcome this additional noise power at the input.

Similar calculations can be carried out for an SSB system. The reader can verify that (Problem 7.7) for an SSB system, the signal-to-noise ratio at the destination point is given by

$$\left(\frac{S}{N}\right)_d = \frac{S_r}{\eta f_X} \tag{7.29}$$

Equation (7.29) indicates that the $(S/N)_d$ for the SSB system is the same as the $(S/N)_d$ for the DSB and baseband systems. However, the transmission bandwidth for the SSB system is $B_T = f_X$, which is one half the transmission bandwidth of DSB modulation.

The reader can also verify that $(S/N)_d$ for VSB modulation using coherent demodulation will be the same as $(S/N)_d$ for the SSB and DSB modulation schemes (Problem 7.8).

7.3.3 Noise in AM Systems

Since AM is usually demodulated by an envelope or square law detector, we will analyze the performance of AM systems assuming envelope or square law demodulators. We will show that such noncoherent demodulation schemes yield satisfactory performance as long as the signal power at the receiver input is considerably higher (10 dB or more) than the inband noise power at the input. We will also show that at lower levels of signal power the performance of AM systems using noncoherent demodulators deteriorates rapidly.

Envelope Demodulation and Threshold Effect. With reference to Figure 7.3, the transmitted signal $X_c(t)$ in an AM system has the form

$$X_c(t) = A_c[1 + mX(t)] \cos \omega_c t \tag{7.30}$$

The signal $X(t)$ is assumed to have values in the range of -1 to 1 and $0 < m < 1$. The parameter "m" is the modulation index of the AM signal.

The input to the detector is then

$$\begin{aligned} Y(t) &= KX_c(t) + n(t) \\ &= \{a_c[1 + mX(t)] + n_c(t)\} \cos \omega_c t - n_s(t) \sin \omega_c t \end{aligned}$$

We can express $Y(t)$ in the form

$$Y(t) = R_Y(t) \cos[\omega_c t + \theta_Y(t)]$$

where

$$R_Y(t) = \sqrt{\{a_c[1 + mX(t)] + n_c(t)\}^2 + [n_s(t)]^2} \tag{7.31a}$$

and

$$\theta_Y(t) = \tan^{-1} \left(\frac{n_s(t)}{a_c[1 + mX(t)] + n_c(t)} \right) \tag{7.31b}$$

The output of the envelope detector will be proportional to $R_Y(t)$. The expression for $R_Y(t)$ given in Equation (7.31a) can be considerably simplified if we assume that the signal power is either very large or very small compared to the noise power. Furthermore, the analysis is easier if we use phasor diagrams.

The phasor diagram for the strong signal case is shown in Figure 7.6a using

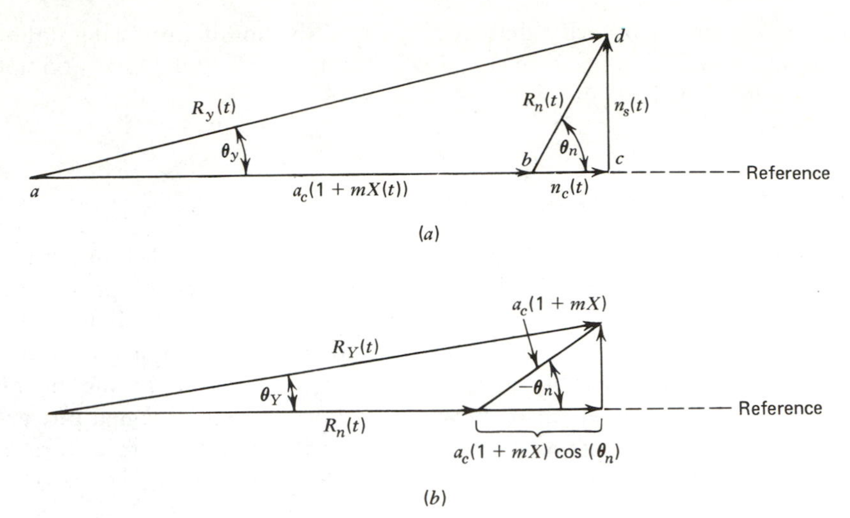

Figure 7.6 Phasor diagram for an AM signal plus noise. (*a*) $a_c^2 \gg E\{n^2(t)\}$. (*b*) $a_c^2 \ll E\{n^2(t)\}$

the signal component as the reference phasor. One of the two quadrature components of the noise adds *in phase* to the signal component, whereas the other noise component adds to the signal component at 90°. The envelope $R_Y(t)$ is represented by the hypotenuse $\overrightarrow{ad}$ of the right-angled triangle a–d–c. Now, since a_c^2 is assumed to be much greater than $E\{n^2(t)\}$, we can expect $a_c[1 + mX(t)]$ to be much larger than $n_s(t)$ and $n_c(t)$ most of the time. That is, the length of $\overrightarrow{ab}$ will be much larger than the lengths of $\overrightarrow{bc}$ and $\overrightarrow{cd}$. Hence we can approximate $R_Y(t)$ as

$$\begin{aligned} R_Y(t) &= \text{length of } \overrightarrow{ad} \approx \text{length of } \overrightarrow{ac} \\ &\approx a_c[1 + mX(t)] + n_c(t) \end{aligned} \tag{7.32}$$

Hence, the demodulator output is given by

$$Z_0(t) \approx a_c mX(t) + n_c(t)$$

assuming that the post-detection filter rejects the DC component due to a_c.

The output signal component is $X_0(t) = a_c mX(t)$ and the noise component is $n_0(t) = n_c(t)$. The output signal-to-noise ratio is given by

$$\left(\frac{S}{N}\right)_d = \frac{a_c^2 m^2 E\{X^2(t)\}}{E\{n_c^2(t)\}} = \frac{a_c^2 m^2 E\{X^2(t)\}}{2\eta f_X}$$

For the AM signal, the received power is given by

$$S_r = E\{K^2X_c^2(t)\}$$
$$= \frac{a_c^2}{2}[1 + m^2E\{X^2(t)\}]$$
$$= \frac{a_c^2}{2}[1 + m^2S_X]$$

Using S_r given in the preceding equation, we can write $(S/N)_d$ as

$$\left(\frac{S}{N}\right)_d = \left(\frac{S_r}{\eta f_X}\right)\left(\frac{m^2S_X}{1 + m^2S_X}\right) \tag{7.33}$$

In Equation (7.33), $m^2S_X/[1 + m^2S_X]$ is the power efficiency of the AM signal that can have a maximum value of $\frac{1}{2}$ (see Section 6.3.2). Hence,

$$\left(\frac{S}{N}\right)_d \leq \frac{1}{2}\left(\frac{S_r}{\eta f_X}\right)$$

which implies that, for a given value of *average* transmitted power, $(S/N)_d$ for AM is at least 3 dB poorer than $(S/N)_d$ for SSB and DSB modulation using coherent demodulation. If we compare these schemes in terms of equal *peak* powers, $(S/N)_d$ for AM is 6 dB poorer than $(S/N)_d$ for DSB modulation since the peak powers are proportional to A_c^2 and $4A_c^2$. In terms of equal peak powers, SSB has been shown to be 2 to 3 dB better than DSB and 8 to 10 dB better than AM.

The output signal-to-noise ratio for the AM system given in (7.33) was derived under the assumption that the signal power at the receiver input is considerably higher than the noise power. If the signal power is much lower than the noise power, the situation is quite different. The phasor diagram in Figure 7.6*b* depicts this hopeless situation. We can now approximate $R_Y(t)$ as

$$R_Y(t) \approx R_n(t) + a_c[1 + mX(t)] \cos \theta_n(t)$$

where $R_n(t)$ and $\theta_n(t)$ are the envelope and phase of $n(t)$.

The output of the demodulator is now

$$Z_0(t) = R_n(t) + a_cmX(t) \cos \theta_n(t) - E\{R_n\}$$

where the term $-E\{R_n\}$ represents the DC component removed by the post-detection filter.

The principal component of the output is $R_n(t)$—a noise term—and the signal component is *multiplied* by a random noise term $\cos \theta_n(t)$. That is, the message term is *captured* by the noise and it is meaningless to talk about output signal-to-noise ratio.

The loss of message signal at low predetection signal-to-noise ratios is called the *threshold effect*. The name comes from the fact that there is some value of $(S/N)_i$ above which signal distortion due to noise is negligible and below which the system performance deteriorates rapidly. The threshold effect does not occur when coherent demodulation is used.

The threshold value of $(S/N)_i$ is usually defined as that value of $(S/N)_i$ for which $R_n < a_c$ with probability 0.99. From Section 3.7 (Equation (3.110)) we know that

$$f_{R_n}(r_n) = \frac{r_n}{N_0}\exp(-r_n^2/2N_0), \quad \text{for } r_n > 0$$

where $N_0 = E\{n^2(t)\} = 2\eta f_X$ and

$$P\{R_n < a_c\} = 0.99 \rightarrow \exp(-a_c^2/2N_0) = 0.01$$

Now, if we take $S_r \approx a_c^2$, then $(S/N)_i = S_r/N_0 = a_c^2/N_0$ and

$$\exp(-a_c^2/2N_0) = \exp[-(S_r/2N_0)] = \exp[-\tfrac{1}{2}(S/N)_i] = 0.01$$

or

$$(S/N)_i \text{ at threshold} = 2\log_e(100) \approx 10\text{ dB}$$

If $(S/N)_i < 10$ dB, message mutilation must be expected and the output will be completely dominated by noise.

For voice and music transmission, $(S/N)_d$ has to be in the range of 30 to 60 dB, which requires that $(S/N)_i$ be considerably above the threshold level. As $(S/N)_i$ drops considerably below 30 dB, additive noise will obscure the signal completely, long before the multiplicative noise captures the signal. Thus for audio transmission, threshold effect is not a serious problem.

If AM is used for transmitting digital information (Chapter 8), it is necessary to use synchronous demodulation when $(S/N)_i$ is low in order to avoid threshold effects.

Square Law Demodulation. The calculation of signal-to-noise ratio at the output of a square law demodulator is rather straightforward. The detector input is as $Y(t) = KX_c(t) + n(t)$, and the output is given by $Z(t) = [Y(t)]^2$, or

$$Z(t) = \frac{a_c^2}{2}[1 + m^2X^2(t) + 2mX(t)][1 + \cos 2\omega_c t] + n^2(t) + 2n(t)a_c[1 + mX(t)]\cos\omega_c t$$

Recalling the requirement $|mX(t)| \ll 1$ for distortionless square law demodulation (Section 6.3 and Problem 6.20), we can approximate $Z(t)$ as

$$Z(t) \approx \frac{a_c^2}{2}[1 + 2mX(t)][1 + \cos 2\omega_c t] + n^2(t) + 2a_cn(t)\cos\omega_c t$$

The signal component at the output of the post-detection filter is (assuming that the filter removes DC components)

$$X_0(t) = a_c^2 mX(t) \quad \text{and} \quad E\{X_0^2(t)\} = a_c^4 m^2 S_X$$

The noise power at the post-detection filter output can be computed by examining the noise components at the input to the filter which consist of $n^2(t)$ and $2a_c n(t)\cos\omega_c t$. The noise component at the output results from filtering $n^2(t)$ and $2a_c n(t)\cos\omega_c t$ and removing DC components.

From the results derived in Chapter 3 we have

$$G_{n^2}(f) = \underbrace{[E\{n^2(t)\}]\delta(f)}_{\text{DC term}} + 2G_n(f) * G_n(f)$$

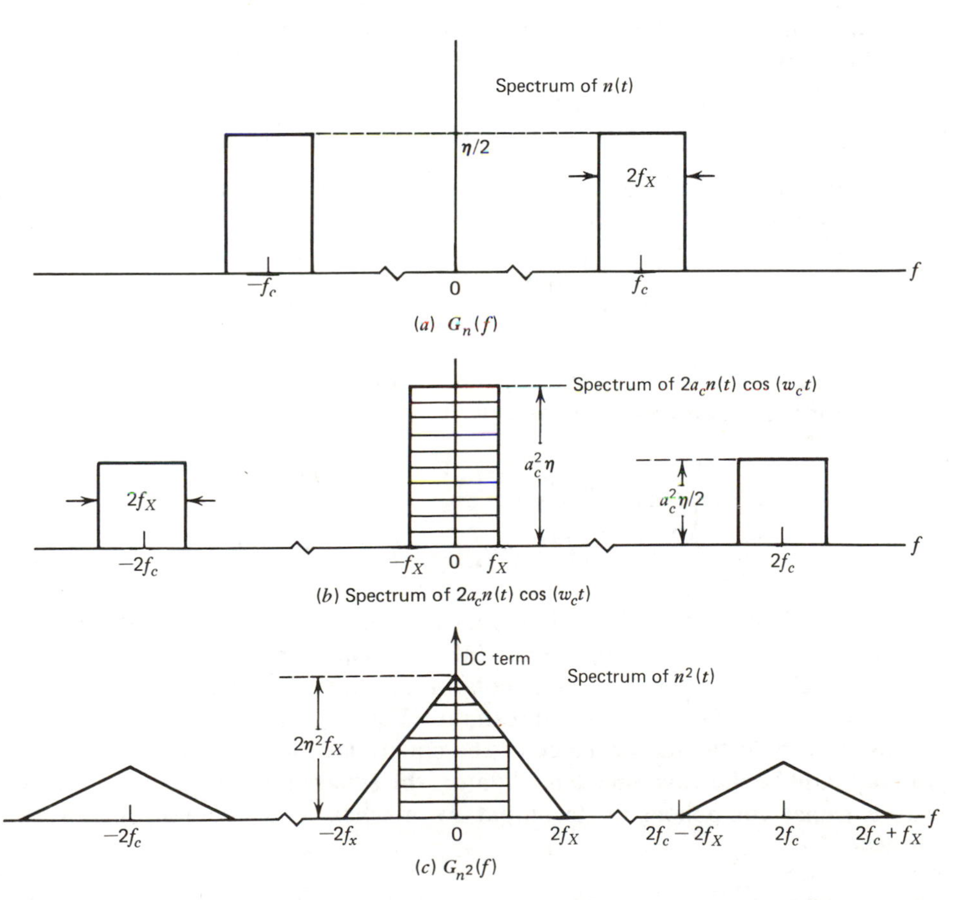

Figure 7.7 Noise power spectral densities. (a) $G_n(f)$. (b) Spectrum of $2a_c n(t)\cos(\omega_c t)$. (c) $G_{n^2}(f)$.

and the psd of $2a_c n(t) \cos \omega_c t$, denoted by say $G_{n_1}(f)$, is given by

$$G_{n_1}(f) = a_c^2[G_n(f - f_c) + G_n(f + f_c)]$$

These spectral densities are shown in Figure 7.7. The post-detection filter removes the DC component in $G_{n^2}(f)$ and passes only those portions of $G_{n^2}(f)$ and $G_{n_1}(f)$ that lie within $-f_X$ to f_X (shaded areas in Figure 7.7). The output noise power is easily computed from Figure 7.7 as

$$E\{n_0^2(t)\} = 2a_c^2\eta f_X + 3\eta^2 f_X^2 \tag{7.34}$$

and

$$\left(\frac{S}{N}\right)_d = \frac{E\{X_0^2(t)\}}{E\{n_0^2(t)\}} = \frac{a_c^4 m^2 S_X}{2a_c^2\eta f_X + 3\eta^2 f_X^2} \tag{7.35}$$

Now, with $|mX(t)| \ll 1$, the average received power is $S_r \approx a_c^2/2$, and the signal-to-noise ratio at the detector input is

$$\left(\frac{S}{N}\right)_i = \frac{S_r}{2\eta f_X} = \frac{S_r}{2N_0} \tag{7.36}$$

where $N_0 = \eta f_X$ is the noise power in the message bandwidth. We can rewrite $(S/N)_d$ in terms of S_r and N_0 as

$$\left(\frac{S}{N}\right)_d = m^2 S_X \left(\frac{S_r}{N_0}\right)\left(\frac{1}{1 + \frac{3}{4}(N_0/S_r)}\right) \tag{7.37}$$

As before, we consider the two special cases when $S_r/N_0 \gg 1$ (high input signal-to-noise power ratio) and $(S_r/N_0) \ll 1$ (low input signal-to-noise power ratio). For these two cases we can rewrite $(S/N)_d$ as

$$\left(\frac{S}{N}\right)_d = \begin{cases} \dfrac{S_r}{\eta f_X}(m^2 S_X), & \text{when } \dfrac{S_r}{\eta f_X} \gg 1 \quad (7.38a) \\[2ex] \left(\dfrac{4}{3}\right) m^2 S_X \left(\dfrac{S_r}{\eta f_X}\right)^2, & \text{when } \dfrac{S_r}{\eta f_X} \ll 1 \quad (7.38b) \end{cases}$$

Equations (7.38a) and (7.38b) show that the square law demodulator exhibits a threshold like the envelope demodulator. When the input signal-to-noise ratio is high, the square law demodulator performs as well as the envelope demodulator operating above threshold (Equation (7.33) with $m^2 S_X \ll 1$). Below threshold the performance of the square law demodulator deteriorates rapidly. Unlike the envelope demodulator, the square law demodulator always has a signal component at its output even when it is operating below the threshold.

Example 7.2. Compare the average transmitter power and channel bandwidth requirements of DSB, SSB, and AM schemes for transmitting an audio signal with a bandwidth of 10 kHz with a destination signal-to-noise ratio of 50 dB. Assume that channel introduces a 50 dB power loss and that the noise power spectral density at the receiver input is 10^{-12} watt/Hz. Assume $m^2 S_X = 0.5$ for AM.

Solution. The bandwidth requirements are easily computed as

$$B_T = \begin{cases} 20\,\text{kHz} & \text{for DSB and AM} \\ 10\,\text{kHz} & \text{for SSB} \end{cases}$$

$(S/N)_d$ for DSB and SSB modulation is given by

$$\left(\frac{S}{N}\right)_d = \frac{S_r}{\eta f_X}$$

With $(S/N)_d = 50$ dB, $\eta f_X = (2)(10^{-12})(10^4)$, we have

$$(S_r)_{\text{DSB,SSB}} = (2)(10^{-8})(10^5) = (2)(10^{-3}) = 3\text{ dBm}$$

Since the channel power loss is 50 dB (or 10^5), we get

$$(S_T)_{\text{DSB,SSB}} = (2)(10^{-3})(10^5) = 200\text{ watts}$$

For the AM, assuming envelope demodulation,

$$\left(\frac{S}{N}\right)_d = \left(\frac{S_r}{\eta f_X}\right)\left(\frac{m^2 S_X}{1 + m^2 S_X}\right) = \frac{1}{3}\frac{S_r}{\eta f_X}$$

and hence $S_T = 600$ watts.

7.4 NOISE IN ANGLE MODULATION SYSTEMS

Having examined noise in linear CW modulation systems, we are now ready to investigate the effects of noise in angle modulated (FM and PM) systems. We will find that the performance of angle modulated systems differs considerably from the performance of linear modulation systems. Furthermore, we will find that there are significant differences between phase and frequency modulation.

Before we proceed to derive expressions for the signal-to-noise ratios at the output of angle modulated systems, let us restate appropriate signal and system models. With reference to Figure 7.3, the transmitted signal $X_c(t)$ has the form

$$X_c(t) = A_c \cos[\omega_c t + \phi(t)]$$

where

$$\phi(t) = \begin{cases} k_p X(t) & \text{for PM} \\ k_f \int_{-\infty}^{t} X(\tau)\, d\tau & \text{for FM} \end{cases}$$

As before, the message signal $X(t)$ is normalized such that $\max|X(t)| = 1$. The phase deviation constant k_p is $\leq \pi$ for PM so that the message signal can be demodulated from the PM waveform without ambiguity.

The detector is assumed to be ideal. With an input of $Y(t) = R_Y(t)\cos[\omega_c t + \theta_Y(t)]$, the output of the detector is

$$Z(t) = \begin{cases} k_d \theta_Y(t) & \text{for phase detection} \quad (7.39a) \\ k_d'\, d\theta_Y(t)/dt & \text{for frequency detection} \quad (7.39b) \end{cases}$$

For convenience, we will assume that the detector gain k_d is such that

$$k_d k_p = k_d' k_f = 1$$

We will use a common approach for analyzing FM and PM cases together and separate the results by replacing $\phi(t)$ by the proper function.

7.4.1 Output Signal-to-Noise Ratios in Angle Modulation Systems

The input to the detector can be written as

$$\begin{aligned} Y(t) &= a_c \cos[\omega_c t + \phi(t)] + n(t) \\ &= a_c \cos[\omega_c t + \phi(t)] + n_c(t)\cos\omega_c t - n_s(t)\sin\omega_c t \end{aligned} \tag{7.40}$$

where $n(t)$ is bandlimited, zero mean Gaussian noise with

$$G_n(f) = \begin{cases} \eta/2, & \text{for } |f - f_c| < B_T/2 \\ 0 & \text{elsewhere} \end{cases} \tag{7.41}$$

The transmission bandwidth of the angle modulated signal is (Equation (6.53))

$$B_T = \begin{cases} 2(D+1)f_X & \text{for FM} \\ 2(k_p+1)f_X & \text{for PM} \end{cases} \tag{7.42}$$

where $D = f_\Delta/f_X = k_f/(2\pi f_X)$ is the deviation ratio for the FM system.

The components that make up $Y(t)$ are shown in the form of a phasor diagram in Figure 7.8. As in the case of envelope demodulation of AM signals, let us first investigate the performance of angle modulation schemes when $a_c^2 \gg E\{n^2(t)\}$, that is, when the signal power at the input to the detector is much higher than the noise power.

The input signal to the detector $Y(t)$ can be written as

$$Y(t) = a_c \cos[\omega_c t + \phi(t)] + R_n(t)\cos[\omega_c t + \theta_n(t)]$$

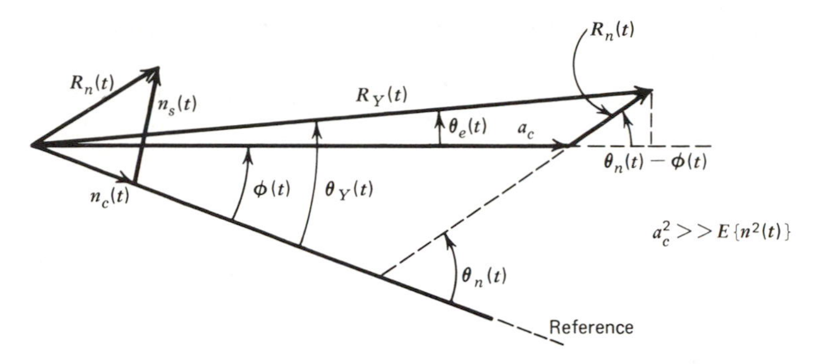

Figure 7.8 Phasor diagram for an angle modulated signal corrupted by additive noise.

where $R_n(t)$ and $\theta_n(t)$ are the envelope and phase of $n(t)$. Using the relationships shown in Figure 7.8, we can express $Y(t)$ in envelope and phase form as

$$Y(t) = R_Y(t)\cos[\omega_c t + \theta_Y(t)]$$

where the angle $\theta_Y(t)$ is

$$\theta_Y(t) = \phi(t) + \theta_e(t) \tag{7.43}$$

We are not interested in the envelope $R_Y(t)$ of $Y(t)$ since the response of the (ideal) detector will depend on $\theta_Y(t)$ only. In a practical situation, a limiter will eliminate any noise-induced variations in the envelope of $Y(t)$.

In Equation (7.43), $\theta_e(t)$ is the perturbation of the carrier phase angle (which contains the signal) due to noise. This perturbation angle can be computed from Figure 7.8 as

$$\theta_e(t) = \tan^{-1}\left(\frac{R_n(t)\sin(\theta_n - \phi)}{a_c + R_n(t)\cos(\theta_n - \phi)}\right) \tag{7.44a}$$

With the assumption $a_c^2 \gg E\{n^2(t)\}$, we can expect $R_n(t) \ll a_c$ most of the time and hence $\theta_e(t)$ can be approximated by

$$\begin{aligned}\theta_e(t) &\approx \tan^{-1}\left(\frac{R_n(t)}{a_c}\sin(\theta_n - \phi)\right)\\ &\approx \frac{R_n(t)}{a_c}\sin(\theta_n - \phi),\end{aligned} \tag{7.44b}$$

the last step being obtained using the approximation $\tan^{-1}\alpha = \alpha$ when $\alpha \ll 1$.

Combining Equations (7.43) and (7.44), we have the following expression for the angle of the signal plus noise at the input to the detector:

$$\theta_Y(t) \approx \underbrace{\phi(t)}_{\substack{\text{signal}\\\text{term}}} + \underbrace{\frac{R_n(t)}{a_c}\sin[\theta_n(t) - \phi(t)]}_{\substack{\text{noise}\\\text{term}}} \tag{7.45}$$

We are now ready to calculate the signal-to-noise ratio at the output of angle modulation systems.

$(S/N)_d$ in PM Systems. Using the detector model stated in Equation (7.39), the output of the phase detector is

$$\begin{aligned} Z(t) &= k_d\theta_Y(t) \\ &= k_d\phi(t) + k_d\frac{R_n(t)}{a_c}\sin[\theta_n(t) - \phi(t)] \end{aligned}$$

If we use $n_p(t)$ to denote the noise term in the preceding equation, then we can express the detector output as

$$Z(t) = k_p k_d X(t) + n_p(t) = X(t) + n_p(t)$$

since $k_p k_d$ is assumed to be unity.

The post-detection filter passes $X(t)$ without distortion and hence the output signal component is $X(t)$ and the output signal power is given by

$$E\{X_0^2(t)\} = E\{X^2(t)\} = S_X \tag{7.46}$$

To calculate the output noise power, we need to derive the spectral characteristics of

$$n_p(t) = k_d\frac{R_n(t)}{a_c}\sin[\theta_n(t) - \phi(t)] \tag{7.47}$$

Unfortunately, the calculation of $G_{n_p}(f)$ is somewhat involved due to the presence of $\sin[\theta_n(t) - \phi(t)]$. To simplify the calculations, let us set the carrier modulation $\phi(t) = 0$ and compute $G_{n_p}(f)$. While setting $\phi(t) = 0$ may seem to be an unreasonable assumption, it has been shown by Downing (1964, Chapter 5) that carrier modulation $\phi(t)$ produces components in $G_{n_p}(f)$ at frequencies above the message band. Such components are rejected by the post-detection filter and do not appear at the output, and hence our assumption $\phi(t) = 0$ does not cause any error in the calculation of inband noise power at the output.

With $\phi(t) = 0$, we have

$$n_p(t) = k_d\frac{R_n(t)}{a_c}\sin\theta_n(t)$$

or

$$n_p(t) = \frac{k_d}{a_c} n_s(t) \tag{7.48}$$

The power spectral density of $n_p(t)$ can be obtained as

$$G_{n_p}(f) = \frac{k_d^2}{a_c^2} G_{n_s}(f) \tag{7.49}$$

where

$$\begin{aligned} G_{n_s}(f) &= G_n(f - f_c) + G_n(f + f_c) \\ &= \begin{cases} \eta & \text{for } |f| < B_T/2 \\ 0 & \text{elsewhere} \end{cases} \end{aligned} \tag{7.50}$$

The post-detection filter passes the spectral components of $n_p(t)$ that lie within $|f| < f_X$, and hence the output noise power is easily computed as

$$\begin{aligned} E\{n_0^2(t)\} &= \int_{-f_X}^{f_X} G_{n_p}(f)\, df \\ &= \frac{k_d^2}{a_c^2} 2\eta f_X \end{aligned} \tag{7.51}$$

The output signal-to-noise ratio for the PM system is given by (using the results stated in Equations (7.46) and (7.51))

$$\left(\frac{S}{N}\right)_d = \frac{E\{X_0^2(t)\}}{E\{n_0^2(t)\}} = \frac{a_c^2}{k_d^2} \frac{S_X}{(2\eta f_X)}$$

Substituting $k_p = 1/k_d$ and $S_r = a_c^2/2$, we have

$$\left(\frac{S}{N}\right)_d = (k_p^2) S_X \left(\frac{S_r}{\eta f_X}\right) \tag{7.52}$$

$(S/N)_d$ in FM Systems. From Equations (7.45) and (7.39), the FM detector output may be obtained as

$$\begin{aligned} Z(t) &= k_d \frac{d}{dt} [\theta_Y(t)] \\ &= k_f k_d X(t) + k_d \frac{d}{dt} \left(\frac{R_n(t)}{a_c} \sin[\theta_n(t) - \phi(t)] \right) \end{aligned}$$

Once again, setting $\phi(t) = 0$ and recognizing that $R_n(t) \sin[\theta_n(t)] = n_s(t)$, we get

$$\begin{aligned} Z(t) &= k_f k_d X(t) + \frac{k_d}{a_c} \frac{d}{dt} [n_s(t)] \\ &= X(t) + n_1(t) \end{aligned} \tag{7.53}$$

since $k_f k_d = 1$. In the preceding equation, $n_1(t)$ denotes the noise term. The psd of $n_1(t)$ is given by

$$G_{n_1}(f) = \frac{k_d^2}{a_c^2} G_{n_s}(f)(2\pi f)^2$$

$$= \begin{cases} \frac{k_d^2}{a_c^2}(2\pi f)^2\eta, & \text{for } |f| < \frac{B_T}{2} \\ 0 & \text{elsewhere} \end{cases} \tag{7.54}$$

The post-detection filter rejects the spectral components of $G_{n_1}(f)$ that lie outside the message band and the output noise psd is

$$G_{n_0}(f) = \begin{cases} \frac{k_d^2}{a_c^2}(2\pi f)^2\eta & \text{for } |f| < f_X \\ 0 & \text{elsewhere} \end{cases} \tag{7.55}$$

This spectrum is illustrated in Figure 7.9. The parabolic shape of the spectrum results from the differentiation action of the FM discriminator. It is clear from Figure 7.9 that low-frequency message components are subjected to lower noise levels than higher frequency message components. More will be said about this feature of FM systems later on.

From Equations (7.53) and (7.55) we obtain the signal power and the noise power at the output of the post-detection filter as

$$E\{X_0^2(t)\} = E\{X^2(t)\} = S_X$$

and

$$E\{n_0^2(t)\} = \int_{-f_X}^{f_X} \frac{k_d^2}{a_c^2}(2\pi f)^2\eta \, df = \frac{2k_d^2}{3a_c^2}(2\pi)^2\eta f_X^3$$

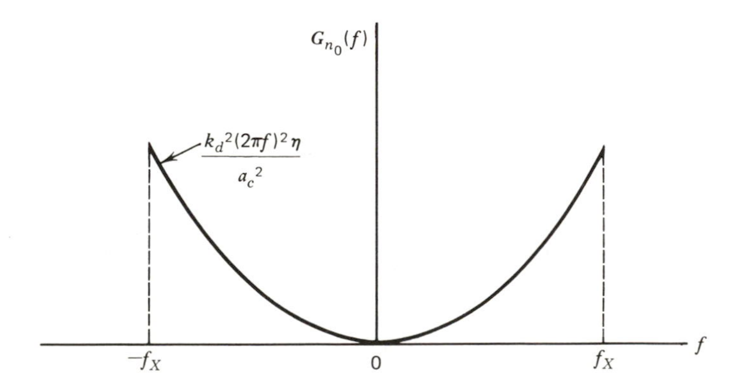

Figure 7.9 Output noise psd in FM systems.

Thus, the signal-to-noise ratio at the output of an FM system is given by

$$\left(\frac{S}{N}\right)_d = \frac{3a_c^2 S_X}{2(2\pi k_d)^2 \eta f_X^3}$$

Our assumptions $\max|X(t)| = 1$, $k_f = 1/k_d$, and $\max|k_f X(t)| = 2\pi f_\Delta$ allow us to write

$$f_\Delta = \frac{k_f}{2\pi} = \frac{1}{(2\pi k_d)}$$

Hence we can rewrite $(S/N)_d$ for the FM system as

$$\left(\frac{S}{N}\right)_d = 3\left(\frac{f_\Delta}{f_X}\right)^2 S_X \left(\frac{S_r}{\eta f_X}\right) \tag{7.56}$$

In Equation (7.56), (f_Δ/f_X) is the deviation ratio (D) of the FM system. Comparing $(S/N)_d$ for PM with FM, we find that FM is superior to PM insofar as noise performance is concerned since $(S/N)_d$ for FM can be made as large as desired by increasing f_Δ whereas in PM this increase is limited by the requirement that $k_p \leqslant \pi$.

The spectral characteristics of the noise at the output of FM and PM systems are quite different. The noise power in an FM system increases as f^2 over the message band. Thus the message spectral components at higher frequencies are corrupted by more noise than the message spectral components at lower frequencies. Now, if the baseband signal is an FDM signal, then message channels occupying the high end of the baseband will be subjected to high values of noise power. In such situations, PM is preferred over FM. The noise spectral characteristics in an FM system can be altered by preemphasis/deemphasis filtering, which we will discuss in a later section of this chapter.

Comparison of $(S/N)_d$ for PM and FM systems with linear modulation reveals an important fact: In angle modulation systems, $(S/N)_d$ can be increased by increasing the modulator sensitivity (k_p for PM and k_f for FM) without having to increase the transmitted power. Increasing k_f or k_p will, of course, increase the transmission bandwidth. *Thus, it is possible to trade off bandwidth for signal-to-noise ratio in angle modulation systems.* For this reason, WBFM systems are used in most of the low power applications such as in space communications.

Example 7.3. Compare the transmitted power and bandwidth requirements of an FM system, an AM system, and a DSB system designed to produce $(S/N)_d =$ 60 dB. The message signal is a video signal with the following characteristics:

$$\max|X(t)| = 1, \quad E\{X^2(t)\} = \tfrac{1}{2}, \quad \text{and} \quad f_X = 5\text{ MHz}$$

Assume that the channel noise has a psd $(0.5)(10^{-14})$ watt/Hz and that the channel introduces a 60-dB power attenuation. Assume a modulation index of 100% for the AM and a deviation ratio of 5 for the FM.

Solution

(a) *FM*: We are given that the FM system operates with a deviation ratio $f_\Delta/f_X = 5$. Hence, $B_T = 2[f_\Delta + f_X] = 60$ MHz. From Equation (7.56), we have

$$\left(\frac{S}{N}\right)_d = 10^6 = \frac{(3)(5)^2(\frac{1}{2})S_r}{(10^{-14})(5)(10^6)}$$

or

$$S_r = 1/750$$

with a channel attenuation of 60 dB, $S_T = (S_r)(10^6) = 1333$ watts.

(b) *AM*: $B_T = 2f_X = 10$ MHz. With 100% modulation ($m = 1$) and envelope demodulation we have

$$\left(\frac{S}{N}\right)_d = 10^6 = \frac{(\frac{1}{3})S_r}{(10^{-14})(5)(10^6)} \quad \text{or} \quad S_r = \frac{15}{100}$$

Hence,

$$S_T = (S_r)(10^6) = 150 \text{ kW}$$

(c) The reader can verify that for a DSB system, $B_T = 10$ MHz and $S_T = 50$ kW.

7.4.2 Threshold Effects in Angle Modulation Systems

The expressions for $(S/N)_d$ for PM and FM systems (Equations (7.52) and (7.56)) were derived under the assumption that the signal power at the detector input was much larger than the noise power. That is, a_c^2 was assumed to be $\gg E\{n^2(t)\}$. When the signal level at the detector input decreases, that is, when $a_c^2 \ll E\{n^2(t)\}$, the phase of the detector input can be shown to be equal to (Problem 7.22)

$$\theta_Y(t) \approx \theta_n(t) + \frac{a_c}{R_n(t)} \sin[\phi(t) - \theta_n(t)] \tag{7.57}$$

Equation (7.57) indicates that the message has been "captured" and mutilated by the noise beyond all hope of recovery—an effect similar to the one we observed in AM systems using envelope demodulation. We will now take a closer look at this threshold effect and discuss techniques for improving the

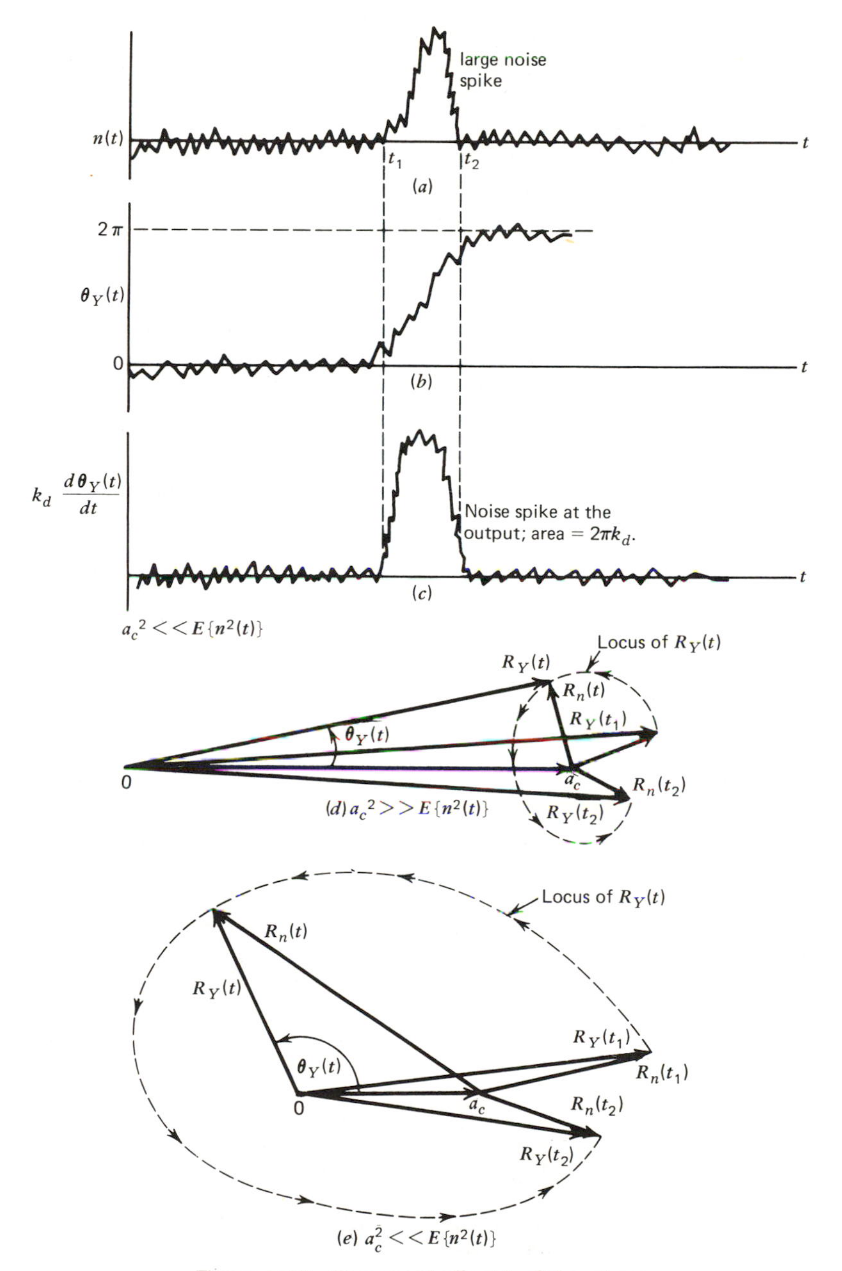

Figure 7.10 Threshold effect in FM systems.

performance of angle modulated systems by lowering the value of $S_r/\eta B_T$ at which threshold effects set in. We will illustrate threshold effects using FM rather than PM as our system model since threshold effects have more adverse effects on WBFM systems.

The mutilation of signal in FM systems takes place as $S_r/\eta B_T \to 1$. When the signal power and noise power at the detector input are of the same order of magnitude, then the signal and noise phasors are of nearly equal length, and variations of the noise phase cause comparable variations of $\theta_Y(t)$. This effect is shown in Figure 7.10, omitting the modulation $\phi(t)$. Figure 7.10*a* shows the noise waveform at the detector input. If the noise power $E\{n^2(t)\}$ is small in comparison to the carrier power, we may expect $R_n(t) \ll a_c$ most of the time and the resultant phasor $R_Y(t)$ will never wander far from the end of the carrier phasor, as illustrated in Figure 7.10*d*. Thus perturbations in the carrier phase angle due to the noise are small.

However, as $S_r/\eta B_T$ decreases, the likelihood of $R_n(t)$ being much less than a_c also decreases. Occasionally, $R_n(t)$ may indeed become much larger compared to a_c (Figure 7.10*e*). When $R_n(t)$ and a_c have comparable magnitude, the resultant phasor $R_Y(t)$ moves away from the end point of the carrier phasor and may even rotate about the origin. If the rotation takes place due to a noise spike in the interval between t_1 and t_2, then $\theta_Y(t)$ changes by 2π during this interval. The output of detector, which is $k_d(d\theta_Y/dt)$, changes by $k_d(2\pi)$ during this interval (t_1, t_2) (see Problem 7.23) and the change appears as a spike in the output.

The duration of the output noise spikes will be of the order of $2/B_T$ and the spacing between the spikes will depend on $S_r/\eta B_T$. These noise spikes can be clearly seen by observing the detector output on an oscilloscope. Aurally, these spikes can be heard as crackling or clicking sounds.

The signal-to-noise ratio at the output of the detector can be calculated by examining the statistical characteristics of the noise spikes or "clicks." The analysis is complicated by the fact that the spike characteristics change when the carrier is modulated. We will simply state* the result for tone modulation $(X(t) = \cos 2\pi f_X t)$ as

$$\left(\frac{S}{N}\right)_d = \frac{\left(\frac{3}{2}\right)\beta^2\left(\frac{S_r}{\eta f_X}\right)}{1+\left(\frac{12\beta}{\pi}\right)\left(\frac{S_r}{\eta f_X}\right)\exp\left[-\frac{1}{2}\left(\frac{1}{1+\beta}\right)\frac{S_r}{\eta f_X}\right]} \tag{7.58}$$

where β is the modulation index. (The reader can verify that Equation (7.58) reduces to Equation (7.56) when $S_r/\eta f_X \gg 1$.)

*The interested reader is referred to the classic papers by Rice (1948) and Stumpers (1948) or the detailed tutorial treatments of Panter (1965) or Taub and Schilling (1971).

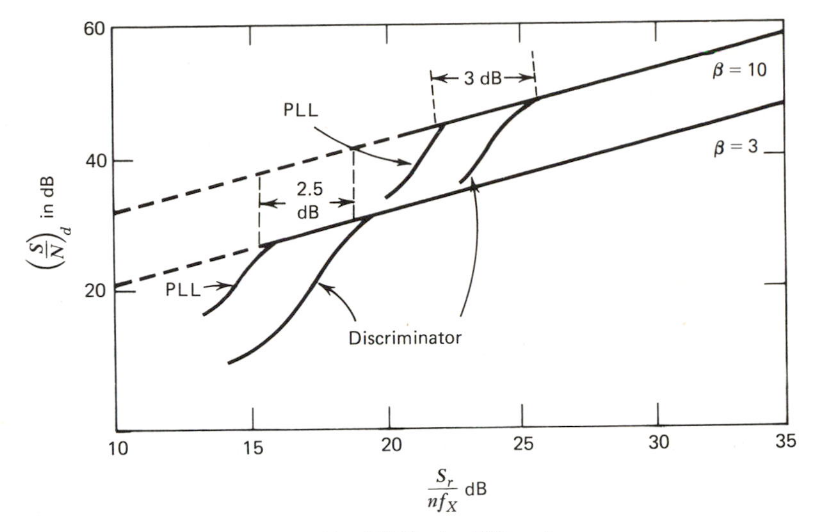

Figure 7.11 $(S/N)_d$ for FM systems.

Plots of $(S/N)_d$ versus $S_r/\eta f_X$ are shown for $\beta = 3$ and 10 in Figure 7.11. These plots show that the performance of FM systems deteriorates rapidly as $S_r/\eta f_X$ falls below certain levels. The threshold value of $S_r/\eta f_X$, below which performance degrades rapidly, depends on the modulation index. Larger values of β require larger values of $S_r/\eta f_X$ if the system is to operate above the threshold. While larger values of β yield an impressive performance above the threshold, larger values of β will require increased signal power if the system is to operate above the threshold. This effect works against the full realization of reduced transmitted signal power at the expense of increased bandwidth. As a matter of fact, one cannot achieve an unlimited exchange of bandwidth for signal-to-noise ratio, and system performance may even deteriorate with large values of β if transmitted power is not appropriately increased. Thus threshold effect is a matter of grave concern in FM systems. However, if the FM system is made to operate above threshold, it does perform much better than other CW modulation schemes.

7.4.3 Threshold Extension in FM Systems

In low power applications, it is desirable to lower the value of the predetection signal-to-noise ratio at which threshold occurs in FM systems. Several techniques have been developed for lowering this value and these schemes use some form of feedback demodulation (Chapter 6). The analysis of noise performance of the feedback demodulator is rather involved and we will simply note that feedback demodulators lower the threshold by as much as 7 dB. The PLL, discussed in Chapter 6, lowers the threshold by about 2 to 3 dB as shown in Figure 7.11. Even a 2 dB power savings is particularly important in low power digital applications such as in space communications.

Above the threshold, discriminators and feedback demodulators yield the same output signal-to-noise ratio.

7.5 PREEMPHASIS/DEEMPHASIS FILTERING IN CW MODULATION SYSTEMS

In the preceding sections, we saw that the output of a CW modulation scheme operating above threshold consists of an undistorted signal component $X(t)$ plus an additive noise component $n_0(t)$. Hence a CW modulation scheme can be modeled by the *baseband-to-baseband model* shown in Figure 7.12. The psd of the noise, $n_0(t)$, depends on the modulation and demodulation process used in the system. Two cases of interest are

$$G_{n_0}(f) = k_1\eta, \quad \text{for } |f| < f_X \tag{7.59a}$$

and

$$G_{n_0}(f) = k_2\eta f^2, \quad \text{for } |f| < f_X \tag{7.59b}$$

Equation (7.59*a*) represents the noise psd at the output of a linear modulation scheme using coherent demodulation, and (7.59*b*) represents the noise psd at the output of an FM system using a discriminator for demodulation.

The reader should note the similarity between the CW system model shown in Figure 7.12 and the baseband system model shown in Figure 7.1. This similarity suggests the possibility of using preemphasis/deemphasis filtering in

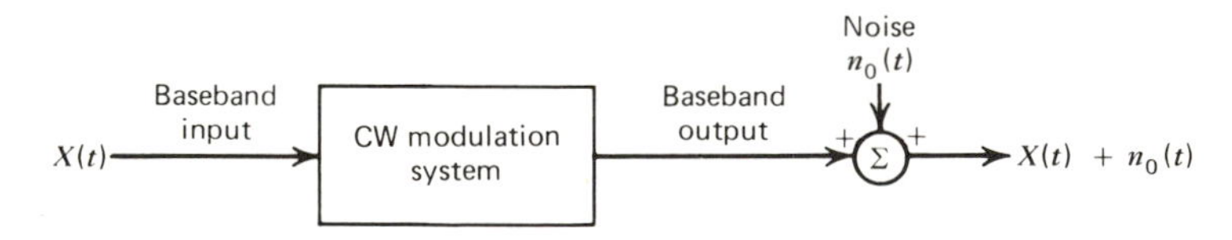

Figure 7.12 Baseband-to-baseband model of a communication system.

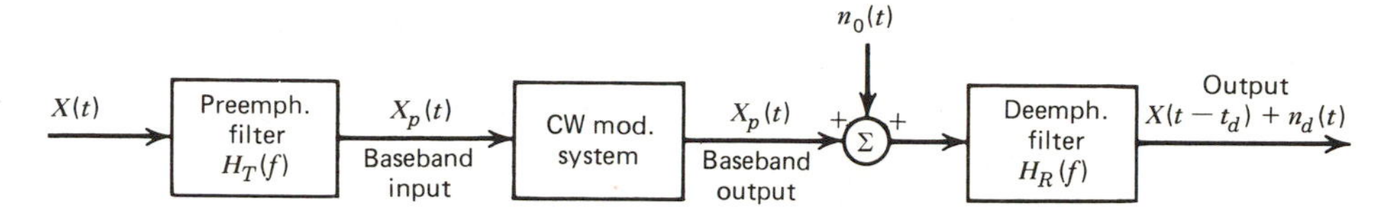

Figure 7.13 Preemphasis/deemphasis filtering in CW communication systems.

$H_T(f)H_R(f) = \exp(-j2\pi f t_d)$

CW modulation systems for improving the output signal-to-noise ratio. The preemphasis/deemphasis filter arrangement is shown in Figure 7.13.

Following the development in Section 7.2.3, we obtain the transfer function of these filters as

$$|H_T(f)|^2 = K_1 G_{n_0}^{1/2}(f)/G_X^{1/2}(f) \tag{7.60a}$$

$$|H_R(f)|^2 = G_X^{1/2}(f)/K_1 G_{n_0}^{1/2}(f) \tag{7.60b}$$

$$H_T(f)H_R(f) = \exp(-j2\pi f t_d) \tag{7.60c}$$

The constant K_1 is chosen to satisfy the constraint

$$\int_{-f_X}^{f_X} G_X(f)\, df = \int_{-f_X}^{f_X} G_X(f)|H_T(f)|^2\, df \tag{7.61}$$

Equation (7.61) requires that the normalized average power of the baseband signal $X(t)$ is the same as the normalized power of the preemphasized signal $X_p(t)$. This constraint assures that for linear modulation schemes the transmitted power is not altered by preemphasis filtering. For FM systems, equal power requirement ensures that the bandwidth of the FM signal remains the same.

The signal-to-noise ratio improvement that results from preemphasis/deemphasis filtering can be calculated as follows. The signal power at the output is the same with or without preemphasis/deemphasis filtering. The noise power at the output without filtering is

$$E\{n_0^2(t)\} = \int_{-f_X}^{f_X} G_{n_0}(f)\, df$$

With the deemphasis filter, the noise psd is given by

$$E\{n_d^2(t)\} = \int_{-f_X}^{f_X} G_{n_0}(f)|H_R(f)|^2\, df$$

$$= \int_{-f_X}^{f_X} [G_{n_0}(f)/|H_T(f)|^2]\, df$$

The signal-to-noise ratio improvement due to preemphasis/deemphasis filtering is given by

$$\gamma = \frac{E\{n_0^2(t)\}}{E\{n_d^2(t)\}} = \frac{\displaystyle\int_{-f_X}^{f_X} G_{n_0}(f)\, df}{\displaystyle\int_{-f_X}^{f_X} [G_{n_0}(f)|H_R(f)|^2]\, df} \tag{7.62}$$

The preemphasis/deemphasis filters specified in Equation (7.60) provide the maximum possible improvement in the output signal-to-noise ratio. In practice, these filters may often be hard to implement and only approximate versions of these filters are used. Even these approximations yield a significant improvement in the output signal quality.

Example 7.4. The circuits shown in Figure 7.14 are used as preemphasis/deemphasis filters in commercial FM broadcasting and reception. The standard value for the time constant is $RC = 75\ \mu\text{sec}$. Assuming a noise psd of $G_{n_0}(f) = k_2 f^2$ and the following signal psd,

$$G_X(f) = \begin{cases} [1 + (f/f_1)^2]^{-1}, & |f| < 15\text{ kHz} \\ 0 \text{ elsewhere}, & f_1 = 2.1\text{ kHz} \end{cases}$$

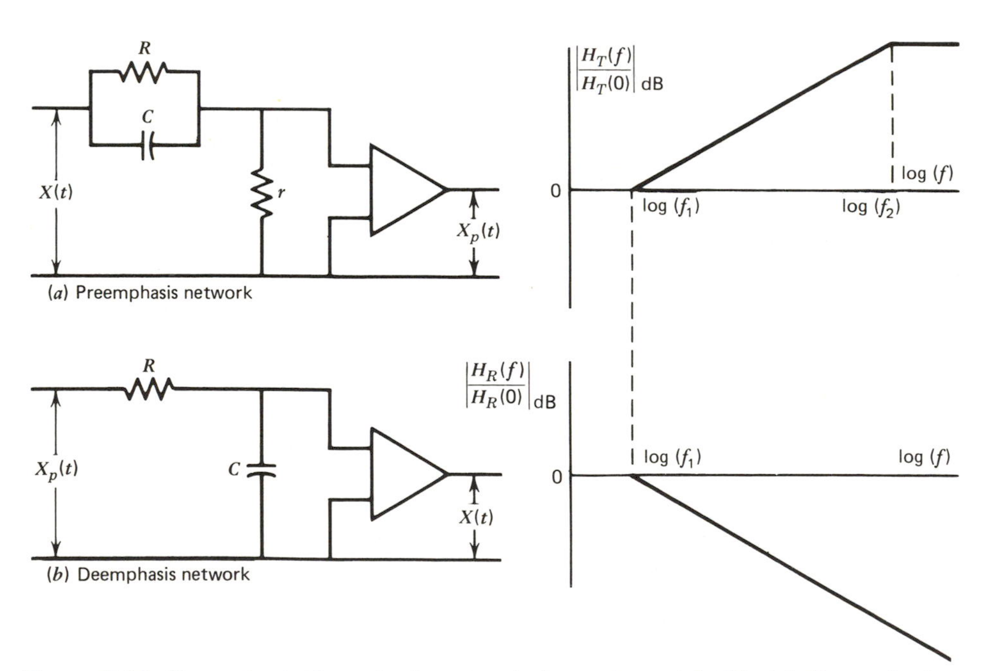

Figure 7.14 Preemphasis/deemphasis network for commercial FM. (a) Preemphasis network. (b) Deemphasis network; $f_1 = 1/2\pi RC$, $f_2 = 1/2\pi rC$, $f_2 \gg f_1$.

find the signal-to-noise ratio improvement that results from the use of the filters shown in Figure 7.14.

Solution. The transfer function of the filters are

$$H_T(f) \approx \sqrt{K_1}\left(1 + j\frac{f}{f_1}\right), \quad (K_1 \text{ is the power gain of the filter})$$

$$H_R(f) = \frac{1}{\sqrt{K_1}}\left(1 + j\frac{f}{f_1}\right)^{-1}$$

The gain K_1 is chosen to satisfy the equal power constraint given in Equation (7.61) as

$$\int_{-f_X}^{f_X} \frac{df}{1 + (f/f_1)^2} = \int_{-f_X}^{f_X} K_1\, df \quad \text{or} \quad K_1 = \frac{f_1}{f_X}\tan^{-1}\frac{f_X}{f_1}$$

Substituting the value of K_1 and $H_R(f)$ in Equation (7.62) with $G_{n_0}(f) = k_2 f^2$ (where k_2 is a positive constant), we obtain the signal-to-noise ratio improvement as

$$\gamma = \frac{K_1 \int_{-f_X}^{f_X} k_2 f^2\, df}{\int_{-f_X}^{f_X} k_2 f^2 \left[1 + \left(\frac{f}{f_1}\right)^2\right]^{-1} df}$$

$$= \tan^{-1}\left(\frac{f_X}{f_1}\right)\left[3\frac{f_1}{f_X}\left(1 - \frac{f_1}{f_X}\tan^{-1}\frac{f_X}{f_1}\right)\right]^{-1}$$

With $f_X = 15$ kHz and $f_1 = 2.1$ kHz, we obtain

$$\gamma \approx 4 = 6.0 \text{ dB}$$

Since the output signal-to-noise ratios are typically 40 to 60 dB, this represents an improvement of over 10%, a significant improvement indeed.

Example 7.5. Consider an FM system for transmitting a large number of multiplexed voice signals (see Chapter 6, Section 6.5). The psd of the composite message signal and the noise at the output of the FM receiver are shown in Figure 7.15. Assume that the signal power is the same in all message channels. Design appropriate preemphasis/deemphasis filters which will yield the same $(S/N)_d$ for all message channels.

Solution. The signal power is the same for all message channels. Hence, to maintain the same $(S/N)_d$ for all channels, we need to have a deemphasis filter (after the FM demodulator) that will yield a flat noise psd. Such a filter

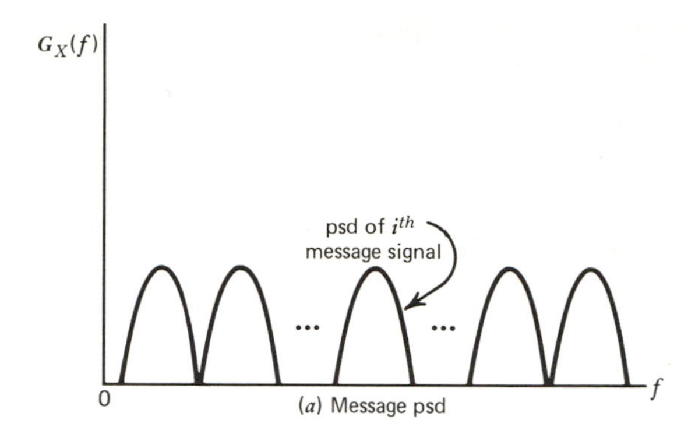

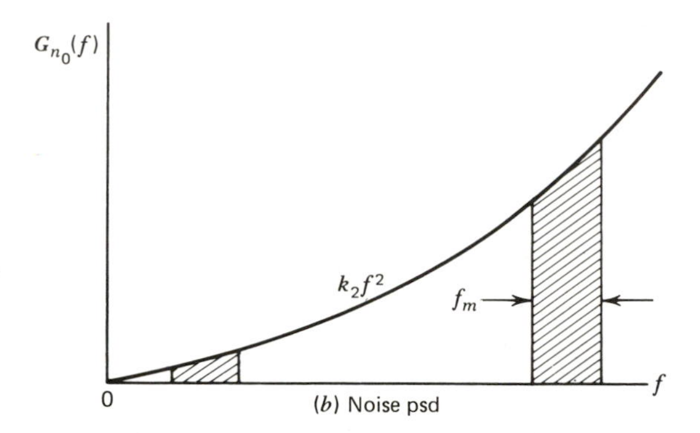

Figure 7.15 Message and noise psd's in an FDM/FM system. Note that the message channels occupying the high frequency end of the baseband are subjected to more noise power than channels located at lower frequencies. (*a*) Message psd. (*b*) Noise psd.

should have a transfer function

$$|H_R(f)|^2 = \frac{1}{K_1^2(2\pi f)^2}$$

where K_1 is a constant. Now, to avoid signal distortion we need a preemphasis filter with a transfer function

$$|H_T(f)|^2 = K_1^2(2\pi f)^2$$

where K_1 is chosen to satisfy the equal power constraint. Since the choice of phase response for the filters is somewhat arbitrary (Equation (7.60c)), let us take

$$H_T(f) = jK_1 f \quad \text{and} \quad H_R(f) = 1/jK_1 f$$

Now, $H_T(f)$ is a differentiator and $H_R(f)$ is an integrator! From Chapter 6, we know that differentiating the input to an FM system along with integrating the output results in a PM system! Thus, it is better to use PM rather than FM for transmitting a large number of multiplexed signals.

Example 7.4 illustrates that preemphasis is effective in FM systems for transmission of audio signals. The effectiveness results from the fact that the spectral density of the noise is greatest where the spectral density of the message signal is the smallest, and vice versa. The advantage of using preemphasis may be somewhat less pronounced in AM systems and in FM stereo broadcasting (see Problems 7.25 and 7.26).

Preemphasis/deemphasis is also used in phonographic and tape recording applications to suppress the noise that is encountered in the recording and replaying process. The Dolby (Dolby Labs, London) system is most widely used in recording and replaying audio signals. The interested reader is referred to Audio Engineering Journals for details of the Dolby and other noise reduction schemes.

7.6 INTERFERENCE IN CW MODULATION

In the preceding sections we investigated the effect of additive noise that accompanies the modulated carrier in CW modulation systems. Additive random noise is not the only unwanted electrical signal that accompanies the modulated message signal at the receiver input. The modulated message signal is sometimes contaminated by other extraneous signals, often manmade, of a form similar to the desired signal. This problem is particularly common in broadcasting when two or more signals are picked up at the same time by the receiver. Oftentimes the strength or average power of the desired signal may be lower than the average power of the interfering signal. When this happens, the presence of the interfering signals causes more problems than the presence of low level additive noise at the receiver input. In the following sections, we present a brief analysis of the effects of interference in linear and exponential CW modulation systems.

7.6.1 Interference in CW Modulation

Let us consider a very simple case of an unmodulated carrier $a_c \cos \omega_c t$ with an interfering signal of the form

$$V(t) = a_c \cos \omega_c t + a_i \cos(\omega_c + \omega_i)t, \quad \omega_i \ll \omega_c$$
$$= R_V(t) \cos[\omega_c t + \theta_V(t)] \tag{7.63}$$

where

$$R_V(t) = \sqrt{(a_c + a_i \cos \omega_i t)^2 + (a_i \sin \omega_i t)^2} \tag{7.64}$$

$$\theta_V(t) = \tan^{-1}\left(\frac{a_i \sin \omega_i t}{a_c + a_i \cos \omega_i t}\right) \tag{7.65}$$

Now, if $a_i \ll a_c$, then $R_V(t) \approx a_c + a_i \cos \omega_i t$ and $\theta_V(t) = (a_i/a_c) \sin \omega_i t$, and hence

$$V(t) \approx a_c[(1 + m_i \cos \omega_i t) \cos(\omega_c t + m_i \sin \omega_i t)] \tag{7.66}$$

where $m_i = a_i/a_c \ll 1$.

At the other extreme, if the interfering signal amplitude is larger, that is, $a_i \gg a_c$, we have (see Problem 7.28)

$$V(t) \approx a_i[1 + m_i^{-1} \cos \omega_i t] \cos[(\omega_c + \omega_i)t - m_i^{-1} \sin \omega_i t] \tag{7.67}$$

Equation (7.66) shows that the interfering signal modulates the *amplitude* and *phase* of the carrier just like a modulating tone of frequency f_i with a modulation index m_i. When the interfering signal is strong, the effective carrier frequency becomes $\omega_c + \omega_i$, and the modulation is at frequency f_i.

7.6.2 Interference in Linear Modulation

Consider an AM system in which there is small amplitude interference as given in Equation (7.63). If the signal is demodulated using envelope (or coherent detection with DC removal), the output of the demodulator is

$$X_0(t) = a_i \cos \omega_i t \tag{7.68}$$

If the interference frequency ω_i is smaller than $2\pi f_X$, then the interference will appear at the detector output with an amplitude a_i.

The result stated in Equation (7.68) is rather an obvious one. The main reason for deriving this simple result is for comparing the effect of interference in linear and exponential modulation schemes. The strong interference case of Equation (7.67) indicates that considerable signal degradation will result from interference. The derivation of an expression for the strong interference case is left as an exercise for the reader (Problem 7.29).

7.6.3 Interference in Angle Modulation

In angle modulation systems, the unmodulated carrier plus a small amplitude interference signal will produce an output (from Equations (7.66) and (7.17))

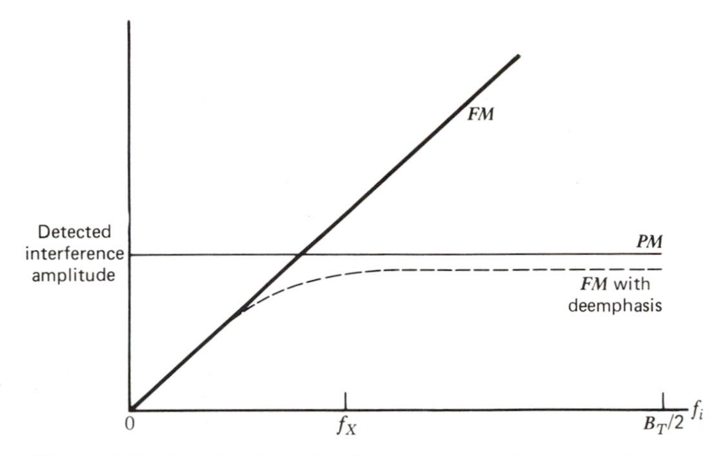

Figure 7.16 Amplitude of the detector output due to interference.

$$X_0(t) = \begin{cases} k_3 \dfrac{a_i}{a_c} \sin \omega_i t & \text{phase detector} \quad (7.69a) \\ k_4 \dfrac{a_i}{a_c} f_i \cos \omega_i t & \text{frequency detector} \quad (7.69b) \end{cases}$$

Comparing Equations (7.69) and (7.68), and remembering that $a_i/a_c \ll 1$, we see that angle modulation is less vulnerable to a small amplitude interference than linear modulation. Furthermore, FM is less vulnerable than PM when $|f_i|$ is small.

For the case of angle modulation, the strength of the detected signal as a function of the frequency of the interfering signal is shown in Figure 7.16. The highest value of the interference frequency f_i is shown to be equal to $B_T/2$ since the predetection filter will remove frequencies beyond $f_c \pm (B_T/2)$. Figure 7.15 shows that when f_i is small, the interfering signal has less effect on the FM system than the PM system, while the opposite is true for large values of f_i. In either case a lowpass filter of bandwidth f_X should follow the demodulator to eliminate detected interference components that are outside of the message band (i.e., components at $f_X < f_i < B_T/2$). The severe effect of interference in FM systems can be reduced by preemphasis/deemphasis filtering and by the use of a limiter before the demodulator.

7.7 COMPARISON OF CW MODULATION SYSTEMS

We are now ready to compare the various types of CW modulation systems. The comparison will be done in terms of transmission bandwidth, destination

Table 7.1. Comparison of CW modulation schemes.

Type	Transmission Bandwidth (B_T)	$\frac{1}{\alpha}\left(\frac{S}{N}\right)_d$	DC Resp	Equipment Complexity	Applications
DSB	$B_T = 2f_X$	1	yes	Moderate: coherent demodulation is required. Small pilot carrier is often transmitted along with the DSB signal.	Analog data; multiplexing low bandwidth signals
AM	$B_T = 2f_X$	1/3	no	Minor: simple envelope detection is used. Modulation is simple. DC isolation is required at the receiver.	Broadcast radio
SSB	$B_T = f_X$	1	no	Major: coherent demodulation is required. Modulators are very complex.	Point to point voice communication (amateur and CB radio), multiplexing

VSB	$f_X < B_T < 2f_X$	1	yes	Major: coherent demodulation is necessary. Modulator requires symmetric filters.	Digital data and other wideband applications
VSB + C	same as VSB	1/3	yes	Moderate: envelope demodulation is used. Symmetric filters are required. DC restoration is required.	Television video and other large bandwidth signals
FM	$B_T \approx 2f_\Delta + 2f_X$	$(3/2)(f_\Delta/f_X)^2$	yes	Moderate: modulation is somewhat complicated. Demodulation is simple using discriminator or PLL	Broadcast radio; microwave relay
PM	$B_T \approx 2(k_p + 1)f_X$	$k_p^2/2$	yes	Moderate: same as FM	Data transmission; multiplexed voice transmission
Baseband	$B_T = f_X$	1	yes	Minor: (no modulation)	Short point-to-point communications

Legend: f_X—message bandwidth. f_Δ—peak frequency deviation. k_p—phase modulator sensitivity, $k_p \leq \pi$. $(S/N)_d$—destination signal-to-noise ratio. $\alpha = S_r/\eta f_X$. $E\{X^2(t)\}$ is assumed to be $\frac{1}{2}$ and 100% modulation is assumed for AM and VSB + C. Effects of preemphasis/deemphasis filtering are not included.

signal-to-noise ratio, and equipment complexity. We will assume that system components such as modulators and demodulators have ideal characteristics and that all systems operate with the same signal power at the receiver input and that the message signal in all the systems is such that

$$E\{X(t)\} = 0, \quad E\{X^2(t)\} = \tfrac{1}{2}, \quad \text{and} \quad \max|X(t)| = 1$$

Furthermore, we will assume that the noise at the receiver input in all systems is additive zero mean and Gaussian with a psd of $\eta/2$ watt/Hz.

The performance characteristics of the various CW modulation schemes

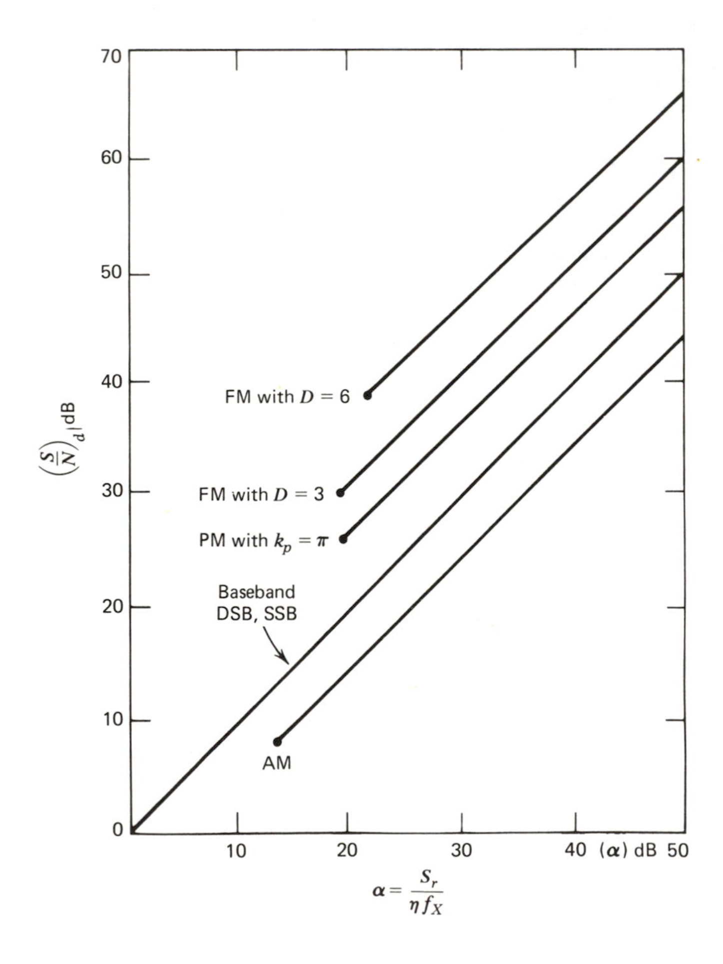

Figure 7.17 $(S/N)_d$ for CW modulation schemes.

are shown in Table 7.1 and Figure 7.17. The baseband system is included in the table for comparison purposes.

Among the linear modulation schemes, suppressed carrier methods (SSB, DSB, and VSB) are better than conventional AM in that signal-to-noise ratios are better and there are no threshold effects. SSB and VSB are attractive when bandwidth conservation is desired. However, these schemes require complex modulation and demodulation equipment. For point to point communication the price of the complex equipment may be worthwhile. But for broadcast systems involving millions of receivers, economic considerations might require the use of AM or VSB + C since these schemes use simple envelope demodulation.

Compared to linear modulation, exponential modulation provides substantially improved signal quality at the expense of bandwidth. WBFM with preemphasis/deemphasis provides the best noise performance of all systems. Phase modulation provides better noise performance than linear modulation schemes, but the performance improvement of PM cannot be made arbitrarily large since k_p is limited to values $\leq \pi$.

When power conservation is important, FM offers considerable savings over other schemes. However, threshold effects in FM may somewhat limit the amount of improvement possible. Still, FM is widely used in a large number of applications such as in commercial broadcasting, mobile radio, microwave links, and in low power satellite and space communication systems (with threshold extension).

In terms of equipment complexity, AM is the least complex, while VSB, which requires synchronous demodulation and special sideband filters, is the most complex. DSB and SSB modulation schemes are also somewhat complex because of the need for synchronous demodulation. FM and PM, in comparison to DSB, SSB, and VSB, require moderately complex instrumentation. VSB + C requires only moderately complex instrumentation since it requires only simple envelope demodulation.

For transmitting signals with significant low-frequency components and large bandwidth, VSB + C offers the best compromise in terms of transmission bandwidth and equipment complexity. While FM systems may require larger bandwidth, the low-frequency response of FM systems is quite good and FM systems are widely used for high quality magnetic-tape recording of signals having significant low-frequency (including DC) components.

The preceding discussion is by no means complete. In many applications, more than one type of system might meet the requirements. In such cases, cost of the equipment and compatibility with equipment already in use might dictate the ultimate choice of a particular system over other comparable systems.

7.8 SUMMARY

The performance of several practical communication systems operating in the presence of additive noise was discussed. The signal-to-noise ratio at the output was used as a measure of output signal quality and a number of communication systems were analyzed to determine the effect of noise on system operation.

It was shown that if the noise power at the receiver input is low, the output of the receiver consists of a signal component plus an additive noise component. If the noise power at the receiver input is the same order of magnitude as the signal power, then many practical systems exhibit a "threshold effect" resulting in a loss of the signal component and/or vastly inferior performance.

Angle modulation schemes operating above threshold were shown to provide improved performance over linear modulation schemes in the presence of noise. It was also shown that angle modulation systems allow a trade off of transmission bandwidth for output signal-to-noise ratio. FM systems provide a wider range of bandwidth to signal-to-noise ratio trade off than PM systems.

Use of preemphasis/deemphasis filtering in both baseband and CW modulation schemes was discussed. It was shown that significant improvement in the signal-to-noise ratio is possible in many applications.

REFERENCES

The noise performance of analog communication systems is dealt with in great detail in a number of graduate and undergraduate texts. Graduate level texts include Panter (1965), Schwartz, Bennet, and Stein (1966). Undergraduate texts such as the ones by Carlson (1975), Taub and Schilling (1971) contain good treatments of noise in analog communication systems. The book by Davenport and Root (1958) contains excellent background material (on random processes) that is applicable to the analysis of the effects of noise in communication systems.

The analysis of noise in FM systems is dealt with, rather nicely, in Taub and Schilling's book. Historical references on this topic include the papers by Rice (1948) and Stumpers (1948). Finally, additional material on commercial preemphasis/deemphasis systems may be found in the articles by Burwen (1971), and Dolby (1967).

1. P. F. Panter. *Modulation, Noise, and Spectral Analysis.* McGraw-Hill, New York (1965).
2. M. Schwartz, W. R. Bennett, and S. Stein. *Communication Systems and Techniques.* McGraw-Hill, New York (1966).

3. A. B. Carlson. *Communication Systems.* McGraw-Hill, New York (1968, 1975).
4. H. Taub and D. L. Schilling. *Principles of Communication Systems.* McGraw-Hill, New York (1971).
5. W. B. Davenport, Jr. and W. L. Root. *Introduction to Random Signals and Noise.* McGraw-Hill, New York (1958).
6. S. O. Rice. "Mathematical Analysis of Random Noise", *Bell System Tech. J.*, vol. 23 (1944), 282–332, and vol. 24 (1945), 46–156. "Statistical Properties of a Sine-Wave Plus Random Noise", *Bell System Tech. J.*, vol. 27 (1948), 109–157.
7. F. L. Stumpers. "Theory of Frequency-Modulation Noise," *Proc. IRE*, vol. 36 September (1948), 1081–1902.
8. D. L. Schilling and J. Billig. "Threshold Extention Capability of the PLL and the FMFB," *Proc. IEEE*, vol. 52 (1964).
9. R. S. Burwen. "A Dynamic Noise Filter," *Journal of the Audio Engineering Society*, vol. 19 (1971).
10. R. M. Dolby. "An Audio Noise Filter," *Journal of the Audio Engineering Society*, vol. 15 (1967).
11. J. J. Downing, *Modulation Systems and Noise.* Prentice Hall, Englewood Cliffs, N.J. (1964).

PROBLEMS

Sections 7.1 *and* 7.2

7.1. An analog baseband communication system is designed for transmitting audio signals with 4 kHz bandwidth. No preemphasis filter is used in the system. At the receiving end an RC lowpass filter with a 3 dB bandwidth of 8 kHz is used to limit the noise power at the output. Assuming the channel noise to be white with a psd of $\eta/2$, calculate the noise power at the output.

7.2. Repeat Problem 7.1 when the filter at the receiving end has an amplitude response $|H(f)| = [1 + (f/f_c)^2]^{-1}$ ($f_c = 4$ kHz).

7.3. Consider a baseband system with the following characteristics:

$$H_c(f) = [1 + j2f/f_X]^{-1} \quad ; \quad G_{n_i}(f) = \eta/2 \text{ watt/Hz}$$

$$G_X(f) = \begin{cases} 1/2f_X & \text{for } |f| < f_X \\ 0 & \text{elsewhere} \end{cases}$$

$$H_{eq}(f) = \begin{cases} 1 + j(2f/f_X) & \text{for } |f| < f_X \\ 0 & \text{for } |f| > f_X \end{cases}$$

(No preemphasis filter.) Calculate $(S/N)_d$ for this system.

7.4. Design optimum preemphasis/deemphasis filters for the system described in Problem 7.3 and compute $(S/N)_d$ for the optimum system. Assume S_T to be the same and compare the results.

7.5. Consider a baseband system in which

$$G_X(f) = \begin{cases} \dfrac{1}{1+(f/f_1)^2} & \text{for } |f| < 15\text{ kHz} \\ 0 & \text{elsewhere, } f_1 = 2\text{ kHz} \end{cases}$$

$$H_c(f) = \begin{cases} 1 & \text{for } |f| < 15\text{ kHz} \\ 0 & \text{elsewhere} \end{cases}$$

and $G_{n_i}(f) = kf^2$, where k is a constant >0.

(a) Calculate $(S/N)_d$ if the system uses an ideal lowpass filter at the receiving end with a cutoff frequency of 15 kHz (no filter at the transmitting end).

(b) Design optimum preemphasis/deemphasis filters for this system and calculate $(S/N)_{d_{max}}$.

(c) Compare the results obtained in (a) and (b) assuming S_T to be the same in both cases.

7.6. In a baseband system, the channel is ideal lowpass with a bandwidth of 1 MHz. The signal psd and noise psd are shown in Figure 7.18.

(a) If the system uses an ideal lowpass filter at the receiving end without a preemphasis filter, calculate $(S/N)_d$.

(b) Design optimum preemphasis/deemphasis filters and compute $(S/N)_{d_{max}}$. Compare the result of parts (a) and (b) assuming S_T to be the same.

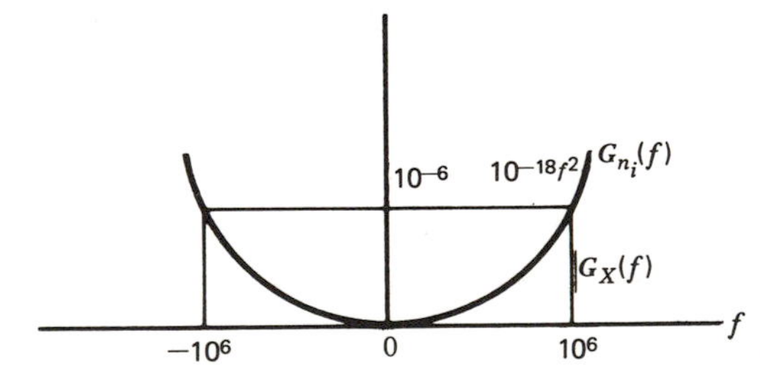

Figure 7.18 Signal and noise psd for Problem 7.6.

Section 7.3

7.7. Show that the $(S/N)_d$ for an SSB system using coherent demodulation is given by Equation (7.29).

7.8. Show that the $(S/N)_d$ for a VSB system using coherent demodulation is the same as $(S/N)_d$ for SSB and DSB modulation schemes using coherent demodulation.

7.9. Show that $(S/N)_d$ for an AM system in which the signal $a_c[1 + mX(t)] \cos w_c t$ is demodulated *coherently* is

$$\left(\frac{S}{N}\right)_d = \frac{m^2 E\{X^2(t)\}}{1 + m^2 E\{X^2(t)\}} \left(\frac{S_r}{\eta f_X}\right)$$

7.10. A carrier of amplitude 1 mv is 50% amplitude modulated by a sinusoidal waveform of frequency $f_X < 1000$ Hz. The signal is accompanied by thermal noise of two-sided spectral density $\eta/2 = 10^{-12}$ watt/Hz. The signal plus noise is passed through a predetection filter having the response shown in Figure 7.19. The signal is demodulated by multiplication with a local carrier of amplitude 2 volts and filtering the product with an ideal lowpass filter having a bandwidth of 1 kHz.

(a) Find the output signal power.
(b) Find the output noise power.
(c) Find (S/N) at the demodulator input and $(S/N)_d$.

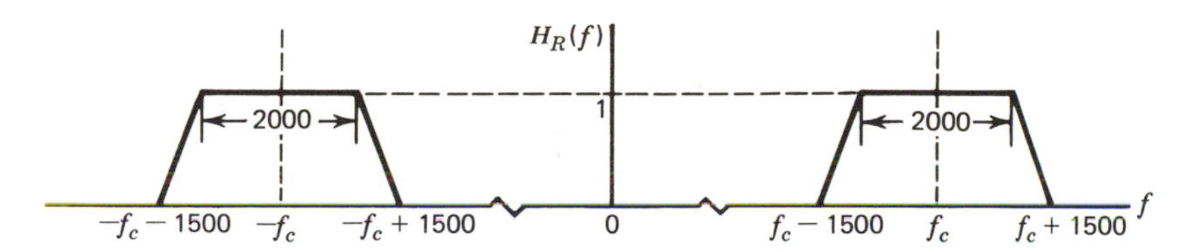

Figure 7.19 Predetection filter for Problem 7.10.

7.11. The signal $X(t)$, whose pdf is shown in Figure 7.20, amplitude modulates a carrier with $m = 0.5$. Calculate $(S/N)_d$ in terms of $S_r/\eta f_X$ for envelope demodulation, assuming white noise with a psd of $\eta/2$ watt/Hz at the input.

7.12. Assume that the baseband message signal in an AM system has a bandwidth of 4 kHz and the amplitude distribution shown in Figure 7.20. The AM signal is accompanied by thermal noise with a psd $(\eta/2) = 10^{-12}$ watt/Hz. The signal is demodulated by envelope detection and

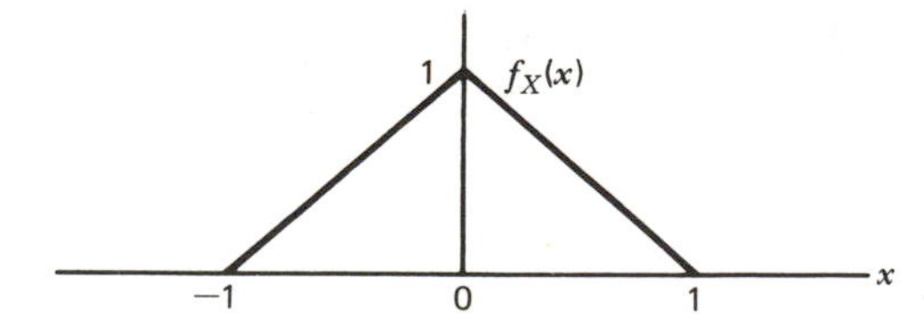

Figure 7.20 Pdf of the signal for Problems 7.11., and 7.12.

appropriate postdetection filtering. It is required that the output signal-to-noise ratio be greater than 40 dB. Assume $m = 1$.
(a) Find the minimum value of the carrier amplitude that will yield $(S/N)_d \geqslant 40$ dB.
(b) Find the threshold value of a_c.

7.13. Plot $(S/N)_d$ versus $(S_r/\eta f_X)$ for an AM system using the square law demodulator. Use Equations (7.38a) and (7.38b). (Assume $m^2 S_X = \frac{1}{2}$.)

7.14. Define the threshold value of $(S_r/\eta f_X)$ for the square law demodulator as the value of $(S_r/\eta f_X)$ for which $(S/N)_d$ is 1 dB below $(S_r/\eta f_X)m^2 S_X$. Find this value from the plot obtained in Problem 7.13 with $m^2 S_X = \frac{1}{2}$. Compare the threshold values of S_r for envelope and square law demodulators.

7.15. An RF communication system has a channel power loss of 90 dB and channel noise psd $(\eta/2) = (0.5)(10^{-14})$ watt/Hz. The baseband message signal has a bandwidth of 1.5 MHz and a uniform amplitude distribution in the interval ± 1 volt. Calculate the transmission bandwidth and transmitter power required for $(S/N)_d = 30$ dB when the modulation is (a) SSB, (b) AM with $m = 1$, (c) DSB.

Section 7.4

7.16. An unmodulated carrier of amplitude 10 mv is accompanied by narrowband Gaussian white noise with a psd of $(\eta/2) = 10^{-12}$ watt/Hz and a bandwidth of 50 kHz. Find the rms value of the carrier phase perturbation due to the noise [i.e., find $E\{\theta_e^2(t)\}$, where $\theta_e(t)$ is defined in Equation (7.44a); if Equation (7.44b) is used, be sure to verify that $a_c^2 \gg E\{n^2(t)\}$].

7.17. Verify the psd of $n_1(t)$ given in Equation (7.54).

7.18. An audio signal $X(t)$ is to be transmitted over an RF channel. It is required that the destination signal-to-noise ratio be >50 dB. Assume the following characteristics for the channel and the baseband signal:

$$\text{Power loss in the channel} = 60 \text{ dB}$$

$$\text{noise psd} = \eta/2 = 10^{-12} \text{ watt/Hz}$$

and

$$E\{X(t)\} = 0, \max|X(t)| = 1, E\{X^2(t)\} = \tfrac{1}{2}, f_X = 15 \text{ kHz}$$

Calculate the bandwidth and average transmitter power requirements

for:
(a) DSB modulation;
(b) AM with 100% modulation;
(c) PM with $k_p = 3$;
(d) FM with a deviation ratio of 5.

7.19. An FDM/FM system uses upper SSB (USSB) subcarrier modulation and FM carrier modulation. There are N independent signals $X_1(t)$, $X_2(t), \ldots, X_N(t)$ each bandlimited to f_X Hz. The subcarrier frequency for the ith signal is $f_{c_i} = (i-1)f_X$, $i = 1, 2, \ldots, N$. The composite baseband signal $X(t)$ is

$$X(t) = \sum_{k=1}^{N} a_k X_{c_k}(t)$$

where a_k are constants and $X_{c_k}(t)$ is the kth USSB signal, and $E\{X_{c_k}^2(t)\} = 1$. At the receiver, the FM signal is demodulated using a discriminator. The discriminator output is passed through a bank of bandpass filters and the USSB signals are coherently demodulated. Assume the channel noise to be white with a psd of $\eta/2$ watt/Hz.

(a) Show that the $(S/N)_d$ for the kth channel is

$$(S/N)_d = f_\Delta^2 a_k^2 / b_k,$$

where

$$b_k = (3k^2 - 3k + 1)\eta f_X^3 / 3S_r$$

(*b*) If a_k's are chosen such that $E\{X^2(t)\} = 1$ and $(S/N)_d$ is the same for all channels, find the value of $(S/N)_d$.

7.20. Six hundred voice channels, each having a bandwidth of 4 kHz, are multiplexed using USSB subcarrier modulation and the multiplexed signal frequency modulates a carrier. The lowest SSB carrier frequency is 10 kHz and the subcarriers are spaced 4 kHz apart. Assume that each voice signal has the same average power. The channel at 10 kHz has an output signal-to-noise ratio of 60 dB.

(a) If $(S/N)_d < 30$ dB is not acceptable, how many channels in the system are usable? (Assume white noise at the FM receiver input.)
(b) If the FDM signal phase modulates the carrier, calculate $(S/N)_d$ for any channel assuming that the transmitted signal power and noise psd are the same as in the FM case.

7.21. One thousand 4 kHz audio signals are multiplexed with a 5 MHz video signal. The video signal is left at baseband and the audio signals are added on using USSB subcarrier modulation with 5 kHz subcarrier spacing. the first subcarrier is at 5.5 MHz. The power spectral density of

the composite baseband signal is constant throughout the baseband. Assuming the carrier modulation is FM, and the demodulation is done using a discriminator followed by coherent linear demodulation:
(a) calculate $(S/N)_d$ for the TV channel;
(b) calculate $(S/N)_d$ for the top audio channel. (The channel noise may be assumed to be white.)

7.22. Verify Equation (7.57).

7.23. Show that the area of the spike shown in Figure 7.10*c* is $2\pi k_d$.

7.24. Define the threshold value of $(S_r/\eta f_X)$ in an FM system to be the value of $(S_r/\eta f_X)$ for which $(S/N)_d$ is 1 dB below $(\frac{3}{2})\beta^2 S_r/\eta f_X$ (Equation (7.58)). Plot β versus the threshold value of $S_r/\eta f_X$ and show that the threshold value of $(S_r/\eta f_X)$ is given by

$$\left(\frac{S_r}{\eta f_X}\right)_{\text{th}} \approx k\left(\frac{B_T}{f_X}\right)$$

where k is a constant. (Assume sinusoidal modulation at a tone frequency of f_X.)

Sections 7.5 *and* 7.6

7.25. The message signal in a DSB system has the psd

$$G_X(f) = \begin{cases} [1 + (f/f_1)^2]^{-1} & \text{for } |f| < 10 \text{ kHz} \\ 0 & \text{elsewhere, } f_1 = 2.1 \text{ kHz} \end{cases}$$

The channel noise is white with a psd of $\eta/2$ watts/Hz, and the DSB signal is coherently demodulated after appropriate bandpass filtering at the receiver. Calculate the signal-to-noise ratio improvement that will result if the preemphasis/deemphasis filters given in Example 7.4 are used in the DSB system.

7.26. In stereo FM (Chapter 6) the sum and difference signals are passed through identical preemphasis filters. At the receiver the sum and difference signals are passed through identical deemphasis filters before they are combined to generate the individual (left and right) signals. Assuming signal model and filters specified in Example 7.4, show that stereophonic FM is 22 dB noisier than monophonic FM transmission (assuming that the transmitted power and bandwidth are the same in both cases).

7.27. With reference to the FDM/FM system described in Problem 7.19, assume that preemphasis/deemphasis filtering is used on the composite baseband signal. The preemphasis filter has a transfer function

$$|H_p(f)|^2 = 4\pi^2\tau^2 f^2$$

where the value of τ is chosen to satisfy the equal power constraint (Equation (7.61)). (You may assume that the psd of the baseband signal is constant over the message bandwidth.)

(a) Find the value of τ.

(*b*) Find the improvement in the signal-to-noise ratio for the top channel.

7.28. Verify Equation (7.67).

7.29. In an AM system using envelope demodulation, the signal at the input to the demodulator has the form

$$V(t) = a_c \cos \omega_c t + a_m \cos \omega_m t \cos \omega_c t + a_i \cos(\omega_c + \omega_i)t$$

where $a_i \cos(\omega_c + \omega_i)t$ is the interference. Show that the envelope detector output is:

(a) $Z(t) = a_m \cos \omega_m t + a_i \cos \omega_i t$ when $a_c \gg a_i$, and

(b) $Z(t) = a_c \cos \omega_i t + a_m \cos \omega_m t \cos \omega_i t$ when $a_c \ll a_i$.

(c) Sketch the envelope detector output spectra for (a) and (b).

[*Hint*: (a) Write $V(t)$ as $V(t) = V_1(t) \cos \omega_c t - V_2(t) \sin \omega_c t$, and then make use of $a_c \gg a_i$. (b) Write $V(t)$ as $V(t) = V_3(t) \cos(\omega_c + \omega_i)t - V_4(t) \sin(\omega_c + \omega_i)t$ and then make use of $a_c \ll a_i$.]

Section 7.7

7.30. Calculate the detection gain (Equation (7.28)) for PM and FM systems.

7.31. One thousand voice signals are to be frequency division multiplexed and transmitted over a microwave RF link that introduces a power loss of 60 dB and additive white noise of 10^{-14} watt/Hz. The composite baseband signal is to have a normalized power of 1 mwatt, and the amplitude distribution may be assumed to be uniform. The psd of the voice signals may be assumed to be constant over the 4 kHz bandwidth. At the receiver output, $(S/N)_d$ has to be maintained at 40 dB. Design a suitable subcarrier and RF carrier modulation scheme for this application. Specify the power and bandwidth requirement of your RF carrier modulation scheme.

7.32. Data has to be transmitted from a remote space probe via an RF link. The baseband data signal has a bandwidth of 1.5 MHz and a normalized power of $\frac{1}{2}$ watt. CW modulation is used to transfer the data signal on to an RF carrier. The RF channel introduces a power loss of 60 dB and adds white noise with a psd of $\eta/2 = 10^{-12}$ watt/Hz. For proper decoding of the data, it is required that (S/N) at the output of the receiver be >20 dB. The power source on the space probe can supply an average power of 10 watts to the RF transmitter on board. Design a suitable CW modulation/demodulation scheme for this application.

8

DIGITAL CARRIER MODULATION SCHEMES

8.1 INTRODUCTION

In Chapter 5 we described several methods of transmitting digital information over baseband channels using discrete baseband PAM techniques. Most real communication channels have very poor response in the neighborhood of zero frequency and hence are regarded as bandpass channels. In order to transmit digital information over bandpass channels, we have to transfer the information to a carrier wave of appropriate frequency. Digital information can be impressed upon a carrier wave in many different ways. In this chapter, we will study some of the most commonly used digital modulation techniques wherein the digital information modifies the amplitude, the phase, or the frequency of the carrier in discrete steps.

Figure 8.1 shows four different modulation waveforms for transmitting binary information over bandpass channels. The waveform shown in Figure 8.1*a* corresponds to discrete amplitude modulation or an amplitude-shift keying (ASK) scheme where the amplitude of the carrier is switched between two values (on and off). The resultant waveform consists of "on" (mark) pulses representing binary 1 and "off" (space) pulses representing binary 0. The waveform shown in Figure 8.1*b* is generated by switching the frequency of the carrier between two values corresponding to the binary information to be transmitted. This method, where the frequency of the carrier is changed, is

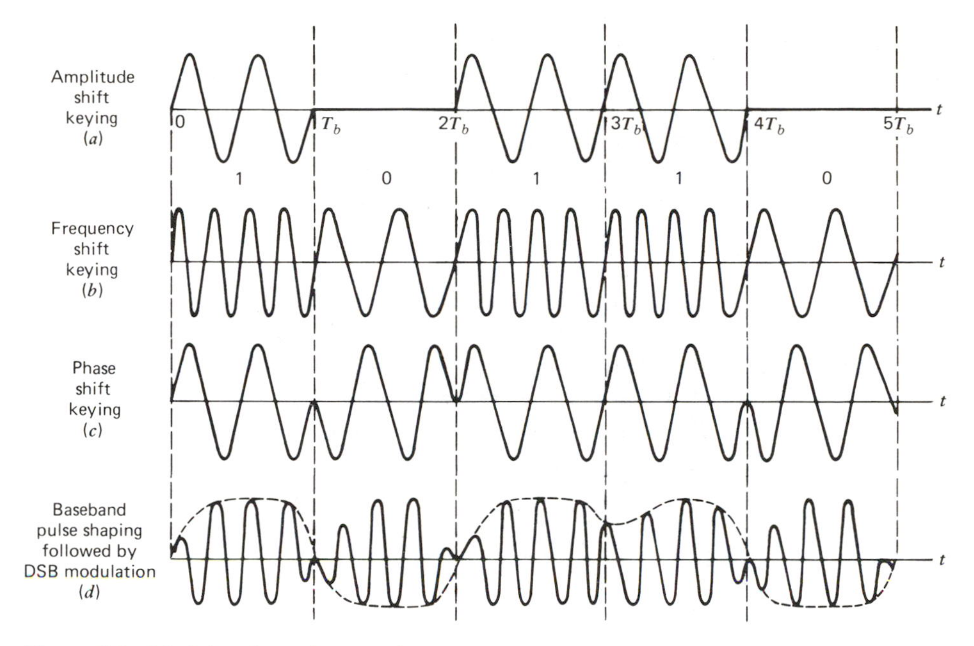

Figure 8.1 Modulated carrier waveforms used in binary data transmission schemes. (*a*) Amplitude-shift keying. (*b*) Frequency-shift keying. (*c*) Phase-shift keying. (*d*) Baseband pulse shaping followed by DSB modulation.

called frequency-shift keying (FSK). In the third method of digital modulation shown in Figure 8.1*c*, the carrier phase is shifted between two values and hence this method is called phase-shift keying (PSK). It should be noted here that in PSK and FSK methods, the amplitude of the carrier remains constant. Further, in all cases the modulated waveforms are continuous at all times. Finally, Figure 8.1*d* shows a modulation waveform generated by amplitude modulating the carrier with a baseband signal generated by the discrete PAM scheme described in the previous chapter.

The modulation scheme using baseband pulse shaping followed by analog modulation (DSB or VSB) requires the minimum transmission bandwidth. However, the equipment required to generate, transmit, and demodulate the waveform shown in Figure 8.1*d* is quite complex. In contrast, the digital modulation schemes are extremely simple to implement. The price paid for this simplicity is excessive bandwidth and possible increase in transmitter power requirements. When bandwidth is not the major consideration, then digital modulation schemes provide very good performance with minimum equipment complexity and with a good degree of immunity to certain channel impairments. In the following sections we will study digital modulation schemes. Primary emphasis will be given to the study of system performance in the presence of additive noise as measured by the probability of error.

We will begin our study of digital modulation schemes with the derivation of an optimum receiver for demodulating ASK, PSK, and FSK signals with minimum probability of error. We will show that such a receiver consists of a matched filter (or a correlation receiver) if the additive noise is white. We will derive expressions for the probability of error for various modulation schemes in terms of the average signal power at the receiver input, power spectral density of the noise at the receiver input, and the signaling rate.

In Section 8.3 we will study the amplitude shift-keying (ASK) method. We will look at optimum and suboptimum methods of demodulating binary ASK signals. In Sections 8.4 and 8.5, we will deal with optimum and suboptimum binary PSK and FSK schemes. Finally, in Section 8.6 we will compare the performance of these binary schemes in terms of power-bandwidth requirements and probability of error. In Section 8.7 we will discuss M-ary PSK, DPSK, and FSK schemes. The problem of synchronizing the receiver to the transmitter will be considered in Section 8.8.

8.2 OPTIMUM RECEIVER FOR BINARY DIGITAL MODULATION SCHEMES

The function of a receiver in a binary communication system is to distinguish between two transmitted signals $s_1(t)$ and $s_2(t)$ in the presence of noise. The performance of the receiver is usually measured in terms of the probability of error and the receiver is said to be optimum if it yields the minimum probability of error. In this section, we will derive the structure of an optimum receiver that can be used for demodulating binary ASK, PSK, and FSK signals.

We will show that the optimum receiver takes the form of a matched filter when the noise at the receiver input is white. We will also show that the matched filter can be implemented as an integrate and dump correlation receiver. The integrate and dump correlation receiver is a *coherent* or synchronous receiver that requires a local carrier reference having the same phase and frequency as the transmitted carrier. Elaborate circuitry is required at the receiver to generate the coherent local carrier reference.

The binary ASK, PSK, and FSK signals can also be demodulated using suboptimal *noncoherent* demodulation schemes. Such schemes are simpler to implement and are widely used in low speed data transmission applications. We will deal with suboptimal (noncoherent) methods of demodulating binary ASK, PSK, and FSK signals in Sections 8.3, 8.4, and 8.5.

8.2.1 Description of Binary ASK, PSK, and FSK Schemes

The block diagram of a bandpass binary data transmission scheme using digital modulation is shown in Figure 8.2. The input to the system is a binary bit sequence $\{b_k\}$ with a bit rate r_b and bit duration T_b. The output of the modulator during the kth bit interval depends on the kth input bit b_k. The modulator output $Z(t)$ during the kth bit interval is a shifted version of one of two basic waveforms $s_1(t)$ or $s_2(t)$, and $Z(t)$ is a random process defined by

$$Z(t) = \begin{cases} s_1[t-(k-1)T_b] & \text{if } b_k = 0 \\ s_2[t-(k-1)T_b] & \text{if } b_k = 1 \end{cases} \tag{8.1}$$

for $(k-1)T_b \leq t \leq kT_b$. The waveforms $s_1(t)$ and $s_2(t)$ have a duration of T_b and have finite energy, that is, $s_1(t)$ and $s_2(t) = 0$ if $t \notin [0, T_b]$ and

$$\begin{aligned} E_1 &= \int_0^{T_b} [s_1(t)]^2\, dt < \infty \\ E_2 &= \int_0^{T_b} [s_2(t)]^2\, dt < \infty \end{aligned} \tag{8.2}$$

The shape of the waveforms depends on the type of modulation used, as shown in Table 8.1. The output of the modulator passes through a bandpass channel $H_c(f)$, which, for purposes of analysis, is assumed to be an ideal channel with adequate bandwidth so that the signal passes through without suffering any distortion other than propagation delay. The channel noise $n(t)$ is assumed to be a zero mean, stationary, Gaussian random process with a known power spectral density $G_n(f)$. The received signal plus noise then is

$$V(t) = \begin{cases} s_1[t-(k-1)T_b - t_d] + n(t) \\ \quad \text{or} \\ s_2[t-(k-1)T_b - t_d] + n(t) \end{cases} \qquad (k-1)T_b + t_d \leq t \leq kT_b + t_d$$

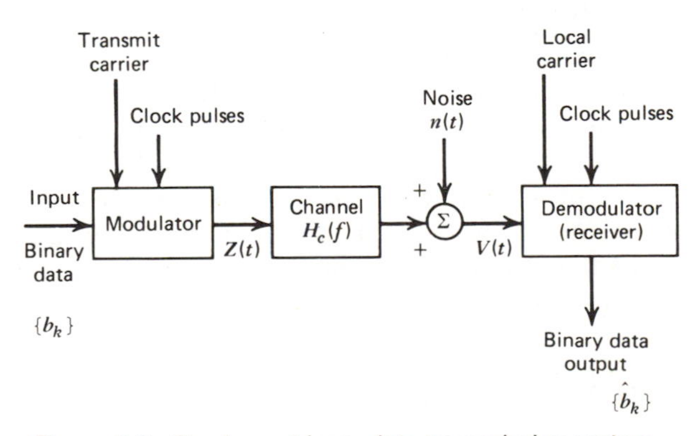

Figure 8.2 Bandpass binary data transmission system.

Table 8.1. Choice of signaling waveforms for various types of digital modulation schemes. $s_1(t)$, $s_2(t) = 0$ for $t \notin [0, T_b]$; $f_c = \omega_c/2\pi$. The frequency of the carrier f_c is assumed to be a multiple of r_b.

$s_1(t)$; $0 \leq t \leq T_b$	$s_2(t)$; $0 \leq t \leq T_b$	Type of Modulation
0	$A \cos \omega_c t$ (or $A \sin \omega_c t$)	Amplitude-shift keying (ASK)
$-A \cos \omega_c t$ (or $-A \sin \omega_c t$)	$A \cos \omega_c t$ ($A \sin \omega_c t$)	Phase-shift keying (PSK)
$A \cos\{(\omega_c - \omega_d)t\}$ (or $A \sin\{(\omega_c - \omega_d)t\}$)	$A \cos\{(\omega_c + \omega_d)t\}$ ($A \sin\{(\omega_c + \omega_d)t\}$)	Frequency-shift keying (FSK)

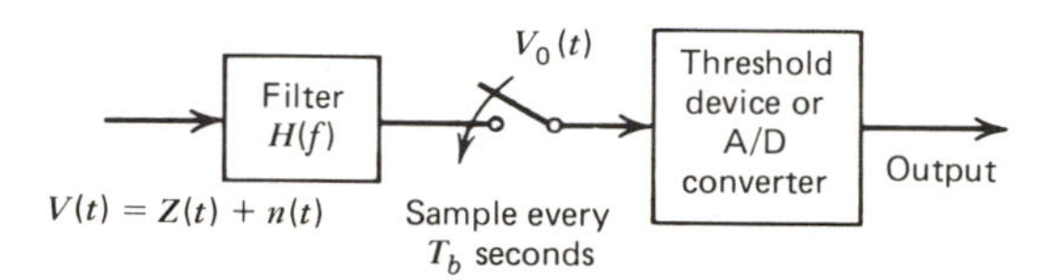

Figure 8.3 Receiver structure.

where t_d is the propagation delay, which can be assumed to be zero without loss of generality.

The receiver shown in Figure 8.3 has to determine which of the two *known waveforms* $s_1(t)$ or $s_2(t)$ was present at its input during each signaling interval. The actual receiver consists of a filter, a sampler, and a threshold device. The signal plus noise $V(t)$ is filtered and sampled at the end of each bit interval. The sampled value is compared against a predetermined threshold value T_0 and the transmitted bit is decoded (with occasional errors) as 1 or 0 depending on whether $V_0(kT_b)$ is greater or less than the threshold T_0.

The receiver makes errors in the decoding process due to the noise present at its input. The error probability will depend on the signal power at the receiver input, noise power spectral density at the input, signaling rate, and receiver parameters such as the filter transfer function $H(f)$ and threshold setting.

8.2.2 Probability of Error

The measure of performance used for comparing digital modulation schemes is the probability of error. The receiver parameters such as $H(f)$ and threshold setting are chosen to minimize the probability of error. In this section, we will derive expressions for the probability of error in terms of the signal parameters, noise power spectral density, and the receiver parameters. These expressions will be used in the following sections for the analysis and design of digital modulation schemes.

We will make the following assumptions while deriving the expressions for the probability of error:

1. We will assume that $\{b_k\}$ is an equiprobable, independent sequence of bits. Hence, the occurrence of $s_1(t)$ or $s_2(t)$ during a bit interval does not influence the occurrence of $s_1(t)$ or $s_2(t)$ during any other non-overlapping bit interval; further, $s_1(t)$ and $s_2(t)$ are equiprobable.
2. The channel noise will be assumed to be a zero mean Gaussian random process with a power spectral density $G_n(f)$.
3. We will assume that the intersymbol interference generated by the filter is small.*

The output of the filter at $t = kT_b$ can be written as

$$V_0(kT_b) = s_0(kT_b) + n_0(kT_b) \tag{8.3}$$

where $s_0(t)$ and $n_0(t)$ denote the response of the filter due to signal and noise components at its input. The signal component in the output at $t = kT_b$ is given by

$$s_0(kT_b) = \int_{-\infty}^{kT_b} Z(\zeta)h(kT_b - \zeta)\, d\zeta \tag{8.4}$$

$$= \int_{(k-1)T_b}^{kT_b} Z(\zeta)h(kT_b - \zeta)\, d\zeta + \text{ISI terms} \tag{8.5}$$

where $h(\zeta)$ is the impulse response of the filter. Since we have assumed the ISI terms to be zero, we can rewrite Equation (8.5) as

$$s_0(kT_b) = \int_{(k-1)T_b}^{kT_b} Z(\zeta)h(kT_b - \zeta)\, d\zeta$$

Substituting $Z(t)$ from Equation (8.1) and making a change of variable, we can write the signal component as

$$s_0(kT_b) = \begin{cases} \displaystyle\int_0^{T_b} s_1(\zeta)h(T_b - \zeta)\, d\zeta = s_{01}(kT_b) \text{ when } b_k = 0 \\ \displaystyle\int_0^{T_b} s_2(\zeta)h(T_b - \zeta)\, d\zeta = s_{02}(kT_b) \text{ when } b_k = 1 \end{cases} \tag{8.6}$$

*We will see later that the optimum filter for the white noise case generates zero ISI. For colored noise case the optimum filter generates nonzero ISI, which can be minimized by making $s_1(t)$ and $s_2(t)$ to have a duration $\ll T_b$ so that the filter response settles down to a negligible value before the end of each bit interval.

The noise component $n_0(kT_b)$ is given by

$$n_0(kT_b) = \int_{-\infty}^{kT_b} n(\zeta)h(kT_b - \zeta)\,d\zeta \tag{8.7}$$

The output noise $n_0(t)$ is a stationary zero mean Gaussian random process. The variance of $n_0(t)$ is

$$N_0 = E\{n_0^2(t)\} = \int_{-\infty}^{\infty} G_n(f)|H(f)|^2\,df \tag{8.8}$$

and the probability density function of $n_0(t)$ is

$$f_{n_0}(n) = \frac{1}{\sqrt{2\pi N_0}} \exp\left(\frac{-n^2}{2N_0}\right), \quad -\infty < n < \infty \tag{8.9}$$

The receiver decodes the kth bit by comparing $V_0(kT_b)$ against the threshold T_0. If we assume that $s_1(t)$ and $s_2(t)$ are chosen such that $s_{01}(T_b) < s_{02}(T_b)$, and that the receiver decodes the kth bit as 0 if $V_0(kT_b) < T_0$ and as 1 if $V_0(kT_b) \geqslant T_0$, then the probability that the kth bit is incorrectly decoded is given by P_e, where

$$\begin{aligned} P_e &= P\{b_k = 0 \text{ and } V_0(kT_b) \geqslant T_0;\text{ or } b_k = 1 \text{ and } V_0(kT_b) < T_0\} \\ &= \tfrac{1}{2}P\{V_0(kT_b) \geqslant T_0 | b_k = 0\} \\ &\quad + \tfrac{1}{2}P\{V_0(kT_b) < T_0 | b_k = 1\} \end{aligned} \tag{8.10}$$

If the kth transmitted bit is 0, then $V_0 = s_{01} + n_0$ where s_{01} is a constant and n_0 is a zero mean Gaussian random variable with the variance given in Equation (8.8). Hence, the conditional pdf of V_0 given $b_k = 0$ is given by

$$f_{V_0|b_k=0}(v_0) = \frac{1}{\sqrt{2\pi N_0}} \exp\left(\frac{-(v_0 - s_{01})^2}{2N_0}\right), \quad -\infty < v_0 < \infty \tag{8.11a}$$

Similarly, when b_k is 1, the conditional pdf of V_0 has the form

$$f_{V_0|b_k=1}(v_0) = \frac{1}{\sqrt{2\pi N_0}} \exp\left\{\frac{-(v_0 - s_{02})^2}{2N_0}\right), \quad -\infty < v_0 < \infty \tag{8.11b}$$

Combining Equations (8.10) and (8.11), we obtain an expression for the probability of error P_e as

$$\begin{aligned} P_e &= \tfrac{1}{2}\int_{T_0}^{\infty} \frac{1}{\sqrt{2\pi N_0}} \exp\left(\frac{-(v_0 - s_{01})^2}{2N_0}\right) dv_0 \\ &\quad + \tfrac{1}{2}\int_{-\infty}^{T_0} \frac{1}{\sqrt{2\pi N_0}} \exp\left(\frac{-(v_0 - s_{02})^2}{2N_0}\right) dv_0 \end{aligned} \tag{8.12}$$

Because of equal probabilities of occurrences of 0's and 1's in the input bit stream and the symmetrical shapes of $f_{V_0|b_k=0}$ and $f_{V_0|b_k=1}$ shown in Figure 8.4, it

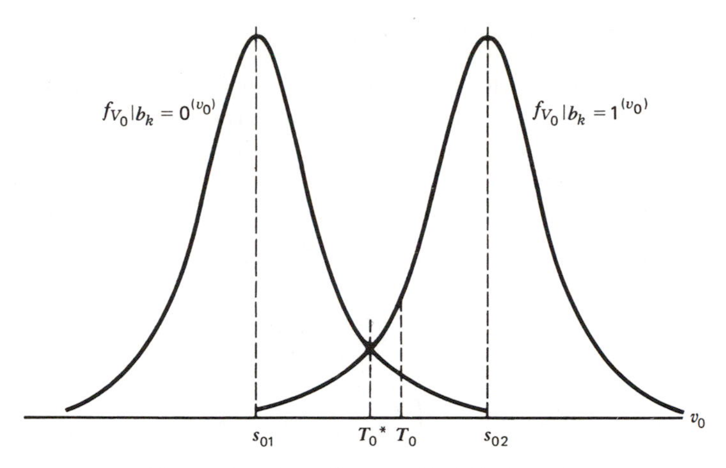

Figure 8.4 Conditional pdf of V_0 given b_k.

can be shown that the optimum choice for the threshold is the value of v_0 at which the conditional pdf's intersect (see Problem 8.5). This optimum value of the threshold T_0^* is

$$T_0^* = \frac{s_{01} + s_{02}}{2}$$

Substituting the value of T_0^* for T_0 in (8.12), we can rewrite the expression for the probability of error as

$$P_e = \int_{(s_{02}+s_{01})/2}^{\infty} \frac{1}{\sqrt{2\pi N_0}} \exp\left(-\frac{(v_0 - s_{01})^2}{2N_0}\right) dv_0$$

$$= \int_{(s_{02}-s_{01})/2\sqrt{N_0}}^{\infty} \frac{1}{\sqrt{2\pi}} \exp\left(-\frac{z^2}{2}\right) dz \tag{8.13}$$

The above expression for the probability of error is a monotonically decreasing function in its argument, that is, P_e becomes smaller as $(s_{02} - s_{01})/\sqrt{N_0}$ becomes larger. Equations (8.6), (8.7) and (8.8) indicate that s_{01}, s_{02}, and $\sqrt{N_0}$ depend on the choice of the filter impulse response (or the transfer function). The optimum filter then is the filter that maximizes the ratio

$$\gamma = \frac{s_{02}(T_b) - s_{01}(T_b)}{\sqrt{N_0}} \tag{8.14}$$

or the square of the ratio γ^2. Observe that maximizing γ^2 eliminates the requirement $s_{01} < s_{02}$.

8.2.3 Transfer Function of the Optimum Filter

The essential function of the receiver shown in Figure 8.3 is that it has to determine which of the two *known waveforms* $s_1(t)$ or $s_2(t)$ was present at its input during each signaling interval. The optimum receiver distinguishes between $s_1(t)$ and $s_2(t)$ from the noisy versions of $s_1(t)$ and $s_2(t)$ with minimum probability of error. We have seen in the preceding section that the probability of error is minimized by an appropriate choice of $h(t)$ which maximizes γ^2, where

$$\gamma^2 = \frac{[s_{02}(T_b) - s_{01}(T_b)]^2}{N_0} \tag{8.15}$$

$$s_{02}(T_b) - s_{01}(T_b) = \int_0^{T_b} [s_2(\zeta) - s_1(\zeta)]h(T_b - \zeta)\, d\zeta$$

and

$$N_0 = \int_{-\infty}^{\infty} G_n(f)|H(f)|^2\, df$$

If we let $p(t) = s_2(t) - s_1(t)$, then the numerator of the quantity to be maximized is

$$\begin{aligned} s_{02}(T_b) - s_{01}(T_b) = p_0(T_b) &= \int_0^{T_b} p(\zeta)h(T_b - \zeta)\, d\zeta \\ &= \int_{-\infty}^{\infty} p(\zeta)h(T_b - \zeta)\, d\zeta \end{aligned} \tag{8.16}$$

since $p(t) = 0$ for $t < 0$ and $h(\lambda) = 0$ for $\lambda < 0$.
If we let $P(f)$ be the Fourier transform of $p(t)$, then we can obtain the Fourier transform $P_0(f)$ of $p_0(t)$ from Equation (8.16) as

$$P_0(f) = P(f)H(f)$$

or

$$p_0(T_b) = \int_{-\infty}^{\infty} P(f)H(f)\exp(j2\pi f T_b)\, df$$

Hence γ^2 can be written as

$$\gamma^2 = \frac{\left|\int_{-\infty}^{\infty} H(f)P(f)\exp(j2\pi f T_b)\, df\right|^2}{\int_{-\infty}^{\infty} |H(f)|^2 G_n(f)\, df} \tag{8.17}$$

We can maximize γ^2 by applying Schwarz's inequality, which has the form

$$\frac{\left|\int_{-\infty}^{\infty} X_1(f)X_2(f)\,df\right|^2}{\int_{-\infty}^{\infty} |X_1(f)|^2\,df} \leq \int_{-\infty}^{\infty} |X_2(f)|^2\,df \tag{8.18}$$

where $X_1(f)$ and $X_2(f)$ are arbitrary complex functions of a common variable f. The equal sign in (8.18) applies when $X_1(f) = KX_2^*(f)$, where K is an arbitrary constant and $X_2^*(f)$ is the complex conjugate of $X_2(f)$. Applying Schwarz's inequality to Equation (8.17) with

$$X_1(f) = H(f)\sqrt{G_n(f)}$$

and

$$X_2(f) = \frac{P(f)\exp(j2\pi fT_b)}{\sqrt{G_n(f)}}$$

we see that $H(f)$, which maximizes γ^2, is given by

$$H(f) = K\,\frac{P^*(f)\exp(-j2\pi fT_b)}{G_n(f)} \tag{8.19}$$

where K is an arbitrary constant. Substituting Equation (8.19) in (8.17), we obtain the maximum value of γ^2 as

$$\gamma^2_{\max} = \int_{-\infty}^{\infty} \frac{|P(f)|^2}{G_n(f)}\,df \tag{8.20}$$

and the minimum probability of error is given by

$$\begin{aligned} P_e &= \int_{\gamma_{\max}/2}^{\infty} \frac{1}{\sqrt{2\pi}}\exp\left(-\frac{z^2}{2}\right)dz \\ &= Q\left(\frac{\gamma_{\max}}{2}\right) \end{aligned} \tag{8.21}$$

Special Case I: Matched Filter Receiver. If the channel noise is white, that is, $G_n(f) = \eta/2$, then the transfer function of the optimum receiver is given by

$$H(f) = P^*(f)\exp(-j2\pi fT_b) \tag{8.22}$$

(from Equation (8.19) with the arbitrary constant K set equal to $\eta/2$). The impulse response of the optimum filter is

$$h(t) = \int_{-\infty}^{\infty} [P^*(f)\exp(-2\pi jfT_b)]\exp(2\pi jft)\,df \tag{8.23}$$

Recognizing the fact that the inverse Fourier transform of $P^*(f)$ is $p(-t)$ and that $\exp(-2\pi jfT_b)$ represents a delay of T_b, we obtain $h(t)$ as

$$h(t) = p(T_b - t)$$

Since $p(t) = s_2(t) - s_1(t)$, we have

$$h(t) = s_2(T_b - t) - s_1(T_b - t) \tag{8.24}$$

The impulse response $h(t)$ in Equation (8.24) is matched to the signal $s_1(t)$ and $s_2(t)$ and for this reason the filter is called a *matched filter*. An example is shown in Figure 8.5 to illustrate the significance of the result stated in Equation (8.24).

Figures 8.5a and 8.5b show $s_2(t)$ and $s_1(t)$ of duration T_b. The waveform $p(t) = s_2(t) - s_1(t)$ is shown in Figure 8.5c and $p(-t)$, which is the waveform $p(t)$ reflected around $t = 0$ is shown in 8.5d. Finally, the impulse response of the filter $h(t) = p(T_b - t)$, which is $p(-t)$ translated in the positive t direction by T_b, is shown in Figure 8.5e. We note here that the filter is causal ($h(t) = 0$ for $t < 0$) and the impulse response has a duration of T_b. The last fact ensures that the signal component of the output at the end of the kth-bit interval is due to signal component at the input during the kth-bit interval only. Thus, there is no intersymbol interference. The probability of error for the matched filter receiver can be obtained from Equations (8.20) and (8.21).

In general, it is very hard to synthesize physically realizable filters that would closely approximate the transfer function specified in Equation (8.22). In the following section we will derive an alternate form for the matched filter that is easier to implement using very simple circuitry.

Special Case II: Correlation Receiver. We will now derive a form of the optimum receiver, which is different from the matched filter implementation. We start with the output of the receiver at $t = T_b$,

$$V_0(T_b) = \int_{-\infty}^{T_b} V(\zeta)h(T_b - \zeta)\, d\zeta$$

where $V(\zeta)$ is the noisy input to the receiver. Substituting $h(\zeta) = s_2(T_b - \zeta) - s_1(T_b - \zeta)$ and noting that $h(\zeta) = 0$ for $\zeta \not\in (0, T_b)$, we can rewrite the preceding expression as

$$\begin{aligned} V_0(T_b) &= \int_0^{T_b} V(\zeta)[s_2(\zeta) - s_1(\zeta)]\, d\zeta \\ &= \int_0^{T_b} V(\zeta)s_2(\zeta)\, d\zeta - \int_0^{T_b} V(\zeta)s_1(\zeta)\, d\zeta \end{aligned} \tag{8.25}$$

Equation (8.25) suggests that the optimum receiver can be implemented as shown in Figure 8.6. This form of the receiver is called a *correlation receiver*.

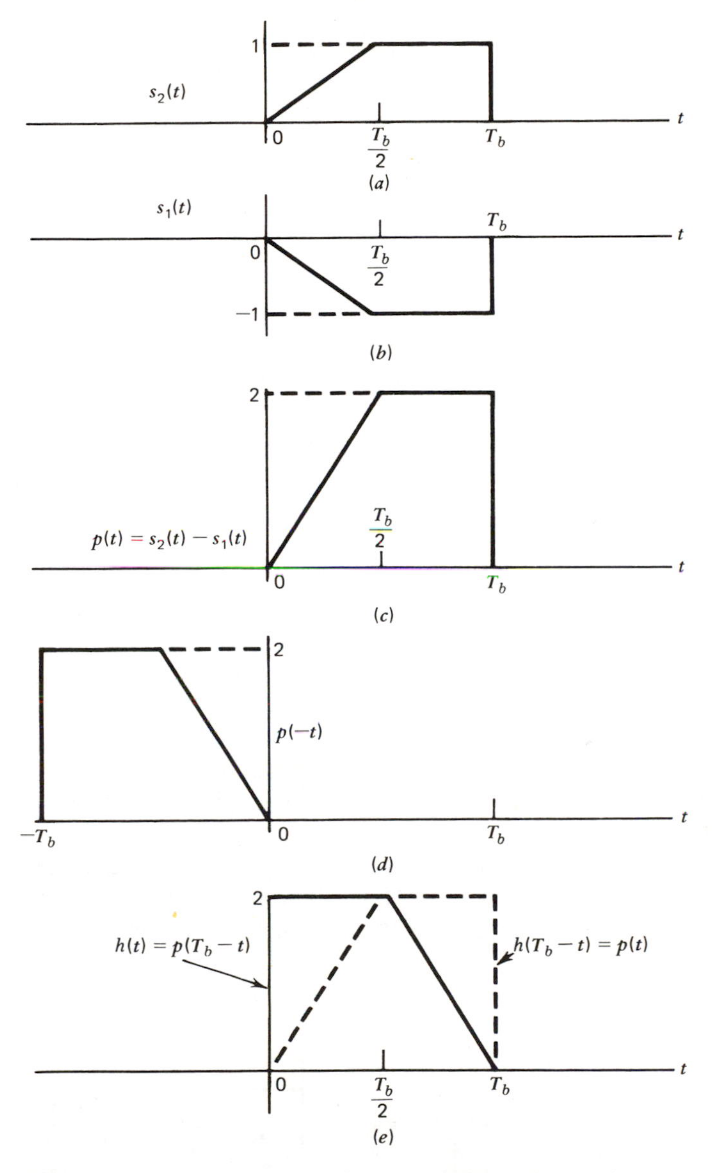

Figure 8.5 Impulse response of a matched filter. (*a*) $s_2(t)$. (*b*) $s_1(t)$. (*c*) $p(t) = s_2(t) - s_1(t)$. (*d*) $p(-t)$. (*e*) $h(t) = p(T_b - t)$.

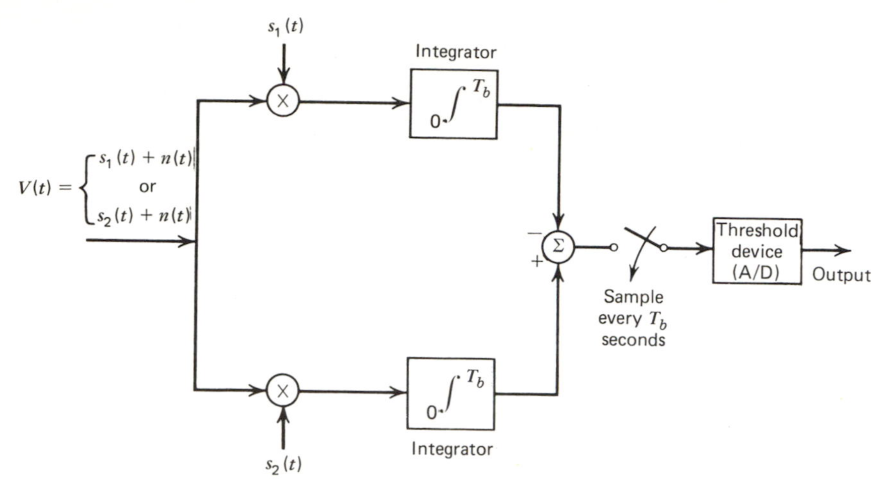

Figure 8.6 Correlation receiver.

It must be pointed out that Equation (8.25) and the receiver shown in Figure 8.6 require that the integration operation be ideal with zero initial conditions. In actual practice, the receiver shown in Figure 8.6 is actually implemented as shown in Figure 8.7. In this implementation, the integrator has to be reset (i.e., the capacitor has to be discharged or dumped) at the end of each signaling interval in order to avoid intersymbol interference. If $RC \gg T_b$, the circuit shown in Figure 8.7 very closely approximates an ideal integrator and operates with the same probability of error as the ideal receiver shown in Figure 8.6.

Needless to say, the sampling and discharging of the capacitor (dumping) must be carefully synchronized. Furthermore, the *local reference signal* $s_2(t) - s_1(t)$ must be in "phase" with the signal component at the receiver

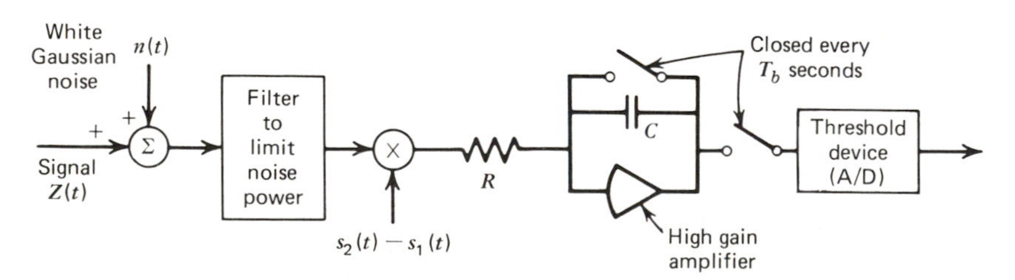

Figure 8.7 Integrate and dump correlation receiver. The bandwidth of the filter preceding the integrator is assumed to be wide enough to pass $Z(t)$ without distortion.

input, that is, the correlation receiver performs *coherent* demodulation. The correlation receiver, also known as an *integrate and dump filter*, represents one of the few cases where matched filtering is closely approximated in practice.

Example 8.1. A bandpass data transmission scheme uses a PSK signaling scheme with

$$s_2(t) = A\cos\omega_c t, \quad 0 \leq t \leq T_b, \quad \omega_c = 10\pi/T_b$$

$$s_1(t) = -A\cos\omega_c t, \quad 0 \leq t \leq T_b, \quad T_b = 0.2\text{ msec}$$

The carrier amplitude at the receiver input is 1 mvolt and the psd of the additive white Gaussian noise at the input is 10^{-11} watt/Hz. Assume that an ideal correlation receiver is used. Calculate the average bit error rate of the receiver.

Solution

$$\text{Data rate} = 5000\text{ bits/sec}, \quad G_n(f) = \eta/2 = 10^{-11}\text{ watt/Hz}$$

$$\begin{aligned}\text{Receiver impulse response} &= h(t)\\ &= s_2(T_b - t) - s_1(T_b - t)\\ &= 2A\cos\omega_c(T_b - t)\end{aligned}$$

Threshold setting is 0 and

$$\begin{aligned}\gamma_{\max}^2 &= \int_{-\infty}^{\infty} \frac{|P(f)|^2}{G_n(f)}\,df \quad \text{(from Equation (8.20))}\\ &= \left(\frac{2}{\eta}\right)\int_{-\infty}^{\infty} |P(f)|^2\,df\\ &= \left(\frac{2}{\eta}\right)\int_0^{T_b} [s_2(t) - s_1(t)]^2\,dt \quad \text{(by Parseval's theorem)}\\ &= \left(\frac{2}{\eta}\right)\int_0^{T_b} 4A^2(\cos\omega_c t)^2\,dt\\ &= \left(\frac{2}{\eta}\right)(2A^2T_b) = \frac{4A^2T_b}{\eta} = 40\end{aligned}$$

$$\begin{aligned}\text{Probability of error} = P_e &= \int_{\frac{1}{2}\gamma_{\max}}^{\infty} \frac{1}{\sqrt{2\pi}}\exp\left(\frac{-z^2}{2}\right)dz\\ &= Q(\sqrt{10})\end{aligned}$$

From the table of Gaussian probabilities, we get $P_e \simeq 0.0008$ and

$$\text{Average error rate} \simeq (r_b)P_e/\text{sec} = 4\text{ bits/sec}$$

8.3 BINARY ASK SIGNALING SCHEMES

The binary ASK signaling scheme was one of the earliest forms of digital modulation used in wireless (radio) telegraphy at the beginning of this century. While amplitude-shift keying is no longer widely used in digital communications, for reasons that will be discussed later, it is the simplest form of digital modulation and serves as a useful model for introducing certain concepts. The binary ASK waveform can be described as

$$Z(t) = \begin{cases} s_1[t-(k-1)T_b] & \text{if } b_k = 0 \\ s_2[t-(k-1)T_b] & \text{if } b_k = 1 \end{cases} \quad (k-1)T_b \leq t \leq kT_b$$

where $s_2(t) = A \cos \omega_c t$ $(0 \leq t \leq T_b)$ and $s_1(t) = 0$. We will assume that the carrier frequency $\omega_c = 2n\pi/T_b$, where n is an integer.

We can represent $Z(t)$ as

$$Z(t) = D(t)(A \cos \omega_c t) \tag{8.26}$$

where $D(t)$ is a lowpass pulse waveform consisting of (often but not necessarily) rectangular pulses. For purposes of analysis we will assume that $D(t)$ is a rectangular random binary waveform with bit duration T_b. The model for $D(t)$ is (Chapter 3, Section 3.5)

$$d(t) = \sum_{k=-\infty}^{\infty} b_k g[t-(k-1)T_b], \quad b_k = 0 \text{ or } 1$$

$$g(t) = \begin{cases} 1 & 0 \leq t \leq T_b \\ 0 & \text{elsewhere} \end{cases} \tag{8.27}$$

$$D(t) = d(t-T)$$

where T represents a random delay with a uniform pdf in the interval $[0, T_b]$. The form of the modulated waveform $Z(t)$ suggests that the ASK signal can be generated by product modulation, that is, by multiplying the carrier with the rectangular waveform $D(t)$ or using $D(t)$ to turn the carrier oscillator on and off.

The bandwidth requirements for transmitting and processing the ASK signal can be obtained from the power spectral density of $Z(t)$, which can be computed as follows: From Equation (8.26) we see that the power spectral density $G_Z(f)$ of $Z(t)$ is related to the power spectral density $G_D(f)$ of $D(t)$ by*

$$G_Z(f) = \frac{A^2}{4}[G_D(f-f_c) + G_D(f+f_c)] \tag{8.28}$$

*Strictly speaking, we need to include a random phase for the carrier in Equation (8.26) so that $Z(t)$ is a stationary random process.

The waveform $D(t)$ is a random binary waveform with levels 0 and 1. The autocorrelation function and the power spectral density of $D(t)$ are (from Example 3.9 and Equations (3.68) and (3.69))

$$R_{DD}(\zeta) = \begin{cases} \frac{1}{4} + \frac{T_b - |\zeta|}{4T_b} & \text{for } |\zeta| \leq T_b \\ 0 \text{ for } |\zeta| > T_b \end{cases}$$

$$G_D(f) = \tfrac{1}{4}\left(\delta(f) + \frac{\sin^2 \pi f T_b}{\pi^2 f^2 T_b}\right) \tag{8.29}$$

Substituting Equation (8.29) into (8.28), we obtain the psd of $Z(t)$ as

$$G_Z(f) = \frac{A^2}{16}\left(\delta(f - f_c) + \delta(f + f_c) + \frac{\sin^2 \pi T_b(f - f_c)}{\pi^2 T_b(f - f_c)^2} + \frac{\sin^2 \pi T_b(f + f_c)}{\pi^2 T_b(f + f_c)^2}\right) \tag{8.30}$$

A sketch of $G_Z(f)$, shown in Figure 8.8, indicates that $Z(t)$ is an infinite bandwidth signal. However, for practical purposes, the bandwidth of $Z(t)$ is often defined as the bandwidth of an ideal bandpass filter centered at f_c whose output (with $Z(t)$ as its input) contains, say, 95% of the total average power content of $Z(t)$. It can be shown that such a bandwidth would be approximately $3r_b$ Hz for the ASK signal.

The bandwidth of the ASK signal can be reduced by using smoothed versions of the pulse waveform $D(t)$ instead of rectangular pulse waveforms. For example, if we use a pulse waveform $D(t)$ in which the individual pulses $g(t)$ have the shape,

$$g(t) = \begin{cases} (a/2)[1 + \cos(2\pi r_b t - \pi)], & 0 \leq t < T_b \\ 0 & \text{elsewhere} \end{cases}$$

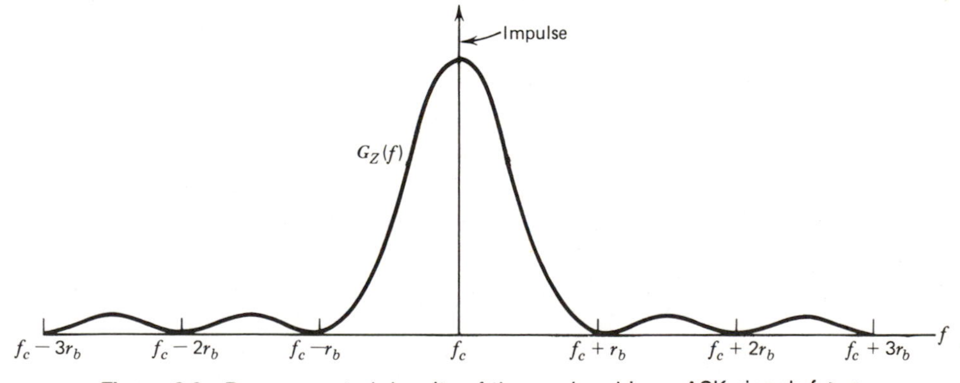

Figure 8.8 Power spectral density of the random binary ASK signal; $f_c \gg r_b$.

the effective bandwidth of the ASK signal will be of the order of $2r_b$. (The magnitude of the Fourier transform of $g(t)$ drops off as $1/f^3$.) Of course, the gain in bandwidth is somewhat offset by the complexity of the pulse shaping networks needed to generate $g(t)$ given above. Depending on the shape of the pulse waveform, we need a channel with a bandwidth of approximately $2r_b$ to $3r_b$ Hz to transmit an ASK signal.

The transmitted bit sequence $\{b_k\}$ can be recovered from the noisy version of $Z(t)$ at the receiver in one of two ways. The first method of demodulation we will study is the integrate and dump-type coherent demodulation; the second method is the noncoherent envelope demodulation procedure. The principal reason for using the ASK signaling method is its simplicity. Hence, coherent demodulation is seldom used in conjunction with ASK schemes because of the complex circuits needed for maintaining phase coherence between the transmitted signal and the local carrier. Nevertheless, we will investigate the performance of coherent ASK schemes for comparison purposes.

8.3.1 Coherent ASK

The receiver shown in Figure 8.7 can be used for coherent demodulation of an ASK signal. As before, we will assume that the input to the receiver consists of an ASK signal that is corrupted by additive, Gaussian white noise. The receiver integrates the product of the signal plus noise and a copy of the noise free signal over one signaling interval. We will assume that the local signal $s_2(t) - s_1(t) = A \cos \omega_c t$ is carefully synchronized with the frequency and phase of the received carrier. The output of the integrator is compared against a set threshold and at the end of each signaling interval the receiver makes a decision about which of the two signals $s_1(t)$ or $s_2(t)$ was present at its input during the signaling interval. Of course, errors will occur in the demodulation process because of the noise. We can derive an expression for the probability of incorrectly decoding the input waveform using the results derived in Section 8.2.

We start with $s_2(t) = A \cos \omega_c t$, $s_1(t) = 0$, and $s_2(t) - s_1(t) = A \cos \omega_c t$. The signal components of the receiver output at the end of a signaling interval are

$$s_{01}(kT_b) = \int_0^{T_b} s_1(t)[s_2(t) - s_1(t)]\, dt = 0$$

and

$$s_{02}(kT_b) = \int_0^{T_b} s_2(t)[s_2(t) - s_1(t)]\, dt$$
$$= \frac{A^2}{2} T_b$$

In the preceding step, we made use of our assumption that $\omega_c T_b = 2n\pi$, n a positive integer. The optimum threshold setting in the receiver is

$$T_0^* = \frac{s_{01}(kT_b) + s_{02}(kT_b)}{2} = \frac{A^2}{4} T_b$$

The receiver decodes the kth transmitted bit as 1 if the output at the kth signaling interval is greater than T_0^*, and as 0 otherwise.

The probability of error P_e can be computed using Equations (8.20) and (8.21) as

$$\begin{aligned}\gamma_{\max}^2 &= \int_{-\infty}^{\infty} \frac{|P(f)|^2}{G_n(f)}\, df \\ &= \frac{2}{\eta} \int_0^{T_b} p^2(t)\, dt \\ &= \frac{2}{\eta} \int_0^{T_b} A^2 \cos^2 \omega_c t\, dt \\ &= \frac{A^2 T_b}{\eta}\end{aligned}$$

and

$$P_e = \int_{\frac{1}{2}\gamma_{\max}}^{\infty} \frac{1}{\sqrt{2\pi}} \exp\left(-\frac{z^2}{2}\right) dz = Q\left(\sqrt{\frac{A^2 T_b}{4\eta}}\right) \tag{8.31}$$

The signal $s_2(t)$ is present at the receiver input only one half the time on the average, and for the remaining half there is no signal since $s_1(t) = 0$. Hence the average signal power at the receiver input is given by

$$S_{av} = A^2/4$$

We can express the probability of error in terms of the average signal power as

$$P_e = Q\left(\sqrt{\frac{S_{av} T_b}{\eta}}\right) \tag{8.32}$$

The probability of error is sometimes expressed in terms of the average signal energy per bit, $E_{av} = (S_{av})T_b$, as

$$P_e = Q(\sqrt{E_{av}/\eta}) \tag{8.33}$$

Example 8.2. Binary data has to be transmitted over a telephone link that has a usable bandwidth of 3000 Hz and a maximum achievable signal-to-noise power ratio of 6 dB at its output. (a) Determine the maximum signaling rate and P_e if a coherent ASK scheme is used for transmitting binary data through this channel. (b) If the data rate is maintained at 300 bits/sec, calculate the error probability.

Solution

(a) If we assume that an ASK signal requires a bandwidth of $3r_b$ Hz, then the maximum signaling rate permissible is $r_b = 1000$ bits/sec. The probability of error can be computed as follows:

$$\text{Average signal power} = A^2/4$$

$$\text{Noise power} = (2)(\eta/2)(3000)$$

$$\frac{\text{Average signal power}}{\text{Noise power}} = 4 = \frac{A^2}{12{,}000\eta} \quad \text{or} \quad \frac{A^2}{\eta} = 48{,}000$$

Hence, $A^2/4\eta r_b = 12$ and

$$P_e = Q(\sqrt{12}) = Q(3.464) \simeq 0.0003$$

(b) If the bit rate is reduced to 300 bits/sec, then

$$\frac{A^2}{4\eta r_b} = 40$$

and

$$P_e = Q(\sqrt{40}) = Q(6.326) \simeq 10^{-10}.$$

8.3.2 Noncoherent ASK

In ideal coherent detection of ASK signals, we assume that there is available at the receiver an exact replica of the arriving signal. That is, we have assumed that a phase coherent local carrier can be generated at the receiver. While this may be possible by the use of very stable oscillators in both the transmitter and receiver, the cost may be excessive.

Noncoherent detection schemes do not require a phase-coherent local oscillator signal. These schemes involve some form of rectification and lowpass filtering at the receiver. The block diagram of a noncoherent receiver for the ASK signaling scheme is shown in Figure 8.9. The computation of the error probability for this receiver is more difficult because of the nonlinear operations that take place in the receiver. In the following analysis we will rely heavily on the results derived in Chapter 3. The input to the receiver is

$$V(t) = \begin{cases} A\cos\omega_c t + n_i(t) & \text{when } b_k = 1 \\ n_i(t) & \text{when } b_k = 0 \end{cases}$$

where $n_i(t)$ is the noise at the receiver input, which is assumed to be zero mean, Gaussian, and white. Now, if we assume the bandpass filter to have a bandwidth of $2/T_b$ centered at f_c, then it passes the signal component without

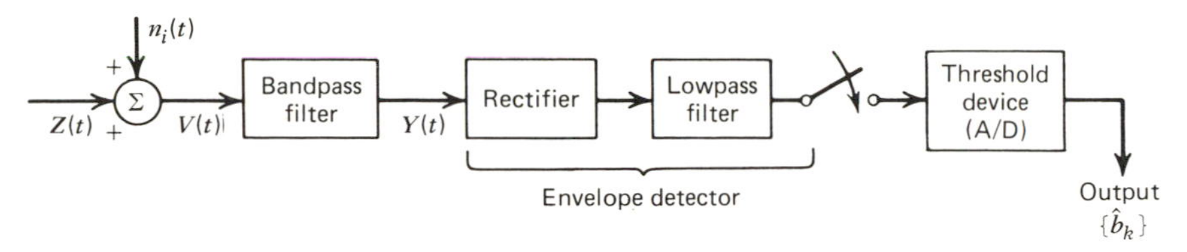

Figure 8.9 Noncoherent ASK receiver.

much distortion. At the filter output we have

$$\begin{aligned} Y(t) &= A_k \cos \omega_c t + n(t) \\ &= A_k \cos \omega_c t + n_c(t) \cos \omega_c t - n_s(t) \sin \omega_c t \end{aligned} \tag{8.34}$$

where $A_k = A$ when the kth transmitted bit $b_k = 1$ and $A_k = 0$ when $b_k = 0$. $n(t)$ is the noise at the output of the bandpass filter and $n_c(t)$, $n_s(t)$ are the quadrature components of the narrowband noise $n(t)$ (Chapter 3). We can rewrite Equation (8.34) in envelope and phase form as

$$Y(t) = R(t) \cos[\omega_c t + \theta(t)]$$

where

$$R(t) = \sqrt{[A_k + n_c(t)]^2 + [n_s(t)]^2} \tag{8.35}$$

Assuming ideal operation, the output of the envelope detector is $R(t)$ and the transmitted bit sequence $\{b_k\}$ is recovered from $R(kT_b)$.

To calculate the probability of error, we need to determine the conditional probability density functions $f_{R|b_k=0}(r)$ and $f_{R|b_k=1}(r)$, and the optimum value of the threshold. Using the results derived in Chapter 3, we obtain these conditional pdfs as

$$f_{R|b_k=0}(r) = \frac{r}{N_0} \exp\left(-\frac{r^2}{2N_0}\right), \quad r > 0 \tag{8.36a}$$

$$f_{R|b_k=1}(r) = \frac{r}{N_0} I_0\left(\frac{Ar}{N_0}\right) \exp\left(-\frac{r^2 + A^2}{2N_0}\right), \quad r > 0 \tag{8.36b}$$

where N_0 is the noise power at the output of the bandpass filter

$$N_0 = \eta B_T \approx 2\eta / T_b$$

and $I_0(x)$ is the modified Bessel function of the first kind and zero order defined by

$$I_0(x) = \frac{1}{2\pi} \int_0^{2\pi} \exp(x \cos(u))\, du$$

In order for the envelope detector to operate above the noise threshold

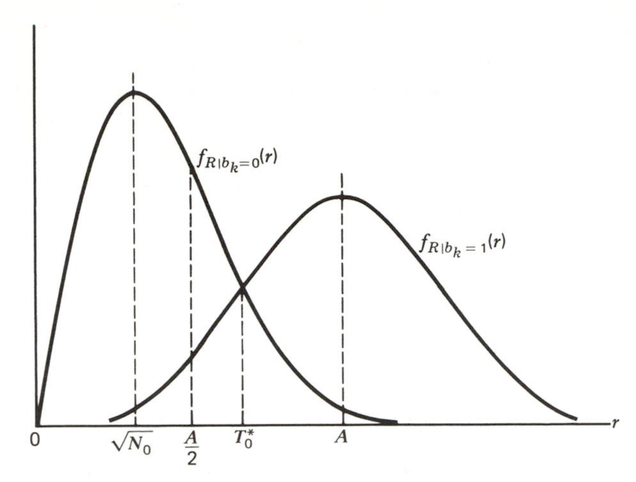

Figure 8.10 Pdf's of the envelope of the noise and the envelope of the signal plus noise.

(Chapter 6, Section 6.3), the carrier amplitude at the receiver input should be such that $A^2 \gg N_0$. If we assume $A^2 \gg N_0$, then we can approximate the Bessel function by

$$I_0\left(\frac{Ar}{N_0}\right) \simeq \sqrt{\frac{N_0}{2\pi Ar}} \exp\left(\frac{Ar}{N_0}\right)$$

Hence,

$$f_{R|b_k=1}(r) \simeq \sqrt{\frac{r}{2\pi AN_0}} \exp\left(-\frac{(r-A)^2}{2N_0}\right), \quad r > 0$$

which is essentially a Gaussian distribution with mean A and variance N_0 since $r/2\pi AN_0 \simeq 1/2\pi N_0$ in the vicinity of $r = A$, where the pdf has the bulk of its area. Sketches of the pdfs are given in Figure 8.10.

The receiver compares the output of the envelope detector $R(t)$ with a threshold value T_0 and decodes the received signal as $s_1(t)$ if $R(kT_b) \leq T_0$ and as $s_2(t)$ if $R(kT_b) > T_0$. Clearly, the threshold T_0 should be set between 0 and A such that the probability of error is minimized. It can be shown that the value of T_0, say T_0^*, which minimizes the probability of error has to satisfy

$$f_{R|b_k=0}(T_0^*) = f_{R|b_k=1}(T_0^*) \tag{8.37a}$$

The relationship

$$T_0^* \simeq \frac{A}{2}\sqrt{1 + \frac{8N_0}{A^2}} \tag{8.37b}$$

is an excellent analytic approximation to the solution of Equation (8.37a). When the carrier amplitude at the receiver input is such that $A^2 \gg N_0$, then $T_0^* \approx A/2$.

The probability of error P_e is given by

$$P_e = \tfrac{1}{2}P(\text{error}|b_k = 0) + \tfrac{1}{2}P(\text{error}|b_k = 1)$$
$$= \tfrac{1}{2}P_{e0} + \tfrac{1}{2}P_{e1}$$

where

$$P_{e0} = \int_{A/2}^{\infty} \frac{r}{N_0} \exp\left(-\frac{r^2}{2N_0}\right) dr = \exp\left(-\frac{A^2}{8N_0}\right) \tag{8.38a}$$

and

$$P_{e1} \approx \int_{-\infty}^{A/2} \frac{1}{\sqrt{2\pi N_0}} \exp\left(-\frac{(r-A)^2}{2N_0}\right) dr$$
$$= Q\left(\frac{A}{2\sqrt{N_0}}\right) \tag{8.38b}$$

Using the approximation

$$Q(x) = \frac{\exp(-x^2/2)}{x\sqrt{2\pi}}$$

for large x, we can reduce P_{e1} to the form

$$P_{e1} \approx \sqrt{\frac{4N_0}{2\pi A^2}} \exp\left(-\frac{A^2}{8N_0}\right) \tag{8.39a}$$

Hence,

$$P_e \approx \tfrac{1}{2}\left[1 + \sqrt{\frac{4N_0}{2\pi A^2}}\right] \exp\left(-\frac{A^2}{8N_0}\right)$$
$$\approx \tfrac{1}{2}\exp\left(-\frac{A^2}{8N_0}\right) \quad \text{if } A^2 \gg N_0 \tag{8.39b}$$

where $N_0 = \eta B_T$ and B_T is the bandwidth of the bandpass filter.

The probability of error for the noncoherent ASK receiver will always be higher than the error probability of a coherent receiver operating with the same signal power, signaling speed, and noise psd. However, the noncoherent receiver is much simpler than the coherent receiver.

In order to obtain optimum performance, the threshold value at the receiver should be adjusted according to Equation (8.37b) as the signal level changes. Furthermore, the filters used in the receiver should be discharged, via auxiliary circuitry, at the end of each bit interval in order to reduce inter symbol interference. While the resulting circuit is no longer a linear time-invariant filter, it does act like a liner time-invariant filter in between discharge intervals. For such a rapidly discharging filter, the filter bandwidth is no longer critical with respect to intersymbol interference.

In the noncoherent receiver shown in Figure 8.9, we have assumed that timing information is available at the receiver. This timing information is usually extracted from the envelope of the received signal using a technique similar to the one described in Chapter 5, Section 5.7.

It is worth noting here that in the noncoherent ASK scheme, the probability of incorrectly decoding "1" is different from the probability of incorrectly decoding "0." Thus the noncoherent ASK scheme results in a nonsymmetric binary channel (Chapter 4, Section 4.5).

Example 8.3. Binary data is transmitted over an RF bandpass channel with a usable bandwidth of 10 MHz at a rate of (4.8) (10^6) bits/sec using an ASK signaling method. The carrier amplitude at the receiver antenna is 1 mv and the noise power spectral density at the receiver input is 10^{-15} watt/Hz. (a) Find the error probability of a coherent receiver, and (b) find the error probability of a noncoherent receiver.

Solution

(a) The bit error probability for the coherent demodulator is

$$P_e = Q\left(\sqrt{\frac{A^2T_b}{4\eta}}\right); \qquad A = 1\text{ mv}, \quad T_b = 10^{-6}/4.8$$

$$\eta/2 = 10^{-15}\text{ watt/Hz}$$

Hence, $P_e = Q(\sqrt{26}) \simeq 2(10^{-7})$.

(b) The noise power at the filter output is $N_0 = 2\eta r_b = 1.92(10^{-8})$ and $A^2 = 10^{-6}$. Hence, $A^2 \gg N_0$ and we can use the approximations given in the preceding paragraphs for P_{e0} and P_{e1}

$$P_{e1} = Q\left(\sqrt{\frac{A^2}{4N_0}}\right) = Q(3.61) = 0.0002$$

$$P_{e0} = \exp\left(-\frac{A^2}{8N_0}\right) \approx 0.0015$$

Hence, $P_e = \frac{1}{2}(P_{e0} + P_{e1}) = 0.00085$.

8.4 BINARY PSK SIGNALING SCHEMES

Phase-shift keying, or discrete phase modulation, is another technique available for communicating digital information over bandpass channels. In PSK signaling schemes the waveforms $s_1(t) = -A \cos \omega_c t$ and $s_2(t) = A \cos \omega_c t$ are used to convey binary digits 0 and 1, respectively. The binary PSK waveform $Z(t)$ can be described by

$$Z(t) = D(t)(A \cos \omega_c t)$$

where $D(t)$ is a random binary waveform with period T_b and levels -1 and 1. The only difference between the ASK and PSK waveform is that in the ASK scheme the carrier is switched on and off whereas in the PSK scheme the carrier is switched between levels $+A$ and $-A$. The power spectral density of the PSK signal can be shown to be

$$G_Z(f) = \frac{A^2}{4}[G_D(f - f_c) + G_D(f + f_c)]$$

where

$$G_D(f) = \frac{\sin^2 \pi f T_b}{\pi^2 f^2 T_b} \tag{8.40}$$

Comparison of Equation (8.40) with Equation (8.30) reveals that the shapes of the psd of the binary PSK signal and the ASK signal are similar. The only difference is that the PSK spectrum does not have an impulse at the carrier frequency. The bandwidth requirement of the PSK signal is the same as that of the ASK signal. The similarity between the ASK and PSK is somewhat misleading. The ASK is a linear modulation scheme whereas the PSK, in the general case, is a nonlinear modulation scheme.

The primary advantage of the PSK signaling scheme lies in its superior performance over the ASK scheme operating under the same peak power limitations and noise environment. In the following sections we will derive expressions for the probability of error for coherent and noncoherent PSK signaling schemes.

8.4.1 Coherent PSK

The transmitted bit sequence $\{b_k\}$ can be recovered from the PSK signal using the integrate and dump correlation receiver shown in Figure 8.7 with a local reference signal $s_2(t) - s_1(t) = 2A \cos \omega_c t$ that is synchronized in phase and frequency with the incoming signal. The signal components of the receiver output at $t = kT_b$ are

$$s_{01}(kT_b) = \int_{(k-1)T_b}^{kT_b} s_1(t)[s_2(t) - s_1(t)]\, dt = -A^2 T_b$$

$$s_{02}(kT_b) = \int_{(k-1)T_b}^{kT_b} s_2(t)[s_2(t) - s_1(t)]\, dt = A^2 T_b$$

The optimum threshold setting is $T_0^* = 0$, which is independent of the carrier strength at the receiver input. The probability of error P_e is given by

$$P_e = Q(\gamma_{\max}/2)$$

where

$$\gamma_{\max}^2 = \frac{2}{\eta} \int_0^{T_b} (2A \cos \omega_c t)^2 \, dt = \frac{4A^2 T_b}{\eta}$$

or

$$P_e = Q(\sqrt{A^2 T_b/\eta}) \tag{8.41}$$

The average signal power S_{av} and the signal energy per bit E_{av} for the PSK scheme are

$$S_{av} = A^2/2$$

and

$$E_{av} = (A^2/2)T_b$$

We can express the probability of error in terms of S_{av} and E_{av} as

$$P_e = (\sqrt{2S_{av}T_b/\eta}) \tag{8.42}$$

$$= Q(\sqrt{2E_{av}/\eta}) \tag{8.43}$$

Comparing the probability of error for the coherent PSK (Equation (8.42)) with the probability of error for the coherent ASK (Equation (8.32)), we see that for equal probability of error the average signal power of the ASK signal should be twice the average power of the PSK signal. That is, the coherent PSK scheme has a 3-dB power advantage over the coherent ASK scheme.

8.4.2 Differentially Coherent PSK

The differentially coherent PSK (DPSK) signaling scheme makes use of a clever technique designed to get around the need for a coherent reference signal at the receiver. In the DPSK scheme, the phase reference for demodulation is derived from the phase of the carrier during the preceding signaling interval, and the receiver decodes the digital information based on the differential phase. If the channel perturbations and other disturbances are slowly varying compared to the bit rate, then the phase of the RF pulses $s(t)$ and $s(t - T_b)$ are affected by the same manner, thus preserving the information contained in the phase difference. If the digital information had been *differentially encoded* in the carrier phase at the transmitter, the decoding at the receiver can be accomplished without a coherent local oscillator

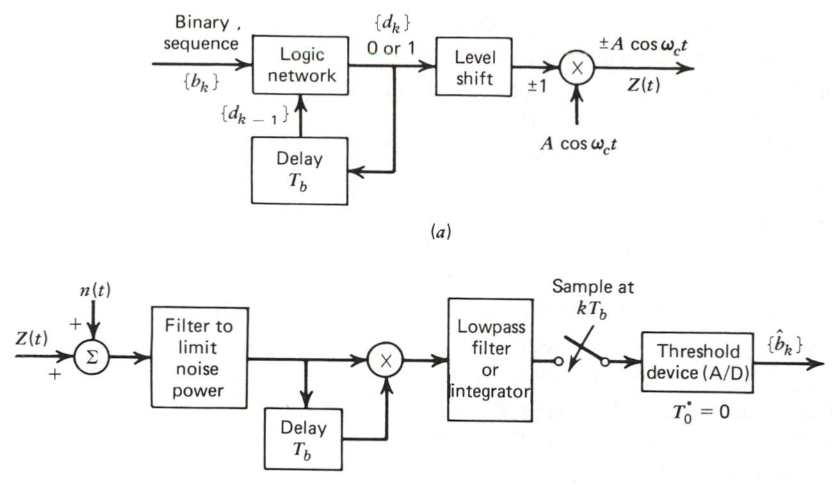

Figure 8.11 (a) DPSK modulator. (b) DPSK demodulator.

signal. The DPSK scheme may be thought of as the noncoherent version of the PSK scheme discussed in the preceding section.

Block diagrams of a DPSK modulator and demodulator are shown in Figures 8.11*a* and 8.11*b*, respectively. The differential encoding operation performed by the modulator is explained in Table 8.2. The encoding process starts with an arbitrary first bit, say 1, and thereafter the encoded bit stream d_k is generated by

$$d_k = d_{k-1}b_k \oplus \bar{d}_{k-1}\bar{b}_k$$

Table 8.2. Differential encoding and decoding

Input sequence (b_k)		1	1	0	1	0	0	0	1	1
Encoded sequence (d_k)	1[a]	1	1	0	0	1	0	1	1	1
Transmitted phase	0	0	0	π	π	0	π	0	0	0
Phase comparison output		+	+	−	+	−	−	−	+	+
Output bit sequence		1	1	0	1	0	0	0	1	1

[a] Arbitrary starting reference bit.

The differential sequence d_k then phase-shift keys a carrier with the phases 0 and π, as shown in Table 8.2.

The DPSK receiver correlates the received signal plus noise with a delayed version (delay = 1-bit duration) of the signal plus noise. The output of the correlator is compared with zero and a decision is made in favor of 1 or 0 depending on whether the correlator output is + or −, respectively. The reader can easily verify that the receiver recovers the bit sequence $\{b_k\}$ correctly, in the absence of noise, by assuring himself that the receiver essentially checks to see if the phase angles of the received carrier during two successive bit intervals are the same or different. With an initial angle of 0 (for the reference bit), the receiver output is 1 at the end of the kth signaling interval if the carrier phase is the same during the $(k-1)$st and the kth signaling intervals. If the phase angles are different, then the receiver output is 0. The last two rows in Table 8.2 illustrate that phase comparison detection at the receiver works correctly.

The noise performance of the DPSK might appear to be inferior compared to coherent PSK because the phase reference is contaminated by noise in the DPSK scheme. However, the perturbations in phase reference due to noise tend to cancel out and the degradation in performance is not too great. In the following paragraphs we will derive an expression for the probability of error for the DPSK scheme.

For purposes of analysis, let us assume that the carrier phase during the $(k-1)$st and the kth signaling intervals is 0, that is, $\phi_{k-1} = \phi_k = 0$. An error in decoding the kth bit occurs if the phase comparator output is negative. The input to the lowpass filter in Figure 8.11b can be written as

$$q(t) = [A \cos \omega_c t + n_f(t)][A \cos \omega_c t' + n_f(t')], \quad (k-1)T_b \leq t \leq kT_b$$

where $t' = t - T_b$, and $n_f(t)$ is the response of the front-end filter to $n(t)$. Substituting the quadrature representation

$$n_f(t) = n_c(t) \cos \omega_c t - n_s(t) \sin \omega_c t$$

in the preceding equation, we obtain

$$\begin{aligned} q(t) = {} & [A + n_c(t)] \cos \omega_c t [A + n_c(t')] \cos \omega_c t' \\ & - [A + n_c(t)] \cos \omega_c t [n_s(t') \sin \omega_c t'] \\ & - n_s(t) \sin \omega_c t [A + n_c(t') \cos \omega_c t'] \\ & + n_s(t) n_s(t') \sin \omega_c t \sin \omega_c t' \end{aligned}$$

The reader can verify that the lowpass filter output $V_0(kT_b)$ is given by [remember that $\omega_c T_b = k\pi (k \geq 2)$, hence $\sin \omega_c t = \sin \omega_c t'$ and $\cos \omega_c t = \cos \omega_c t'$]

$$V_0(kT_b) = c[A + n_c(t)][A + n_c(t')] + n_s(t) n_s(t')$$

where c is a positive constant $t = kT_b$, and $t' = (k-1)T_b$. The probability of error P_e is given by

$$P_e = P[V_0(kT_b) < 0] = P\left(\frac{1}{c} V_0(kT_b) < 0\right)$$

In order to simplify the expression for the probability of error, let us define

$$\alpha = A + \frac{n_c(t) + n_c(t')}{2}, \qquad \beta = \frac{n_c(t) - n_c(t')}{2}$$

$$\nu = \frac{n_s(t) + n_s(t')}{2}, \qquad \delta = \frac{n_s(t) - n_s(t')}{2}$$

We now have

$$\frac{1}{c} V_0(kT_b) = (\alpha^2 + \beta^2) - (\nu^2 + \delta^2)$$

$$P\left(\frac{1}{c} V_0(kT_b) < 0\right) = P(\alpha^2 + \beta^2 < \nu^2 + \delta^2)$$

$$= P(\sqrt{\alpha^2 + \beta^2} < \sqrt{\nu^2 + \delta^2})$$

If we denote $\sqrt{\alpha^2 + \beta^2}$ by X_1 and $\sqrt{\nu^2 + \delta^2}$ by X_2, then X_1 has a Rice pdf and X_2 has a Raleigh pdf, and the probability of error P_e has the form

$$P_e = P(X_1 < X_2) = \int_0^\infty P(X_2 > x_1 | X_1 = x_1) f_{X_1}(x_1)\, dx_1$$

where

$$P(X_2 > x_1 | X_1 = x_1) = \int_{x_1}^\infty f_{X_2}(x_2)\, dx_2$$

since X_1 and X_2 are statistically independent. The pdf's involved in the preceding expressions have the forms given in Equations (8.36*a*) and (8.36*b*) and the probability of error can be shown to be equal to (see Problem 8.17)

$$P_e = \tfrac{1}{2} \exp(-A^2 T_b / 2\eta) \tag{8.44}$$

The example given below shows that the DPSK scheme requires 1 dB more power than the coherent PSK scheme when the error probabilities of both systems are of the order of 10^{-4}. The slight increase in power requirements for the DPSK signaling method is more than offset by the fact that DPSK does not require a coherent reference signal at the receiver. Because of the fixed delay in the DPSK receiver, the system is locked on a specific signaling speed, thus precluding asynchronous data transmission. Another minor problem in DPSK schemes is that errors tend to propagate, at least to adjacent bits, due to the correlation between signaling waveforms and the noise over adjacent signaling intervals.

Example 8.4. Binary data is transmitted at a rate of 10^6 bits/sec over a microwave link having a bandwidth of 3 MHz. Assume that the noise power spectral density at the receiver input is $\eta/2 = 10^{-10}$ watt/Hz. Find the average carrier power required at the receiver input for coherent PSK and DPSK signaling schemes to maintain $P_e \leqslant 10^{-4}$.

Solution

The probability of error for the PSK scheme is

$$(P_e)_{\text{PSK}} = Q(\sqrt{2S_{av}T_b/\eta}) \leqslant 10^{-4},$$

This requires

$$\sqrt{2S_{av}T_b/\eta} \geqslant 3.75$$

or

$$(S_{av})_{\text{PSK}} \geqslant (3.75)^2(10^{-10})(10^6) = 1.48\ \text{dBm}$$

For the DPSK scheme we have

$$(P_e)_{\text{DPSK}} = \tfrac{1}{2}\exp[-(A^2T_b/2\eta)] \leqslant 10^{-4},$$

hence

$$S_{av}T_b/\eta \geqslant 8.517 \quad \text{or} \quad (S_{av})_{\text{DPSK}} \geqslant 2.313\ \text{dBm}$$

This example illustrates that the DPSK signaling scheme requires about 1 dB more power than the coherent PSK scheme when the error probability is of the order of 10^{-4}.

8.5 BINARY FSK SIGNALING SCHEMES

FSK signaling schemes find a wide range of applications in low-speed digital data transmission systems. Their appeal is mainly due to hardware advantages that result principally from the use of a noncoherent demodulation process and the relative ease of signal generation. As we will see later in this section, FSK schemes are not as efficient as PSK schemes in terms of power and bandwidth utilization. In the binary FSK signaling scheme, the waveforms $s_1(t) = A\cos(\omega_c t - \omega_d t)$ and $s_2(t) = A\cos(\omega_c t + \omega_d t)$ are used to convey binary digits 0 and 1, respectively. The information in an FSK signal is essentially in the frequency of the signal.

The binary FSK waveform is a continuous phase constant envelope FM waveform. The binary FSK waveform can be mathematically represented as follows:

$$Z(t) = A\cos\left(\omega_c t + \omega_d \int_{-\infty}^{t} D(t')\,dt' + \theta\right) \tag{8.45}$$

where $D(t)$ is a random binary waveform with levels $+1$ when $b_k = 1$ and -1 when $b_k = 0$, and θ is the phase angle of the carrier at time $t = 0$. The instantaneous frequency of the binary FSK signal is given by

$$\begin{aligned} f_i &= \frac{d}{dt}\,[\text{phase of } Z(t)] \\ &= \omega_c + \omega_d D(t) \end{aligned}$$

Since $D(t) = \pm 1$, the instantaneous frequency ω_i has two values: $\omega_i = \omega_c \pm \omega_d$.

The derivation of the power spectral density of the digital FM waveform is rather involved and hence we will look only at the results of the derivation shown in Figure 8.12 (see Lucky's book, Chapter 8, for a detailed derivation).

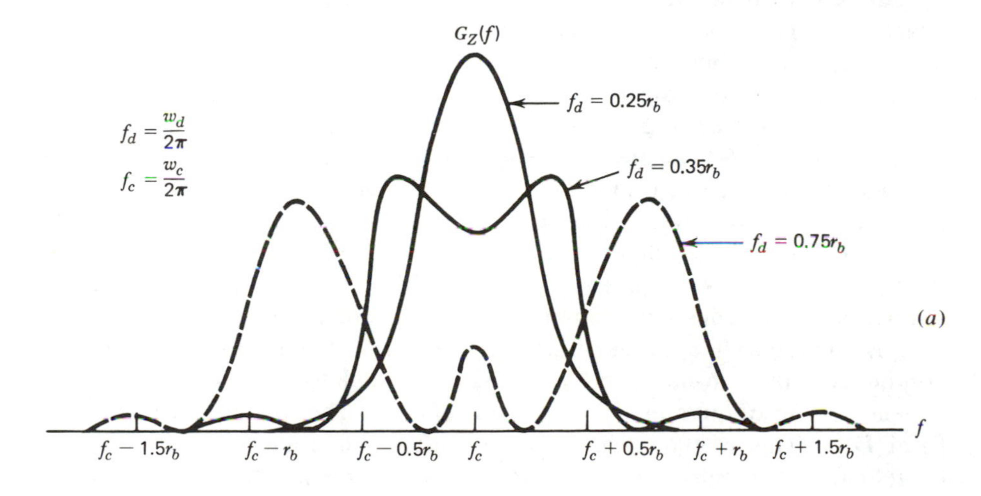

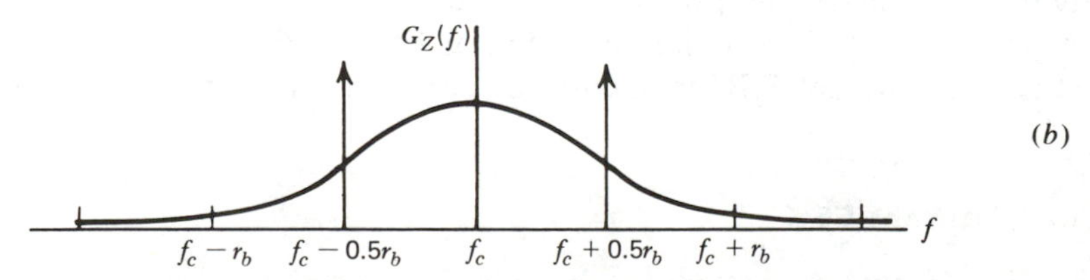

Figure 8.12 (*a*) Power spectral density of FSK signals. (*b*) Power spectral density of a binary FSK signal with $2f_d = r_b$.

The power spectral density curves displayed in Figure 8.12*a* exhibit the following characteristics: For low values of f_d/r_b the curve has a smooth roll off with a peak at the carrier frequency. The FSK signal bandwidth in this case is of the order of $2r_b$ Hz, which is the same as the order of bandwidth of the PSK signal. As f_d/r_b increases, major peaks in the power spectral density curve occur at the transmit frequencies $f_c + f_d$ and $f_c - f_d$ and the bandwidth of the signal exceeds $2r_b$, the bandwidth of the PSK signal. For large values of f_d/r_b, the FSK signal essentially consists of two interleaved ASK signals of differing carrier frequencies, say $f_c + f_d$ and $f_c - f_d$. Further, when $2f_d = mr_b$, m an integer, the psd has impulses corresponding to discrete frequency sinusoidal components as shown in Figure 8.12*b*. In general, we can say that the bandwidth of the FSK signal is greater than the bandwidth of the ASK and the PSK signals.

As mentioned earlier, the binary FSK waveform given in Equation (8.45) is a continuous phase waveform. In order to maintain phase continuity, the phase at every transition is made to be dependent on the past data sequence. To visualize how one could generate a continuous phase constant envelope FSK signal, consider the following waveform construction procedure: The sequence $\{b_k\}$ is used to generate a sequence of segmented cosine waveforms $A\cos(\omega_c t + \omega_k t + \theta_k)$, where $\omega_k = +\omega_d$ if $b_k = 1$ and $\omega_k = -\omega_d$ if $b_k = 0$. The FM waveform given in Equation (8.45) is then constructed by specifying the sequence $\{\theta_k\}$ as follows: Let θ_1 be arbitrarily set equal to some value θ. Then $\theta_2 = \theta + (\omega_1 + \omega_c)T_b$, $\theta_3 = \theta + (\omega_1 + \omega_c)T_b + (\omega_2 + \omega_c)T_b, \ldots,$ and $\theta_n = \theta + (\omega_1 + \omega_c)T_b + \cdots + (\omega_{n-1} + \omega_c)T_b$. By shifting the phase of the different segments, one obtains a continuous phase constant envelope FM wave.

It is also possible to generate a digital FM wave ignoring the phase continuity. This can be done by having two oscillators tuned to $\omega_c + \omega_d$ and $\omega_c - \omega_d$ whose outputs are directly controlled by the digital baseband waveform $D(t)$. This method gives rise to undesirable transients in addition to complicating the transmitter. For these reasons, this method is rarely used in practice.

The FSK signal $Z(t)$ can be demodulated using a coherent correlation receiver or using a suboptimal scheme consisting of bandpass filters and envelope detectors. Correlation detection of FSK signals is very seldom used; our study of coherent FSK is mainly for comparison purposes.

8.5.1 Coherent FSK

If the FSK signal is demodulated using the correlation receiver shown in Figure 8.7, the local carrier signal required is

$$s_2(t) - s_1(t) = A\cos(\omega_c t + \omega_d t) - A\cos(\omega_c t - \omega_d t)$$

The input to the A/D converter at sampling time $t = kT_b$ is $s_{01}(kT_b)$ or $s_{02}(kT_b)$, where

$$s_{02}(kT_b) = \int_0^{T_b} s_2(t)[s_2(t) - s_1(t)]\, dt$$

$$s_{01}(kT_b) = \int_0^{T_b} s_1(t)[s_2(t) - s_1(t)]\, dt$$

If the signal energy E_1 and E_2 are the same, then $s_{02}(kT_b) = -s_{01}(kT_b)$ and hence the receiver threshold setting is at 0. The probability of error P_e for the correlation receiver is given by (from Equation (8.21))

$$P_e = Q(\gamma_{\max}/2)$$

where

$$\gamma_{\max}^2 = \frac{2}{\eta}\int_0^{T_b} [s_2(t) - s_1(t)]^2\, dt$$

Substituting $s_2(t) = A\cos(\omega_c t + \omega_d t)$ and $s_1(t) = A\cos(\omega_c t - \omega_d t)$ and performing the integration, we get

$$\gamma_{\max}^2 = \frac{2A^2T_b}{\eta}\left(1 - \frac{\sin 2\omega_d T_b}{2\omega_d T_b} + \frac{1}{2}\frac{\sin[2(\omega_c + \omega_d)T_b]}{2(\omega_c + \omega_d)T_b} - \frac{1}{2}\frac{\sin[2(\omega_c - \omega_d)T_b]}{2(\omega_c - \omega_d)T_b} - \frac{\sin 2\omega_c T_b}{2\omega_c T_b}\right) \tag{8.46a}$$

If we make the following assumptions:

$$\omega_c T_b \gg 1, \qquad \omega_c \gg \omega_d$$

which are usually encountered in practical systems, then the last three terms in Equation (8.46a) may be neglected. We now have

$$\gamma_{\max}^2 = \frac{2A^2T_b}{\eta}\left(1 - \frac{\sin 2\omega_d T_b}{2\omega_d T_b}\right) \tag{8.46b}$$

The quantity $\gamma_{\max}^2$ in the preceding equation attains the largest value when the frequency offset ω_d is selected so that $2\omega_d T_b = 3\pi/2$. For this value of ω_d we find

$$\gamma_{\max}^2 = (2.42)(A^2T_b/\eta)$$

and

$$P_e = Q(\sqrt{0.61(A^2T_b/\eta)}) \tag{8.47}$$

Once again, if we define $S_{av} = A^2/2$ and $E_{av} = A^2T_b/2$, then we can express P_e as

$$P_e = Q(\sqrt{1.2S_{av}T_b/\eta})$$
$$= Q(\sqrt{1.2E_{av}/\eta}) \tag{8.48}$$

Comparison of the probability of error for the coherent FSK scheme with the error probability for coherent PSK scheme (Equation (8.42)) shows that coherent FSK requires about 2.2 dB more power than the coherent PSK scheme. The FSK signal also uses more bandwidth than the PSK signal. Thus coherent FSK does not have any advantages over the coherent PSK scheme.

8.5.2 Noncoherent FSK

Since the FSK scheme can be thought of as the transmission of two interleaved ASK signals (assuming that $2f_d = mr_b$, m an integer), the first with a carrier frequency $f_c - f_d$ and the second with carrier frequency $f_c + f_d$, it should be possible to detect the signal using two bandpass filters with center frequencies $f_c + f_d$ and $f_c - f_d$. Such a detection scheme is shown in Figure 8.13. The probability of error for the noncoherent FSK receiver can be derived easily using the results derived in Section 8.3.2 for the noncoherent ASK receiver. As a matter of fact, the derivation is somewhat simpler since we do not have to face the problem of calculating the optimum threshold setting. Because of symmetries, the threshold is set at zero in noncoherent FSK receivers.

Assuming that $s_1(t) = A\cos(\omega_c - \omega_d)t$ has been transmitted during the kth signaling interval, the pdf of the envelope $R_1(kT_b)$ of the bottom filter is

$$f_{R_1|s_1(t)}(r_1) = \frac{r_1}{N_0} I_0\left(\frac{Ar_1}{N_0}\right) \exp\left(-\frac{r_1^2 + A^2}{2N_0}\right), \quad r_1 > 0$$

where $N_0 = \eta B_T$, and B_T is the filter bandwidth. The top filter responds to noise alone and therefore $R_2(kT_b)$ has a Rayleigh pdf given by

$$f_{R_2|s_1(t)}(r_2) = \frac{r_2}{N_0} \exp\left(\frac{-r_2^2}{2N_0}\right), \quad r_2 > 0$$

An error occurs when $R_2 > R_1$, and this error probability is obtained as

$$\begin{aligned} P\,[\text{error}|s_1(t)\text{sent}] \\ &= P(R_2 > R_1) \\ &= \int_0^\infty f_{R_1|s_1}(r_1)\left[\int_{r_1}^\infty f_{R_2|s_1}(r_2)\,dr_2\right] dr_1 \end{aligned} \tag{8.49}$$

since the random variables $R_1(kT_b)$ and $R_2(kT_b)$ will be independent if $f_d = mr_b/4$, where m is an integer (Problem 8.24). By symmetry, we have $P[\text{error}|s_1(t)\text{ sent}] = P[\text{error}|s_2(t)\text{ sent}]$ so that

$$P(\text{error}) = P_e = P[\text{error}|s_1(t)\text{ sent}]$$

Substituting the appropriate pdf's in Equation (8.49) and carrying out the

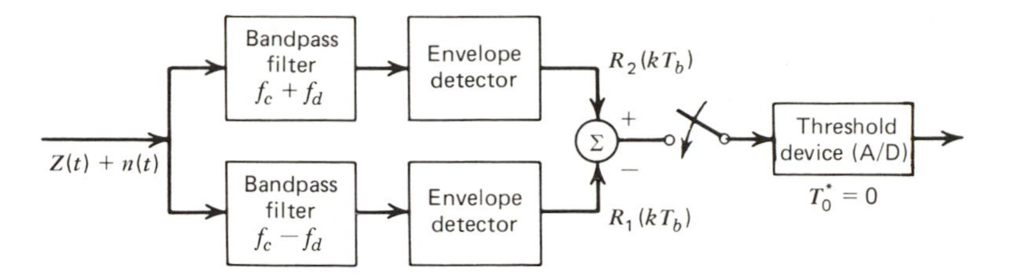

Figure 8.13 Noncoherent demodulation of binary FSK signals.

integration (with the help of a table of definite integrals), we obtain

$$P_e = \tfrac{1}{2}\exp(-A^2/4N_0) \tag{8.50}$$

The filter bandwidth is usually of the order of $2/T_b$ and hence N_0 in Equation (8.50) is approximately equal to $2\eta/T_b$.

The error probability for a noncoherent FSK receiver will be higher than the error probability of a coherent FSK receiver. However, because of its simplicity, the noncoherent FSK scheme is widely used in practice.

Example 8.5. Binary data is transmitted over a telephone line with usable bandwidth of 2400 Hz using the FSK signaling scheme. The transmit frequencies are 2025 and 2225 Hz, and the data rate is 300 bits/sec. The average signal-to-noise power ratio at the output of the channel is 6 dB. Calculate P_e for the coherent and noncoherent demodulation schemes.

Solution. We are given $r_b = 300$, $f_c + f_d = 2225$ Hz, and $f_c - f_d = 2025$ Hz. Hence, $f_c = 2125$ Hz and $f_d = 100$ Hz. Before we use Equation (8.48) to obtain P_e, we need to make sure that the assumptions $\omega_c T_b \gg 1$, $\omega_c \gg \omega_d$, and $2\omega_d T_b \approx 3\pi/2$ are valid. The reader can verify that all these assumptions are satisfied. Now, we are given that $S/N = (A^2/2)/(2400\eta) = 4$ or $A^2T_b/\eta = 64$. Using Equation (8.48), we obtain

$$(P_e)_{\substack{\text{coh}\\ \text{FSK}}} = Q(\sqrt{(0.61)64}) \approx 10^{-9}$$

For the noncoherent scheme, P_e is given by Equation (8.50) as

$$(P_e)_{\substack{\text{noncoh}\\ \text{FSK}}} = \tfrac{1}{2}\exp\left(-\frac{A^2T_b}{8\eta}\right) = \frac{e^{-8}}{2} = 1.68(10^{-4})$$

8.6 COMPARISON OF DIGITAL MODULATION SYSTEMS

We have developed formulas in the preceding sections that relate the performance of various modulation schemes, as measured by the probability of error, to parameters of the system, such as signaling rate, noise power spectral density, and signal power. We also discussed the complexity of equipment required to generate, transmit, and demodulate the different types of modulated signals. In this section we will attempt to compare the performance of various digital modulation schemes.

We begin our comparison by emphasizing that the choice of a modulation method depends on the specific application. The choice may be based on relative immunity to noise and channel impairments (such as nonlinearities, phase jitter, fading, and frequency offset), simplicity of equipment, and compatibility with other equipment already in place in the system. While it is not our intent to compare modulation systems under all conditions cited above, we will however attempt to provide the reader with some guidelines that might be useful in comparing and selecting a modulation scheme for a particular application. We will compare systems operating at the same data rate (r_b), probability of error (P_e), and noise environment.

8.6.1 Bandwidth Requirements

If one is interested in high speed data transmission over a noisy bandpass channel, then vestigial-sideband (VSB) modulation with baseband signal shaping is a better choice than ASK, PSK, or FSK schemes for efficient bandwidth utilization. Bandwidth requirements of VSB schemes with baseband signal shaping are of the order of r_b. The bandwidth requirements of ASK and PSK schemes are of the order of $2r_b$, whereas the bandwidth of the FSK signal is somewhat larger than $2r_b$. Thus if bandwidth is of primary concern, the FSK scheme is generally not considered.

8.6.2 Power Requirements

The power requirements of various schemes can be compared using the relationships derived in the preceding sections. These relationships are summarized in Table 8.3 and plots of the probability of error P_e versus $A^2T_b/2\eta$ are shown in Figure 8.14. The horizontal axis in Figure 8.14 should be read as peak received (or transmitted) power and the peak power A^2 is the same for all schemes. The error probability in most practical systems is in the range of 10^{-4} to 10^{-7} and hence we will do our comparison of power requirements assuming that $10^{-7} < P_e < 10^{-4}$.

Table 8.3. Comparison of Binary digital modulation schemes

Scheme	$s_1(t)$, $s_2(t)$	BW	P_e	S/N for $P_e = 10^{-4}$ (dB)	Equipment complexity	Comments
Coherent ASK	$s_1(t) = A\cos\omega_c t$ $s_2(t) = 0$ $\omega_c = k2\pi r_b$ k-integer	$\simeq 2r_b$	$Q\left(\sqrt{\frac{A^2T_b}{4\eta}}\right)$	14.45	Moderate	Rarely used $T_0^* = A^2T_b/4$
Noncoh. ASK	Same as above	$\simeq 2r_b$	$\frac{1}{2}\exp\left\{-\frac{A^2T_b}{16\eta}\right\}$	18.33	Minor	$T_0^* = A/2$ $P_{e0} \neq P_{e1}$
Coherent FSK	$s_1(t) = A\cos(\omega_c - \omega_d)t$ $s_2(t) = A\cos(\omega_c + \omega_d)t$ $2\omega_d = 1.5\pi r_b$	$>2r_b$	$Q\left(\sqrt{\frac{0.61A^2T_b}{\eta}}\right)$	10.6	Major	Seldom used; performance does not justify complexity $T_0^* = 0$
Noncoh. FSK	Same as above $2\omega_d = (k2\pi)r_b$	$>2r_b$	$\frac{1}{2}\exp\left\{-\frac{A^2T_b}{8\eta}\right\}$	15.33	Minor	Used for slow speed data transmission; poor utilization of power and bandwidth $T_0^* = 0$.
Coherent PSK	$s_1(t) = A\cos\omega_c t$ $s_2(t) = -A\cos\omega_c t$ $\omega_c = k2\pi r_b$	$\simeq 2r_b$	$Q\left(\sqrt{\frac{A^2T_b}{\eta}}\right)$	8.45	Major	Used for high speed data transmission. $T_0^* = 0$; best overall performance, but requires complex equipment
DPSK	Same as above with differential coding	$\simeq 2r_b$	$\frac{1}{2}\exp\left(-\frac{A^2T_b}{2\eta}\right)$	9.30	Moderate	Most commonly used in medium speed data transmission. $T_0^* = 0$; errors tend to occur in pairs

P_e—Prob. of error; A—carrier amplitude at receiver input; $\eta/2$—two-sided noise psd; T_b—bit duration; r_b—bit rate; $f_c = \omega_c/2\pi$ = carrier frequency; T_0^*—threshold setting; $S/N = A^2/2\eta r_b$; $P_{e0} = P$ (error|0 sent); $P_{e1} = P$ (error|1 sent).

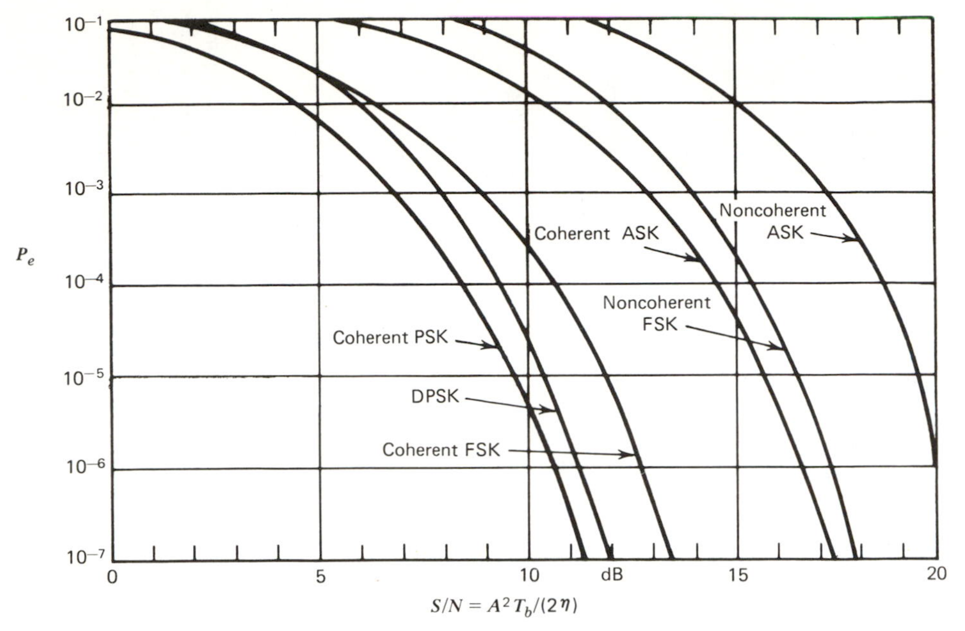

Figure 8.14 Probability of error for binary digital modulation schemes. (Note that the average signal power for ASK schemes is $A^2/4$, whereas it is $A^2/2$ for other schemes).

Plots in Figure 8.14 reveal that a coherent PSK signaling scheme requires the least amount of power followed by DPSK, coherent FSK, coherent ASK, noncoherent FSK, and noncoherent ASK signaling schemes. If the comparison is done in terms of average power requirements, then the ASK schemes require about the same amount of power as the FSK schemes. Since the cost of transmitting and receiving equipment depends more upon the peak power requirements than average power requirements, the comparison is usually made on the basis of peak power requirements. Thus, if the peak power requirement is of primary concern, then ASK schemes are not used.

It must be pointed out here that three of the most widely used digital modulation schemes are PSK, DPSK and noncoherent FSK. The power requirement of DPSK is approximately 1 dB more than coherent PSK, and the noncoherent FSK requires about 7 dB more power than coherent PSK. The reader may at this point ask the significance of say a 1 to 2 dB increase in power requirements. Industry sources claim that, in a large communication network, every 1 dB saving in power will result in savings of many millions of dollars annually.

8.6.3 Immunity to Channel Impairments

In selecting a signaling scheme, one should consider if the scheme is to some degree immune to channel impairments such as amplitude nonlinearities and fading (slow random variations in channel characteristics.) The FSK and PSK schemes are constant amplitude signals, and the threshold setting in the receiver does not depend on the received signal level. In the ASK scheme the receiver threshold setting depends on the received signal level and has to be changed as the received signal level changes. Thus ASK schemes are more sensitive to variations in received signal level due to changes in channel characteristics.

If the communication channel has fading, then noncoherent schemes have to be used because of the near impossibility of establishing a coherent reference at the receiver under fading channel conditions. However, if the transmitter has serious power limitations (as in the case of remote data transmission from space vehicles with limited energy storage and power generation capabilities), then a coherent scheme may have to be considered since coherent schemes use less power than noncoherent schemes for a given data rate and probability of error.

8.6.4 Equipment Complexity

There is very little difference in the complexity of transmitting equipment for the PSK, FSK, and ASK signals. At the receiver, the complexity depends on whether a coherent or noncoherent demodulation method is used. Hardware implementations of coherent demodulation schemes are more complex. Among the noncoherent schemes, DPSK is more complex than noncoherent FSK, which is more complex than noncoherent ASK. Complexity of equipment will increase the cost.

Summary. It must be obvious to the reader by now that there are a large number of factors that must be taken into account in the selection of a particular type of signaling scheme for a specific application. However, the following broad guidelines could be used to simplify the selection procedure:

1. If bandwidth is a premium quantity, then the most desirable signaling scheme is VSB with baseband signal shaping, and the least desirable scheme is FSK.
2. If power requirements are most important, then coherent PSK or DPSK is most desirable while ASK schemes are least desirable.

3. If equipment complexity is a limiting factor, then noncoherent demodulation schemes are preferrable to coherent schemes.

8.7 *M*-ARY SIGNALING SCHEMES

In Section 5.4 of Chapter 5, we saw that M-ary signaling schemes can be used for reducing the bandwidth requirements of baseband PAM data transmission systems. M-ary signaling schemes can be used in conjunction with digital modulation techniques also. Here, one of M $(M > 2)$ signals $s_1(t), s_2(t), \ldots, s_M(t)$ is transmitted during each signaling interval of duration T_s. These signals are generated by changing the amplitude, phase, or frequency of a carrier in M discrete steps. Thus we can have M-ary ASK, M-ary PSK, and M-ary FSK digital modulation schemes. M-ary digital modulation schemes are preferred over binary digital modulation schemes for transmitting digital information over bandpass channels when one wishes to conserve bandwidth (at the expense of increasing power requirements), or to conserve power (at the expense of increasing bandwidth requirements).

In practice, we seldom find a channel that has the exact bandwidth required for transmitting the output of a source using binary signaling schemes. When the bandwidth of the channel is less, M-ary digital modulation schemes are used to transmit the information over the bandpass channel. If the channel has a bandwidth much larger than the bandwidth required for transmitting the source output using binary modulation techniques, M-ary schemes may be used to utilize the additional bandwidth to provide increased immunity to channel noise. In this section, we will look at M-ary PSK schemes that are used for conserving bandwidth, and wideband M-ary FSK schemes that can be used for conserving power in digital modulation schemes.

In our discussion of M-ary schemes, we will assume that the input to the modulator is an independent sequence of equiprobable binary digits. We will further assume that the modulator takes blocks of λ binary digits and assigns one of M possible waveforms to each block ($M = 2^\lambda$).

8.7.1 *M*-ary Coherent PSK

In M-ary PSK systems, the phase of the carrier is allowed to take on one of M possible values $\phi_k = k2\pi/M$ $(k = 0, 1, 2, \ldots, M-1)$. Thus the M possible signals that would be transmitted during each signaling interval of duration T_s are

$$s_k(t) = A\cos(\omega_c t + k2\pi/M), \quad k = 0, 1, \ldots, M-1, 0 \leq t \leq T_s \tag{8.51}$$

We will assume that ω_c, the carrier frequency, is a multiple of $r_s (r_s = 1/T_s)$. The digital M-ary PSK waveform can be represented in the form

$$Z(t) = A \sum_{k=-\infty}^{\infty} g(t - kT_s) \cos(\omega_c t + \phi_k) \tag{8.52}$$

where $g(t)$ is a rectangular unit amplitude pulse with a duration T_s. The sequence of phase angles $\{\phi_k\}$ carries the digital information. We can rewrite Equation (8.52) as

$$\begin{aligned} Z(t) = A \cos \omega_c t \sum_{k=-\infty}^{\infty} (\cos \phi_k) g(t - kT_s) \\ - A \sin \omega_c t \sum_{k=-\infty}^{\infty} (\sin \phi_k) g(t - kT_s) \end{aligned} \tag{8.53}$$

which shows that the waveform $Z(t)$ is the difference of two AM signals using $\cos \omega_c t$ and $\sin \omega_c t$ as carriers. The power spectral density of $Z(t)$ is a shifted version of the power spectral density of the M-ary rectangular waveforms $\Sigma \cos \phi_k g(t - kT_s)$ and $\Sigma \sin \phi_k g(t - kT_s)$. The psd of these waveforms has a $(\sin x/x)^2$ form with zero crossings at $\pm kr_s$ Hz. Thus the bandwidth requirement of an M-level PSK signal will be of the order of $2r_s$ to $3r_s$ Hz.

If the information to be transmitted is an independent binary sequence with a bit rate of r_b, then the bandwidth required for transmitting this sequence using binary PSK signaling scheme is of the order of $2r_b$. Now, if we take blocks of λ bits and use an M-ary PSK scheme with $M = 2^\lambda$ and $r_s = r_b/\lambda$, the bandwidth required will be of the order of $2r_s = 2r_b/\lambda$. Thus the M-ary PSK signaling scheme offers a reduction in bandwidth by a factor of λ over the binary PSK signaling scheme.

The M-ary PSK signal can be demodulated using a coherent demodulation scheme if a phase reference is available at the receiver. For purposes of illustration we will discuss the demodulation of four-phase PSK (also known as QPSK or quadrature PSK) in detail and then present the results for the general M-ary PSK.

In four-phase PSK, one of four possible waveforms is transmitted during each signaling interval T_s. These waveforms are:

$$\left.\begin{aligned} s_1(t) &= A \cos \omega_c t \\ s_2(t) &= -A \sin \omega_c t \\ s_3(t) &= -A \cos \omega_c t \\ s_4(t) &= A \sin \omega_c t \end{aligned}\right\} \text{ for } 0 \leq t \leq T_s \tag{8.54}$$

These waveforms correspond to phase shifts of 0°, 90°, 180°, and 270° as shown in the phasor diagram in Figure 8.15. The receiver for the system is shown in Figure 8.16. The receiver requires two local reference waveforms

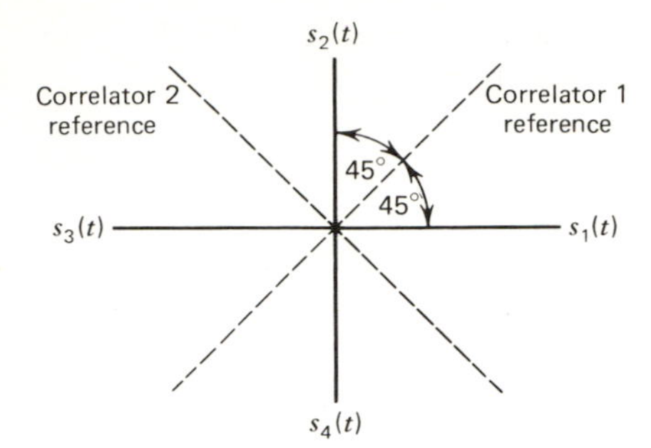

Figure 8.15 Phasor diagram for QPSK.

$A\cos(\omega_c t + 45°)$ and $A\cos(\omega_c t - 45°)$ that are derived from a coherent local carrier reference $A\cos\omega_c t$.

For purposes of analysis, let us consider the operation of the receiver during the signaling interval $(0, T_s)$. Let us denote the signal component at the output of the correlators by s_{01} and s_{02}, respectively, and the noise component by $n_0(t)$. If we assume that $s_1(t)$ was the transmitted signal during the signaling interval $(0, T_s)$, then we have

$$s_{01}(T_s) = \int_0^{T_s} (A\cos\omega_c t) A\cos\left(\omega_c t + \frac{\pi}{4}\right) dt$$
$$= \frac{A^2}{2} T_s \cos\frac{\pi}{4} = L_0$$

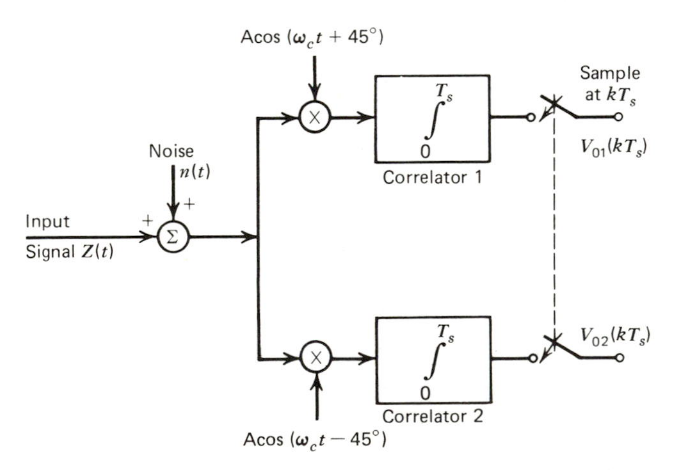

Figure 8.16 Receiver for QPSK scheme. Polarities of $V_{01}(kT_s)$ and $V_{02}(kT_s)$ determine the signal present at the receiver input during the kth signaling interval as shown in Table 8.4.

$$s_{02}(T_s) = \int_0^{T_s} (A \cos \omega_c t) A \cos\left(\omega_c t - \frac{\pi}{4}\right) dt$$
$$= \frac{A^2}{2} T_s \cos \frac{\pi}{4} = L_0$$

Table 8.4 shows s_{01} and s_{02} corresponding to each of the four possible signals $s_1(t)$, $s_2(t)$, $s_3(t)$, and $s_4(t)$.

Output signal levels shown in Table 8.4 indicate that the transmitted signal can be recognized from the polarities of the outputs of both correlators (i.e., the threshold levels are zero). In the presence of noise, there will be some probability that an error will be made by one or both correlators. An expression for the probability of incorrectly decoding the transmitted signal can be derived as follows.

The outputs of the correlators at time $t = T_s$ are

$$V_{01}(T_s) = s_{01}(T_s) + n_{01}(T_s)$$
$$V_{02}(T_s) = s_{02}(T_s) + n_{02}(T_s)$$

where $n_{01}(T_s)$ and $n_{02}(T_s)$ are zero mean Gaussian random variables defined by

$$n_{01}(T_s) = \int_0^{T_s} n(t) A \cos(\omega_c t + 45°)\, dt$$

$$n_{02}(T_s) = \int_0^{T_s} n(t) A \cos(\omega_c t - 45°)\, dt,$$

and $n(t)$ is a zero mean Gaussian random process with a power spectral density of $\eta/2$. With our assumption that $\omega_c = k2\pi r_s$ (k an integer > 0), we can show that $n_{01}(T_s)$ and $n_{02}(T_s)$ are independent Gaussian random variables with equal variance N_0 given by (see Problems 8.1, and 8.24)

$$N_0 = \frac{\eta}{4} A^2 T_s \tag{8.55}$$

Let us now calculate the probability of error assuming that $s_1(t)$ was the

Table 8.4. Output signal levels at sampling times.

	Input			
Output	$s_1(t)$	$s_2(t)$	$s_3(t)$	$s_4(t)$
$s_{01}(kT_s)$	L_0	$-L_0$	$-L_0$	L_0
$s_{02}(kT_s)$	L_0	L_0	$-L_0$	$-L_0$

transmitted signal. If we denote the probability that correlator 1 (Figure 8.16) makes an error by P_{ec1}, then

$$\begin{aligned} P_{ec1} &= P(n_{01}(T_s) < -L_0) \\ &= P(n_{01}(T_s) > L_0) \\ &= Q\left(\frac{L_0}{\sqrt{N_0}}\right) = Q\left(\sqrt{\frac{A^2T_s}{2\eta}}\right) \end{aligned} \tag{8.56}$$

By symmetry, the probability that the correlator 2 makes an error is

$$P_{ec2} = P_{ec1} = Q(\sqrt{A^2T_s/2\eta}) \tag{8.57}$$

The probability P_c that the transmitted signal is received correctly is

$$\begin{aligned} P_c &= (1 - P_{ec1})(1 - P_{ec2}) \\ &= 1 - 2P_{ec1} + P_{ec1}^2 \end{aligned}$$

We have made use of the fact that the noise outputs of the correlators are statistically independent at sampling times, and that $P_{ec1} = P_{ec2}$. Now, the probability of error P_e for the system is

$$\begin{aligned} P_e &= 1 - P_c \\ &= 2P_{ec1} - P_{ec1}^2 \\ &\simeq 2P_{ec1} \end{aligned}$$

since P_{ec1} will normally be $\ll 1$. Thus for the QPSK system we have

$$P_e = 2Q(\sqrt{A^2T_s/2\eta}) \tag{8.58}$$

We can extend this result to the M-ary PSK signaling scheme when $M > 4$. In the general case of this scheme, the receiver consists of a phase discriminator—a device whose output is directly proportional to the phase of the incoming carrier plus noise measured over a signaling interval. The phase of the signal component at the receiver input is determined as θ_k if the phase discriminator output $\theta(t)$ at $t = kT_s$ is within $\pm\pi/M$ of θ_k. Thus the receiver makes an error when the magnitude of the noise-induced phase perturbation exceeds π/M (see Figure 8.17). A detailed derivation of an expression for the probability of error in an M-ary PSK scheme using ideal phase detection is given in References 1 and 8.

We will simply note here that the probability of error in an optimum M-ary PSK signaling scheme can be approximated by (see Reference 8, Chapter 14)

$$P_e \simeq 2Q\left(\sqrt{\frac{A^2T_s}{\eta}\sin^2\frac{\pi}{M}}\right), \quad M \geq 4 \tag{8.59}$$

when the signal-to-noise power ratio at the receiver input is large. We are now

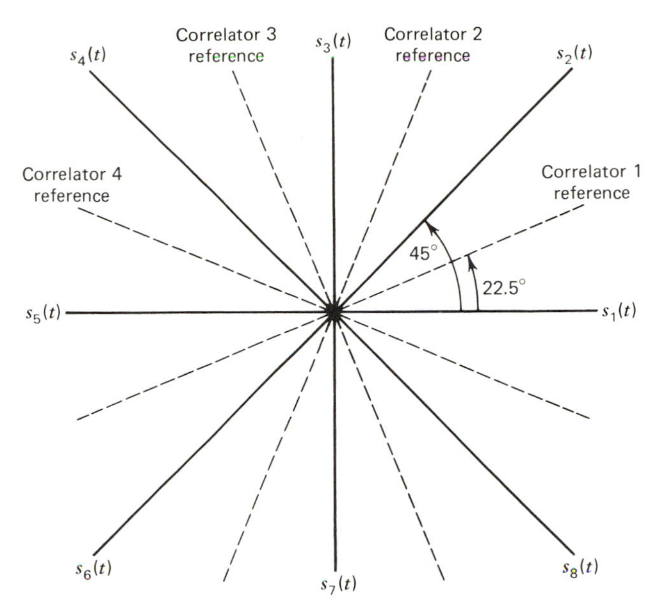

Figure 8.17 Phasor diagram for M–ary PSK; $M = 8$.

ready to compare the power-bandwidth trade off when we use an M-ary PSK to transmit the output of a source emitting an independent sequence of equiprobable binary digits at a rate of r_b. We have already seen that if $M = 2^\lambda$ (λ an integer), the M-ary PSK scheme reduces the bandwidth by a factor of λ over the binary PSK scheme. It is left as an exercise for the reader to show that the ratio of the average power requirements of an M-ary PSK scheme $(S_{av})_M$ and the average power requirement of a binary PSK scheme $(S_{av})_b$ are given by

$$\frac{(S_{av})_M}{(S_{av})_b} = \left(\frac{z_1^2}{z_2^2}\right)\frac{1}{\lambda \sin^2(\pi/M)} \tag{8.60a}$$

where z_1 and z_2 satisfy $Q(z_1) = P_e/2$ and $Q(z_2) = P_e$, respectively. If P_e is very small, then z_1 will be approximately equal to z_2 and we can rewrite Equation (8.60a) as

$$\frac{(S_{av})_M}{(S_{av})_b} \approx \frac{1}{\lambda \sin^2(\pi/M)} \tag{8.60b}$$

Typical values of power bandwidth requirements for binary and M-ary schemes are shown in Table 8.5, assuming that the probability of error is equal to 10^{-4} and that the systems are operating in identical noise environments.

Table 8.5. Comparison of power-bandwidth requirements for M-ary PSK scheme. $P_e = 10^{-4}$.

Value of M	$\frac{(\text{Bandwidth})_M}{(\text{Bandwidth})_b}$	$\frac{(S_{av})_M}{(S_{av})_b}$
4	0.5	0.34 dB
8	0.333	3.91 dB
16	0.25	8.52 dB
32	0.2	13.52 dB

Values shown in the table indicate that the QPSK scheme offers the best trade off between power and bandwidth requirements. For this reason QPSK is very widely used in practice. For $M > 8$, power requirements become excessive and hence PSK schemes with $M > 8$ are very seldom used in practice. It must be pointed out here that M-ary PSK schemes require considerably more complex equipment than binary PSK schemes for signal generation and demodulation.

The results shown in Table 8.5 were derived under the assumption that the binary PSK and the M-ary PSK schemes operate with the *same symbol error probability* P_e. If the comparison is to be done with the same *bit error probability* P_{eb} for all schemes, then P_e should be modified according to Equation (5.57a) or (5.57b).

8.7.2 *M*-ary Differential PSK

M-ary PSK signals can be differentially encoded and demodulated using the phase comparison detection scheme discussed in Section 8.5. As an example of an M-ary differential PSK signaling scheme, let us consider the case where $M = 4$. The PSK signal with $M = 4$ given in Equation (8.53) can be thought of as two binary PSK signals using $\sin \omega_c t$ and $\cos \omega_c t$ as carriers. The four-phase PSK signal can be differentially encoded by encoding its two constituent binary PSK signals differentially, as explained in Table 8.2. The receiver for a four-phase differential PSK scheme consists of essentially two biphase comparison detectors, as shown in block diagram form in Figure 8.18.

Comparison of Figures 8.18 and 8.19 reveals that the receiver for the differential four-phase PSK signaling scheme uses a delayed version of the received signal as its phase reference. This principle was discussed earlier when we were discussing the binary DPSK signaling scheme in Section 8.4.2.

The performance of the four-phase differential PSK can be analyzed using

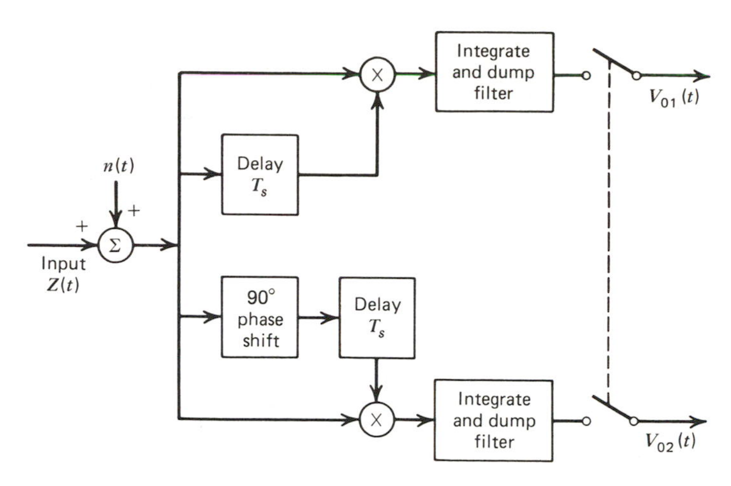

Figure 8.18 Receiver for four-phase differential PSK.

a procedure that combines the results of Sections 8.4.2 and 8.5. The derivation of an expression for the probability of error for M-ary DPSK is rather involved and we will simply state the following expression (Reference 1) for the probability of error in an M-ary differential PSK scheme:

$$P_e \simeq 2Q\left(\sqrt{\frac{A^2T_s}{\eta} 2\sin^2\left(\frac{\pi}{2M}\right)}\right) \tag{8.61}$$

Comparison of P_e for the M-ary DPSK scheme with P_e for the M-ary PSK scheme shows that differential detection increases power requirements by a factor of

$$\frac{\sin^2(\pi/M)}{2\sin^2(\pi/2M)} \approx 2 \text{ for large values of } M$$

With $M = 4$, the increase in power requirement is about 2 dB. This slight increase is more than offset by the simplicity of the equipment needed to handle the four-phase DPSK signal.

Figure 8.19 shows block diagrams of one of the earliest and most successful modems (modulator and demodulator) that uses a four-phase DPSK signaling scheme for transmitting data over (equalized) voice grade telephone lines. The binary data to be transmitted is grouped into blocks of two bits called *dibits*, and the resulting four possible combinations 00, 01, 10, and 11 differentially phase modulate the carrier. The data rate is fixed at 2400 bits/sec and the carrier frequency is 1800 Hz. The differential PSK waveform has the form

$$Z(t) = A\sum_k g(t - kT_s)\cos(\omega_c t + \phi_k)$$

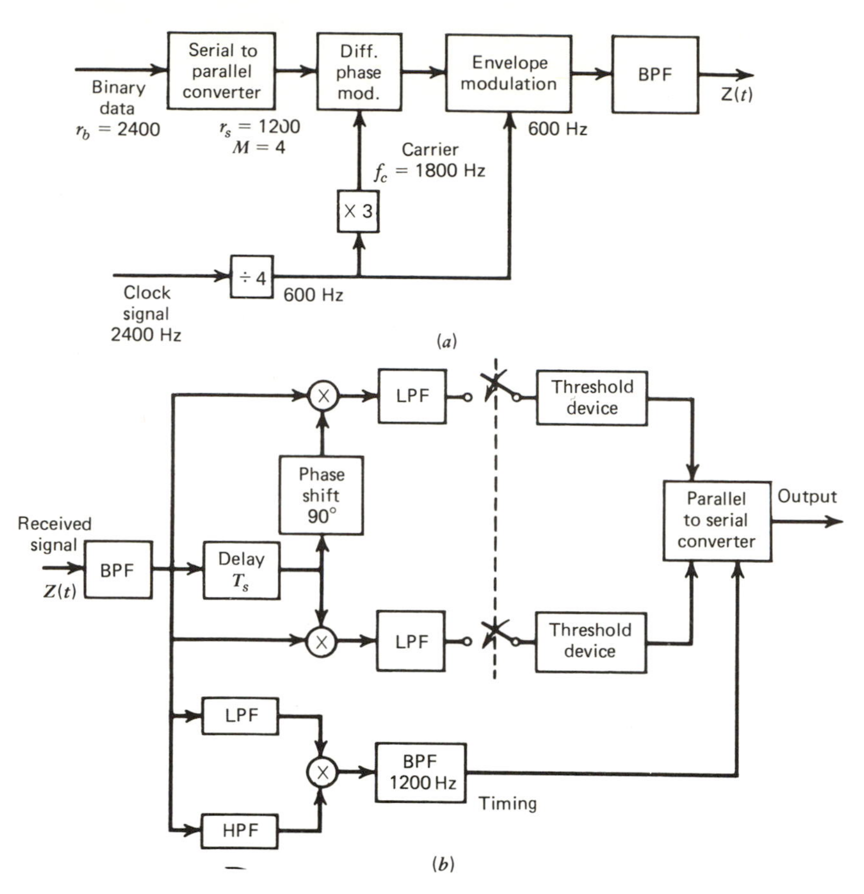

Figure 8.19 (*a*) Transmitter for differential PSK. (*b*) Receiver for differential PSK.

where

$$g(t) = \begin{cases} \frac{1}{2}(1 + \cos \pi r_s t) & \text{for } -T_s \leq t \leq T_s \\ 0 & \text{elsewhere, } T_s = 0.8333 \text{ msec} \end{cases}$$

The pulse shape $g(t)$ described above has negligible power content for $|f| > r_s$, thus the bandwidth of the transmitted signal is of the order of $2r_s$ or 2400 Hz. The non-rectangular shape for $g(t)$ conveys useful timing information to the receiver without introducing excessive ISI. Because of the differential phase-shift scheme (Table 8.6), the modulated waveform undergoes a phase change every T_s seconds. This feature along with the shape of $g(t)$ produces discrete frequency components at $f_c + 600$ and $f_c - 600$ that are used at the receiver to generate the 1200 Hz timing signal.

Table 8.6. Differential coding and decoding of quadrature PSK signals.

Dibit	$\phi_k - \phi_{k-1}$	$\sin(\phi_k - \phi_{k-1})$	$\cos(\phi_k - \phi_{k-1})$
00	$+45°$	$+$	$+$
01	$+135°$	$+$	$-$
10	$-135°$	$-$	$-$
11	$-45°$	$-$	$+$

The receiver produces two outputs—the first output is proportional to $\sin(\phi_k - \phi_{k-1})$ and the second is proportional to $\cos(\phi_k - \phi_{k-1})$. The input bits can be decoded uniquely from the sign of these two outputs as shown in Table 8.6. Finally, the parallel to serial converter interleaves the dibits to yield a serial binary output. Tests have shown that this modem has an error probability less than 10^{-5} when the signal-to-noise power ratio at the output of the channel is about 15 dB.

8.7.3 *M*-ary Wideband FSK Scheme

In this section we look at the possibility of using an M-ary FSK scheme to conserve power at the expense of (increased) bandwidth. Let us consider an FSK scheme where the M transmitted signals $s_i(t)$ $(i = 1, 2, \ldots, M)$ have the following properties:

$$s_i(t) = \begin{cases} A \cos \omega_i t, & 0 \leq t \leq T_s \\ 0 & \text{elsewhere} \end{cases} \tag{8.62}$$

and

$$\int_0^{T_s} s_i(t)s_j(t) = \begin{cases} A^2T_s/2 & \text{if } i = j \\ 0 & \text{if } i \neq j \end{cases} \tag{8.63}$$

The signals are of duration T_s, have equal energy, and are orthogonal to each other over the interval $(0, T_s)$. The minimum bandwidth of this signal set is approximately equal to $Mr_s/2$. In order to achieve this minimum bandwidth, the closest frequency separation $\omega_d = |\omega_n - \omega_m|$, $m \neq n$ must satisfy $\omega_d \geq \pi r_s$. One such choice of frequencies is $\omega_1 = k\pi r_s$, k an integer, and $\omega_m = \omega_1 + (m-1)\pi r_s$ $(m = 2, 3, \ldots, M)$.

The optimum receiver for this orthogonal signal set consists of a bank of M integrate and dump (or matched) filters as shown in Figure 8.20. The optimum receiver samples the filter output at $t = kT_s$ and decides that $s_i(t)$ was present at its input during the kth signaling interval if

$$\max_j [Y_j(kT_s)] = Y_i(kT_s) \tag{8.64}$$

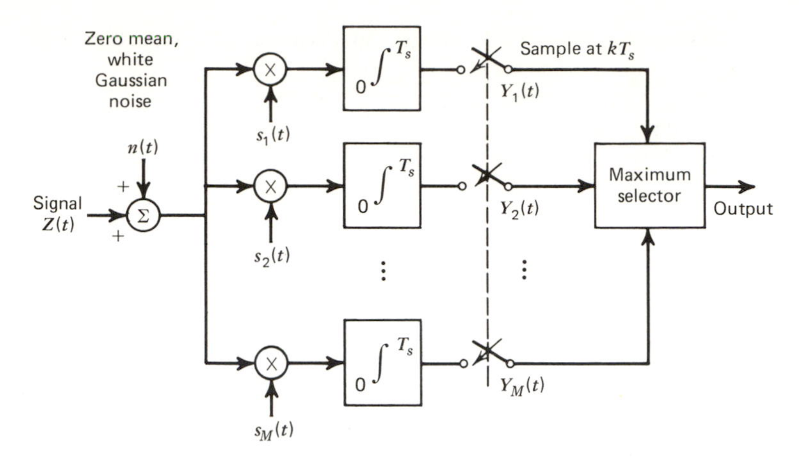

Figure 8.20 Structure of the receiver for an orthogonal (wideband FSK) signaling scheme.

To calculate the probability of error, let us consider the output of the filter at $t = T_s$ under the assumption that $s_1(t)$ was present at the receiver input during the interval $(0, T_s)$. The filter outputs are:

$$Y_j(T_s) = \int_0^{T_s} s_j(t)[n(t) + s_1(t)]\, dt, \quad j = 1, 2, \ldots, M \tag{8.65}$$

$$= \int_0^{T_s} s_j(t)s_1(t)\, dt + \int_0^{T_s} s_j(t)n(t)\, dt \tag{8.66}$$

$$= s_{0j}(T_s) + n_j(T_s)$$

where $s_{0j}(T_s)$ is the signal component of the j-th filter output and $n_j(T_s)$ is the noise component. The reader can verify that (Problems 8.1 and 8.24)

$$s_{0j}(T_s) = \begin{cases} A^2T_s/2 & \text{if } j = 1 \\ 0 & \text{if } j \geq 2 \end{cases} \tag{8.67}$$

and $n_j(T_s)$ $(j = 1, 2, \ldots, M)$ are independent Gaussian random variables with zero means and equal variances N_0 given by

$$N_0 = A^2T_s(\eta/4) \tag{8.68}$$

The receiver correctly decodes that $s_1(t)$ was present during the signaling interval $(0, T_s)$ if $Y_1(T_s)$ is larger than $Y_j(T_s)$ $(j = 2, 3, \ldots, M)$. We calculate this probability of correct decoding as

$$P_{c1} = P\{Y_2 < Y_1, Y_3 < Y_1, \ldots, Y_M < Y_1 | s_1 \text{ sent}\}$$
$$= \int_{-\infty}^{\infty} P\{Y_2 < y_1, \ldots, Y_M < y_1 | {}^{s_1 \text{ sent and}}_{Y_1 = y_1}\} f_{Y_1|s_1}(y_1)\, dy_1 \tag{8.69}$$

In the preceding step we made use of the identity

$$P(X < Y) = \int_{-\infty}^{\infty} P(X < y | Y = y) f_Y(y)\, dy$$

where X and Y are continuous, independent random variables.

We can simplify Equation (8.69) by making use of the fact that when s_1 is present during the signaling interval, $Y_2, Y_3, \ldots, Y_M$ are independent, zero mean Gaussian random variables with variance N_0, and hence the joint pdf of $Y_2, Y_3, \ldots, Y_M$ is given by

$$f_{Y_2, \ldots, Y_M | s_1;\, Y_1 = y_1}(y_2, \ldots, y_M) = \prod_{i=2}^{M} f_{Y_i}(y_i) \tag{8.70}$$

where

$$f_{Y_i}(y_i) = \frac{1}{\sqrt{2\pi N_0}} \exp\left(-\frac{y_i^2}{2N_0}\right), \quad -\infty < y_i < \infty$$

Substituting Equation (8.70) into (8.69), we have

$$P_{c1} = \int_{-\infty}^{\infty} \Bigg\{ \underbrace{\int_{-\infty}^{y_1} \cdots \int_{-\infty}^{y_1}}_{M-1 \text{ integrals}} \prod_{i=2}^{M} f_{Y_i}(y_i)\, dy_i \Bigg\} f_{Y_1|s_1}(y_1)\, dy_1$$

$$= \int_{-\infty}^{\infty} \left[\int_{-\infty}^{y_1} f_Y(y)\, dy \right]^{M-1} f_{Y_1|s_1}(y_1)\, dy_1 \tag{8.71}$$

where

$$f_Y(y) = \frac{1}{\sqrt{2\pi N_0}} \exp\left(-\frac{y^2}{2N_0}\right), \quad -\infty < y < \infty$$

$$f_{Y_1|s_1}(y_1) = \frac{1}{\sqrt{2\pi N_0}} \exp\left(-\frac{(y_1 - s_{01})^2}{2N_0}\right), \quad -\infty < y_1 < \infty$$

and

$$N_0 = \left(\frac{A^2}{2} T_s\right) \frac{\eta}{2}$$
$$s_{01} = \frac{A^2}{2} T_s \tag{8.72}$$

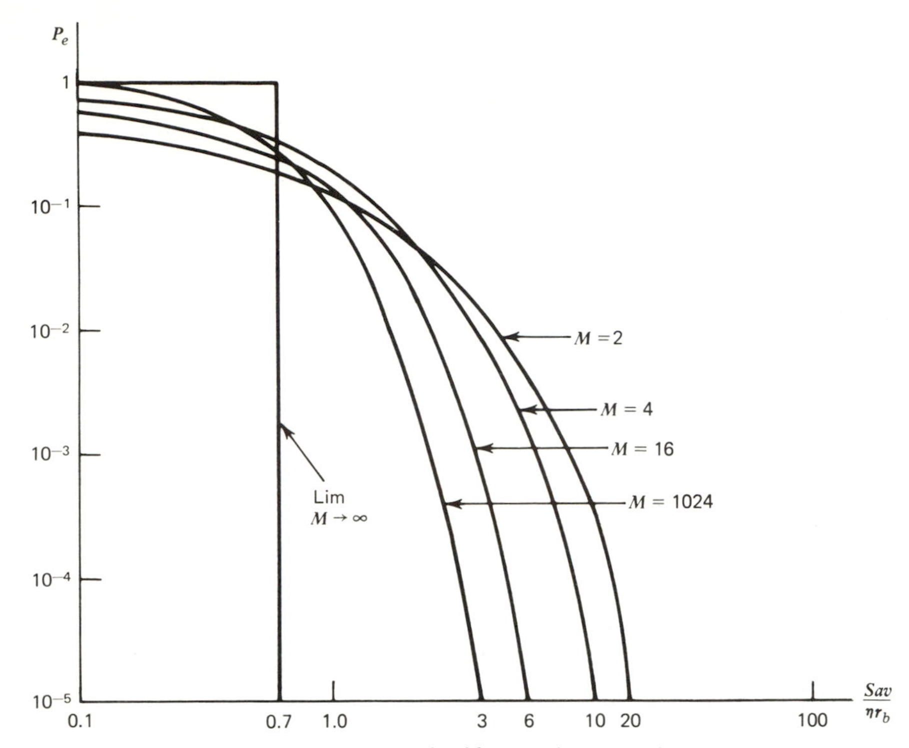

Figure 8.21 Probability of error for M-ary orthogonal signaling schemes.

Now, the probability that the receiver incorrectly decodes the incoming signal $s_1(t)$ is

$$P_{e1} = 1 - P_{c1}$$

and the probability that the receiver makes an error in decoding is

$$P_e = P_{e1}$$

because of symmetry.

The integral in Equation (8.71) cannot be expressed in a closed form for $M > 2$. Numerical integration techniques have been used to evaluate the integral and the results for several values of M are given in Figure 8.21.

The horizontal axis in Figure 8.21 is $S_{av}/\eta r_b$, where r_b is the data rate in bits/per second, S_{av} is the average signal power at the receiver input, and $\eta/2$ is the noise power spectral density at the receiver input. Because of our assumption that $M = 2^\lambda$, we have $r_b = r_s \log_2 M = \lambda r_s$ (λ a positive integer). The plots in Figure 8.21 reveal several interesting points. First, for fixed values of data rate r_b, noise psd $\eta/2$, and probability of error P_e, we see that increasing values of M lead to smaller power requirements. Of course the price paid is the increase in bandwidth since the minimum bandwidth of M-ary orthogonal FSK signal set is $M/2T_s$ and it increases as the value of M increases. Also, large values of M lead to more complex transmitting and receiving equipment.

Figure 8.21 also reveals that in the limiting case as $M \to \infty$ the probability of error P_e satisfies

$$P_e = \begin{cases} 1 & \text{if } S_{av}/\eta r_b < 0.7 \\ 0 & \text{if } S_{av}/\eta r_b > 0.7 \end{cases}$$

The above relationship indicates that the maximum errorless rate r_b at which data can be transmitted using an M-ary orthogonal FSK signaling scheme is

$$r_b = \frac{S_{av}}{0.7\eta} \simeq \frac{S_{av}}{\eta} \log_2 e \tag{8.73}$$

The bandwidth of the signal set $\to \infty$ as $M \to \infty$.

It is interesting to note that the capacity C of a Gaussian channel of infinite bandwidth is $(S_{av}/\eta) \log_2 e$ (see Section 4.6). Equation (8.73) states that if the bit rate r_b is less than channel capacity, the probability of error can be made arbitrarily small. Thus we have indeed constructed a signaling scheme capable of signaling at a rate up to channel capacity with an arbitrarily small probability of error.

Example 8.6. Binary data is to be transmitted over a microwave channel at a rate of $(3)(10^6)$ bits/sec. Assuming the channel noise to be white Gaussian with a psd $\eta/2 = 10^{-14}$ watt/Hz, find the power and bandwidth requirements of four-phase PSK and 16-tone FSK signaling schemes to maintain an error probability of 10^{-4}.

Solution

(a) For the QPSK scheme we have

$$(P_e)_{\text{QPSK}} = 2Q(\sqrt{A^2T_s/2\eta})$$

where $T_s = 2T_b = (0.6667)10^{-6}$ and $\eta/2 = 10^{-14}$ watt/Hz, $P_e = 10^{-4}$, and hence

$$Q\left(\sqrt{\frac{A^2T_s}{2\eta}}\right) = \frac{10^{-4}}{2}$$

which requires

$$A^2T_s/2\eta = (3.9)^2$$

or

$$A^2/2 = S_{av} = -33.41 \text{ dBm}$$

The QPSK scheme requires a bandwidth of $2r_s = 3$ MHz.

(b) For the 16-tone FSK ($M = 16$), we obtain from Figure 8.21

$$S_{av}/\eta r_b = 5$$

or

$$S_{av} = (30)(10^{-8}) = -35.23\ \text{dBm}$$

for $P_e = 10^{-4}$. The bandwidth requirements of the 16-tone FSK scheme is $\geqslant M/2T_s$, where $M = 16$ and $T_s = (T_b)\log_2 16 = 4T_b = (1.333)(10^{-6})$. Hence the bandwidth required is $\geqslant 6$ MHz. Thus the multitone FSK has lower power requirements than QPSK, but requires more bandwidth and a more complex receiver structure.

8.8 SYNCHRONIZATION METHODS

For optimum demodulation of ASK, FSK, and PSK waveforms, timing information is needed at the receiver. In particular, the integrate and dump operation in correlation receivers and the sampling operation in other types of receivers must be carefully controlled and sychronized with the incoming signal for optimum performance. Three general methods are used for synchronization in digital modulation schemes. These methods are:

1. Use of a primary or secondary time standard.
2. Utilization of a separate synchronization signal.
3. Extraction of clock information from the modulated waveform itself, referred to as self-synchronization.

In the first method, the transmitter and receiver are slaved to a precise master timing source. This method is often used in large data communication networks. In point-to-point data communication this method is very seldom used because of its cost.

Separate synchronization signals in the form of pilot tones are widely used in point-to-point data communication systems. In this method, a special synchronization signal or a sinusoidal signal of known frequency is transmitted along with the information carrying modulation waveform. The synchronization signal is sent along with the modulation waveform using one of the following methods:

1. by frequency division multiplexing, wherein the frequency of the pilot tone is chosen to coincide with a null in the psd of the signaling waveform;

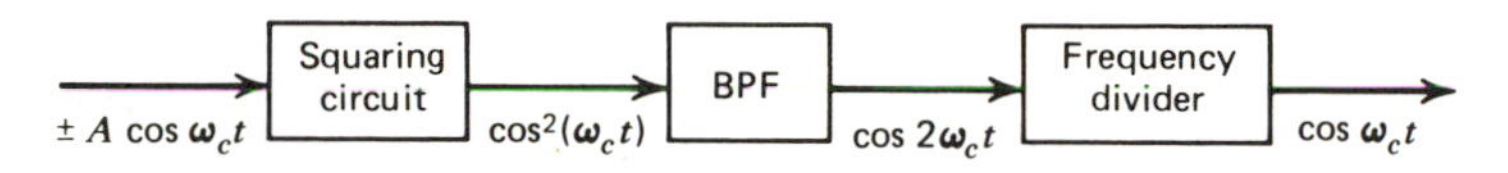

(a) Open loop carrier recovery scheme

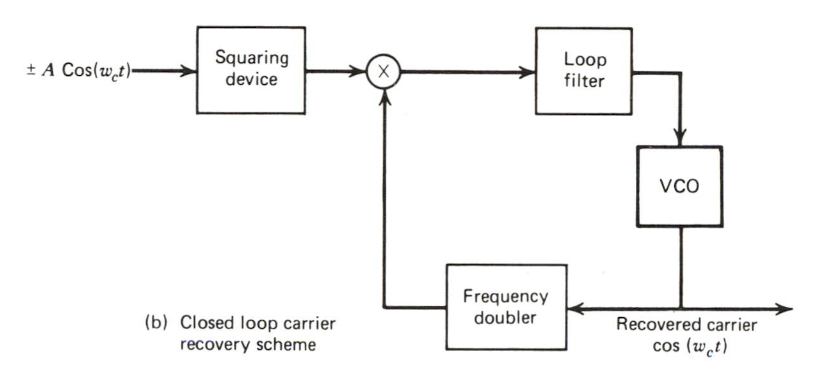

(b) Closed loop carrier recovery scheme

Figure 8.22 Extraction of local carrier for coherent demodulation of PSK signals.

2. by time division multiplexing where the modulated waveform (data stream) is interrupted for a short period of time during which the synchronizing signal is transmitted; or
3. by additional modulation such as the one shown in Figure 8.19.

In all of the above methods, the synchronization signal is isolated at the receiver and the zero crossings of the synchronization signal control the sampling operations at the receiver. All three methods discussed above add overhead (or additional requirements) to the system in terms of an increase in power and bandwidth requirements or a reduction in the data rate in addition to increasing the equipment complexity.

Self-synchronization methods extract a local carrier reference as well as timing information from the received waveforms. The block diagram of a system that derives a coherent local carrier from a PSK waveform is shown below in Figure 8.22a. Similar systems can be used to extract such a reference signal for other types of digital modulation schemes.

A feedback version of the squaring synchronizer is shown in Figure 8.22b. This version makes use of a PLL for extracting the correct phase and the frequency of the carrier waveform. The feedback version tracks the carrier phase more accurately, but its response is slower compared to the open-loop type synchronizing network.

If the carrier frequency is a multiple of the bit rate, then timing information can be derived from the local carrier waveform. Or, one could use self-synchronizing schemes similar to the ones described in Section 5.7.2. For these schemes to work properly, the incoming signal should have frequent symbol or bit changes. In some applications, the data stream might have to be scrambled at the transmitter to force frequent signal transitions (see Section 5.7.3).

8.9 SUMMARY

We developed procedures for analyzing and designing various signaling schemes for transmitting digital information over bandpass channels. Binary ASK, PSK, and FSK schemes were described in detail. Expressions for the probability of error for various schemes were derived in terms of average signal power at the receiver input, psd of the noise and signaling rate. The performances of various binary digital modulation schemes were compared. Finally, some of the commonly used M-ary signaling schemes were presented, and it was shown that a wideband M-ary orthogonal FSK scheme offers good trade-off between power and bandwidth.

It must be pointed out that combined modulation schemes, such as simultaneous amplitude and phase modulation, have also been used in data transmission applications. The treatment of combined modulation schemes is rather involved. The interested reader can find detailed treatment of these schemes in communication systems journals.

The signaling waveforms discussed in this chapter have spectral components that are nonzero for all values of frequencies—that is, the bandwidths of these signals approach infinity. In practical systems, filters are introduced to limit the transmission bandwidth. Such filtering introduces ISI and hence the performance of practical systems will be inferior to the performance of ideal systems discussed in this chapter. The analysis of systems that have ISI and additive noise is very complicated. Interested readers may refer to recent issues of the IEEE Transactions on Communications Technology which contain a large number of papers dealing with the combined effects of ISI and additive noise in digital communication systems.

REFERENCES

A very thorough treatment of digital modulation schemes, written at an advanced level, may be found in *Principles of Data Communications* by Lucky et al. (1968) and in *Modern Communication Principles* by Stein and Jones (1967). Sakrison's and Viterbi's books contain good theoretical treatment of signal detection theory and its application to digital modulation schemes. Many undergraduate texts provide easily readable treatments of several aspects of digital modulation schemes.

1. R. W. Lucky, J. Salz, and E. J. Weldon Jr. *Principles of Data Communications*. McGraw-Hill, New York (1968).
2. D. J. Sakrison. *Communication Theory: Transmission of Waveforms and Digital Information*. Wiley, New York (1968).
3. A. J. Viterbi. *Principles of Coherent Communication*. McGraw-Hill, New York, 1966.
4. Mischa Schwartz. *Information Transmission, Modulation and Noise*, 2nd ed. McGraw-Hill, New York (1970).
5. A. Bruce Carlson. *Communication Systems*. McGraw-Hill, New York (1975).
6. R. E. Ziemer and W. H. Tranter. *Principles of Communications*. Houghton Mifflin, Boston, Mass. (1976).
7. H. Taub and D. L. Schilling. *Principles of Communication Systems*. McGraw-Hill, New York (1971).
8. S. Stein and J. J. Jones. *Modern Communication Principles*. McGraw-Hill, New York (1967).

PROBLEMS

Section 8.2

8.1. $n(t)$ is a zero mean Gaussian white noise with a psd of $\eta/2$. $n_0(T_b)$ is related to $n(t)$ by

$$n_0(T_b) = \int_0^{T_b} n(t)s(t)\,dt$$

where $s(t) \equiv 0$ for t outside the interval $[0, T_b]$ and

$$\int_0^{T_b} s^2(t)\,dt = E_s$$

Show that $E\{n_0(T_b)\} = 0$ and $E\{[n_0(T_b)]^2\} = \eta E_s/2$.

8.2. A statistically independent sequence of equiprobable binary digits is transmitted over a channel having infinite bandwidth using the rectangular signaling waveform shown in Figure 8.23. The bit rate is r_b,

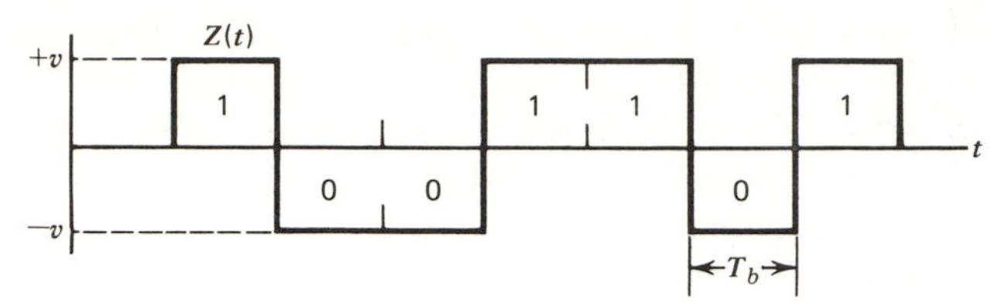

Figure 8.23 Signal waveform at the receiver input, Problem 8.2.

and the channel noise is white Gaussian with a psd of $\eta|2$.

(a) Derive the structure of an optimum receiver for this signaling scheme.

(b) Derive an expression for the probability of error.

8.3. In Problem 8.2, assume that the channel noise has a psd $G_n(f)$ given by

$$G_n(f) = G_0[1 + (f/f_1)^2]^{-1}$$

(a) Find the transfer function of the optimum receiver and calculate P_e.

(b) If an integrate and dump receiver is used instead of the optimum receiver, find P_e and compare with the P_e for the optimum receiver.

8.4. A received signal is ± 1 mv for T_b second intervals with equal probability. The signal is accompanied by white Gaussian noise with a psd of 10^{-10} watt/Hz. The receiver integrates the signal plus noise synchronously for T_b second duration and decodes the signal by comparing the integrator output with 0.

(a) Find the maximum signaling rate (minimum value of T_b) such that $P_e = 10^{-4}$.

(b) If actual signaling takes place at $\frac{1}{2}$ the rate found in (a), what is the signal amplitude required to maintain $P_e = 10^{-4}$?

8.5. Verify that the threshold value T_0^* shown in Figure 8.4 yields the minimum probability of error when $P(b_k = 0) = P(b_k = 1) = \frac{1}{2}$.

8.6. Assume that the ideal integrator in an integrate and dump receiver for Problem 8.2 is replaced by an RC lowpass filter with

$$H(f) = 1/(1 + jf/f_0)$$

where f_0 is the half-power frequency.

(a) Assuming the capacitor is initially discharged, find $s_{01}(T_b)$, $s_{02}(T_b)$, and $E\{n_0^2(t)\}$, where s_{01} and s_{02} are output signal values at $t = T_b$ and $n_0(t)$ is the noise at the filter output.

(b) Find the relationship between T_b and f_0 that will maximize $[s_{01}(T_b) - s_{02}(T_b)]^2/E\{n_0^2(t)\}$, that is, find the value of f_0 that will minimize P_e.
(c) Find the maximum value of $[s_{01} - s_{02}]^2/E\{n_0^2(t)\}$.

8.7. Referring to Equation (8.20), we have the signal-to-noise power ratio at the output of a matched filter receiver as

$$\gamma_{\max}^2 = \frac{2}{\eta}\int_0^T [s_1(t) - s_2(t)]^2\,dt$$

Now suppose that we want $s_1(t)$ and $s_2(t)$ to have the same signal energy. Show that the optimum choice of $s_2(t)$ is

$$s_2(t) = -s_1(t)$$

and that with $s_2(t) = -s_1(t)$

$$\gamma_{\max}^2 = (8/\eta)\int_0^T s_1^2(t)\,dt$$

8.8. An on-off binary system uses the following waveforms:

$$s_2(t) = \begin{cases} 2t/T_b, & 0 < t < T_b/2 \\ (2/T_b)(T_b - t), & T_b/2 \leq t < T_b \end{cases}$$

$s_1(t) = 0$

Assume that $T_b = 20\,\mu$sec and the noise psd is $\eta/2 = 10^{-7}$ watt/Hz. Find P_e for the optimum receiver assuming $P(0 \text{ sent}) = \frac{1}{4}$, $P(1 \text{ sent}) = \frac{3}{4}$.

Sections 8.3, 8.4, *and* 8.5

8.9. The input to a threshold device has the following conditional probabilities:

$$f_{R|0\text{ sent}}(r) = \frac{r}{N_0}\exp\left(-\frac{r^2}{2N_0}\right), \quad r > 0$$

$$f_{R|1\text{ sent}}(r) \simeq \frac{1}{\sqrt{2\pi N_0}}\exp\left(-\frac{(r-A)^2}{2N_0}\right) \quad A \gg 0$$

$P(0 \text{ sent}) = \frac{1}{2}$ and $P(1 \text{ sent}) = \frac{1}{2}$. Find the optimum value of the threshold setting that will minimize the probability of error for $A = 1$ and $N_0 =$ 0.01, 0.2, and 0.5. (*Hint*: Plot the pdf's and find the point where they intersect.) Compare the threshold values you get with the values obtained using the approximation given in Equation (8.37*b*).

8.10. In a binary PSK scheme using correlation receiver, the local carrier waveform is $A\cos(\omega_c t + \phi)$ instead of $A\cos\omega_c t$ due to poor carrier synchronization. Derive an expression for the probability of error and compute the increase in error probability when $\phi = 15°$ and $A^2T_b/\eta = 10$.

8.11. In a coherent binary PSK system, the peak carrier amplitude at the receiver A varies slowly due to fading. Assume that A has a pdf

$$f_A(a) = \frac{a}{A_0^2}\exp\left(-\frac{a^2}{2A_0^2}\right), \quad a \geq 0$$

(a) Find the mean and standard deviation of A.
(b) Find the *average* probability of error P_e. [Use the approximation for $Q(x)$ given in appendix D.]

8.12. In a coherent binary PSK system with $f_c = 5r_b$, the local carrier is in synchronism with the received signal, but the integrate and dump operation in the receiver is not fully synchronized. The sampling takes place at $t = 0.2T_b, 1.2T_b, 2.2T_b, \ldots$.
(a) How much intersymbol interference is generated by the offset in sampling times. (See Problem 5.8.)
(b) Calculate the probability of error and compare it with the probability of error that can be achieved with perfect timing.

8.13. In a coherent binary PSK system the symbol probabilities are $P(0 \text{ sent}) = p$ and $P(1 \text{ sent}) = 1 - p$. The receiver is operating with a signal-to-noise ratio $(A^2T_b)/\eta = 4$.
(a) Find the optimum threshold setting for $p = 0.4, 0.5,$ and 0.6 and find the probability of error P_e for $p = 0.4, 0.5,$ and 0.6.
(b) Suppose that the receiver threshold setting was set at 0 for $p = 0.4,$ $0.5,$ and 0.6. Find P_e and compare with P_e obtained in part (a).

8.14. An ideal analog bandpass channel has a usable bandwidth of 3000 Hz. The maximum average signal power allowed at the input to the channel is 0.001 mW. The channel noise (at the output) can be assumed to be zero mean white Gaussian with a psd of $\eta/2 = 10^{-10}$ watt/Hz.
(a) Find the capacity of the analog channel.
(b) Find the maximum rate at which binary data can be transmitted over this channel using binary PSK and FSK signaling schemes.
(c) Find P_e for coherent PSK and noncoherent FSK, assuming maximum signaling rate.
(d) Using the results of (b) and (c), find the capacity of the discrete channels corresponding to the coherent PSK and noncoherent FSK

signaling schemes.

8.15. Consider a bandpass channel with the response shown in Figure 8.24.

(a) Binary data is transmitted over this channel at a rate of 300 bits/sec using a noncoherent FSK signaling scheme with tone frequencies of 1070 and 1270 Hz. Calculate P_e assuming $A^2/\eta = 8000$.

(b) How fast can a PSK signaling scheme operate over this channel? Find P_e for the PSK scheme assuming coherent demodulation.

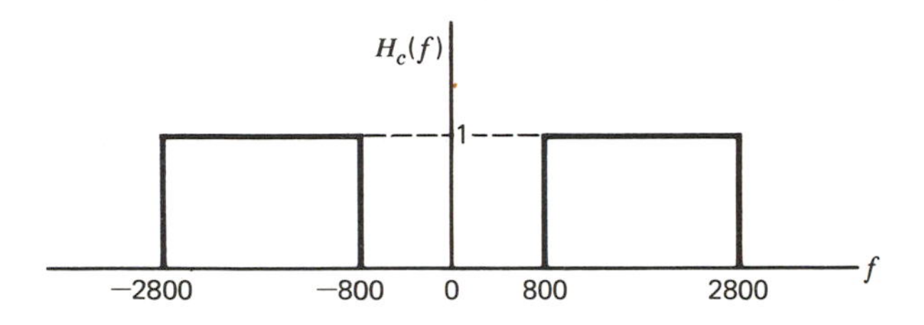

Figure 8.24 Response of a bandpass channel, Problem 8.15.

8.16. Compare the average power requirements of binary noncoherent ASK, coherent PSK, DPSK, and noncoherent FSK signaling schemes operating at a data rate of 1000 bits/sec over a bandpass channel having a bandwidth of 3000 Hz, $\eta/2 = 10^{-10}$ watt/Hz, and $P_e = 10^{-5}$.

8.17. Fill in the missing steps in the derivation of P_e for the DPSK signaling scheme.

8.18. A correlation receiver for a PSK system uses a carrier reference $A \sin \omega_c t$ for detecting

$$s_1(t) = A \cos(\omega_c t + \Delta\theta)$$

$$s_2(t) = A \sin(\omega_c t + \Delta\theta)$$

Assuming that $s_1(t)$ and $s_2(t)$ are equiprobable and the noise is white and Gaussian with a psd of $\eta/2$, find the probability of incorrect decoding.

8.19. The bit stream 11011100101 is to be transmitted using DPSK. Determine the encoded sequence and the transmitted phase sequence. Show that the phase comparison scheme described in Section 8.4.2 can be used for demodulating the signal.

8.20. A high frequency transmitter used in a binary communication system is peak power limited to 1 kW. The power loss in the channel is 60 dB and the noise power at the receiver input (ηr_b) is 10^{-4} watts. Assuming maximum signaling rate and equiprobable message bits, find P_e for noncoherent ASK and coherent PSK signaling schemes.

8.21. In some threshold devices a no-decision zone centered at the optimum threshold level is used such that if the input Y to the threshold device falls in this region, no decision is made, that is, the output is 0 if say $Y < T_1$ and 1 if $Y > T_2$ and no decision is made if $T_1 \leq Y \leq T_2$. Assuming that

$$f_{Y/1\text{ sent}}(y) = \frac{1}{2\sqrt{\pi}} \exp\left(-\frac{(y-1)^2}{4}\right), \quad -\infty < y < \infty$$

$$f_{Y/0\text{ sent}}(y) = \frac{1}{2\sqrt{\pi}} \exp\left(-\frac{(y+1)^2}{4}\right), \quad -\infty < y < \infty$$

$$P(1\text{ sent}) = P(0\text{ sent}) = 0.5$$

$$T_1 = -\epsilon, \quad T_2 = \epsilon, \quad \epsilon > 0$$

Sketch P_e and the probability of no decision versus ϵ. (Use $\epsilon = 0.1$, 0.2, 0.3, 0.4, and 0.5.)

8.22. An ASK signaling scheme uses the noncoherent demodulation scheme shown in Figure 8.25. The center frequency of the filter is f_c and the

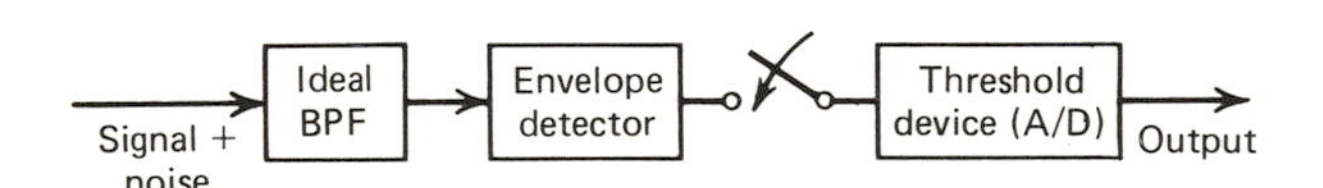

Figure 8.25 Noncoherent ASK receiver.

bandwidth $B = 10r_b$ Hz. Assume that the bandwidth is such that the ASK signal passes through the filter with minimum distortion, and that the filter generates no ISI.

(a) Calculate the P_e for the receiver shown above assuming that $A^2/(\eta r_b) = 200$.

(b) Compare with P_e for a noncoherent ASK scheme if the filter is matched to the mark pulses.

8.23. Repeat Problem 8.22 for the noncoherent FSK scheme.

8.24. Let $n(t)$ be a stationary zero mean Gaussian white noise and let

$$n_{01}(T_b) = \int_0^{T_b} n(t)\cos(\omega_c t + \omega_d t)\, dt$$

$$n_{02}(T_b) = \int_0^{T_b} n(t)\cos(\omega_c t - \omega_d t)\, dt$$

Show that $n_{01}(T_b)$ and $n_{02}(T_b)$ are independent if $\omega_c = 2\pi k/T_b$ and $\omega_d = m\pi/2T_b$, where k and m are (arbitrary) positive integers ($k \gg m$).

Section 8.6

8.25. An M-ary signaling scheme uses the following signals: $s_k(t) = A_k \cos(\omega_c t + \phi_k)$, $(0 \leq t \leq T_s)$, where

$A_k = A$ or $2A$, and

$\phi_k = 45°, 90°, 135°$, or $270°$,

(Observe that $M = 8$ and the signaling scheme is combined ASK/PSK.)

(a) Draw the block diagram of a coherent receiver for this system.

(b) Derive an approximate expression for the probability of error.

8.26. Consider the channel described in Problem 8.15.

(a) Compute the fastest rate at which data can be transmitted over this channel using four-phase PSK signaling schemes.

(b) Compute P_e for QPSK and differential QPSK.

8.27. A microwave channel has a usable bandwidth of 10 MHz. Data has to be transmitted over this channel at a rate of $(1.5)(10^6)$ bits/sec. The channel noise is zero mean Gaussian with a psd of $\eta/2 = 10^{-14}$ watt/Hz.

(a) Design a wideband FSK signaling scheme operating at $P_e = 10^{-5}$ for this problem, that is, find a suitable value of M and $A^2/2$.

(b) If a binary differential PSK signaling scheme is used for this problem, find its power requirement.

8.28. If the value of M obtained in Problem 8.27 is doubled, how will it affect the bandwidth and power requirements if P_e is to be maintained at 10^{-5}?

8.29. The design of a high-speed data communication system calls for a combined ASK/PSK signaling scheme with $M = 16$. Three alternate designs corresponding to three different sets of signaling waveforms are to be comparatively evaluated. The signaling waveforms in each set are shown in Figure 8.26 in a phasor diagram. The important parameters of the signal sets are the minimum distance between the phasors (parameter "a"

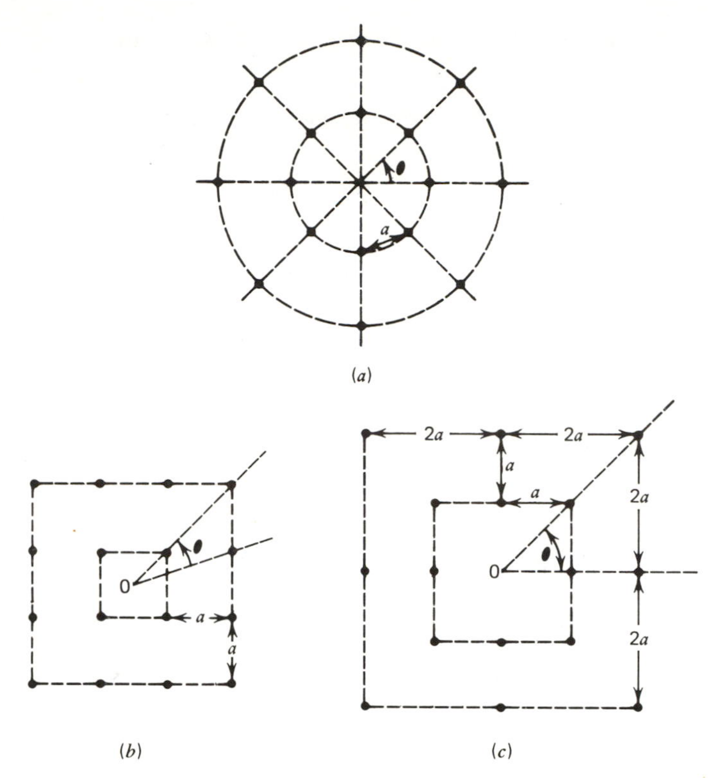

Figure 8.26 Constellations of signals, $M = 16$. Dots denote the tips of signal phasors.

in Figure 8.26 that is a measure of immunity against additive noise), the minimum phase difference (which is a measure of immunity against phase jitter/delay distortion), and the ratio of peak-to-average power (which is a measure of immunity against nonlinear distortion in the channel). Assuming that the average power is to be the same for all three signal sets, compare their robustness against the following channel impairments.

(a) Nonlinear distortion.
(b) Additive noise.
(c) Phase jitter.
(d) Combination of noise, phase jitter, and nonlinear distortion.

8.30. Derive the structure of a carrier recovery network (similar to the one shown in Figure 8.22a) for a QPSK signaling scheme.

9

ERROR CONTROL CODING

9.1 INTRODUCTION

In Chapters 5 and 8 we described signaling schemes for transmitting digital information over noisy channels. We saw that the probability of error for a particular signaling scheme is a function of the signal-to-noise ratio at the receiver input and the data rate. In practical systems the maximum signal power and the bandwidth of the channel are restricted to some fixed values due to governmental regulations on public channels or regulations imposed by private companies if the channel is leased. Furthermore, the noise power spectral density $\eta/2$ is also fixed for a particular operating environment. In addition, parameters of signaling schemes, such as the number and type of signals used, are chosen to minimize the complexity and cost of the equipment. With all of these constraints, it is often not possible to arrive at a signaling scheme which will yield an acceptable probability of error for a given application. Faced with this problem, the only practical alternative for reducing the probability of error is the use of error control coding, also known as channel coding.

In a nutshell, error control coding is the calculated use of *redundancy*. The functional blocks that accomplish error control coding are the channel encoder and the channel decoder. The channel encoder systematically adds digits to the transmitted message digits. These additional digits, while conveying no new information themselves, make it possible for the channel decoder to detect and correct errors in the information bearing digits. Error detection and/or correction lowers the overall probability of error.

The source encoder discussed in Chapter 4 removes redundancy from the source output. The channel encoder adds or puts back just enough redundancy to lower the overall error probability to the desired value. By looking at these two encoders separately, we can isolate their functions and arrive at their optimum design.

In this chapter we will develop procedures for designing efficient coding schemes for controlling various types of errors in a digital communication system. We will look at methods of detecting and/or correcting errors and compare different coding schemes on the basis of their error control capabilities and ease of implementation.

The approach used here differs considerably from the one used in the preceding chapters. In the preceding chapters we developed procedures for arriving at optimum solutions for the problem in hand, such as optimum design of signal sets, optimum receivers for a class of signal sets, and so forth; our approach was one of design or synthesis. In contrast, the approach we will use in this chapter will be directed mainly towards analysis. We will describe various coding procedures and discuss their properties rather than attempting to design an optimum encoding scheme for a given application. The reason for this approach is that optimum encoding schemes do not exist in general, and even in a few special cases where they do exist the mathematical rigor needed to treat them is beyond the scope of this text.

The problem of error control coding can be formulated as follows. With reference to Figure 9.1 we have a digital communication system for transmitting the binary output $\{b_k\}$ of a source encoder over a noisy channel at a rate of r_b bits/sec. Due to channel noise, the bit stream $\{\hat{b}_k\}$ recovered by the

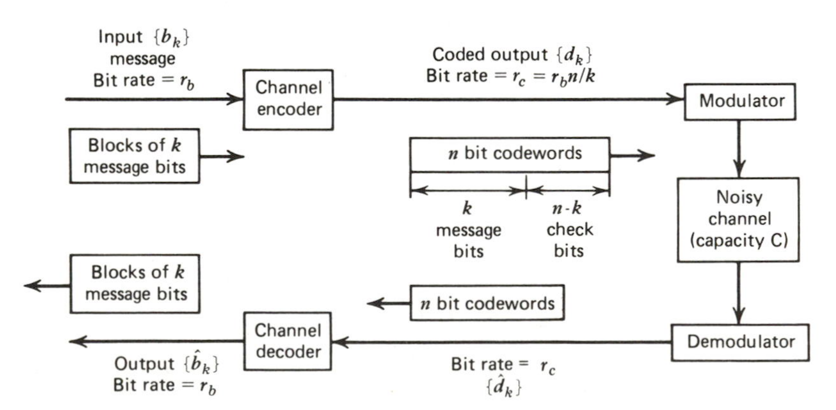

Figure 9.1 Error control coding. Channel bit error probability is $q_c = P\{d_k \neq \hat{d}_k\}$, and message bit error probability is $P_e = P\{b_k \neq \hat{b}_k\}$. It is assumed that $r_c < C$, and $q_c > P_e$.

receiver differs occasionally from the transmitted sequence $\{b_k\}$. It is desired that the probability of error $P(\hat{b}_k \neq b_k)$ be less than some prescribed value. The actual data transmission over the channel is accomplished by a modem (modulator and demodulator) that can operate at data rates r_c ranging from, say, $r_1 \leq r_c \leq r_2$. The probability of error q_c for the modem, defined as $P(\hat{d}_k \neq d_k)$, will depend on the data rate r_c over the channel. We are told that we have to use a particular modem due to cost and other practical constraints, that is, the signaling scheme is specified. Furthermore, we are told that the error probability of the modem, without error control coding, is higher than the desired error probability. Hence, we are asked to design an error control coding scheme so that the overall probability of error is acceptable.

The channel encoder and the decoder are the functional blocks in the system that, by acting together, reduce the overall probability of error. The encoder divides the input message bits into blocks of k message bits and replaces each k bit message block D with an n bit codeword C_w by adding $n-k$ check bits to each message block. The decoder looks at the received version of the codeword C_w, which may occasionally contain errors, and attempts to decode the k message bits. While the check bits convey no new information to the receiver, they enable the decoder to detect and correct transmission errors and thereby lower the probability of error.

The design of the encoder and decoder consists of selecting rules for generating codewords from message blocks and for extracting message blocks from the received version of the codewords. The mapping rules for coding and decoding are to be chosen such that error control coding lowers the overall probability of error.

9.1.1 Example of Error Control Coding

Let us consider the following example that illustrates some of the important aspects of error control coding. Suppose that we want to transmit data over a telephone link that has a usable bandwidth of 3000 Hz and a maximum S/N at the output of 13 dB at a rate of 1200 bits/sec with a probability of error less than 10^{-4}. We are given a DPSK modem that can operate at speeds of 1200, 2400, or 3600 bits/sec with error probabilities $2(10^{-4})$, $4(10^{-4})$, and $8(10^{-4})$, respectively. We are asked to design an error control coding scheme that would yield an overall probability of error $<10^{-4}$. In terms of the notation shown in Figure 9.1, we have $r_b = 1200$, $C \simeq 13{,}000$ $r_c = 1200$, 2400, or 3600, $q_c(r_c) = 2(10^{-4})$, $4(10^{-4})$, and $8(10^{-4})$, respectively. Since $r_b < C$, according to Shannon's theorem we should be able to transmit the data with an arbitrarily small probability of error.

Consider an error control coding scheme for this problem wherein the triplets 000 and 111 are transmitted when $b_k = 0$ and 1, respectively. These triplets, 000 and 111, are called codewords. These triplets of 0's and 1's are certainly redundant and two of the three bits in each triplet carry no new information. Data comes out of the channel encoder at a rate of 3600 bits/sec, and at this data rate the modem has an error probability of $8(10^{-4})$. The decoder at the receiver looks at the received triplets (which might occasionally be different from the transmitted triplet due to noise) and extracts the information-bearing digit using a majority-logic decoding scheme shown below.

Received triplet	000	001	010	100	011	101	110	111
Output message bit $\hat{b}_k$	0	0	0	0	1	1	1	1

Notice that 000 and 111 are the only valid codewords. The reader can verify that the information bearing bit is recovered correctly if no more than one of the bits in the triplet is affected by the channel noise. Now, without error control coding the lowest achievable error probability for the given modem is $(2)(10^{-4})$. However, with error control coding the receiver output bit is different from the input message bit only if two or more of the bits in triplet are in error due to channel noise. Thus,

$$\begin{aligned} P_e &= P(b_k \neq \hat{b}_k) \\ &= P(\text{two or more bits in a triplet are in error}) \\ &= \binom{3}{2} q_c^2(1 - q_c) + \binom{3}{3} q_c^3 \\ &= 3q_c^2 - 2q_c^3 \end{aligned}$$

With the modem bit error probability q_c of $(8)(10^{-4})$ at $r_c = 3600$ bits/sec, we have

$$P_e = (191.9)10^{-8} = (0.01919)10^{-4}$$

which is less than the desired error probability of 10^{-4}. Thus error control coding reduces the overall probability of error.

The preceding example illustrates the following important aspects of error control coding:

1. It is possible to detect and correct errors by adding extra bits called *check bits* to the message stream. Because of the additional bits, not all bit sequences will constitute bonafide messages.

2. It is *not* possible to detect and correct *all* errors.
3. Addition of check bits reduces the effective data rate through the channel. Quantitatively, the *rate efficiency* of a coding scheme is defined as r_b/r_c. (The rate efficiency of the code discussed previously is $\frac{1}{3}$.)

9.1.2 Methods of Controlling Errors

The channel decoder discussed in the preceding example corrected (or at least attempted to correct) possible errors in the received sequence. This method of controlling errors at the receiver through attempts to correct noise-induced errors is called the *forward-acting error correction* method. If system considerations permit, errors can be handled in an entirely different manner as follows. The decoder, upon examining the demodulator output, accepts the received sequence if it matches a valid message sequence. If not, the decoder discards the received sequence and notifies the transmitter (over a reverse channel from the receiver to the transmitter) that errors have occurred and that *the received message must be retransmitted.* Thus the decoder attempts to detect errors but does not attempt to correct the error. This method of error control is called *error detection.*

Error detection schemes yield a lower overall probability of error than error correction schemes. To illustrate this point, let us return to the previous example. If the decoder in the preceding example used error detection only, then it would accept only 000 and 111 and reject all other triplets and request retransmission. Now, an information bit will be incorrectly decoded at the receiver only when all three bits in a received code word are in error. Thus

$$P_e = (q_c)^3 = (5.12)(10^{-10})$$

which is much lower than the probability of error for the error correction scheme. It must be emphasized here that error detection and retransmission of messages require a reverse channel that may not be available in some applications. Furthermore, error detection schemes slow down the effective rate of data transmission since the transmitter has to wait for an acknowledgment from the receiver before transmitting the next message.

Forward-acting error correction and error detection are usually used separately, but occasionally it may be necessary to use both in a particular system. For example, the receiver may decode a sequence only if it closely resembles an allowable sequence and request a retransmission otherwise. In most situations, system considerations determine which of the above error control techniques is to be employed. For instance, if a reverse channel is not available, then forward error correction is the only error control scheme that can be used. On the other hand, if a reverse channel is available and a large reduction in error rate is to be accomplished, then the error detection method with retransmission is preferable over the error correction method.

In the following sections we will investigate coding schemes for detecting and/or correcting transmission errors and analyze their performance.

9.1.3 Types of Errors

Transmission errors in a digital communication system are caused by the noise in the communication channel. Generally, two kinds of noise can be distinguished in a communication channel. The first kind, *Gaussian noise*, had been our chief concern in designing and evaluating modulators and demodulators for data transmission. Sources of Gaussian noise include thermal and shot noise in the transmitting and receiving equipment, thermal noise in the channel, and radiation picked up by the receiving antenna. Often, the power spectral density of the Gaussian noise at the receiver input is white. The transmission errors introduced by white Gaussian noise are such that the occurrence of error during a particular signaling interval does not affect the performance of the system during the subsequent signaling interval. The discrete channel in this case can be modeled by the binary symmetric channel discussed in Chapter 4, and the transmission errors due to *white Gaussian noise* are referred to as *random errors*.

A second type of noise often encountered in a communication channel is called *impulse noise*, which is characterized by long quiet intervals followed by high amplitude noise bursts. This type of noise results from many natural and manmade causes such as lightning and switching transients. When a noise burst occurs, it affects more than one symbol or bit, and there is usually a dependence of errors in successive transmitted symbols. Thus errors occur in *bursts*.

Error control schemes for dealing with random errors are called *random error-correcting codes*, and coding schemes designed to correct burst errors are called *burst error-correcting codes*. We will consider coding schemes for controlling both random and burst errors.

9.1.4 Types of Codes

Error control codes are often divided into two broad categories: *block codes* and *convolutional codes*. In block codes, a block of k information bits is followed by a group of r check bits that are derived from the block of information bits. At the receiver, the check bits are used to verify the information bits in the information block preceding the check bits. In convolutional codes, check bits are continuously interleaved with information bits; the check bits verify the information bits not only in the block immediately preceding them, but in other blocks as well.

While it is possible to unify the mathematical treatment of block and convolutional codes, such a description is often very complex. Hence we will treat these two types of codes separately. Since block codes are better developed and easily understood, we will devote a considerable amount of our efforts to describing them.

9.2 LINEAR BLOCK CODES

In this section, we will deal with block codes* in which each block of k message bits is encoded into a block of $n > k$ bits, as illustrated in Figure 9.2, by adding $n - k$ check bits derived from the k message bits.

The n-bit block of the channel encoder output is called a *codeword*, and codes (or coding schemes) in which the message bits appear at the beginning of a codeword are called *systematic codes*. Furthermore, if each of the 2^k codewords can be expressed as *linear* combinations of k linearly independent code vectors, then the code is called a *systematic, linear block code*.† It is this subclass of block codes that we will treat in this section.

We will first introduce the basic notations for describing linear block codes in a matrix form. Then we will define the minimum distance of a block code and show that the error control capabilities of a linear block code are determined by its minimum distance. Finally, we will discuss the table look-up method of decoding of (correcting or detecting errors in) a linear block code.

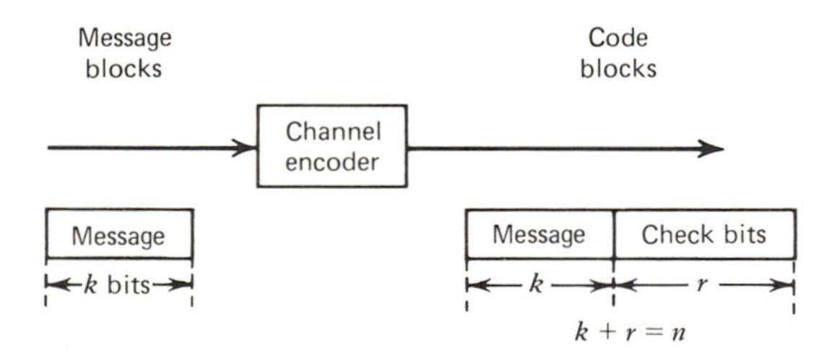

Figure 9.2 Example of a block coder.

*We use "codes" and "coding schemes" synonymously; In a strict sense, the term block "code" refers to all the codewords of a block coding scheme.

†The 2^k codewords of a linear block code form a k-dimensional subspace of the vector space of all n-tuples. Thus the sum of any two codewords of a linear code also belongs to the code.

9.2.1 Matrix Description of Linear Block Codes

The encoding operation in a linear block encoding scheme consists of two basic steps: (1) the information sequence is segmented into message blocks, each block consisting of k successive information bits; (2) the encoder transforms each message block into a larger block of n bits according to some predetermined set of rules. These $n-k$ additional bits are generated from linear combinations of the message bits, and we can describe the encoding operations using matrices.

We will denote the message block as a row vector or k-tuple $D = (d_1, d_2, \ldots, d_k)$, where each message bit can be a 0 or 1. Thus we have 2^k distinct message blocks. Each message block is transformed to a codeword C of length n bits $(C = (c_1, c_2, \ldots, c_n))$ by the encoder and there are 2^k distinct codewords, one unique codeword for each distinct message block. This set of 2^k codewords, also called code vectors, is called a (n, k) block code. The rate *efficiency* of this code is k/n.

In a *systematic linear block code*, the first k bits of the codeword are the message bits, that is,

$$c_i = d_i, \quad i = 1, 2, \ldots, k \tag{9.1}$$

The last $n-k$ bits in the codeword are check bits generated from the k message bits according to some predetermined rule:

$$\begin{aligned} c_{k+1} &= p_{11}d_1 + p_{21}d_2 + \cdots + p_{k,1}d_k \\ c_{k+2} &= p_{12}d_1 + p_{22}d_2 + \cdots + p_{k,2}d_k \\ &\vdots \\ c_n &= p_{1,n-k}d_1 + p_{2,n-k}d_2 + \cdots + p_{k,n-k}d_k \end{aligned} \tag{9.2}$$

The coefficients $p_{i,j}$ in Equation (9.2) are 0's and 1's and the addition operations in Equation (9.2) are performed in modulo-2 arithmetic* $(1+1=0,\ 0+1=1,\ 1+0=1,\ 0+0=0)$ so that c_k's are 0's and 1's. We can combine Equations (9.1) and (9.2) and write them in matrix form as

$$[c_1, c_2, \ldots, c_n] = [d_1, d_2, \ldots, d_k] \begin{bmatrix} 1000\ldots0 & p_{11}p_{12}\cdots p_{1,n-k} \\ 0100\ldots0 & p_{21}p_{22}\cdots p_{2,n-k} \\ 0010\ldots0 & p_{31}p_{32}\cdots p_{3,n-k} \\ \vdots \quad\quad \vdots & \vdots \quad \vdots \quad\quad \vdots \\ 0000\ldots1 & p_{k1}p_{k2}\cdots p_{k,n-k} \end{bmatrix}_{k\times n} \tag{9.3}$$

or

$$C = DG \tag{9.4}$$

*All of our additions from here on will be modulo-2, unless otherwise specified.

where G is the matrix on the right-hand side of Equation (9.3). The $k \times n$ matrix G is called the *generator matrix* of the code and it has the form

$$G = [I_k \mid P]_{k \times n} \tag{9.5}$$

The matrix I_k is the identity matrix of order k and P is an arbitrary k by $n-k$ matrix. When P is specified, it defines the (n, k) block code completely.

An important step in the design of a (n, k) block code is the selection of a P matrix so that the code generated by G has certain desirable properties such as ease of implementation, ability to correct random and burst errors, high rate efficiency, and so forth. While no procedure exists for selecting a P matrix to satisfy all of the above constraints, past research in coding theory has yielded several special classes of linear block codes that possess many desirable properties and a fair amount of mathematical structure. In the remaining sections of this chapter, we will look at constructive procedures for choosing P matrices for different applications.

Example 9.1. The generator matrix for a (6, 3) block code is given below. Find all code vectors of this code.

$$G = \left[\begin{array}{ccc|ccc} 1 & 0 & 0 & 0 & 1 & 1 \\ 0 & 1 & 0 & 1 & 0 & 1 \\ 0 & 0 & 1 & 1 & 1 & 0 \end{array}\right]$$

Solution. The message block size k for this code is 3, and the length of the code vectors n is 6. There are eight possible message blocks: (0, 0, 0), (0, 0, 1), (0, 1, 0), (0, 1, 1), (1, 0, 0), (1, 0, 1), (1, 1, 0), (1, 1, 1). The code vector for the message block $D = (111)$ is given by

$$C = DG = (111)\begin{bmatrix} 1 & 0 & 0 & 0 & 1 & 1 \\ 0 & 1 & 0 & 1 & 0 & 1 \\ 0 & 0 & 1 & 1 & 1 & 0 \end{bmatrix}$$
$$= (1 \quad 1 \quad 1 \quad 0 \quad 0 \quad 0)$$

The reader can verify that the remaining code vectors of this code are:

Messages	Code vectors
0 0 0	0 0 0 0 0 0
0 0 1	0 0 1 1 1 0
0 1 0	0 1 0 1 0 1
0 1 1	0 1 1 0 1 1
1 0 0	1 0 0 0 1 1
1 0 1	1 0 1 1 0 1
1 1 0	1 1 0 1 1 0

The preceding example illustrates how an encoder for the (6, 3) code generates code vectors. The encoder essentially has to store the G matrix (or at least the submatrix P of G) and perform binary arithmetic operations to generate the check bits. The complexity of the encoder increases as the block size n and the number of check bits $n-k$ increase.

Associated with each (n, k) block code is a *parity check* matrix H, which is defined as

$$H = \left[\begin{array}{cccc|cccccc} p_{11} & p_{21} & \cdots & p_{k1} & 1 & 0 & 0 & 0 & \cdots & 0 \\ p_{12} & p_{22} & \cdots & p_{k2} & 0 & 1 & 0 & 0 & \cdots & 0 \\ \vdots & \vdots & & & \vdots & & & & & \vdots \\ p_{1,n-k} & p_{2,n-k} & \cdots & p_{k,n-k} & 0 & 0 & 0 & 0 & \cdots & 1 \end{array}\right]$$

$$= [P^T \mid I_{n-k}]_{(n-k)\times n} \tag{9.6}$$

The parity check matrix can be used to verify whether a codeword C is generated by the matrix $G = [I_k \mid P]$. This verification can be done as follows. C is a codeword in the (n, k) block code generated by $G = [I_k \mid P]$ if and only if

$$CH^T = 0 \tag{9.7}$$

where H^T is the transpose of the matrix H. The proof of (9.7) is very simple and is left as an exercise for the reader. (Caution: Remember that additions are in modulo-2 arithmetic!)

While the generator matrix is used in the encoding operation, the parity check matrix is used in the decoding operation as follows. Consider a linear (n, k) block code with a generator matrix $G = [I_k \mid P]$ and a parity check matrix $H = [P^T \mid I_{n-k}]$. Let C be a code vector that was transmitted over a noisy channel and let R be the noise-corrupted vector that was received. The vector R is the sum of the original code vector C and an error vector E, that is,

$$R = C + E \tag{9.8}$$

The receiver does not know C and E; its function is to decode C from R, and the message block D from C. The receiver does the decoding operation by determining an $(n-k)$ vector S defined as

$$S = RH^T \tag{9.9}$$

The vector S is called the *error syndrome* of R. One can rewrite (9.9) as

$$\begin{aligned} S &= [C + E]\,H^T \\ &= CH^T + EH^T \end{aligned}$$

and obtain

$$S = EH^T \tag{9.10}$$

since $CH^T = 0$. Thus the syndrome of a received vector is zero if R is a valid code vector. If errors occur in transmission, then the syndrome S of the received vector is nonzero. Furthermore, S is related to the error vector E and the decoder uses S to detect and correct errors. Later on in this chapter we will see how this is accomplished by the decoder. Let us now look at an example to clarify some of the notations introduced in this section.

Example 9.2. Consider a (7, 4) block code generated by

$$G = \left[\begin{array}{cccc|ccc} \multicolumn{4}{c|}{\overbrace{\hphantom{1\ 0\ 0\ 0}}^{I_4}} & \multicolumn{3}{c}{\overbrace{\hphantom{1\ 1\ 1}}^{P}} \\ 1 & 0 & 0 & 0 & 1 & 1 & 1 \\ 0 & 1 & 0 & 0 & 1 & 1 & 0 \\ 0 & 0 & 1 & 0 & 1 & 0 & 1 \\ 0 & 0 & 0 & 1 & 0 & 1 & 1 \end{array}\right]$$

According to Equation (9.6), the parity check matrix for this code is

$$H = \left[\begin{array}{cccc|ccc} 1 & 1 & 1 & 0 & 1 & 0 & 0 \\ 1 & 1 & 0 & 1 & 0 & 1 & 0 \\ 1 & 0 & 1 & 1 & 0 & 0 & 1 \\ \multicolumn{4}{c|}{\underbrace{\hphantom{1\ 0\ 1\ 1}}_{P^T}} & \multicolumn{3}{c}{\underbrace{\hphantom{0\ 0\ 1}}_{I_3}} \end{array}\right]$$

For a message block $D = (1011)$, the code vector C is given by

$$\begin{aligned} C &= DG \\ &= (1 \quad 0 \quad 1 \quad 1 \,\vdots\, 0 \quad 0 \quad 1) \end{aligned}$$

For this code vector, the syndrome S is given by

$$S = CH^T = (000)$$

If the third bit of the code vector C suffered an error in the transmission, then the received vector R will be

$$\begin{aligned} R &= (1 \quad 0 \quad \underline{0} \quad 1 \quad 0 \quad 0 \quad 1) \\ &= (1 \quad 0 \quad 1 \quad 1 \quad 0 \quad 0 \quad 1) + (0 \quad 0 \quad 1 \quad 0 \quad 0 \quad 0 \quad 0) \\ &= C + E \end{aligned}$$

and the syndrome of R is

$$\begin{aligned} S = RH^T &= (101) \\ &= EH^T \end{aligned}$$

where E is the error vector $(0 \quad 0 \quad 1 \quad 0 \quad 0 \quad 0 \quad 0)$. Note that the syndrome S for an error in the third bit is the third row of the H^T matrix. The reader can verify that, for this code, a single error in the ith bit of C would lead to a syndrome vector that would be identical to the ith row of the matrix H^T. Thus single errors can be corrected at the receiver by comparing S with the rows of H^T and correcting the ith received bit if S matches with the ith row of H^T. It is easy to see that the error correction scheme would not work if multiple errors occur (consider an error vector 0 1 1 0 0 0 0 containing two errors). The error correction procedure is generally more complicated than what is presented in this simple example. We will discuss a general procedure for error correction later in this chapter.

9.2.2 Error Detection and Error Correction Capabilities of Linear Block Codes

At this point we introduce some of the basic terminology that will be used in defining the error control capabilities of a linear block code. First, the (Hamming) *weight* of a (code) vector C is defined as the number of nonzero components of C. The (Hamming) *distance* between two vectors C_1 and C_2 is defined as the number of components in which they differ. Finally, the *minimum distance* of a block code is the smallest distance between any pair of codewords in the code.

Theorem 9.1

The minimum distance of a linear block code is equal to the minimum weight of any nonzero word in the code.

Proof

First of all it is clear that the minimum distance cannot be greater than the weight of the minimum weight nonzero codeword since the all zero n-tuple is a codeword. But if it is less, the sum of the two closest codewords is a codeword of weight lower than the minimum weight, which is impossible. Thus for linear block codes, minimum weight is equal to minimum distance.

Example 9.3. Consider the code described in Example 9.1. Find the weights of the codewords and the minimum distance of the code.

Solution

The codewords and their weights are listed in the table on page 455. From the set of code vectors listed we see that the minimum weight is 3 and hence the minimum distance of the code is 3. The reader can verify that no two codewords of this code differ in less than three places.

The ability of a linear block code to correct *random* errors can be specified

Codewords	Weight
0 0 0 0 0 0	0
0 0 1 1 1 0	3
0 1 0 1 0 1	3
0 1 1 0 1 1	4
1 0 0 0 1 1	3
1 0 1 1 0 1	4
1 1 0 1 1 0	4
1 1 1 0 0 0	3

in terms of the minimum distance of the code. In defining the error correcting (or detecting) capability of a linear block code, we will assume that the decoder will associate a received vector R with a transmitted code vector C if C is the code vector closest to R in the sense of Hamming distance.

Theorem 9.2

A linear block code with a minimum distance $d_{\min}$ can correct up to $[(d_{\min}-1)/2]$ errors and detect up to $d_{\min}-1$ errors in each codeword, where $[(d_{\min}-1)/2]$ denotes the largest integer no greater than $(d_{\min}-1)/2$.

Proof

Let R be the received codeword and let C be the transmitted codeword. Let C' be any other codeword. Then, the Hamming distance between codewords C and C', $d(C, C')$, and the distances $d(C, R)$ and $d(C', R)$ satisfy

$$d(C, R) + d(C', R) \geq d(C, C')$$

since $d(U, V) =$ weight of $U + V$, where addition is on a bit by bit basis in modulo-2 arithmetic with no carry. If an error pattern of t' errors occurs, then the Hamming distance between the transmitted vector C and the received vector R is $d(C, R) = t'$. Now, since the code is assumed to have a distance of $d_{\min}$, $d(C, C') \geq d_{\min}$. Thus we have

$$d(C, C') \geq d_{\min}$$

$$d(C, R) = t'$$

or

$$d(C', R) \geq d_{\min} - t'$$

The decoder will correctly identify C as the transmitted vector if $d(C, R)$ is less than $d(C', R)$, where C' is any other codeword of the code. For $d(C, R) < d(C', R)$, the number of errors t' should satisfy

$$t' < \frac{d_{\min}}{2}$$

Thus we have shown that a linear block code with a minimum distance d_{min} can correct up to t errors (or identify a received codeword correctly) if

$$t \leq [(d_{min} - 1)/2] \tag{9.11}$$

where $[(d_{min} - 1)/2]$ denotes the largest integer no greater than $(d_{min} - 1)/2$. A similar proof can be developed to show that such a code can detect up to $d_{min} - 1$ or fewer errors.

From the preceding theorem it is obvious that for a given n and k, we would like to design an (n, k) code with d_{min} as large as possible. Unfortunately, there is no systematic way to do it except in some extremely simple cases.

9.2.3 Single Error-Correcting Hamming Codes

From Equation (9.11) it follows that linear block codes capable of correcting single errors must have a minimum distance $d_{min} = 3$. Such codes are extremely simple to construct. We know that when a single error occurs, say in the ith bit of the codeword, the syndrome of the received vector is equal to the ith row of H^T. Hence, if we choose the n rows of the $(n) \times (n - k)$ matrix H^T to be distinct, then the syndrome of all single errors will be distinct and we can correct single errors. There are two points we must remember in choosing the rows of H^T. First, we must not use a row of 0's since a syndrome of 0's corresponds to no error. Second, the last $n - k$ rows of H^T must be chosen so that we have an identity matrix in H as required in Equation (9.6). Once we choose H, the generator matrix can be obtained using Equations (9.5) and (9.6).

Each row in H^T has $(n - k)$ entries, each of which could be a 0 or 1. Hence we can have 2^{n-k} distinct rows of $(n - k)$ entries out of which we can select $2^{n-k} - 1$ distinct rows of H^T (the row of 0's is the only one we cannot use). Since the matrix H^T has n rows, for all of them to be distinct we need

$$2^{n-k} - 1 \geq n$$

or the number of parity bits in this (n, k) code satisfies the inequality

$$(n - k) \geq \log_2(n + 1) \tag{9.12}$$

So, given a message block size k, we can determine the minimum size n for the codewords from

$$n \geq k + \log_2(n + 1) \tag{9.13}$$

(Note that n has to be an integer.)

Example 9.4. Design a linear block code with a minimum distance of three and a message block size of eight bits.

Solution. From (9.13) we have

$$n \geqslant 8 + \log_2(n+1)$$

The smallest value of n that satisfies the preceding inequality is $n = 12$. Thus we need a (12, 8) block code. The transpose of the parity check matrix H will be of size 12 by 4. The last four rows of H^T will be a 4×4 identity matrix (Equation (9.6)). The first eight rows are arbitrarily chosen, with the restrictions that no row is identically zero, and all rows are distinct. One choice for H^T is given below:

$$H^T = \left[\frac{P}{I_{n-k}}\right] = \begin{bmatrix} 1 & 1 & 0 & 0 \\ 0 & 1 & 1 & 0 \\ 0 & 0 & 1 & 1 \\ 1 & 0 & 0 & 1 \\ 1 & 0 & 1 & 0 \\ 0 & 1 & 0 & 1 \\ 1 & 1 & 1 & 0 \\ 0 & 1 & 1 & 1 \\ \hline 1 & 0 & 0 & 0 \\ 0 & 1 & 0 & 0 \\ 0 & 0 & 1 & 0 \\ 0 & 0 & 0 & 1 \end{bmatrix}$$

The generator matrix for this code is

$$G = [I_k \,\vdots\, P] = \begin{bmatrix} 1 & 0 & 0 & 0 & 0 & 0 & 0 & 0 & 1 & 1 & 0 & 0 \\ 0 & 1 & 0 & 0 & 0 & 0 & 0 & 0 & 0 & 1 & 1 & 0 \\ 0 & 0 & 1 & 0 & 0 & 0 & 0 & 0 & 0 & 0 & 1 & 1 \\ 0 & 0 & 0 & 1 & 0 & 0 & 0 & 0 & 1 & 0 & 0 & 1 \\ 0 & 0 & 0 & 0 & 1 & 0 & 0 & 0 & 1 & 0 & 1 & 0 \\ 0 & 0 & 0 & 0 & 0 & 1 & 0 & 0 & 0 & 1 & 0 & 1 \\ 0 & 0 & 0 & 0 & 0 & 0 & 1 & 0 & 1 & 1 & 1 & 0 \\ 0 & 0 & 0 & 0 & 0 & 0 & 0 & 1 & 0 & 1 & 1 & 1 \end{bmatrix}$$

The receiver forms the syndrome $S = RH^T$ and accepts the received code vector if $S \equiv 0$. Otherwise, the receiver decodes as $R = C_i$, if $\min_j d(R, C_j) = d(R, C_i)$, where $C_1, C_2, \ldots C_{2^k}$ are the valid codewords of the code. Single errors are always corrected. Double errors can be detected but cannot be corrected. Multiple errors will usually result in incorrect decoding.

The efficiency of block codes with minimum distance three improves as the message block size is increased. The reader can verify that for $k = 64$ a minimum distance three code with efficiency 0.90 exists and that for $k = 256$ a code with efficiency 0.97 can be found.

So far in our discussion of linear block codes we have assumed that the decoding operation can be performed by finding a codeword C_i that is nearest to the received word R. This procedure would involve storing (or generating) 2^k codewords at the decoder and comparing R with these 2^k codewords. Since each codeword is of length n, the storage required at the decoder would be equal to $n2^k$ bits. Even for moderate sizes of k and n, the storage requirement becomes excessive. Furthermore, the search or comparing requires the processing of a total number of bits of the order of $n2^k$ and this processing requirement also becomes excessive when n and k are large. In the next section, we will look at a procedure that would require considerably less storage and processing for decoding linear block codes.

9.2.4 Table Lookup Decoding Using the Standard Array

Suppose that an (n, k) linear code C is used for error correcting purposes. Let $C_1, C_2, \ldots, C_{2^k}$ be the code vectors of C. Let R be the received vector. R can be any one of 2^n n-tuples and the decoder has the task of associating R with one of 2^k n-tuples that are valid codewords. The decoder can perform this task by partitioning the 2^n n-tuples into 2^k disjoint sets $T_1, T_2, \ldots, T_{2^k}$ such that each subset contains only one code vector C_i. Then if $R \in T_i$, the decoder identifies C_i as the transmitted code vector. Correct decoding results if the error vector E is such that $C_i + E \in T_i$ and an incorrect decoding results if $C_i + E \notin T_i$.

One way of partitioning the set of 2^n n-tuples is shown below in Table 9.1. The code vectors $C_1, C_2, \ldots, C_{2^k}$ are placed in the first row, *with the code vector of all zeros appearing in the leftmost position*. The first element in the second row E_2 is any one of $(2^n - 2^k)$ n-tuples not appearing in the first row. Once E_2 is chosen, the second row is completed by adding E_2 to the

Table 9.1. A Standard array for an (n, k) linear block code.

C_1	C_2	C_3	$\ldots$	C_{2^k}
E_2	$C_2 + E_2$	$C_3 + E_2 \ldots$		$C_{2^k} + E_2$
E_3	$C_2 + E_3$	$C_3 + E_3 \ldots$		$C_{2^k} + E_3$
$\vdots$	$\vdots$	$\vdots$		$\vdots$
$E_{2^{n-k}}$	$C_2 + E_{2^{n-k}}$	$.$	$\ldots$	$C_{2^k} + E_{2^{n-k}}$

codewords as shown in the table. Having completed the second row, an unused n-tuple E_3 is chosen to begin the third row and the sum of $C_i + E_3$ ($i = 1, 2, \ldots, 2^k$) are placed in the third row. The process is continued until all of 2^n n-tuples are used. The resulting array is called the *standard array* for the code and it consists of 2^k columns that are disjoint. Each column has 2^{n-k} n-tuples with the top most n-tuple as a code vector. The jth column is the partition T_j that will be used for decoding as described at the beginning of this section. The rows of the standard array are called *co-sets* and the first element in each row is called a *co-set leader.* The standard array has the following properties.

1. Each element in the standard array is distinct and hence the columns of the standard array T_j are disjoint.
2. If the error pattern caused by the channel coincides with a co-set leader, then the received vector is correctly decoded. On the other hand, if the error pattern is not a co-set leader, then an incorrect decoding will result. This can be easily verified by noting that the received vector corresponding to a code vector C_i will be in T_i if the error pattern caused by the channel is a co-set leader. Thus the co-set leaders are called *correctable error patterns.*

In order to minimize the probability of incorrect decoding, the 2^{n-k} *co-set leaders are chosen to be the error patterns that are most likely to occur for a given channel.* If E_i and E_j are two error patterns with weights W_i and W_j, then for a channel in which only random errors are occurring, E_i is more likely to occur than E_j if $W_j > W_i$. Hence, in constructing the standard array, the co-set leader should be chosen as the vector with minimum weight from the remaining available vectors.

Example 9.5. Construct the standard array for a (6, 3) linear block code whose generator matrix is given below:

$$G = \begin{bmatrix} 1 & 0 & 0 & 1 & 1 & 0 \\ 0 & 1 & 0 & 0 & 1 & 1 \\ 0 & 0 & 1 & 1 & 0 & 1 \end{bmatrix}$$

Solution. The code vectors are (000 000), (100 110), (010 011), (001 101), (110 101), (011 110), (101 011), and (111 000). One co-set decomposition of this code is shown in Table 9.2.

If the standard array shown above is used for decoding, then the decoder output would be the ith code vector (first entry in column i) if the received vector matches one of the vectors in column i. For example, a received vector

Table 9.2. Standard array for the (6, 3) block code of Example 9.5.

Syndrome	Co-set leader							
000	000000	100110	010011	001101	110101	011110	101011	111000
110	100000	000110	110011	101101	010101	111110	001011	011000
011	010000	110110	000011	011101	100101	001110	111011	101000
101	001000	101110	011011	000101	111101	010110	100011	110000
100	000100	100010	010111	001001	110001	011010	101111	111100
010	000010	100100	010001	001111	110111	011100	101001	111010
001	000001	100111	010010	001100	110100	011111	101010	111001
111	100001	000111	110010	101100	010100	111111	001010	011001

0 1 0 1 0 0 will be decoded as 1 1 0 1 0 1 and a received vector 1 1 1 1 0 0 will be decoded as 1 1 1 0 0 0.

A standard array has the following important property that leads to a simpler decoding process.

Theorem 9.3

All the 2^k n-tuples of a co-set have the same syndrome and the syndromes of different co-sets are different.

Proof

Consider the jth co-set leader E_j. An n-tuple in this co-set is $E_j + C_i$ for some i. The syndrome of this n-tuple is

$$[E_j + C_i]H^T = E_jH^T \tag{9.14}$$

since $C_iH^T = 0$. Thus the syndrome of any n-tuple in a co-set is equal to the syndrome of the co-set leader. To prove the second part, suppose that the syndromes of ith and jth co-sets ($i < j$) are equal. From (9.14) we have

$$E_iH^T = E_jH^T$$

or

$$[E_i + E_j]H^T = 0$$

This implies that the n-tuple $E_i + E_j$ is a code vector (only code vectors have 0 syndrome), say C_k. Then $E_j = E_i + C_k$, that is, E_j is in the ith co-set that contradicts the construction rule of the standard array. Thus for $i \neq j$, $E_iH^T \neq E_jH^T$.

The one-to-one correspondence between a co-set leader (correctable error pattern) and a syndrome leads to the following procedure for decoding:

1. Compute the syndrome RH^T for the received vector R. Let $RH^T = S$.
2. Locate the co-set leader E_i that has a syndrome $E_iH^T = S$. Then E_i is assumed to be the error pattern caused by the noisy channel.
3. The code vector C is obtained from R by $C = R + E_i$. Since the most probable error patterns have been chosen as the co-set leaders, this scheme will correct the 2^{n-k} most likely error patterns introduced by the channel.

The table lookup decoding scheme described above requires that we store the 2^{n-k} syndrome vectors of length $n-k$ bits and the 2^{n-k} n-bit error patterns corresponding to these syndromes. (Actually the storage required is one less since we need not store the syndrome of all zeros.) Thus the storage required is of the order of $2^{n-k} \times (2n-k)$ bits. For high efficiency codes $2n - k \simeq n$; hence, the storage required will be of the order of $n2^{n-k}$ bits as compared to the $n2^k$ bits required for the exhaustive search scheme described previously. When $n - k \ll k$, then the table look up scheme described in the preceding paragraph requires considerably less storage than a scheme that stores all 2^k, n-bit code words at the decoder.

In spite of the reduced storage requirements, the table lookup scheme might require many millions of bits of storage even for moderately complex codes. For example, for a (200, 175) code, the storage requirements will be of the order of $2^{25}(225) = (7.6)(10^9)$ bits. Thus the table lookup decoding scheme is also often impractical. The encoding operations for linear block codes are also complex if special care is not taken in the design of the codes. In the next section we will look at special classes of linear block codes that are easier to implement. In particular, we will look at a class of linear block codes, called cyclic codes, which can be implemented using simple shift registers.

9.3 BINARY CYCLIC CODES

Binary cyclic codes form a subclass of linear block codes described in the preceding section. Cyclic codes are attractive for two reasons. First, encoding and syndrome calculations can be easily implemented using simple shift registers with feedback connections. Secondly, these codes have a fair amount of mathematical structure that makes it possible to design codes with useful error-correcting properties.

We saw in the preceding section that linear block codes can be described well using the matrix representation. In this section we will develop a polynomial representation for cyclic codes and use this representation to derive procedures for encoding and syndrome calculations. We will also look at special subclasses of cyclic codes that are suitable for correcting burst-type errors in a system.

9.3.1 Algebraic Structure of Cyclic Codes

We begin our study of cyclic codes with the following definition:

An (n, k) linear block code C is called a cyclic code if it has the following property: If an n-tuple

$$V = (v_0, v_1, v_2, \ldots, v_{n-1}) \tag{9.15}$$

is a code vector of C, then the n-tuple

$$V^{(1)} = (v_{n-1}, v_0, v_1, \ldots, v_{n-2})$$

obtained by shifting V cyclically one place to the right is also a code vector of C.

From the above definition it is clear that

$$V^{(i)} = (v_{n-i}, v_{n-i+1}, \ldots, v_0, v_1, \ldots, v_{n-i-1}) \tag{9.16}$$

is also a code vector of C. This property of cyclic codes allows us to treat the elements of each code word as the coefficients of a polynomial of degree $n-1$. For example, the code word V can be represented by a *code polynomial* as

$$V(x) = v_0 + v_1x + v_2x^2 + \cdots + v_{n-1}x^{n-1} \tag{9.17}$$

The coefficients of the polynomial are 0's and 1's and they belong to a binary field with the following rules of addition and multiplication:

$$\begin{array}{ll} 0+0=0 & 0.0=0 \\ 0+1=1 & 0.1=0 \\ 1+0=1 & 1.0=0 \\ 1+1=0 & 1.1=1 \end{array}$$

The variable x is defined on the real line with $x^2 = x \cdot x$, $x^3 = (x^2)x = x \cdot x \cdot x$, and so on. In this notation if we have

$$p(x) = a_0 + a_1x + \cdots + a_mx^m$$

$$q(x) = b_0 + b_1x + \cdots + b_nx^n, \quad m > n$$

then

$$p(x)q(x) = \sum_{i=0}^{m} \sum_{j=0}^{n} a_ib_jx^{i+j}$$

$$p(x) + q(x) = (a_0 + b_0) + (a_1 + b_1)x + \cdots + (a_n + b_n)x^n + a_{n+1}x^{n+1} + \cdots + a_mx^m$$

The rules of addition and multiplication for the coefficients are as given above.

With the polynomial representation $V(x)$ for the code vector V given in Equation (9.17), we can obtain the code polynomial $V^{(i)}(x)$ for the code vector $V^{(i)}$ as

$$V^{(i)}(x) = v_{n-i} + v_{n-i+1}x + \cdots + v_0 x^i + v_1 x^{i+1} + \cdots + v_{n-i-1}x^{n-1}$$

It can be shown that $V^{(i)}(x)$ is the remainder resulting from dividing $x^i V(x)$ by $x^n + 1$, that is,

$$x^i V(x) = q(x)(x^n + 1) + V^{(i)}(x) \tag{9.18}$$

With the polynomial representation of code vectors, we now state and prove the following important theorem on cyclic codes.

Theorem 9.4

If $g(x)$ is a polynomial of degree $(n-k)$ and is a factor of $x^n + 1$, then $g(x)$ generates an (n, k) cyclic code in which the code polynomial $V(x)$ for a data vector $D = (d_0, d_1, d_2, \ldots, d_{k-1})$ is generated by

$$V(x) = D(x)g(x) \tag{9.19}$$

Proof

This theorem can be proved as follows. Consider k polynomials $g(x)$, $xg(x)$, $x^2g(x), \ldots, x^{k-1}g(x)$, which all have degree $n-1$ or less. Now, any linear combination of these polynomials of the form

$$\begin{aligned} V(x) &= d_0 g(x) + d_1 x g(x) + \cdots + d_{k-1}x^{k-1}g(x) \\ &= D(x)g(x) \end{aligned}$$

is a polynomial of degree $n-1$ or less and is a multiple of $g(x)$. There are a total of 2^k such polynomials corresponding to the 2^k data vectors and the code vectors corresponding to the 2^k polynomials form a linear (n, k) code. To prove that this code is cyclic, let $V(x) = v_0 + v_1 x + \cdots + v_{n-1}x^{n-1}$ be a code polynomial in this code. Consider

$$\begin{aligned} xV(x) &= v_0 x + v_1 x^2 + \cdots + v_{n-1}x^n \\ &= v_{n-1}(x^n + 1) + (v_{n-1} + v_0 x + \cdots + v_{n-2}x^{n-1}) \\ &= v_{n-1}(x^n + 1) + V^{(1)}(x) \end{aligned}$$

where $V^{(1)}(x)$ is a cyclic shift of $V(x)$. Since $xV(x)$ and $x^n + 1$ are both divisible by $g(x)$, $V^{(1)}(x)$ must be divisible by $g(x)$. Thus $V^{(1)}(x)$ is a multiple of $g(x)$ and can be expressed as a linear combination of $g(x)$, $xg(x), \ldots, x^{k-1}g(x)$. This says that $V^{(1)}(x)$ is also a code polynomial. Hence from the definition of cyclic codes, it follows that the linear code generated by $g(x), xg(x), \ldots, x^{k-1}g(x)$ is an (n, k) cyclic code.

The polynomial $g(x)$ is called the *generator polynomial* of the cyclic code. Given the generator polynomial $g(x)$ of a cyclic code, the code can be put into a systematic form as

$$V = (\underbrace{r_0, r_1, r_2 \ldots, r_{n-k-1}}_{\substack{n-k \text{ parity} \\ \text{check bits}}} \quad \underbrace{d_0, d_1, \ldots, d_{k-1}}_{k \text{ message bits}}) \tag{9.20}$$

where

$$r(x) = r_0 + r_1 x + r_2 x^2 + \cdots + r_{n-k-1} x^{n-k-1}$$

is the *parity check polynomial* for the message polynomial $D(x)$. The parity check polynomial $r(x)$ is the remainder from dividing $x^{n-k}D(x)$ by $g(x)$:

$$x^{n-k}D(x) = q(x)g(x) + r(x) \tag{9.21}$$

where $q(x)$ and $r(x)$ are the quotient and remainder, respectively. The code polynomial $V(x)$ is given by

$$V(x) = r(x) + x^{n-k}D(x) \tag{9.22}$$

Example 9.6. The generator polynomial of a $(7, 4)$ cyclic code is $g(x) = 1 + x + x^3$. Find the 16 codewords of this code in the following ways:

(a) by forming the code polynomials using $V(x) = D(x)g(x)$, where $D(x)$ is the message polynomial,

(b) by using the systematic form given in Equations (9.20) and (9.21).

Solution. (a) Consider the message vector

$$D = (d_0, d_1, d_2, d_3) = (1\ 0\ 1\ 0)$$

The message polynomial is

$$D(x) = 1 + x^2$$

The code polynomial $V(x)$ is given by

$$\begin{aligned} V(x) &= D(x)g(x) \\ &= (1 + x^2)(1 + x + x^3) \\ &= 1 + x^2 + x + x^3 + x^3 + x^5 \\ &= 1 + x + x^2 + x^5 \end{aligned}$$

since $x^3 + x^3 = (1 + 1)x^3 = (0)x^3 = 0$. Hence the code vector is $V = (1\ 1\ 1\ 0\ 0\ 1\ 0)$. The remaining code vectors are shown in Table 9.3.

(b) In the systematic form, the first three bits are the check bits and the last four bits are the message bits. The check bits are obtained from the remainder polynomial $r(x)$ given by

$$\frac{x^{n-k}D(x)}{g(x)} = q(x) + r(x)$$

where $q(x)$ is the quotient of the division. For the message vector, say $D = (1\ 1\ 1\ 0)$, the message polynomial is

$$D(x) = 1 + x + x^2,$$

and

$$x^3D(x) = x^3 + x^4 + x^5$$

The division of $x^3D(x)/g(x)$ can be done as:

$$\begin{array}{r} x^2 + x \\ x^3 + x + 1 \overline{\smash{)}\, x^5 + x^4 + x^3} \\ \underline{x^5 + x^3 + x^2} \\ x^4 + x^2 \\ \underline{x^4 + x^2 + x} \\ x \end{array}$$

Note: Subtraction is the same as addition in modulo-2 arithmetic.

Hence, $r(x) = x$, and

$$V = (\underbrace{0\ \ 1\ \ 0}_{r}\ \ \ \underbrace{1\ \ 1\ \ 1\ \ 0}_{D})$$

The remaining code vectors are shown in column 2 of Table 9.3.
Column 3 contains the codewords in systematic form. Column 2 contains the codewords in a non-systematic form. Observe that in both forms the code contains the same set of 16 code vectors.

The main advantage of cyclic codes over other classes of linear block codes is the ease of implementation of the encoding process as described in the following section.

9.3.2 Encoding Using an $(n - k)$ Bit Shift Register

The encoding operations described in Equations (9.20), (9.21), and (9.22) involve the division of $x^{n-k}D(x)$ by the generator polynomial $g(x)$ to calculate the parity check polynomial $r(x)$. This division can be accomplished using the dividing circuit* consisting of a feedback shift register as shown in Figure 9.3.

*A very good treatment on the use of shift registers to perform multiplication and division of polynomials using modulo-2 arithmetic may be found in Peterson's book on error-correcting codes.

Table 9.3. Codewords generated by $g(x) = 1 + x + x^3$.

Messages	Code vectors obtained using $V(x) = D(x)g(x)$	Code vectors obtained using the systematic form Equations (9.20), (9.21)
0 0 0 0	0 0 0 0 0 0 0	0 0 0 0 0 0 0
0 0 0 1	0 0 0 1 1 0 1	1 0 1 0 0 0 1
0 0 1 0	0 0 1 1 0 1 0	1 1 1 0 0 1 0
0 0 1 1	0 0 1 0 1 1 1	0 1 0 0 0 1 1
0 1 0 0	0 1 1 0 1 0 0	0 1 1 0 1 0 0
0 1 0 1	0 1 1 1 0 0 1	1 1 0 0 1 0 1
0 1 1 0	0 1 0 1 1 1 0	1 0 0 0 1 1 0
0 1 1 1	0 1 0 0 0 1 1	0 0 1 0 1 1 1
1 0 0 0	1 1 0 1 0 0 0	1 1 0 1 0 0 0
1 0 0 1	1 1 0 0 1 0 1	0 1 1 1 0 0 1
1 0 1 0	1 1 1 0 0 1 0	0 0 1 1 0 1 0
1 0 1 1	1 1 1 1 1 1 1	1 0 0 1 0 1 1
1 1 0 0	1 0 1 1 1 0 0	1 0 1 1 1 0 0
1 1 0 1	1 0 1 0 0 0 1	0 0 0 1 1 0 1
1 1 1 0	1 0 0 0 1 1 0	0 1 0 1 1 1 0
1 1 1 1	1 0 0 1 0 1 1	1 1 1 1 1 1 1

In Figure 9.3 we use the symbol $\rightarrow\square\rightarrow$ to denote the flip-flops that make up a shift register, $\oplus$ to denote modulo-2 adders, and $\rightarrow\!(g_i)\!\rightarrow$ to denote a closed path if $g_i = 1$ and an open path if $g_i = 0$. We will assume that at the occurrence of a clock pulse, the register inputs are shifted into the register and appear at the output at the end of the clock pulse.

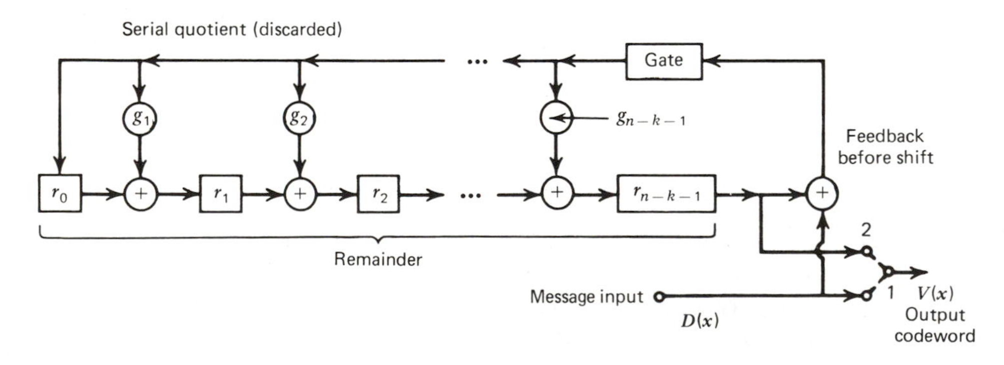

Figure 9.3 An $(n\text{-}k)$ stage encoder for an (n, k) binary cyclic code generated by $g(x) = 1 + g_1x + g_2x^2 + \cdots + g_{n-k-1}x^{n-k-1} + x^{n-k}$.

The encoding operation consists of two steps:

1. With the gate turned on and the switch in position 1, the information digits $(d_0, d_1, \ldots, d_{k-1})$ are shifted into the register (with d_{k-1} first) and simultaneously into the communication channel. As soon as the k information digits have been shifted into the register, the register contains the parity check bits $(r_0, r_1, \ldots, r_{n-k-1})$.
2. With the gate turned off and the switch in position 2, the contents of the shift register are shifted into the channel. Thus the codeword $(r_0, r_1, \ldots, r_{n-k-1}, d_0, d_1, \ldots, d_{k-1})$ is generated and sent over the channel.

The hardware required to implement the encoding scheme consists of

1. an $(n-k)$ bit shift register
2. a maximum of $(n-k)$ modulo-2 adders
3. an AND gate and
4. a counter to keep track of the shifting operations.

This encoder is much simpler than the encoder needed for implementing an (n, k) linear block code in matrix form where portions of the G or H matrix have to be stored.

Example 9.7. Design an encoder for the (7, 4) binary cyclic code generated by $g(x) = 1 + x + x^3$ and verify its operation using the message vector (0 1 0 1).

Solution. The encoder for this code is obtained from Figure 9.4 with $n = 7$, $k = 4$, $g_0 = 1$, $g_1 = 1$, $g_2 = 0$, and $g_3 = 1$. The encoding operation on the message vector 0 1 0 1 is shown in Table 9.4.

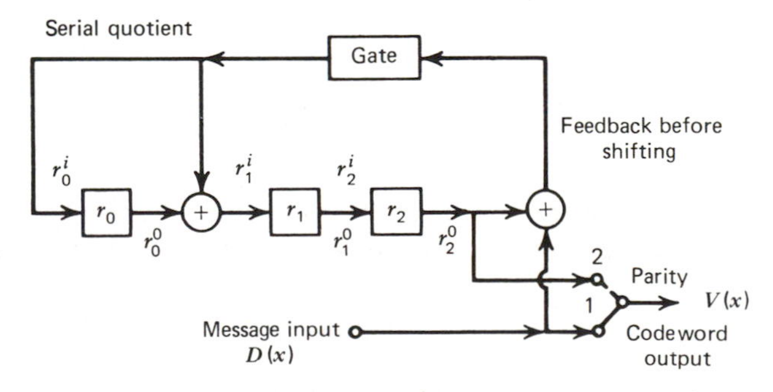

Figure 9.4 Encoder for a (7, 4) cyclic code generated by $g(x) = 1 + x + x^3$.

Table 9.4. Encoding operation with message vector = (0 1 0 1). $r_0^i = d$, $r_1^i = r_0^0 + d$, $r_2^i = r_1^0$, where d is the message bit; shift operation moves input bits into the registers and to the outputs. The code vector for (0 1 0 1) is (1 1 0 0 1 0 1).

Input bit	Register inputs				Register outputs		
d	r_0^i	r_1^i	r_2^i		r_0^0	r_1^0	r_2^0
—	0	0	0		0	0	0
1	1	1	0	→ shift 1 →	1	1	0
0	0	1	1	→ shift 2 →	0	1	1
1	0	0	1	→ shift 3 →	0	0	1
0	1	1	0	→ shift 4 →	1	1	0

9.3.3 Syndrome Calculation, Error Detection, and Error Correction

Suppose that a code vector V is transmitted over a noisy communication channel. The received vector R may or may not be the transmitted code vector; indeed R may not be any one of the 2^k valid code vectors of the code. The function of the decoder is to determine the transmitted code vector based on the received vector.

The decoder first tests whether or not the received vector is a valid code vector by calculating the syndrome of the received word. If the syndrome is zero, the received vector is divisible by the generator polynomial and hence is a code vector. The decoder accepts the received vector as the transmitted code vector. A nonzero syndrome indicates that transmission errors have occurred. The syndrome $S(x)$ of the received vector $R(x)$ is the remainder resulting from dividing $R(x)$ by $g(x)$, that is,

$$\frac{R(x)}{g(x)} = P(x) + \frac{S(x)}{g(x)} \tag{9.23}$$

where $P(x)$ is the quotient of the division. The syndrome $S(x)$ is a polynomial of degree $n-k-1$ or less. If $E(x)$ is the error pattern caused by the channel, then

$$R(x) = V(x) + E(x)$$

and

$$\frac{R(x)}{g(x)} = \frac{V(x)}{g(x)} + \frac{E(x)}{g(x)} \tag{9.24}$$

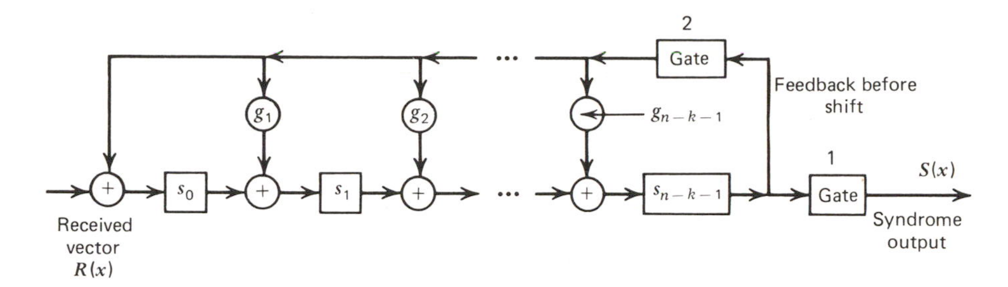

Figure 9.5 An $(n-k)$ syndrome calculation circuit for an (n, k) cyclic code.

Now $V(x) = D(x)g(x)$ where $D(x)$ is the message polynomial. Hence from Equations (9.23) and (9.24) we obtain

$$E(x) = [P(x) + D(x)]g(x) + S(x) \tag{9.25}$$

Hence, the syndrome of $R(x)$ is equal to the remainder resulting from dividing the error pattern by the generator polynomial, and the syndrome contains information about the error pattern that can be used for error correction. The division operation required for calculating the syndrome is implemented using a circuit shown in Figure 9.5, which is similar to the encoding circuit shown in Figure 9.3.

The syndrome calculations are carried out as follows:

1. The register is first initialized. Then, with gate 2 turned on and gate 1 turned off, the received vector $R(x)$ is entered into the shift register.
2. After the entire received vector is shifted into the register, the contents of the register will be the syndrome. Now, gate 2 is turned off, gate 1 is turned on, and the syndrome vector is shifted out of the register. The circuit is ready for processing the next received vector.

Cyclic codes are extremely well suited for error detection. Error detection can be implemented by simply adding an additional flip-flop to the syndrome calculator. If the syndrome is nonzero, the flip-flop sets and an indication of error is provided. Thus if we are interested in error detection only, cyclic codes are much easier to implement than noncyclic linear block codes discussed in Section 9.2.

If we are interested in error correction, then the decoder has to determine a correctable error pattern $E(x)$ from the syndrome $S(x)$ and add $E(x)$ to $R(x)$ to determine the transmitted code vector $V(x)$.

In principle, the decoding operation can be implemented using the system shown in Figure 9.6.

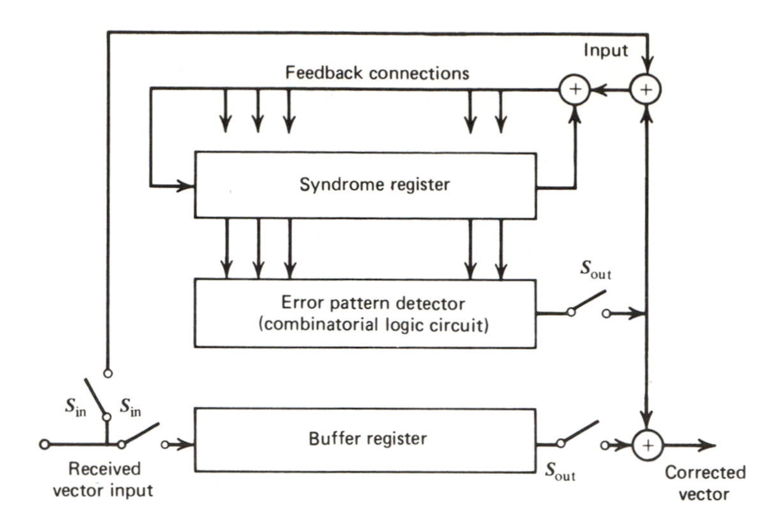

Figure 9.6 General form of a decoder for cyclic codes. Received data is shifted in with S_{in} closed and S_{out} open. Error correction is performed with S_{out} closed and S_{in} open.

The error correction procedure consists of the following steps.

Step 1. The received vector is shifted into the buffer register and the syndrome register.

Step 2. After the syndrome for the received vector is calculated and placed in the syndrome register, the contents of the syndrome register are read into the detector. The detector is a combinatorial logic circuit designed to output a 1 if and only if the syndrome in the syndrome register corresponds to a correctable error pattern with an error at the highest order position x^{n-1}. That is, if the detector output is 1, then the received digit at the right-most stage of the buffer register is assumed to be erroneous and hence is corrected. If the detector output is 0, then the received digit at the right-most stage of the buffer register is assumed to be correct. Thus the detector output is the estimated error value for the digit coming out of the buffer register.

Step 3. The first received digit is shifted out of the buffer and at the same time the syndrome register is shifted right once. If the first received digit is in error, the detector output will be 1, which is used for correcting the first received digit. The output of the detector is also fed to the syndrome register to modify the syndrome. This results in a new syndrome corresponding to the altered-received vector shifted to the right by one place.

Step 4. The new syndrome is now used to check whether or not the second received digit, now at the right-most stage of the buffer register, is an erroneous digit. If so, it is corrected, a new syndrome is calculated as in step 3, and the procedure is repeated.

Step 5. The decoder operates on the received vector digit by digit until the entire received vector is shifted out of the buffer.

At the end of the decoding operation (that is, after the received vector is shifted out of the buffer), errors will have been corrected if they correspond to an error pattern built into the detector, and the syndrome register will contain all zeros. If the syndrome register does not contain all zeros, then an uncorrectable error pattern has been detected.

The decoder shown in Figure 9.5 can be used for decoding any cyclic code. However, whether or not the decoder will be practical will depend on the complexity of the combinatorial logic circuits in the error detector. Special classes of codes have been developed that lead to simpler logic circuits in the decoder. However, this is accomplished at the expense of lowering the efficiency of the code for a given block size. We will briefly describe some important classes of cyclic codes that can be used for error correction without requiring excessively complex decoding circuits.

9.3.4 Special Classes of Cyclic Codes: Bose–Chaudhuri–Hocquenghem (BCH) Codes

Optimum design of error-correcting codes consists of designing a code with the smallest block size (n) for a given size of the message block (k) and for a desirable value of the minimum distance ($d_{\min}$) for the code. Or, for a given code length (n) and efficiency (k/n), we might want to design codes with the largest possible values for $d_{\min}$. That is, we might want to design codes with the best error-correcting capabilities. BCH codes, as a class, are the most extensive and powerful error-correcting cyclic codes known.

Decoding algorithms for BCH codes can be implemented with a reasonable amount of equipment. A detailed mathematical description of BCH codes would require extensive use of modern algebra. To discuss modern algebra is beyond the scope of this book and hence we will not include a mathematical description of BCH codes in this book. However, we will state the following properties of the BCH code that illustrate the power of this code:

For any positive integer m and t ($t < 2^{m-1}$) there exists a BCH code with the following parameters:

$$\text{Block length:} \quad n = 2^{m-1}$$
$$\text{Number of parity check bits:} \quad n - k \leqslant mt$$
$$\text{Minimum distance:} \quad d_{\min} \geqslant 2t + 1 \tag{9.26}$$

Several iterative procedures for decoding BCH codes are available. Many of these procedures can be programmed on a general purpose digital computer. In many practical applications, digital computers form an integral part of data communication networks. In such systems software implementation of decoding algorithms has several advantages over hardware implementation. The reader may find descriptions of BCH codes in the references listed at the end of this chapter.

Majority Logic Decodable Codes. These codes form a smaller sub-class of cyclic codes than do the BCH codes. Also, they are slightly inferior to BCH codes in terms of error-correcting capabilities for most interesting values of code length and efficiency. The main advantage of majority logic decodable codes is that the decoding operation can be implemented using simple circuits. The decoder for these codes has the form shown in Figure 9.6 with the combinatorial portion consisting of modulo-2 adders and a few layers of majority gates. An example is shown in Figure 9.7 for a (7, 4) majority logic decodable code. This code has $d_{\min} = 3$ and the reader can verify (using the decoding steps given below Figure 9.6) that the decoder can correct single errors.

Several classes of cyclic codes have been found recently that could be decoded using layers of majority gates. The construction procedures and the derivation of decoding rules are based on properties of finite geometries. Interested readers may refer to the books listed at the end of this chapter.

Shortened Cyclic Codes. The cyclic codes we have considered so far have generator polynomials that are divisors of $x^n + 1$. In general, the polynomial $x^n + 1$ has relatively few divisors and as a result there are usually very few cyclic codes of a given length. To circumvent this difficulty and to increase the number of pairs (n, k) for which useful codes can be constructed, cyclic codes are often used in shortened form. In the shortened form the last i information digits are always taken to be zeros (i.e., the last i bits of the codeword are padded with zeros). These bits are not transmitted; the decoder for the original cyclic code can decode the shortened code words simply by padding the received $(n - i)$-tuple with i zeros.

Hence, given an (n, k) cyclic code, it is always possible to construct an $(n - i, k - i)$ shortened cyclic code. The shortened cyclic code is a subset of

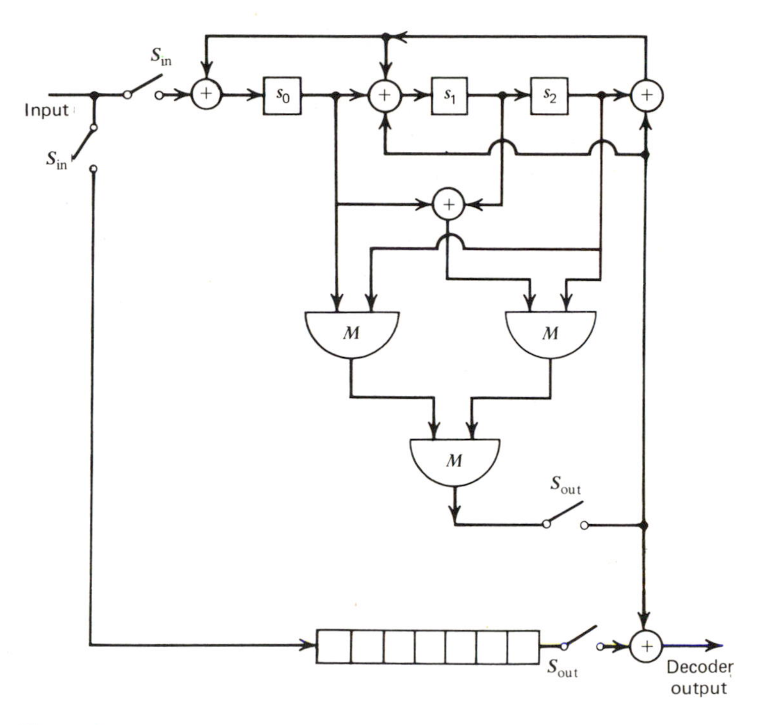

Figure 9.7 Decoder for a (7, 4) majority logic code. ($g(x) = 1 + x + x^3$). M–majority gate.

the cyclic code from which it was derived and hence its minimum distance and error correcting ability is at least as great as that of the original code. The encoding operation, syndrome calculation, and error correction procedures for shortened cyclic codes are identical to the procedures described for cyclic codes. As a result, shortened cyclic codes inherit nearly all of the implementation advantages and much of the mathematical structure of cyclic codes.

9.4 BURST-ERROR-CORRECTING CODES

In the preceding sections we developed coding techniques for channels in which transmission errors occur independently in digit positions within a codeword. We will now describe cyclic and shortened cyclic codes to correct bursts of errors due to impulse noise in the communication channel. Impulse noise causes transmission errors to cluster into bursts. Codes designed for correcting random errors are in general not efficient for correcting burst errors. Special codes have been developed for correcting burst errors.

In this section we will describe codes for correcting a single error burst of length q. We start with the definition of *a burst of length* q which is defined as a vector whose nonzero components are confined to q consecutive digit positions, the first and last of which are nonzero. For example, the vector $V = (0\,0\,1\,0\,1\,0\,0\,0\,1\,0\,0\,0)$ has a burst of length 7. A code that is capable of correcting all burst errors of length q or less is called q burst-error-correcting code, or the code is said to have burst-error-correcting ability q.

It is obvious that for a given n and q we would like to construct an (n, k) code with as small a redundancy $n - k$ as possible. The following theorem states a lower bound on the number of check bits needed:

Theorem 9.4

The number of parity check digits of a q-burst-error-correcting code must be at least $2q$, that is,

$$n - k \geqslant 2q \tag{9.27}$$

Proof

We can prove this theorem by showing that the following statements are true:

1. A necessary condition for an (n, k) linear code to be able to correct all bursts of length q or less is that no burst of length $2q$ or less can be a code vector.
2. The number of parity check digits of an (n, k) linear code that has no burst of length b or less as a code vector is at least b, that is, $n - k \geqslant b$.

To prove statement (1), consider a code vector V with a burst of $2q$ or less. This code vector can be expressed as a vector sum of V_1 and V_2 of length q or less. Thus, in a standard array of the code, V_1 and V_2 must be in the same co-set. If one of these vectors is used as a co-set leader (correctable error pattern), then the other will be an uncorrectable error pattern. As a result, this code will not be able to correct all bursts of length q or less.

To prove statement (2), consider the vectors whose nonzero components are confined to the first b bits. There are 2^b such vectors. No two such vectors can be in the same co-set of a standard array for this code; otherwise, their sum, which is a burst of b or less, would be a code vector. Hence, these 2^b vectors must be in 2^b distinct co-sets. There are a total of 2^{n-k} co-sets for an (n, k) code. Thus $(n - k)$ must be at least equal to b. Combining (1) and (2), we obtain the result stated in Equation (9.27).

Table 9.5. Some burst-error-correcting cyclic and shortened cyclic codes.

$n-k-2l$	Code (n, k)	Burst-correcting ability l	Generator polynomial[a]
0	(7, 3)	2	35
	(15, 9)	3	171
	(19, 11)	4	1151
	(27, 17)	5	2671
	(34, 22)	6	15173
	(38, 24)	7	114361
	(50, 34)	8	224531
	(56, 38)	9	1505773
	(59, 39)	10	4003351
1	(15, 10)	2	65
	(27, 20)	3	311
	(38, 29)	4	1151
	(48, 37)	5	4501
	(67, 54)	6	36365
	(103, 88)	7	114361
	(96, 79)	8	501001
2	(31, 25)	2	161
	(63, 55)	3	711
	(85, 75)	4	2651
	(131, 119)	5	15163
	(169, 155)	6	55725
3	(63, 56)	2	355
	(121, 112)	3	1411
	(164, 153)	4	6255
	(290, 277)	5	24711
4	(511, 499)	4	10451
5	(1023, 1010)	4	22365

[a] Note: Generator polynomials are given in an octal representation. Each digit in the table represents three binary digits according to the following code:

$0 \leftrightarrow 000 \quad 2 \leftrightarrow 010 \quad 4 \leftrightarrow 100 \quad 6 \leftrightarrow 110$

$1 \leftrightarrow 001 \quad 3 \leftrightarrow 011 \quad 5 \leftrightarrow 101 \quad 7 \leftrightarrow 111$

The binary digits are then the coefficients of the polynomial, with the high-order coefficients at the left. For example, the binary representation of 171 is 0 0 1 1 1 1 0 0 1, and the corresponding polynomial is $g(x) = x^6 + x^5 + x^4 + x^3 + 1$.

The bound given in (9.27) says that the burst-error-correcting capability of an (n, k) code is at most $(n-k)/2$. In other words, the upper bound on the burst-error-correcting capability of an (n, k) code is

$$q \leq (n-k)/2 \tag{9.28}$$

The bound given in (9.28) is called the ***Reiger bound*** and we use it to define the burst-correcting efficiency z of an (n, k) code as

$$z = 2q/(n-k) \tag{9.29}$$

Unlike the random error-correcting codes in which the most useful codes have been derived by analytical techniques, the best burst-error-correcting codes have been found by computer aided search procedures. We list in Table 9.5 some of the very efficient cyclic and shortened cyclic codes for correcting single short burst of errors.

If a code is needed for *detecting* burst errors of length $\leq d$, then the number of check bits needed must satisfy

$$n-k \geq d \tag{9.30}$$

The decoding algorithms for correcting burst errors are similar to the algorithms described in preceding sections for cyclic codes designed to correct random errors.

9.5 BURST- AND RANDOM-ERROR-CORRECTING CODES

In the preceding sections we dealt with the design of codes to correct specific well-defined classes of error patterns. We treated the problems of correcting random errors, and burst errors, separately. Unfortunately, in most practical systems, errors occur neither independently, at random, nor in well-defined bursts. Consequently, random-error-correcting codes or single-burst-error-correcting codes will be either inefficient or inadequate for combating a mixture of random and burst errors. For channels in which both types of errors occur, it is better to design codes capable of correcting random errors and/or single or multiple bursts.

Several methods of constructing codes for the correction of random and burst errors have been proposed. The most effective method uses the interlacing technique. Given an (n, k) cyclic code, it is possible to construct a $(\lambda n, \lambda k)$ cyclic *interlaced code* by simply arranging λ code vectors of the original code into λ rows of a rectangular array and transmitting them column by column. The resulting code is called an interlaced code with an interlacing degree λ.

In an interlaced code, a burst of length λ or less will affect no more than one digit in each row since the transmission is done on a column by column fashion. If the original code (whose code words are rows of the two-dimensional array) can correct single errors, then the interlaced code can correct single bursts of length λ or less. If the original code can correct, say, t errors ($t > 1$), then the interlaced code can correct any combination of t bursts of length λ or less. The performance of the $(\lambda n, \lambda k)$ interleaved cyclic code against purely random errors is identical to that of the (n, k) cyclic code from which it was generated. The following example illustrates the concepts of interleaving.

Consider a (15, 7) BCH code generated by $g(x) = x^8 + x^4 + x^2 + x + 1$. This code has a minimum distance 5; and hence it is double-error-correcting. We can construct a (75, 35) interleaved code with $\lambda = 5$ with a burst-error-correcting ability of 10. The arrangement of codewords in an interleaved fashion is shown in Table 9.6. A 35-bit message block is divided into five 7-bit message blocks, and five codewords of length 15 bits are generated using $g(x)$ given above. These codewords are arranged as five rows of a 5×15 matrix. The columns of the matrix are transmitted in the order indicated in Table 9.6 as a 75-bit-long code vector.

To illustrate the burst- and random-error-correcting capabilities of this code, assume that errors have occurred in bit positions 5, 37 through 43, and 69. The decoder operates on the rows of the Table 9.6. Each row has a maximum of two errors, and the (15,7) BCH code from which the rows of the table are obtained is capable of correcting up to two errors per row. Hence, the error pattern shown in Table 9.6 can be corrected. The reader may think of the isolated errors in bits 5 and 69 as random errors, and the cluster of errors in bit positions 37 to 43 as a burst error.

Table 9.6

Each row is a 15 bit code word. (Five code words)

1	6		36	41			66	71
2	7		37	42			67	72
3	8	···	38	43	·	···	68	73
4	9		39	44			69	74
5	10		40	45			70	75

While operating on the rows of the code array may be an obvious way to encode and decode an interlaced code, this is generally not the simplest implementation. The simplest implementation results from the property that if the original code is cyclic, then the interlaced code is also cyclic. Further, the generator polynomial for the interlaced code is $g(x^\lambda)$, where $g(x)$ is the generator polynomial for the original code. Thus, encoding and decoding can be accomplished using shift registers. The decoder for the interlaced code can be derived from the decoder for the original code by replacing each shift register stage of the original decoder by λ stages without changing other connections. This allows the decoder to look at successive rows of the code array on successive decoder cycles. Also, if the decoder for the original code was simple, then the decoder for the interlaced code will also be simple. Thus, the interlacing technique is an effective tool for deriving long powerful codes from short optimal codes.

9.6 CONVOLUTIONAL CODES

The main difference between block codes discussed in the preceding sections and the convolutional (or recurrent) codes is the following: In the block code, a block of n digits generated by the encoder in a particular time unit depends only on the block of k input message digits within that time unit. In a convolutional code, the block of n code digits generated by the encoder in a time unit depends on not only the block of k message digits within that time unit, but also on the preceding $(N-1)$ blocks of message digits $(N > 1)$. Usually the values of k and n will be small.

Like block codes, convolutional codes can be designed to either detect or correct errors. However, because data are usually retransmitted in blocks, block codes are better suited for error detection and convolutional codes are mainly used for error correction. Encoding of convolutional codes can be accomplished using simple shift registers, and several practical procedures for decoding have also been developed. Recent studies have shown that convolutional codes perform as well or better than block codes in many error control applications.

Analysis of the performance of convolutional codes is quite complicated since the encoding and decoding operations are highly dependent. Some analytical results are available, but these are of limited applicability. In general, it is easier to evaluate the performance of convolutional codes by simulating the encoding and decoding operations, and the channel on a digital computer.

The following sections contain an introductory level treatment of convolutional codes. We will first describe the encoding operation for convolutional codes and then look at various methods of decoding convolutional codes.

9.6.1 Encoder for Convolutional Codes

A convolutional encoder, shown in its general form in Figure 9.8, takes sequences of message digits and generates sequences of code digits. In any time unit, a message block consisting of k digits is fed into the encoder and the encoder generates a code block consisting of n code digits ($k < n$). The n-digit code block depends not only on the k-digit message block of the same time unit, but also on the previous $(N-1)$ message blocks. The code generated by the above encoder is called an (n, k) convolutional code of *constraint length* nN digits and rate efficiency k/n. The parameters of the encoder, k, N, and n, are in general small integers and the encoder consists of shift registers and modulo-2 adders. An encoder for a (3, 1) convolution code with a constraint length of nine bits is shown in Figure 9.9.

The operation of the encoder shown in Figure 9.9 proceeds as follows. We assume that the shift register is clear initially. The first bit of the input data is entered into D_1. During this message bit interval the commutator samples the modulo-2 adder outputs c_1, c_2, and c_3. Thus a single message bit yields three output bits. The next message bit in the input sequence now enters D_1, while the content of D_1 is shifted into D_2 and the commutator again samples the

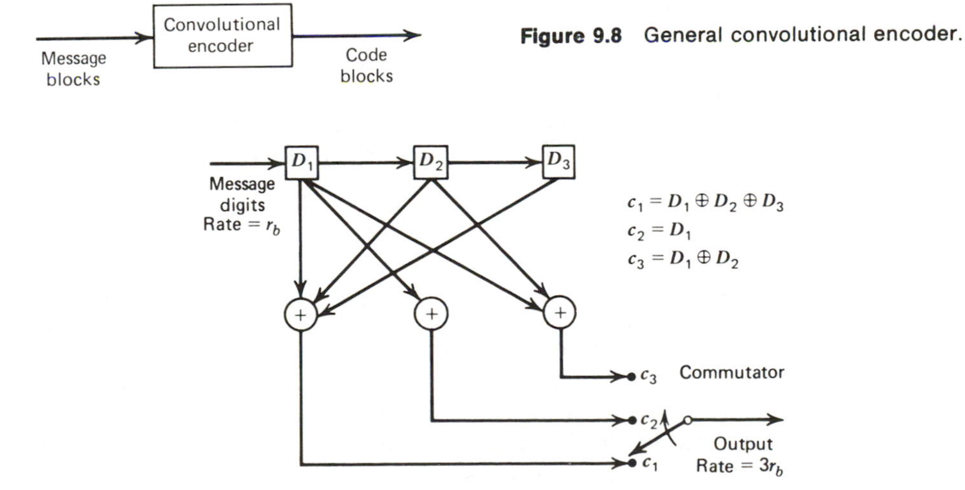

Figure 9.8 General convolutional encoder.

Figure 9.9 An example of convolutional encoder, $N = 3$, $n = 3$, $k = 1$.

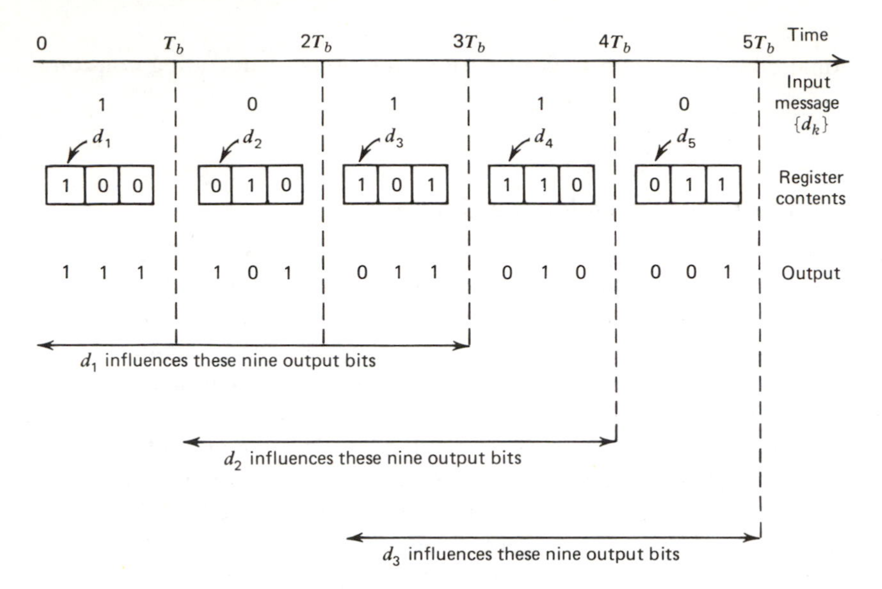

Figure 9.10 Encoding operation of the convolutional encoder shown in Figure 9.9.

three adder outputs. The process is repeated until the last message digit is shifted into D_3. An example of the encoding operation is shown in Figure 9.10.

It must be emphasized here that the convolutional encoder operates on the message stream in a continuous manner; that is, the message stream is run continuously through the encoder unlike in block coding schemes where the message stream is first divided into long blocks and then encoded. Thus the convolutional encoder requires very little buffering and storage hardware. Another point that must be remembered is that each message bit has its influence on $N \times n$ digits, where N is the size of the shift register and n is the number of commutator segments.

9.6.2 Decoder for Convolutional Codes: Exhaustive Search Method

In order to understand the decoding procedures used for convolutional codes, let us consider the code tree shown in Figure 9.11, which applies to the convolutional encoder shown in Figure 9.9. The starting point on the code tree is at left and it corresponds to the situation before the occurrence of the ith message bit d_i. For convenience let us assume that the message bits d_{i-1} and d_{i-2} are zero. The paths shown in the code tree are generated using the convention that we shall diverge upward from a node of the tree when the

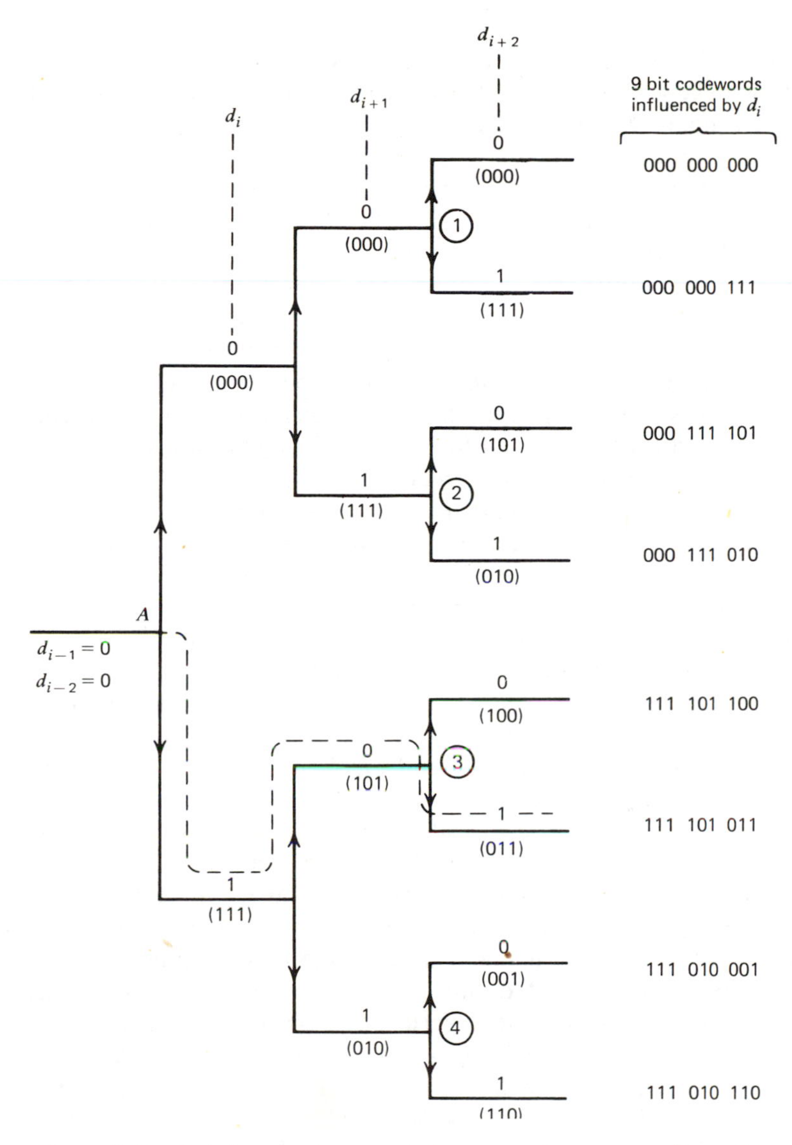

Figure 9.11 Code tree for the convolutional encoder shown in Figure 9.9.

input bit is 0. Since the starting node (A) corresponds to $d_{i-1} = 0$ and $d_{i-2} = 0$, then, if $d_i = 0$ we will move upward from the initial node and the coder output will be 0 0 0. The coder output is shown marked on the upward path from the initial node. Nodes 1, 2, 3, and 4 can be considered as starting nodes for d_{i+2} with $d_i d_{i+1} = 0\,0$, $0\,1$, $1\,0$, and $1\,1$, respectively. For any given starting node, the first message bit influences the code blocks generated from the starting node and the two succeeding nodes. Thus each message bit will have an effect on nine code digits and there are eight distinct 9-bit code blocks associated with each starting node.

The code tree can be used in the decoding operation as follows. In the absence of noise, the codewords or code blocks will be received as transmitted. In this case it is a simple matter to reconstruct the transmitted bit sequence. We simply follow the codeword through the code tree n bits at a time ($n = 3$ for the example considered in this section). The transmitted message is then reconstructed from the path taken through the tree.

The presence of noise introduces transmission errors and hence the codeword received is not always the codeword transmitted. In this situation, the following procedure can be used for reconstructing the transmitted codeword in a manner that hopefully will correct errors.

Consider the ith message digit d_i that has an influence on nN bits in the codeword. For the example shown in Figures 9.9–9.11, $N = 3$ and $n = 3$, and d_i has an effect on nine code digits (see Figure 9.10). Hence, in order to deduce d_i, there is no point in examining the codeword beyond the nine-digit block that has been influenced by d_i. On the other hand, we would not be taking advantage of the redundancy of the code if we decided d_i on the basis of less than nine bits of the received codeword. If we assume that d_{i-1} and d_{i-2} have been correctly decoded, then a starting node is defined on the code tree, and we can identify eight distinct and valid 9-bit code blocks that emerge from this node. We compare the 9-bit received code block we are examining with the eight valid code blocks, and discover the valid code block closest to the received code block (closest in Hamming distance). If this valid code block corresponds to an upward path from the starting node on the code tree, then d_i is decoded as 0; otherwise it is decoded as 1. After d_i is decoded, we use the decoded values of d_{i-1} and d_i to define a starting node for decoding d_{i+1}. This procedure is repeated until the entire message sequence is decoded.

The decoding algorithm just described yields a probability of error that decreases exponentially with N. However, as N gets large each bit requires the examination of 2^N branch sections of the code tree, and the search procedure in the decoding process becomes too lengthy. Another problem associated with the decoding procedure described above is that errors tend to propagate. That is, when d_i is decoded incorrectly, then the digits following d_i will probably be incorrectly decoded since the starting node will be incorrect. Fortunately, it has been found that this is not a serious problem in most of the practical convolutional codes known at this time.

Sequential Decoding. The exhaustive tree-search method of decoding convolutional codes becomes impractical as N becomes large since the decoder has to examine 2^N branch sections of the code tree. Sequential decoding schemes avoid the lengthy process of examining every branch of 2^N code tree branches in decoding a single message bit. In sequential decoding, at the arrival of an n-bit code block, the encoder compares these bits with the code blocks associated with the two branches diverging from the starting node. The encoder follows an upward or downward path (hence decodes a message bit as 0 or 1, respectively) in the code tree depending on which of the code blocks exhibit the fewest discrepancies with the received bits.

If a received code block contains transmission errors, then the decoder might make an error and start out on a wrong path in the code tree. In such a case, the entire continuation of the path taken by the encoder will be in error. Suppose that the decoder keeps a running record of the total number of discrepancies between the received code bits and the code bits encountered along its path. Then, the likelihood is great that, after having made the wrong turn at some node, the total number of errors will grow much more rapidly than would be the case if the decoder were following the correct path. The decoder may be programmed to respond to such a situation by retracing its path to the node at which an apparent error has been made and then taking an alternate branch out of that node. In this way the decoder will eventually find a path through N nodes. When such a path is found, the decoder decides about the first message bit. Similarly, the second message bit is then determined on the basis of the path searched out by the decoder, again N branches long.

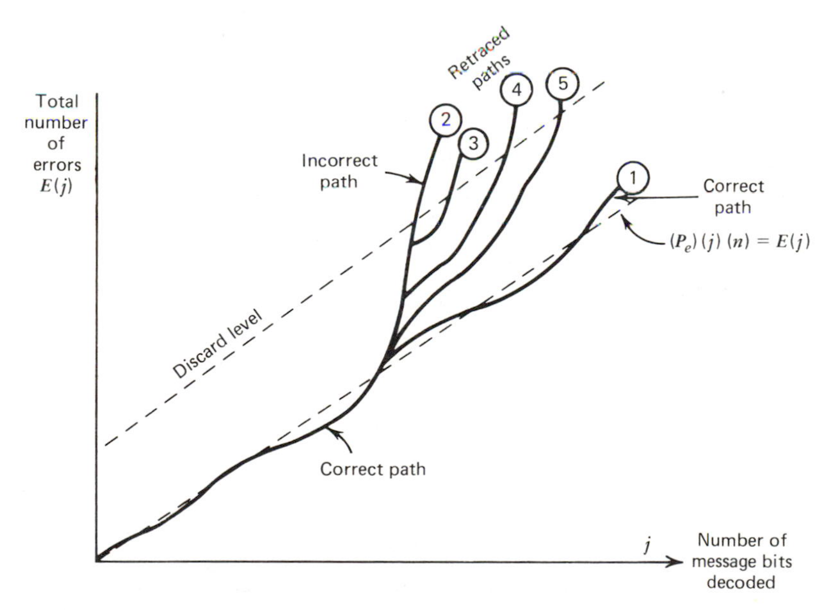

Figure 9.12 Threshold setting in sequential decoding. P_e = probability that a received bit is in error.

The decoder begins retracing when the number of accumulated errors exceeds a threshold, as shown in Figure 9.12. Since every branch in the codetree is associated with n bits, then, on the average over a long path of j branches, we can expect the total number of bit differences between the decoder path and the corresponding received bit sequence to be (P_e) (n) (j) even when the correct path is being followed. The number of bit errors accumulated will oscillate about $E(j)$ if the encoder is following the correct path (path 1 in Figure 9.12). However, the accumulated bit error will diverge sharply from $E(j)$ soon after a wrong decoder decision (path 2 in Figure 9.12). When the accumulated errors exceed a *discard level*, the decoder decides that an error has been made and retraces its path to the nearest unexplored path and starts moving forward again. It is possible that this path may again be reversed if the retracing did not result in proper error correction (paths 3, 4, and 5 in Figure 9.12). After some trial and error, an entire N node section of the code tree is retraced and at this point a decision is made about the message bit associated with this N node section. Thus, the sequential decoder operates on short code blocks most of the time and reverts to trial and error search over long code blocks only when it judges that an error has been made.

Majority Logic (Threshold) Decoding. Like block codes, a subclass of convolutional codes can be decoded using majority logic decoding techniques. To illustrate this technique let us consider the convolutional code generated by the encoder shown in Figure 9.13. The (2, 1) convolutional encoder generates a check bit r_i for each message bit d_i according to

$$r_i = d_i + d_{i-2} + d_{i-4} \tag{9.31}$$

and the output consists of an interleaved sequence of r_i and d_i. Let $v_i^{(m)}$ and $v_i^{(c)}$ be the output of the channel (received bits) corresponding to input bit d_i

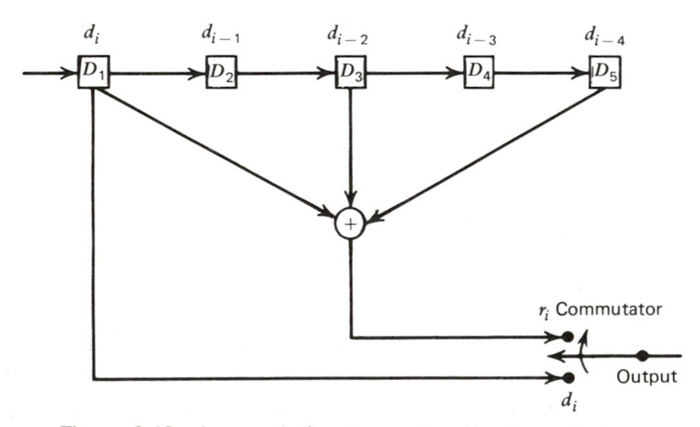

Figure 9.13 A convolutional encoder, $N = 5$, $n = 2$, $k = 1$.

and r_i, respectively. If we denote the errors by $e_i^{(m)}$ and $e_i^{(c)}$, then $v_i^{(m)} = d_i + e_i^{(m)}$ and $v_i^{(c)} = r_i + e_i^{(c)}$. The decoder for this code forms the ith syndrome digit as

$$s_i = v_{i-4}^{(m)} + v_{i-2}^{(m)} + v_i^{(m)} + v_i^{(c)}, \quad i \geq 5$$

It follows from Equation (9.31) that the preceding equation can be written as

$$s_i = e_{i-4}^{(m)} + e_{i-2}^{(m)} + e_i^{(m)} + e_i^{(c)}, \quad i \geq 5 \tag{9.32a}$$

For the first four bits in a message, s_i is given by

$$\begin{aligned} s_1 &= e_1^{(m)} + e_1^{(c)} \\ s_2 &= e_2^{(m)} + e_2^{(c)} \\ s_3 &= e_1^{(m)} + e_3^{(m)} + e_3^{(c)} \\ s_4 &= e_2^{(m)} + e_4^{(m)} + e_4^{(c)} \end{aligned} \tag{9.32b}$$

Equations (9.32*a*) and (9.32*b*) reveal that any transmission error $e_i^{(m)}$ in the ith message digit appears in syndrome bits s_i, s_{i+2}, and s_{i+4}. Table 9.7 shows the effect of error bits on various syndrome bits. It is apparent from what is shown in the table that if there are two or three 1's in s_1 s_3 s_5, then it is most likely that $e_1^{(m)} = 1$, meaning that the first message digit suffered a transmission error and hence it must be corrected. Similarly, we should correct the second message digit if there are more 1's than 0's in s_2 s_4 s_6, and so on. The reader can verify that this majority logic decoding will correct up to four successive errors (in message bits and check bits) if the following eight digits are error free. Thus this convolution code can correct burst errors of length four or less.

A variety of other techniques for decoding convolutional codes have also proved to be workable in general. One such technique is the Viterbi algorithm that has been shown to be optimal in a number of situations. For a summary of the evolution of convolutional codes, decoding methods, and their performance, the interested reader is referred to an excellent tutorial paper written by Viterbi (reference cited at the end of this chapter).

Table 9.7. Decoding procedure for the coder shown in Figure 9.13.

	$e_1^{(m)}$	$e_1^{(c)}$	$e_2^{(m)}$	$e_2^{(c)}$	$e_3^{(m)}$	$e_3^{(c)}$	$e_4^{(m)}$	$e_4^{(c)}$	$e_5^{(m)}$	$e_5^{(c)}$	$e_6^{(m)}$	$e_6^{(c)}$
s_1	X	X										
s_2			X	X								
s_3	X				X	X						
s_4			X				X	X				
s_5	X				X				X	X		
s_6			X				X				X	X
s_7					X				X			
s_8							X				X	

9.6.3 Performance of Convolutional Codes

Like block codes, convolutional codes can be designed to detect and/or correct errors. Since error detection involves retransmission of blocks of data, block codes are better suited for error detection applications. Convolutional codes are most commonly used for correcting errors. Practical convolutional codes exist that are capable of correcting random errors, bursts, or both. It is very difficult to derive a meaningful expression for the probability of incorrectly decoding a convolutional code since successive decodings are dependent, and as a result a past decoding error tends to cause subsequent decodings to be erroneous also. The performance of convolutional codes are usually evaluated by simulating the encoding and decoding operations along with a suitable channel model on the digital computer.

Although the theory of convolutional codes is not as well developed as that of block codes, several practical convolutional encoders/decoders have been built for use in various applications. Convolutional codes have many of the basic properties of block codes: (1) they can be encoded using simple shift registers and modulo-2 adders; (2) while decoding is difficult, several classes of convolutional codes have been found for which the amount of decoding equipment required is not excessive; and (3) practical convolutional codes exist that are capable of correcting random and burst types of errors.

Convolutional codes have the following advantages over block codes:

1. since convolutional codes operate on smaller blocks of data, the decoding delay is small;
2. because of the smaller block size, convolutional codes require a smaller amount of storage hardware;
3. loss of synchronization is not as serious a problem in convolutional codes as it is in systems using block codes with large block size.

In summary, convolutional codes, though not as well developed and more difficult to analyze than block codes, are competitive with block codes in many applications.

9.7 PERFORMANCE OF BLOCK CODES–ERROR CORRECTION

In this section, we compare the performance of systems using block codes for error correction with systems using no error control coding. We will use the probability of incorrectly decoding a message bit, and the probability of incorrectly decoding a block (or word) of message digits as our measure of performance. We will do the comparison under the condition that the rate of information transmission (message digits or words per unit time) is the same for the coded and uncoded systems, and that both systems are operating with the same average signal power (S_{av}) and noise power spectral density. Coded or uncoded, a block or word of, say, k message digits must be transmitted in a duration of time $T_w = k/r_b$, where r_b is the message bit rate. If the system uses an (n, k) block code, then the bit rate going into the channel will be $r_c = r_b(n/k) > r_b$. Thus the transmitted bit duration in the coded case T_c must be less than the bit duration T_b in the uncoded case. If the transmitter power S_{av} is held constant, then the energy per bit is decreased by the use of coding and the probability of decoding a transmitted bit (channel bit error probability) increases. We must determine whether coding results in a significant reduction in the probability of incorrectly decoding a message bit even though the channel bit error probability is now somewhat higher. We will use the following notation:

- r_b message bit rate
- r_c channel bit rate; $r_c = r_b(n/k)$ for the coded case, and $r_c = r_b$ for the uncoded case
- q_u channel bit error probability for the uncoded system
- q_c channel bit error probability for the coded system
- P_{be}^u probability of incorrectly decoding a message bit in the uncoded system
- P_{be}^c probability of incorrectly decoding a message bit in the coded system
- P_{we}^u probability of incorrectly decoding a word (block) of message bits in the uncoded system
- P_{we}^c probability of incorrectly decoding a word (block) of message bits in the coded system
- t error-correcting capability of the block code

Now, in the uncoded system we have

$$P_{be}^u = q_u \tag{9.33}$$

and the probability that a word of k message bits is correctly received is given by

$$\begin{aligned} P_{we}^u &= 1 - P(\text{all } k \text{ bits are correctly received}) \\ &= 1 - (1 - q_u)^k \approx kq_u \text{ when } kq_u \ll 1 \end{aligned} \tag{9.34}$$

since *transmission errors are assumed to be independent.*

In the coded case, we can calculate the word error rate as follows. A word or block of k message digits will be incorrectly decoded when more than t errors occur in the n-bit codeword. A maximum of t errors per codeword will be corrected by the decoder since the block code is assumed to be able to correct up to t errors. Thus,

$$\begin{aligned} P_{we}^{c} &= P((t+1) \text{ or more errors occur in an } n\text{-bit codeword}) \\ &= \sum_{i=t+1}^{n} \binom{n}{i} (q_c)^i (1-q_c)^{n-i} \\ &= \sum_{i=t+1}^{n} P(n, i) \end{aligned} \tag{9.35}$$

where $P(n, i)$ is defined as

$$P(n, i) \triangleq \binom{n}{i} (q_c)^i (1-q_c)^{n-i}$$

and

$$\binom{n}{i} = \frac{n!}{i!(n-i)!}$$

For $nq_c \ll 1$, $P(n, i+1) \ll P(n, i)$ and we can approximate P_{we}^{c} by

$$P_{we}^{c} \approx \binom{n}{t+1} (q_c)^{t+1} (1-q_c)^{n-t-1} \tag{9.36}$$

We can use the approximation given in Equation (9.36) to compute the message bit error probability P_{be}^{c} as follows. Equation (9.36) implies that the majority of decoding errors are due to $(t+1)$ bit errors in the n-bit codeword. Out of these $(t+1)$ errors, the fraction k/n represents erroneous message bits. Hence the average message bit error rate is given by

$$r_b P_{be}^{c} = r_w (t+1) \frac{k}{n} P_{we}^{c}$$

or

$$P_{be}^{c} = \frac{(t+1)}{n} P_{we}^{c} \tag{9.37}$$

where r_w is the word rate. Let us use the following example to illustrate error probabilities in a coded and uncoded system.

Example 9.8. The probability of error over a binary symmetric channel is given by

$$P_e = Q\left(\left[\frac{2S_{av}}{\eta r_c}\right]^{1/2}\right)$$

where r_c is the bit rate over the channel. Assume that information is transmitted over this channel in blocks of seven message bits. The uncoded system transmits the message bits directly and the coded system uses a (15, 7) BCH code that can correct up to two errors in each 15-bit codeword. Compare the word and message bit error probabilities for the coded and uncoded systems. Assume S_{av}, η, and the message bit rate r_b to be the same for both systems.

Solution. We are given that $n = 15$, $k = 7$, $t = 2$, and

$$q_u = Q\left(\left[\frac{2S_{av}}{\eta r_b}\right]^{1/2}\right)$$

$$q_c = Q\left(\left[\frac{2S_{av}}{\eta r_b}\frac{7}{15}\right]^{1/2}\right)$$

Hence for the uncoded system we have

$$P_{be}^u = q_u = Q\left(\left[\frac{2S_{av}}{\eta r_b}\right]^{1/2}\right)$$

and

$$P_{we}^u = 1 - (1 - q_u)^7$$

For the coded system,

$$P_{we}^c \approx \binom{15}{3} q_c^3 (1 - q_c)^{12}$$

and

$$P_{be}^c = \tfrac{3}{15} P_{we}^c$$

Plots of q_u, q_c, P_{we}^u, and P_{we}^c are shown in Figure 9.14.

Plots shown in Figure 9.14 reveal several interesting points. First, coded transmission always leads to higher channel bit error probability since coding increases the bit rate. However, for $S_{av}/\eta r_b > 2$, coded transmission yields a lower word error probability than uncoded transmission. The difference in word error probabilities becomes significant as $S_{av}/\eta r_b$ exceeds 8.

The preceding example also illustrates that coding is not a cure-all. The additional expense of coding and decoding equipment can be justified only if coding results in a significant improvement. For example, if r_b and η are fixed and an overall word error probability of 10^{-5} is desired, coding results in a 3 dB savings in power requirements. However, at an error probability of, say, 10^{-3}, coding saves only about 1 dB of power. Thus, before deciding to employ coding, one should carefully analyze the performance of the system with and without coding to determine if the improvement obtained is worth the complication and expense. In many applications, the message bit error probabilities can be lowered by signaling at a slower speed. That is, with S_{av} and η fixed, error probabilities can be reduced by lowering r_b, the message bit rate.

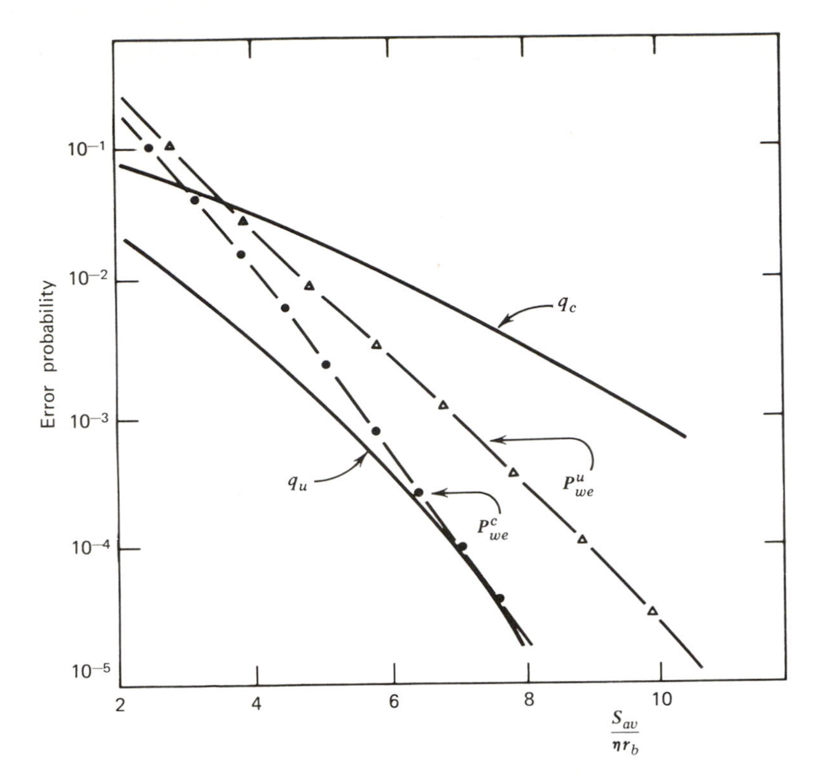

Figure 9.14 Channel bit error probabilities and word error probabilities for the coded and uncoded systems discussed in Example 9.8.

9.8 PERFORMANCE OF BLOCK CODES–ERROR DETECTION

In this section we will compare the performance of data transmission systems using block codes for error detection with systems using direct transmission without error control. We will do the comparison under the assumption that S_{av}, η, and r_b remain the same for the coded and uncoded systems, and we will use the probability of incorrectly decoding a block of k message bits as our measure of comparative performance.

In systems using block codes for error detection, a word or message block of k bits is transmitted using an n-bit codeword. We will assume that the (n, k) block code used is capable of detecting up to $2t$ errors per block. The decoder checks the received codewords for errors, and when an error is detected the decoder may either discard the message block or request the transmitter to retransmit the message block. We will assume that a reverse channel is available between the receiver and the transmitter for sending requests for retransmission.

In the preceding sections we saw that the data rate r_c over the channel is $r_b(n/k)$ when an (n, k) error-correcting block code is used. The data rate r_c will have to be higher than $r_b(n/k)$ for (n, k) error-detecting block codes because of retransmission. The actual value of r_c will depend on the method of retransmission scheme used. In the following sections we will derive expressions for r_c for the *stop and wait* and *continuous retransmission* schemes. After we calculate r_c, we can use expressions similar to the ones developed in the preceding section for comparing the performance of coded and uncoded systems.

9.8.1 Stop and Wait Transmission Method

In the stop and wait method of transmission, the transmitter begins transmission at time, say, t_0 and completes the transmission of a block of n bits (a codeword) at time, say, $t_0 + t_n$. The decoder starts receiving the message block at time $t_0 + \Delta$, where Δ is the propagation delay. At time $t_0 + \Delta + t_n$ the decoder checks the n-bit block that was received and sends a positive acknowledgment (ACK) or a negative acknowledgment (NACK) to the transmitter depending on whether or not it detected an error in the received block. The transmitter receives an ACK or NACK at time $t_0 + t_n + 2\Delta$ and begins another transmission. If the acknowledgment is positive, then the next message block is transmitted; if not, the previous message block is retransmitted. In either case, the transmitter is waiting from $t_0 + t_n$ to $t_0 + t_n + 2\Delta$ for an acknowledgment from the receiver. (See Figure 9.15.)

To determine the channel bit rate, let us consider the transmission of N blocks of data over the channel at a rate of r_c bits per second. The total time needed to complete this transmission is $N(n/r_c + 2\Delta)$. Out of these N blocks, NP_{we}^c blocks will be in error on the average. Hence, on the average, $N(1 - P_{we}^c)$ blocks will be accepted by the receiver, and the number of message bits transmitted during the time interval $N(n/r_c + 2\Delta)$ seconds is $Nk(1 - P_{we}^c)$. With the message bit rate of r_b, the time allowed for transmitting $Nk(1 - P_{we}^c)$ bits is $Nk(1 - P_{we}^c)/r_b$. Hence we have

$$\frac{Nk(1 - P_{we}^c)}{r_b} = N\left(\frac{n + 2\Delta r_c}{r_c}\right) \tag{9.38}$$

or

$$r_b = \left(\frac{k}{n}\right)\left(\frac{nr_c}{n + 2\Delta r_c}\right)(1 - P_{we}^c) \tag{9.39}$$

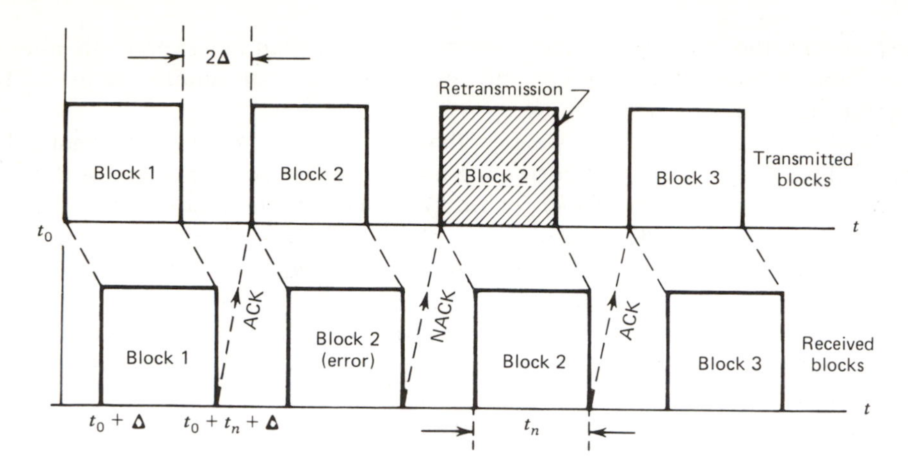

Figure 9.15 Stop and wait method of transmission.

In Equation (9.39) P_{we}^{c} is the word error probability we are trying to calculate and hence is not known. However, $P_{we}^{c} \approx 0$ and hence

$$r_b \approx \left(\frac{k}{n}\right) \frac{nr_c}{(n + 2\Delta r_c)} \tag{9.40}$$

In Equation (9.40), the values of r_b, k, n, and Δ are assumed to be known. The value of r_c obtained from (9.40) is used to calculate the channel bit error probability q_c which in turn is used to calculate the word error probability P_{we}^{c} as follows:

$$\begin{aligned} P_{we}^{c} &= P((2t+1) \text{ or more bits in an } n\text{-bit codeword are in error}) \\ &= \sum_{i=2t+1}^{n} \binom{n}{i} (q_c)^i (1 - q_c)^{n-i} \\ &\approx \binom{n}{2t+1} (q_c)^{2t+1} (1 - q_c)^{n-2t-1} \end{aligned} \tag{9.41}$$

The approximation given in Equation (9.41) is valid when $nq_c < 0.1$, q_c being the probability of channel bit error for the coded system at channel bit rate r_c computed from Equation (9.40).

The stop and wait method of block transmission becomes very inefficient when the propagation delay is large. For example, if the loop delay 2Δ is such that $2\Delta r_c \to n$, then $r_b \to (k/n)\ (r_c/2)$, which is $\frac{1}{2}$ the message bit rate that can be maintained by using an (n, k) error-correcting code. On very long data links such as the ones involving satellites, the loop delay may be several hundred milliseconds and long block lengths are necessary to maintain a reasonably effective data rate.

9.8.2 Continuous Transmission Method

In the continuous transmission method illustrated in Figure 9.16, message blocks are transmitted continuously until an error is detected. Then, the transmitter returns to the last correctly acknowledged message block and

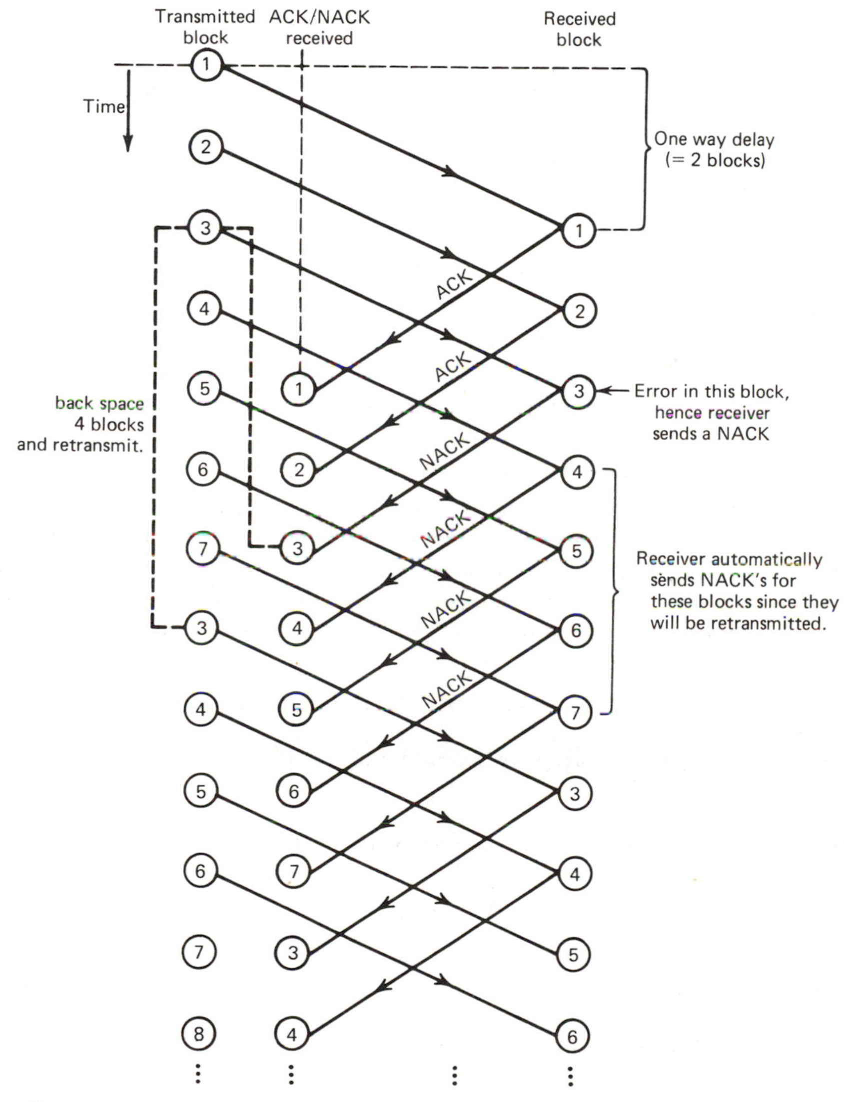

Figure 9.16 Example of a continuous transmission scheme. Loop delay is assumed to be equal to $4(n/r_c)$.

repeats it together with all succeeding blocks. A running account of message blocks sent is maintained at the transmitter, and message blocks are stored until a positive acknowledgment for each block is received before the storage is released for a new message block. When noise corrupts a message block on the forward link, the receiver will return a negative acknowledgment requesting retransmission of the message block. However, because of the delay through the loop, several other message blocks will follow before the transmitter can be stopped. Hence these blocks must be retransmitted as well. Of course the noise on the reverse channel may corrupt the acknowledgment in both methods of block transmission schemes discussed thus far.

The reader can verify that for the continuous transmission scheme, r_c and r_b are related by

$$r_b \approx \left(\frac{k}{n}\right) r_c \tag{9.42}$$

when P_{we}^{c} is very small.

The word error probability is given by

$$P_{we}^{c} \approx \binom{n}{2t+1} (q_c)^{2t+1} (1 - q_c)^{n-2t-1} \tag{9.43}$$

Comparison of Equation (9.42) with (9.39) indicates that in the continuous transmission method the loop delay does not affect data transmission rate when the overall error probabilities are small. However, it must be pointed out that the equipment needed to implement the continuous retransmission scheme is quite complex. Combinations of the stop and wait method and the continuous retransmission method are also used in some systems.

Example 9.9. A digital communication system has the following parameters:

Message bit rate = 24,000 bits/sec
Word size = 120 bits
Channel bit rate = r_c; variable up to 48,000 bits/sec
Channel bit error probability = $Q\left(\sqrt{\frac{2S_{av}}{\eta r_c}}\right)$,
$S_{av}/\eta = 150{,}000$
Loop delay = 2 msec

For the system described above, calculate:

(a) The word error probability for direct transmission (P_{we}^{u}).

(b) P_{we}^{c} for a (127,120) single error-correcting BCH code.

(c) P_{we}^{c} for a (127, 120) double error-detecting code using stop and wait transmission method.

(d) P_{we}^{c} for a (127, 120) double error-detecting code using continuous retransmission method.

Solution. (a) With $r_b = 24{,}000$ bits/sec and no coding we have a channel bit rate $r_c = 24{,}000$ and hence the bit error probability is

$$q_u = Q(\sqrt{12.5}) = (2)(10^{-4})$$

and the word error probability is

$$P_{we}^{u} = 1 - (1 - q_u)^{120} = 1 - (0.9998)^{120} = 0.0237$$

(b) With a (127, 120) error-correcting code, the channel bit rate $r_c = (24{,}000)(127/120) = 25{,}400$. Hence the channel bit error probability is equal to

$$q_c = Q(\sqrt{11.81}) = (3)(10^{-4})$$

and the word error probability is equal to

$$\begin{aligned} P_{we}^{c} &= \binom{127}{2}(0.0003)^2(0.9997)^{125} \\ &= 0.00069 \end{aligned}$$

(c) With a double error-detecting (127, 120) code using stop and wait method of transmission, we have (from Equation (9.40))

$$24{,}000 \approx \frac{120}{127}\frac{(127)r_c}{(127 + 0.002r_c)}$$

or

$$r_c \approx \frac{(24{,}000)127}{72} = 42{,}333$$

Hence, q_c is given by

$$q_c = Q(\sqrt{7.087}) = 0.0047$$

The word error probability is equal to

$$\begin{aligned} P_{we}^{c} &\approx \binom{127}{3}(0.0047)^3(0.9953)^{124} \\ &\approx 0.01930 \end{aligned}$$

(d) With continuous transmission, $r_c = 25{,}400$ and $q_c = (3)(10^{-4})$, and

$$P_{we}^{c} = \binom{127}{3}(0.0003)^3(0.9997)^{124}$$
$$= (8.67)(10^{-6})$$

This example illustrates the reduction in word error probability that can be achieved with various methods of error control coding and block transmission.

9.9 SUMMARY

We discussed several methods of error control using digital coding techniques. Codes for detecting and/or correcting random and burst types of errors were discussed and coding schemes were compared on the basis of their error control capabilities. Considerable attention was given to the complexity of encoding and decoding operations.

Two techniques, forward acting error correction and error detection, were described for handling transmission errors in a digital communication system. Substantially large reduction in error rate may be accomplished with a modest amount of equipment by using coding schemes for error detection. However, error detection requires a reverse channel. If a reverse channel is not available, then forward acting error correction has to be used for error control. Convolutional encoders provide excellent performance in forward error correction applications.

REFERENCES

Easily readable treatments of error-correcting codes may be found in the books written by Peterson (1961) and Lin (1970). For advanced level treatments of topics in error control coding, the reader may refer to the books *Principles of Data Communication* (by Lucky et al. 1968) and *Principles of Communication Engineering* (by Wozencraft and Jacobs 1965). An excellent tutorial treatment of convolutional codes may be found in a recent article by Viterbi. Practical schemes for implementing error control coding are discussed in *Communication Networks for Computers* by Davies and Barber (1975).

1. W. W. Peterson. *Error-Correcting Codes.* M.I.T. Press, Cambridge, Mass. (1961, 1970).
2. S. Lin. *An Introduction to Error-Correcting Codes.* Prentice-Hall, Englewood Cliffs, N.J. (1970).

3. R. W. Lucky, J. Salz, and E. J. Weldon, Jr. *Principles of Data Communications.* McGraw-Hill, New York (1968).
4. J. M. Wozencraft and I. M. Jacobs. *Principles of Communication Engineering.* Wiley, New York (1965).
5. D. W. Davies and D. L. Barber. *Communication Networks for Computers.* Wiley, New York (1975).
6. A. J. Viterbi. Convolutional Codes and their Performance in Communication Systems, *IEEE Transactions on Communications Technology,* vol 5, October, 1971.

PROBLEMS

Section 9.1

9.1. Suppose that we want to transmit binary data over a microwave link at a rate of (1.5) (10^6) bits/sec using a coherent PSK signaling scheme. Assume that the system is operating with $(S_{av}/\eta) = 2(10^7)$. Error control coding used in the system consists of repeating each binary digit twice. The decoding algorithm used at the receiver is given below:

Received pair	Output
00	0
11	1
01	Reject
10	Reject

For the above system:

(a) Find the channel bit error probability for the coded system.
(b) Find the message bit error probability for the coded system.
(c) Find the average number of message bits per sec that are rejected by the receiver.

Section 9.2

9.2. Consider a (7, 4) linear code whose generator matrix is

$$G = \left[\begin{array}{cccc|ccc} 1 & 0 & 0 & 0 & 1 & 0 & 1 \\ 0 & 1 & 0 & 0 & 1 & 1 & 1 \\ 0 & 0 & 1 & 0 & 1 & 1 & 0 \\ 0 & 0 & 0 & 1 & 0 & 1 & 1 \end{array}\right]$$

(a) Find all the code vectors of this code.
(b) Find the parity check matrix for this code.
(c) Find the minimum weight of this code.

9.3. Prove Equation (9.7).

9.4. In a repeated code, a binary 0 is encoded as a sequence of $(2t + 1)$ zeros and a binary 1 as a similar number of ones. Thus $k = 1$ and $n = 2t + 1$. Find the generator matrix and the parity check matrix for a repeated code with $t = 1$.

9.5. The parity check bits of a (8, 4) block code are generated by

$$c_5 = d_1 + d_2 + d_4$$
$$c_6 = d_1 + d_2 + d_3$$
$$c_7 = d_1 + d_3 + d_4$$
$$c_8 = d_2 + d_3 + d_4$$

where d_1, d_2, d_3, and d_4 are the message digits.
(a) Find the generator matrix and the parity check matrix for this code.
(b) Find the minimum weight of this code.
(c) Find the error-detecting capabilities of this code.
(d) Show through an example that this code can detect three errors/codeword.

9.6. Let H be the parity check matrix of an (n, k) linear block code C with an odd minimum weight d. Construct a new code C_1 whose parity check matrix is

$$H_1 = \left[\begin{array}{ccc|c} & & & 0 \\ & & & 0 \\ & H & & 0 \\ & & & \vdots \\ & & & 0 \\ \hline 1 & 1 \cdots & 1 & 1 \end{array}\right]$$

(a) Show that C_1 is an $(n + 1, k)$ linear block code.
(b) Show that every code vector in C_1 has even weight.
(c) Show that the minimum weight of C_1 is $d + 1$.

9.7. Design a single-error-correcting code with a message block size = 11 and show by an example that the code can correct single errors.

Section 9.3

9.8. A (15, 5) linear cyclic code has a generator polynomial

$$g(x) = 1 + x + x^2 + x^4 + x^5 + x^8 + x^{10}$$

(a) Draw block diagrams of an encoder and syndrome calculator for this code.

(b) Find the code polynomial for the message polynomial $D(x) = 1 + x^2 + x^4$ (in a systematic form).

(c) Is $V(x) = 1 + x^4 + x^6 + x^8 + x^{14}$ a code polynomial? If not, find the syndrome of $V(x)$.

9.9. The generator polynomial for a (15, 7) cyclic code is

$$g(x) = 1 + x^4 + x^6 + x^7 + x^8$$

(a) Find the code vector (in systematic form) for the message polynomial $D(x) = x^2 + x^3 + x^4$.

(b) Assume that the first and last bits of the code vector $V(x)$ for $D(x) = x^2 + x^3 + x^4$ suffer transmission errors. Find the syndrome of $V(x)$.

9.10. The decoder for a class of single-error-correcting cyclic codes (called Hamming codes) is shown in Figure 9.17. Show, by way of an example,

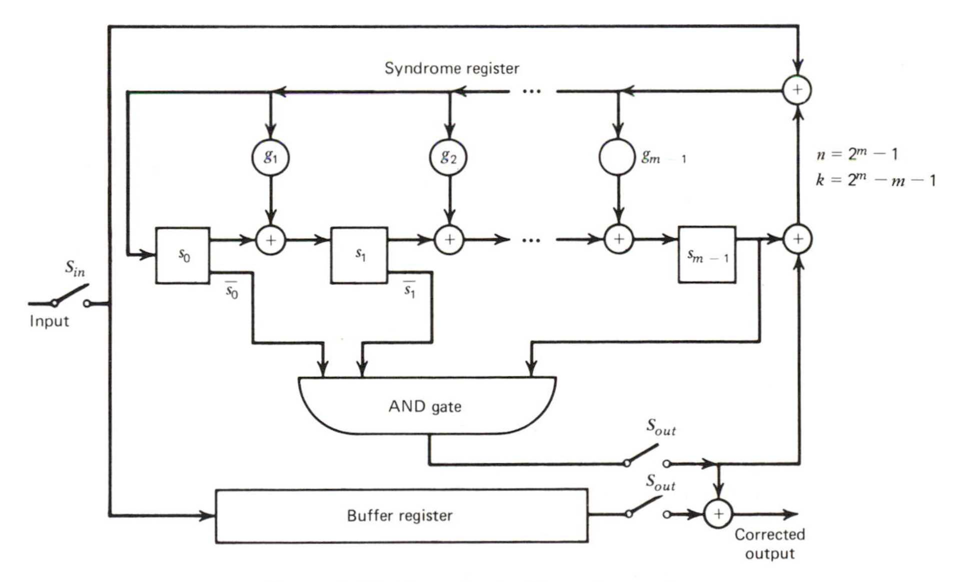

Figure 9.17 Decoder for Hamming codes.

that single errors in a (15, 11) Hamming code generated by $g(x) = 1 + x + x^4$ can be decoded using the decoder shown in the figure.

9.11. Verify the operation of the decoder shown in Figure 9.7 assuming that the transmitted code vector (1 1 0 0 1 0 1) was received as (1 1 0 0 1 1 1).

9.12. A t-error-correcting code is said to be a *perfect code* if it is possible to form a standard array with all error patterns of t or fewer errors and no others as coset leaders. Show that a (7, 4) linear block code generated by $g(x) = 1 + x + x^3$ is a perfect code.

9.13. When a cyclic code is used for error detection, the receiver is equipped with a block check register (BCCR) (see Figure 9.18) that is identical to the encoder used at the transmitter (see Figure 9.20). The BCR is initialized and the received code vector is shifted in, beginning with the message bits. When all of the received message bits are shifted in, the BCCR will contain the check bits that correspond to the *received message bits*. Now, the received check bits are shifted in to the BCCR.

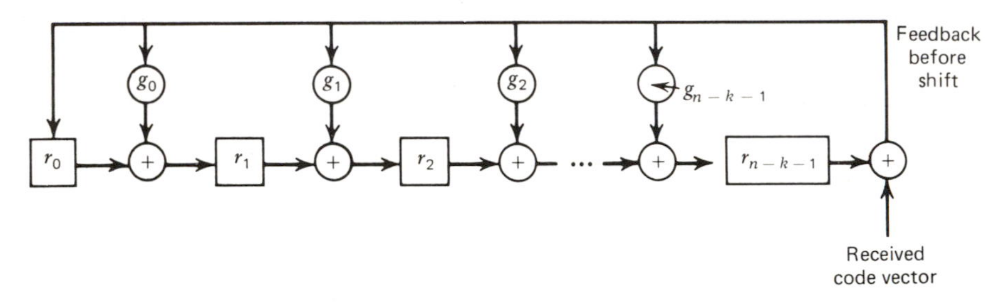

Figure 9.18 Block check register (BCCR).

(a) Verify that the BCCR will contain all zeros if there are no transmission errors.
(b) Verify that a 1 appearing in the feedback path while the check bits are being shifted in indicates a transmission error.

9.14. Verify the error detection operation of the encoder shown in Figure 9.19

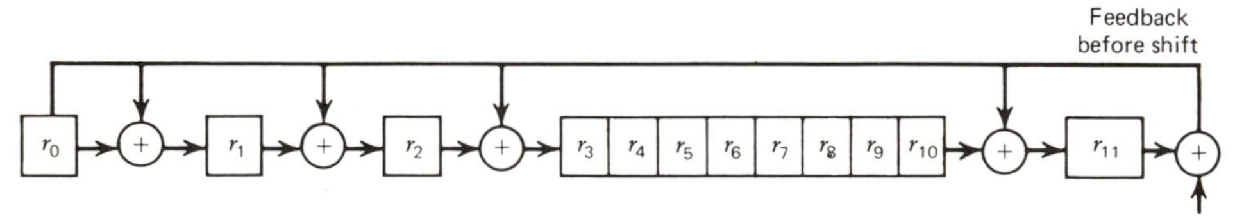

Figure 9.19 Block check register for the cyclic code generated by $g(x) = 1 + x + x^2 + x^3 + x^{11} + x^{12}$.

using the principle stated in Problem 9.13. Assume that the transmitted message block is (1 0 0 0 0 0 0 0 0 0 0 1) and that the first two message bits suffer transmission errors.

Section 9.4

9.15. Consider a (15, 9) cyclic code generated by

$$g(x) = 1 + x^3 + x^4 + x^5 + x^6$$

This code has a burst-error-correcting ability $q = 3$. Find the burst-correcting efficiency of this code.

9.16. Consider a (7, 3) cyclic code generated by

$$g(x) = 1 + x^2 + x^3 + x^4$$

(a) Find the burst-error-correcting ability of this code.
(b) Find the burst-error-correcting efficiency of this code.
(c) Construct an interlaced code derived from the (7, 3) code given above that can correct bursts of length up to 10 using a message block size of 15 bits.

9.17. A source produces binary symbols at a rate of 10,000 symbols/sec. The channel is subjected to error bursts lasting up to 1 msec. Devise an encoding scheme using an interlaced (n, k) single error-correcting code that will allow full correction of a burst. What is the minimum time between bursts if the system is to operate properly?

Section 9.6

9.18. Consider the convolutional encoder shown in Figure 9.20. The message bits are shifted into the encoder two bits at a time.

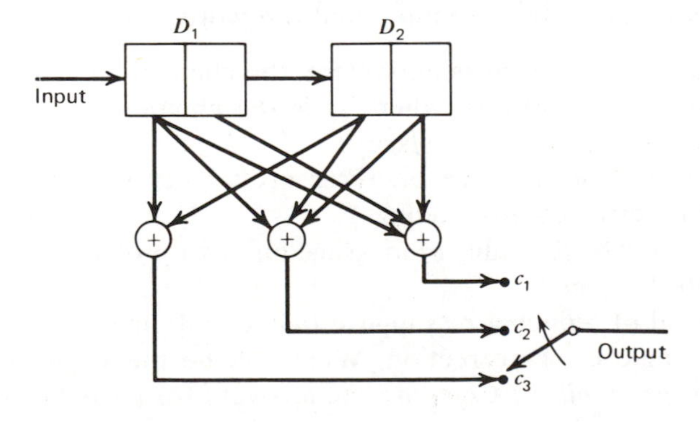

Figure 9.20 Convolutional encoder for Problem 9.18.

(a) Find the constraint length and the rate efficiency of the code.
(b) Assume the initial content of the registers to be zero and find the code block for the input message block (110101).

9.19. Assume that the convolutional encoder shown in Figure 9.9 is initially loaded with zeros. Assume that a 6-bit message sequence (followed by zeros) is transmitted.

(a) Draw the tree diagram that can be used for decoding.
(b) How would you decode the messages

111 101 001 000 100 and

001 101 001 000 100 ?

Sections 9.7 and 9.8

9.20. Consider a data communication system with the following characteristics:

message bit rate = 1000 bits/sec

word size = 8 bits

channel bit error probability $= \frac{1}{2}\exp(-10{,}000/r_c)$,
where r_c is the channel bit rate that can vary from 1000 to 3000 bits/sec

For this system:
(a) Calculate the word error probability for uncoded transmission.
(b) Design a single error-correcting code with $k = 8$ and calculate the word error probability for the coded system.
(c) If a repeated code in which each message bit is repeated three times is used for error correction, find the word error probability.

9.21. In a binary data transmission system, the channel bit error probability is given by $\frac{1}{2}\exp(-8000/r_c)$, where r_c is the channel bit rate. The normal message bit rate is 1000 bits/sec.
(a) Calculate the message bit error probability with $r_b = 1000$ bits/sec and no error control coding.
(b) What will be the value of message bit error probability if r_b is reduced to 500 bits/sec?
(c) Instead of reducing r_b, suppose that a (7, 4) linear block code is used for single error correction. What will be the value of *message bit error probability*? Compare the answers for parts (a), (b), and (c).

9.22. A digital communication system has the following requirements:

message block size = 40 bits
word error probability $\leq 10^{-3}$
channel bit error probability $= \frac{1}{2}\exp(-20{,}000/r_c)$
where r_c is the channel bit rate and r_c has to be ≤ 5000 bits/sec

(a) What is the highest message bit rate possible for an uncoded system? (P^u_{we} has to be $\leq 10^{-3}$).
(b) What is the highest message bit rate possible for a single error-correcting system? (Design the error-correcting block code and then find the maximum message bit rate.)

9.23. A digital communication system has the following parameters:

message bit rate r_b = variable from 900 to 4800 bits/sec
channel bit rate r_c = variable from 900 to 9600 bits/sec
loop delay = 10 msec
channel bit error probability $= \frac{1}{2}\exp(-12{,}000/r_c)$

Two error control coding schemes are being considered for this system: (1) A single error-correcting scheme using 16 message bits/block. (2) A (255, 231) BCH code with $d_{\min} = 7$ for error detection using stop and wait method of transmission.
(a) Sketch P^u_{be} and P^c_{be} versus r_b for the two schemes ($900 \leq r_b \leq 4800$).
(b) Which coding scheme would you prefer and why?

9.24. The state diagram of a discrete information source is shown in Figure 9.21. The output of the source is to be transmitted over a bandpass channel with a bandwidth of 15,000 Hz. The modem available for transmitting the source output in binary form will accept bit rates up to 15,000 bits/sec and yield a channel bit error probability q given by

$$q = \frac{1}{2}\exp(-60{,}000/r_c)$$

where r_c is the channel bit rate.

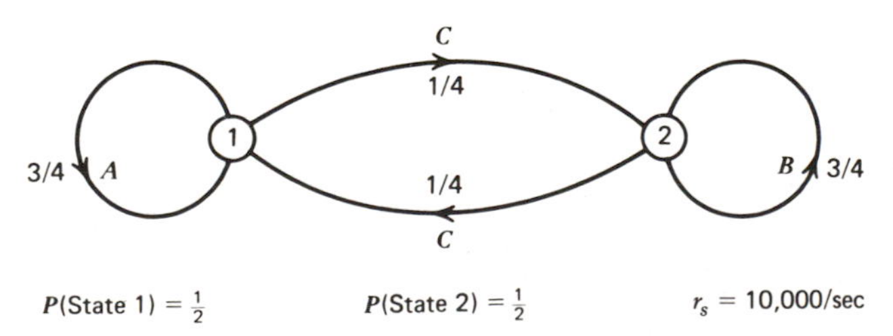

Figure 9.21 State diagram of a source, Problem 9.24.

Design a source encoding and a channel encoding scheme for this system that would yield an average symbol (letter) error rate of 50 symbols/sec. The source encoder is to use fixed length codewords with a block size ≤ 3. Assume that one or more bit errors in the codeword will result in the incorrect decoding of all the symbols contained in the codeword (worst case).

10

DIGITAL TRANSMISSION OF ANALOG SIGNALS

10.1 INTRODUCTION

Communication systems are designed to handle the output of a variety of information sources. In the preceding chapters we considered analog communication systems using CW modulation schemes (AM, DSB, SSB, PM, and FM) for transmitting the output of analog information sources. We also discussed digital communication systems that used digital modulation schemes (discrete PAM, ASK, FSK, and PSK) for transmitting the output of discrete information sources. Simplified block diagrams of analog and digital communication systems are shown in Figure 10.1. In this chapter we will consider the use of digital communication systems such as the one shown in Figure 10.1*b* for transmitting the output of analog information sources.

Digital transmission of analog signals is possible by virtue of the sampling theorem which tells us that an analog signal can be reproduced from an appropriate set of its samples and hence we need transmit only the sample values as they occur rather than the analog signal itself. Samples of the analog signal can be transmitted using analog pulse modulation schemes wherein the amplitude, width, or position of a pulse waveform is varied in proportion to the values of the samples. The key distinction between analog pulse modulation and CW modulation is as follows: In CW modulation, some parameter of the modulated wave varies *continuously with the message.* In analog pulse modulation, some parameter of each pulse is modulated by a *particular sample value* of the message.

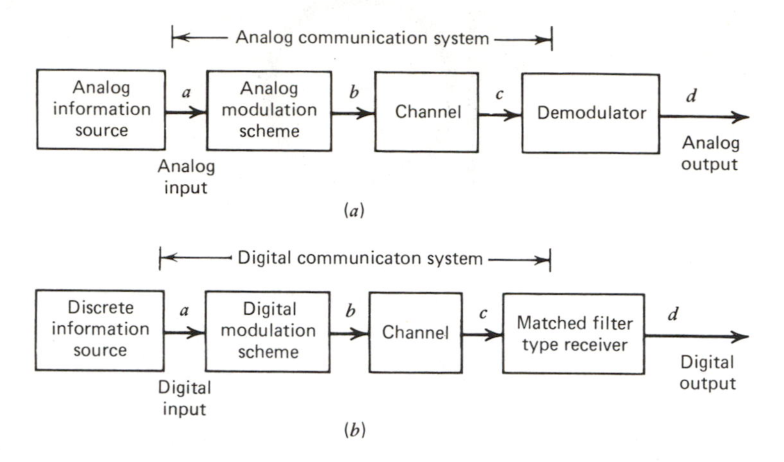

Figure 10.1 Communication systems. (*a*) Analog communication system for transmitting the output of an analog information source. (*b*) Digital communication system for transmitting the output of a discrete information source.

Another method of transmitting the sampled values of an analog signal is to round off (quantize) the sampled values to one of Q predetermined values and then transmit the sampled and quantized signal using digital modulation schemes. The block diagram of a system that uses this scheme is shown in Figure 10.2. Here, the output $X(t)$ of the analog information source is converted to an M-ary symbol sequence $\{S_k\}$ through the processes of sampling, quantizing, and encoding. The M-ary sequence $\{S_k\}$ is transmitted using a digital communication system. The receiver output $\{\hat{S}_k\}$ will differ occasionally from the input $\{S_k\}$ due to transmission errors caused by channel noise and ISI in the digital communication system. An estimate $\hat{X}(t)$ of $X(t)$ is obtained from $\{\hat{S}_k\}$ through the process of decoding, and digital to analog (D/A) conversion. The reconstructed waveform $\hat{X}(t)$ will be a noisy version of the transmitted signal. The noise is due to sampling and quantizing of $X(t)$,

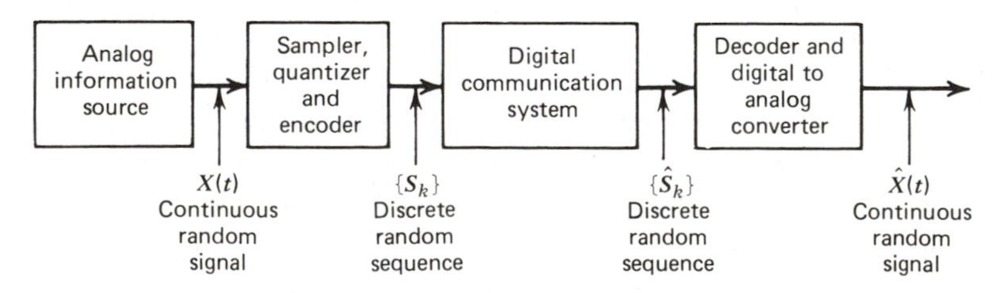

Figure 10.2 Digital transmission of analog signals.

and due to symbol errors that occur in the digital communication system. The overall performance of this system is measured by the signal-to-noise power ratio at the receiver output. A major portion of this chapter is devoted to the analysis of sampling, quantizing, and encoding procedures that are used to convert the analog output of an information source into a discrete symbol sequence suitable for transmission over a digital communication system.

At the beginning of this chapter, we will review the sampling theorem and discuss how analog signals are sampled and reconstructed in practical systems. We will then discuss methods of quantizing and encoding the sampled values for transmission over a digital communication system. Finally, we will derive expressions for the signal-to-noise power ratio at the receiver output and use these expressions for comparing the performance of digital and analog transmission schemes. We will also point out the advantages of using digital schemes for transmitting analog information.

In pulse communication systems, both analog and digital pulse modulation schemes are used. Analog pulse modulation, such as continuous pulse amplitude modulation and pulse position modulation, are similar to linear (AM) or exponential CW (PM or FM) modulation schemes. Digital or coded pulse modulation schemes such as pulse code modulation (PCM) and Delta modulation (DM) have no CW equivalent. We will treat only the digital pulse modulation schemes in this chapter. We begin our study with a review of sampling techniques.

10.2 SAMPLING THEORY AND PRACTICE

In many applications (such as in sample data control systems, digital computers, and in discrete pulse and CW modulation systems that we are currently dealing with) it is necessary and useful to represent an analog signal in terms of its sampled values taken at appropriately spaced intervals. In this section, we will first consider the representation of a low pass (bandlimited) deterministic signal $x(t)$ by its sampled values $x(kT_s)$ $(k = \ldots, -2, -1, 0, 1, 2, \ldots)$, where T_s is the time between samples. We will then extend the concept of sampling to include bandpass deterministic signals as well as to random signals. Finally, we will point out how the sampling and reconstruction of analog signals are carried out in practice.

10.2.1 Sampling Theory

The principle of sampling can be explained using the switching sampler shown in Figure 10.3. The switch periodically shifts between two contacts at a rate of

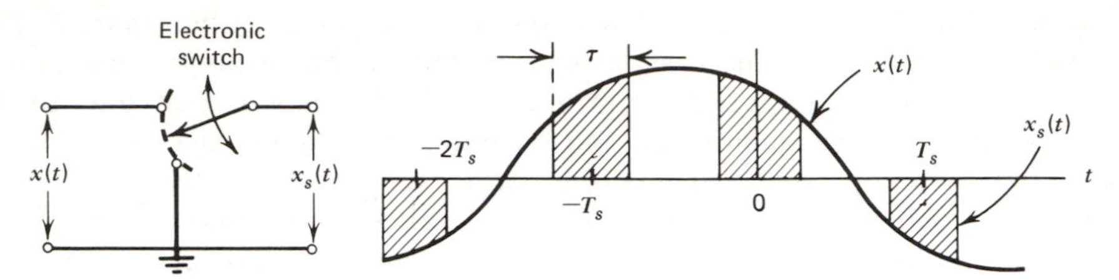

Figure 10.3 Switching sampler.

$f_s = 1/T_s$ Hz staying on the input contact for τ seconds and on the grounded contact for the remainder of each sampling period. The output $x_s(t)$ of the sampler consists of segments of $x(t)$, and $x_s(t)$ can be represented as

$$x_s(t) = x(t)s(t) \tag{10.1}$$

where $s(t)$ is the *sampling* or *switching function* shown in Figure 10.4.

Two questions that need to be answered with the sampling scheme shown in Figure 10.3 are: (1) Are the sampled segments sufficient to describe the original signal $x(t)$? (2) If so, how can we reconstruct $x(t)$ from $x_s(t)$? These questions can be answered by looking at the spectra (Fourier transforms) $X(f)$ and $X_s(f)$ of $x(t)$ and $x_s(t)$. Using the results derived in Chapter 2, we can express $s(t)$ as a Fourier series of the form

$$s(t) = C_0 + \sum_{n=1}^{\infty} 2C_n \cos n\omega_s t \tag{10.2}$$

where

$$C_0 = \tau/T_s, \quad C_n = f_s\tau \operatorname{sinc}[nf_s\tau], \quad \text{and} \quad \omega_s = 2\pi f_s$$

Combining Equations (10.2) and (10.1), we can write $x_s(t)$ as

$$x_s(t) = C_0x(t) + 2C_1x(t)\cos\omega_s t + 2C_2x(t)\cos 2\omega_s t + \cdots \tag{10.3}$$

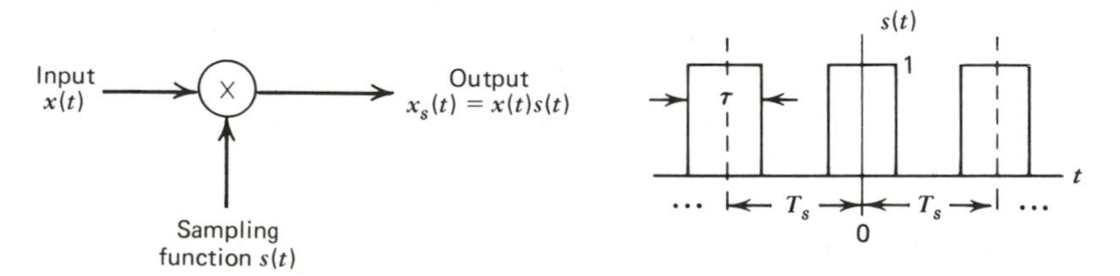

Figure 10.4 Sampling interpreted as multiplication. This type of sampling is often called *natural sampling*.

The Fourier transform of Equation (10.3) yields

$$X_s(f) = C_0X(f) + C_1[X(f - f_s) + X(f + f_s)] + C_2[X(f - 2f_s) + X(f + 2f_s)] + \cdots \tag{10.4a}$$

$$= C_0X(f) + \sum_{\substack{n=-\infty \\ n\neq 0}}^{\infty} C_nX(f - nf_s) \tag{10.4b}$$

We can use Equation (10.4*a*) to find the spectrum of $x_s(t)$ given the spectrum of $x(t)$. Figure 10.5 shows the spectrum of the sampler output when the input $x(t)$ is bandlimited to f_x Hz.

It follows from Equation (10.4*a*) and from Figure (10.5*b*) that if $f_s > 2f_x$, then the sampling operation leaves the message spectrum intact, merely repeating it periodically in the frequency domain with a period of f_s. We also note that the first term in Equation (10.4*a*) (corresponding to the first term in Equation (10.3)) is the message term attenuated by the duty cycle C_0 of the sampling pulse. Since the sampling operation has not altered the message

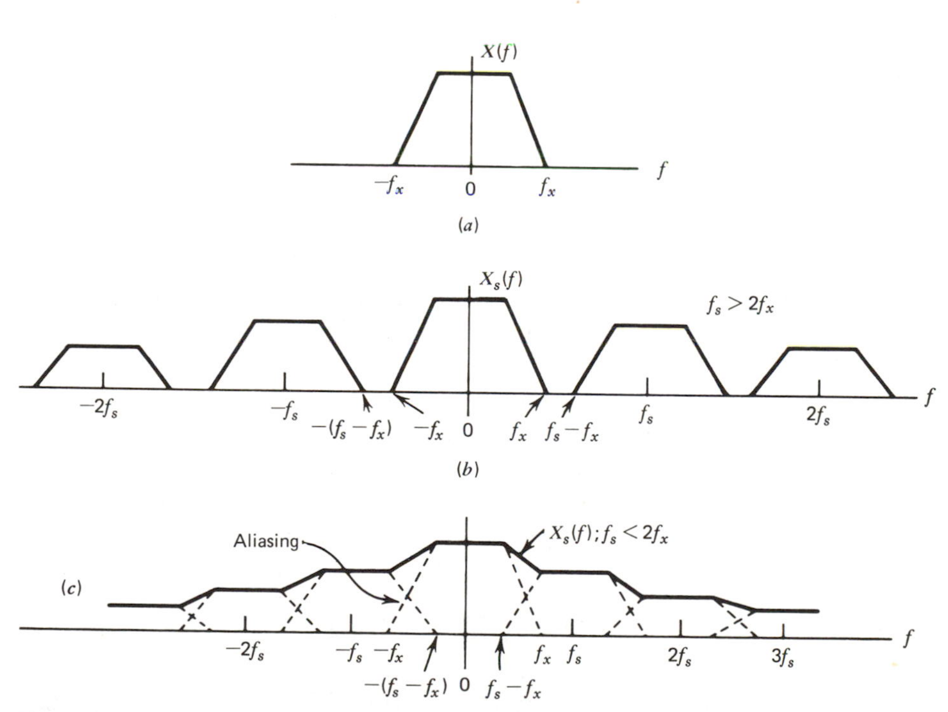

Figure 10.5 Sampling operation shown in frequency domain. (*a*) Message. (*b*) Sampled output, $f_s > 2f_x$. (*c*) Sampled output, $f_s < 2f_x$. (The width of the sampling pulse τ is assumed to be much smaller than T_s.)

spectrum, it should be possible to reconstruct $x(t)$ from the sampled waveform $x_s(t)$. While the procedure for reconstruction is not obvious from time domain relationships, it can be seen from Figure 10.5*b* that $X(f)$ can be separated from $X_s(f)$ by lowpass filtering. If we can filter $X(f)$ from $X_s(f)$, then we have recovered $x(t)$. Of course, such recovery is possible only when $x(t)$ is bandlimited and $f_s > 2f_x$. If the sampling rate $f_s < 2f_x$, then the sidebands of the signal overlap (Figure 10.5*c*) and $x(t)$ cannot be recovered without distortion from $X_s(f)$. This distortion is referred to as *aliasing*.

Thus the sampling frequency f_s must satisfy

$$f_s \geq 2f_x \quad \text{or} \quad T_s \leq 1/2f_x \tag{10.5}$$

The minimum sampling frequency $f_{s_{\min}} = 2f_x$ is called the *Nyquist rate*. When Equation (10.5) is satisfied, $x(t)$ can be recovered by passing $x_s(t)$ through an ideal lowpass filter with a bandwidth B, where B satisfies

$$f_x \leq B \leq f_s - f_x \tag{10.6}$$

At this point, we restate our reason for studying sampling theory: namely, we want to represent an analog signal by a sequence of sampled values. So far we have seen how an analog signal can be represented by a sequence of *segments*; now we proceed to show that indeed it is sufficient to have *instantaneous values* of $x(t)$ rather than segments of $x(t)$ for adequate representation of $x(t)$.

10.2.2 Ideal Sampling and Reconstruction of Lowpass Signals

Ideal sampling, by definition, is *instantaneous sampling*, and is accomplished by using a train of impulses $s_\delta(t)$ as the sampling function. Thus we have, for ideal sampling,

$$x_\delta(t) = x(t)s_\delta(t) \tag{10.7}$$

where

$$s_\delta(t) = \sum_{k=-\infty}^{\infty} \delta(t - kT_s) \tag{10.8}$$

Using the properties of the uniformly spaced impulse train, the reader can verify that

$$\begin{aligned} x_\delta(t) &= x(t) \sum_{k=-\infty}^{\infty} \delta(t - kT_s) \\ &= \sum_{k=-\infty}^{\infty} x(kT_s)\delta(t - kT_s) \end{aligned} \tag{10.9}$$

and

$$X_\delta(f) = X(f) * S_\delta(f) \tag{10.10}$$

where

$$S_\delta(f) = f_s \sum_{k=-\infty}^{\infty} \delta(f - nf_s) \tag{10.11}$$

or

$$X_\delta(f) = f_s \sum_{n=-\infty}^{\infty} X(f - nf_s) \tag{10.12}$$

Comparing Equation (10.12) and (10.4*b*), we see that the only difference is that the constants C_n in Equation (10.4*b*) are equal to f_s in Equation (10.12). Thus for perfect reconstruction of $x(t)$ from $x_\delta(t)$ we invoke the same conditions as we had for recovering $x(t)$ from $x_s(t)$, that is, $x(t)$ must be bandlimited to f_x and $f_s \geqslant 2f_x$. Then, we can reconstruct $x_\delta(t)$ from $x(t)$ by passing $x_\delta(t)$ through an ideal lowpass filter $H_R(f)$ with a bandwidth B satisfying $f_x \leqslant B \leqslant f_s - f_x$ as shown in Figure 10.6. Next, if we let the filter gain $K = 1/f_s$, then, from Equation (10.12) and Figure 10.6, we have

$$X(f) = X_\delta(f)H_R(f)$$

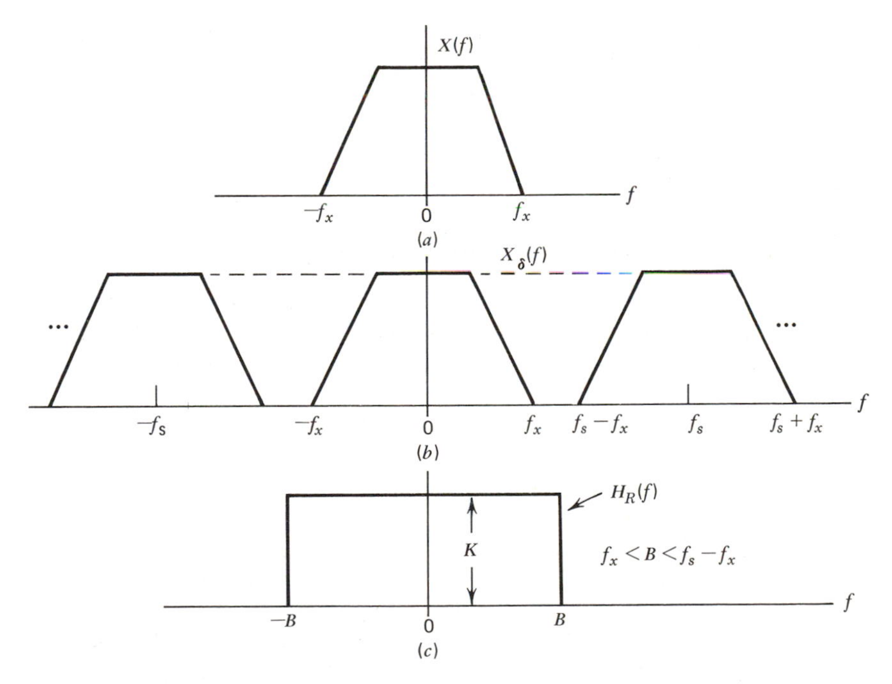

Figure 10.6 Spectra of ideally sampled signals. (*a*) Message. (*b*) Ideally sampled message. (*c*) Reconstruction filter. $X_\delta(f)H_R(f) = CX(f)$, where $C = Kf_s$.

In the time domain, we can represent the reconstruction process by

$$\begin{aligned} x(t) &= F^{-1}\{X_\delta(f)H_R(f)\} \\ &= [h_R(t) * x_\delta(t)] \end{aligned} \tag{10.13}$$

where $h_R(t)$ is the impulse response of the reconstruction filter. We know that $h_R(t)$ is $2BT_s \operatorname{sinc}(2Bt)$ and hence we have

$$\begin{aligned} x(t) &= [2BT_s \operatorname{sinc}(2Bt)] * \left[\sum_{k=-\infty}^{\infty} x(kT_s)\delta(t-kT_s)\right] \\ &= 2BT_s \sum_{k=-\infty}^{\infty} x(kT_s) \operatorname{sinc} 2B(t-kT_s) \end{aligned} \tag{10.14}$$

Equation (10.14) gives us the result we were seeking; namely, a bandlimited signal $x(t)$ can be represented by a sequence of sampled values $\{x(kT_s)\}$ if the sampling is done such that $f_s > 2f_x$. We state this result in the following theorem:

The Uniform Sampling Theorem for Lowpass Signals

If a signal $x(t)$ contains no frequency components for $|f| > f_x$, then it is completely described by instantaneous values $x(kT_s)$ uniformly spaced in time with period $T_s \leq 1/2f_x$. If the sampling rate f_s is equal to the Nyquist rate or greater ($f_s \geq 2f_x$), and if the sampled values are represented by weighted impulses, then the signal can be exactly reconstructed from its samples by an ideal lowpass filter of bandwidth B, where $f_x \leq B \leq f_s - f_x$ and $f_s = 1/T_s$.

10.2.3 Ideal Sampling and Reconstruction of Bandpass Signals

Signals with bandpass spectra can also be represented by their sampled values. Consider a signal $x(t)$ with the spectrum shown in Figure 10.7. The following sampling theorem gives the conditions for representing $x(t)$ by its sampled values.

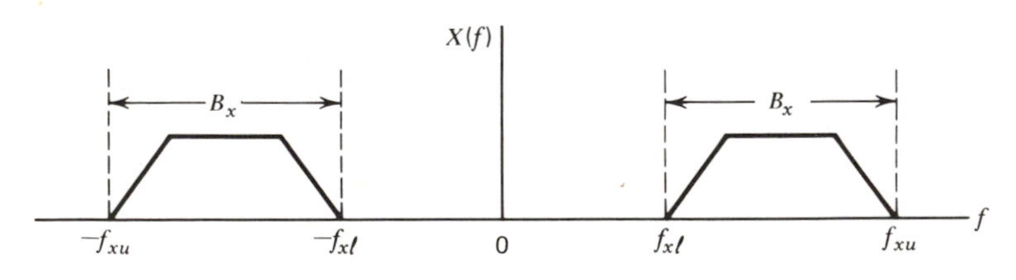

Figure 10.7 Spectrum of a bandpass signal.

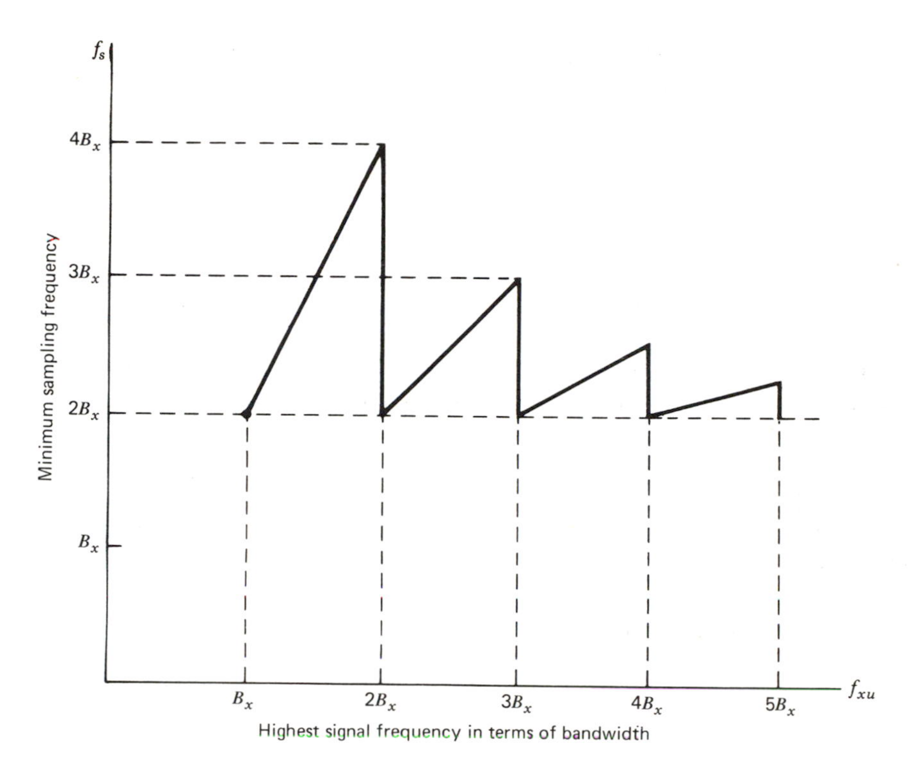

Figure 10.8 Minimum sampling frequency for a signal occupying a bandwidth B_x.

The Uniform Sampling Theorem for Bandpass Signals

If a bandpass signal $x(t)$ has a spectrum of bandwidth B_x and upper frequency limit f_{xu}, then $x(t)$ can be represented by instantaneous values $x(kT_s)$ if the sampling rate f_s is $2f_{xu}/m$, where m is the largest integer not exceeding f_{xu}/B_x. (Higher sampling rates are not always usable unless they exceed $2f_{xu}$.) If the sample values are represented by impulses, then $x(t)$ can be exactly reproduced from its samples by an ideal bandpass filter $H(f)$ with the response

$$H(f) = \begin{cases} 1 & f_{xl} < |f| < f_{xu} \\ 0 & \text{elsewhere} \end{cases}$$

The sampling rate for a bandpass signal depends on the ratio f_{xu}/B_x. If $f_{xu}/B_x \gg 1$, then the minimum sampling rate approaches $2B_x$. A sketch of f_{xu}/B_x versus f_s/B_x is shown in Figure 10.8. The reader can easily verify that $f_s > 2f_{xu}$ will result in exact reconstruction. Proof of exact reconstruction when

$f_s = 2f_{xu}/m$, where m is an integer satisfying $(f_{xu}/B_x) - 1 < m \leq f_{xu}/B_x$, is left as an exercise for the reader (see Problems 10.3 and 10.4).

10.2.4 Sampling Theorem for Random Signals

Having looked at sampling methods for deterministic signals, we now turn our attention to the sampling of random processes. The message waveform $X(t)$ in communication systems is often modeled as a bandlimited stationary random process at the baseband level. The power spectral density $G_X(f)$ of a bandlimited random process $X(t)$ is zero for $|f| > f_x$. Hence the autocorrelation function $R_{XX}(\tau)$ can be written as (see Equation (10.14))

$$R_{XX}(\tau) = 2BT_s \sum_{k=-\infty}^{\infty} R_{XX}(kT_s) \operatorname{sinc} 2B(\tau - kT_s) \tag{10.15}$$

where $1/T_s = f_s > 2f_x$ and $f_x < B < f_s - f_x$. It is convenient to state two different versions of Equation (10.15). With a an arbitrary constant, the transform of $R_{XX}(\tau - a)$ is equal to $G_X(f) \exp(-2\pi jfa)$. This function is also bandlimited, and hence Equation (10.15) can be applied to $R_{XX}(\tau - a)$ as

$$R_{XX}(\tau - a) = 2BT_s \sum_{n=-\infty}^{\infty} R_{XX}(nT_s - a) \operatorname{sinc} 2B(\tau - nT_s) \tag{10.16}$$

Changing $(\tau - a)$ to τ in Equation (10.16), we have

$$R_{XX}(\tau) = 2BT_s \sum_{n=-\infty}^{\infty} R_{XX}(nT_s - a) \operatorname{sinc} 2B(\tau + a - nT_s) \tag{10.17}$$

We will now state and prove the sampling theorem for bandlimited random processes using Equations (10.15) and (10.16).

The Uniform Sampling Theorem for Bandlimited Random Signals

If a random process $X(t)$ is bandlimited to f_x Hz, then $X(t)$ can be represented using the instantaneous values $X(kT_s)$ as

$$X(t) \overset{MS}{=} \hat{X}(t) = 2BT_s \sum_{-\infty}^{\infty} X(nT_s) \operatorname{sinc}[2B(t - nT_s)] \tag{10.18}$$

(where $\overset{MS}{=}$ stands for equality in the mean squared sense*) if the sampling rate f_s is equal to or greater than the Nyquist rate $2f_x$. If the sampled values are represented by weighted impulses, then $X(t)$ can be reconstructed from its samples by an ideal lowpass filter of bandwidth B, where $f_x \leq B \leq f_s - f_x$ and $f_s = 1/T_s$.

*$X(t) \overset{MS}{=} \hat{X}(t)$ if $E\{[X(t) - \hat{X}(t)]^2\} = 0$.

To prove Equation (10.18), we need to show that

$$E\{[X(t) - \hat{X}(t)]^2\} = 0 \tag{10.19}$$

where

$$\hat{X}(t) = 2BT_s \sum_{n=-\infty}^{\infty} X(nT_s) \operatorname{sinc}[2B(t - nT_s)]$$

Now,

$$E\{[X(t) - \hat{X}(t)]^2\} = E\{[X(t) - \hat{X}(t)]X(t)\} - E\{[X(t) - \hat{X}(t)]\hat{X}(t)\} \tag{10.20}$$

The first term on the right-hand side of the previous equation may be written as

$$E\{[X(t) - \hat{X}(t)]X(t)\} = R_{XX}(0) - 2BT_s \sum_{n=-\infty}^{\infty} R_{XX}(nT_s - t) \operatorname{sinc}[2B(t - nT_s)]$$

From Equation (10.17) with $\tau = 0$ and $a = t$, we have

$$2BT_s \sum_{n=-\infty}^{\infty} R_{XX}(nT_s - t) \operatorname{sinc}[2B(t - nT_s)] = R_{XX}(0)$$

and hence

$$E\{[X(t) - \hat{X}(t)]X(t)\} = 0 \tag{10.21}$$

The second term in Equation (10.20) can be written as

$$E\{[X(t) - \hat{X}(t)]\hat{X}(t)\} = \sum_{m=-\infty}^{\infty} E\{[X(t) - \hat{X}(t)]X(mT_s)\}2BT_s \operatorname{sinc}[2B(t - mT_s)]$$

Now,

$$E\{[X(t) - \hat{X}(t)]X(mT_s)\} = R_{XX}(t - mT_s) - \sum_{n=-\infty}^{\infty} 2BT_s R_{XX}(nT_s - mT_s) \operatorname{sinc}[2B(t - nT_s)]$$

and from Equation (10.16) with $\tau = t$ and $a = mT_s$, we have

$$R_{XX}(t - mT_s) = 2BT_s \sum_{n=-\infty}^{\infty} R_{XX}(nT_s - mT_s) \operatorname{sinc}[2B(t - nT_s)]$$

Hence,

$$E\{[X(t) - \hat{X}(t)]\hat{X}(t)\} = 0 \tag{10.22}$$

Substitution of Equations (10.21) and (10.22) in (10.20) completes the proof of Equation (10.19).

The proof of the second part of the theorem dealing with the reconstruction follows the steps outlined in Section 10.2.2. If the random process $X(t)$ is a bandpass process, then a theorem similar to the uniform sampling theorem for deterministic bandpass signals can be developed.

The sampling theorems for random signals tell us that the output of analog information sources can be adequately represented by the sampled values of the signals. Thus, rather than transmitting an analog signal, we need to transmit only the sampled values. At the receiver, the analog signal can be reconstructed from the received sequence of sampled values by appropriate filtering.

10.2.5 Practical Sampling

There are a number of differences between the ideal sampling and reconstruction techniques described in the preceding sections and the actual signal sampling as it occurs in practice. The major differences are:

1. The sampled wave in practical systems consists of finite amplitude and finite duration pulses rather than impulses.
2. Reconstruction filters in practical systems are not ideal filters.
3. The waveforms that are sampled are often timelimited signals and hence are not bandlimited.

Let us look at the effects of these differences on the quality of the reconstructed signals.

The sampled waveform produced by practical sampling devices, especially the sample and hold variety, has the form

$$\begin{aligned} x_s(t) &= \sum_{k=-\infty}^{\infty} x(kT_s)p(t-kT_s) \\ &= [p(t)] * \left[\sum_{k=-\infty}^{\infty} x(kT_s)\delta(t-kT_s) \right] \end{aligned}$$

where $p(t)$ is a flat topped pulse of duration τ. (This type of sampling is called *flat topped sampling.*) The spectrum $X_s(f)$ of $x_s(t)$ is given by

$$X_s(f) = P(f)X_\delta(f) = P(f)\left[f_s \sum_{n=-\infty}^{\infty} X(f-nf_s) \right] \tag{10.23}$$

where $P(f)$ is the Fourier transform of $p(t)$ and $X_\delta(f)$ is the Fourier transform of the ideal sampled wave. $P(f)$ is a sinc function and hence we can say from Equation (10.23) that the primary effect of flat topped sampling is an attenuation of high-frequency components. This effect, sometimes called an *aperture effect*, can be compensated by an equalizing filter with a transfer function $H_{eq}(f) = 1/P(f)$. However, if the pulsewidth is chosen to be small compared to the time between samples (i.e., $\tau \ll 1/f_x$) then $P(f)$ is essentially constant over the message band and no equalization may be needed. Thus, effects of pulse shape are often unimportant and flat topped sampling is a good approximation to ideal impulse sampling.

The effect of nonideal reconstruction filters is shown in Figure 10.9. To recover the sampled signal shown in the figure, we need an ideal lowpass filter. However, such filters can only be approximated in practice. The output of the filter shown in Figure 10.9*a* will consist of $x(t)$ plus spurious frequency components at $|f| > f_x$ that lie outside the message band. While these components are considerably attenuated compared to $x(t)$, their presence may be annoying in some applications such as in audio systems. Good filter design will minimize this effect. Alternatively, for a given filter response, the high-frequency spurious components can be suppressed or eliminated by increasing the sampling frequency as shown in Figure 10.9*b*. Increasing the sampling frequency produces a *guard band* of width $(f_s - 2f_x)$ Hz.

In many practical applications, the waveform to be sampled might last for only a finite amount of time. Such message waveforms are not strictly bandlimited, and when such a message is sampled, there will be unavoidable overlapping of spectral components at frequencies $f > f_s/2$ (see Figure 10.10). The effect of this overlapping (also called aliasing) is far more serious than spurious high-frequency components passed by nonideal reconstruction filters, for the latter fall outside the message band. Aliasing effects can be minimized by bandlimiting the signal by filtering before sampling and sampling at a rate moderately higher than the nominal Nyquist rate.

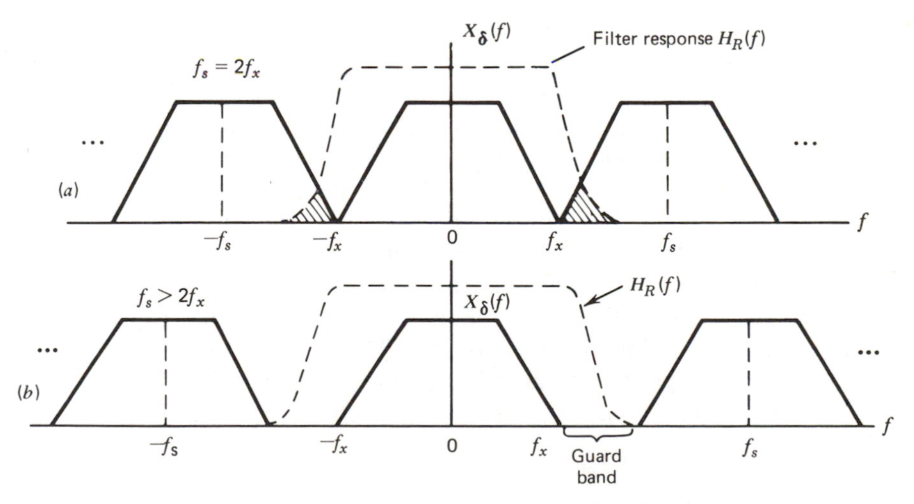

Figure 10.9 Reconstruction of sampled signals.

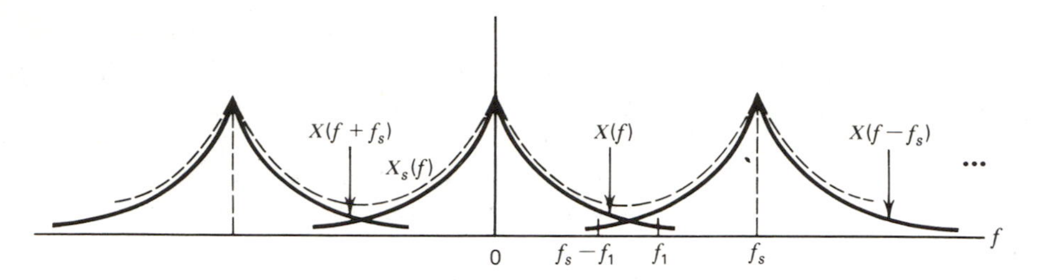

Figure 10.10 Sampling of nonbandlimited signals.

10.3 QUANTIZING OF ANALOG SIGNALS

In the preceding sections we established the fact that an analog message signal can be adequately represented by its sampled values. Message signals such as speech waveforms or video waveforms have a continuous amplitude range and hence the samples are also continuous in amplitude. When these continuous amplitude samples are transmitted over a noisy channel, the receiver cannot discern the exact sequence of transmitted values. The effect of the noise in the system can be minimized by representing the samples by a finite number of predetermined levels and transmitting the levels using a discrete signaling scheme such as discrete PAM. Now, if the separation between the levels is large compared to the noise perturbations, it will be a simple matter for the receiver to decide precisely which specific value was transmitted. Thus the effect of random noise can be virtually eliminated.

Representing the analog sampled values by a finite set of levels is called *quantizing*. While sampling converts a continuous time signal to a discrete time signal, quantizing converts a continuous amplitude sample to a discrete amplitude sample. Thus sampling and quantizing operations convert the output of an analog information source into a sequence of levels (or symbols), that is, the analog source is transformed to a discrete (digital) source. The sequence of levels can be transmitted using any one of the many digital signaling schemes discussed in the preceding chapters. An example of the quantizing operation is shown in Figure 10.11.

The input to the quantizer is a random process $X(t)$ that represents the output of an analog information source. The random waveform $X(t)$ is sampled at an appropriate rate and the sampled values $X(kT_s)$ are converted to one of Q allowable levels, $m_1, m_2, \ldots, m_Q$, according to some predetermined rule:

$$X_q(kT_s) = m_i \quad \text{if} \quad x_{i-1} \leq X(kT_s) < x_i \qquad x_0 = -\infty,\ x_Q = +\infty \tag{10.24}$$

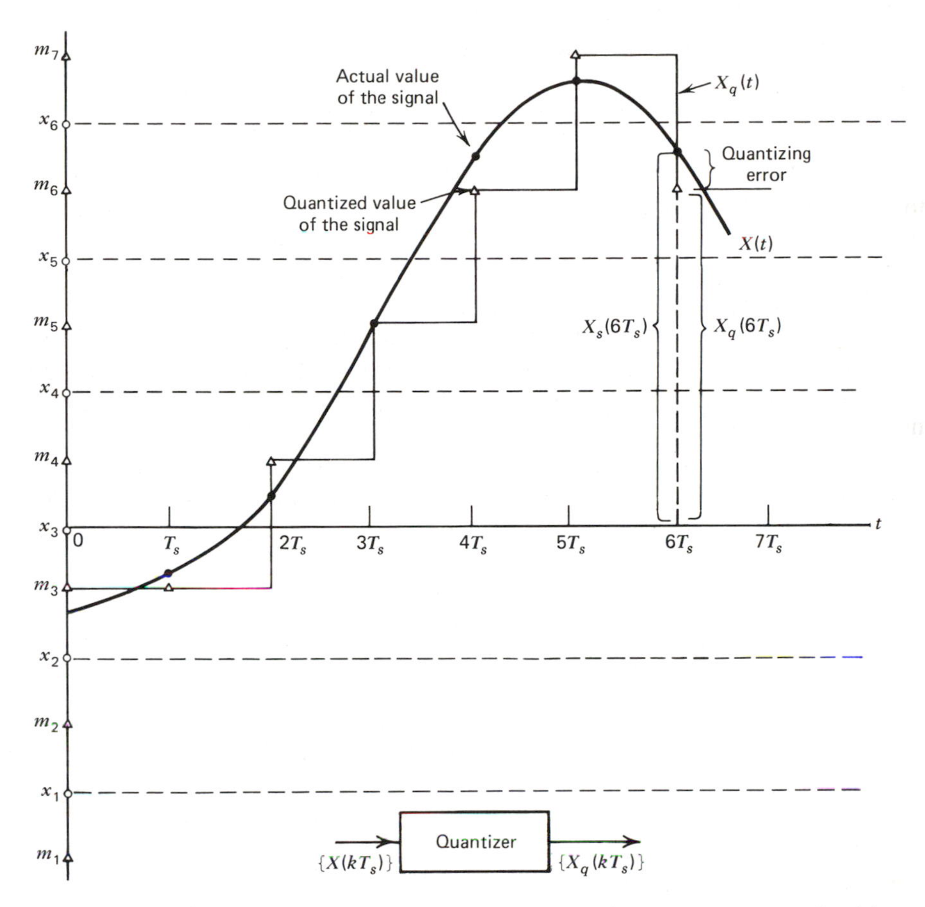

Figure 10.11 Quantizing operation; $m_1, m_2, \ldots, m_7$ are the seven output levels of the quantizer.

The output of the quantizer is a sequence of levels, shown in Figure 10.11 as a waveform $X_q(t)$, where

$$X_q(t) = X_q(kT_s), \qquad kT_s \leq t < (k+1)T_s$$

We see from Figure 10.11 that the quantized signal is a good approximation to the original signal. The quality of the approximation may be improved by a careful choice of x_i's and m_i's such that some measure of performance is optimized. The measure of performance that is most commonly used for evaluating the performance of a quantizing scheme is the output signal to quantizing noise power ratio defined as

$$\frac{S_q}{N_q} = \frac{E\{[X_q(kT_s)]^2\}}{E\{[X(kT_s) - X_q(kT_s)]^2\}} \tag{10.25}$$

Since the overall performance of digital transmission schemes for analog signals will be measured by signal-to-noise power ratios, and since the overall received signal quality will depend on the accuracy of representation of the sample values, the signal to quantizer noise power ratio, defined in Equation (10.25) is an appropriate measure of signal quality.

We will now consider several methods of quantizing the sampled values of a random process $X(t)$. For convenience, we will assume $X(t)$ to be a zero mean stationary random process with a pdf $f_X(x)$. We will use the abbreviated notation X to denote $X(kT_s)$ and X_q to denote $X_q(kT_s)$. The problem of quantizing consists of approximating the continuous random variable X by a discrete random variable X_q. We will use the mean squared error $E\{(X - X_q)^2\}$ as a measure of quantizing error.

10.3.1 Uniform Quantizing

In this method of quantizing, the range of the continuous random variable X is divided into Q intervals of equal length, say Δ. If the value of X falls in the ith *quantizing interval*, then the quantized value of X is taken to be the midpoint of the interval (see Figure 10.12). If a and b are the minimum and maximum values of X, respectively, then the *step size* or interval length Δ is given by

$$\Delta = (b - a)/Q \tag{10.26a}$$

The quantized output X_q is generated according to

$$X_q = m_i \quad \text{if} \quad x_{i-1} < X \leq x_i \tag{10.26b}$$

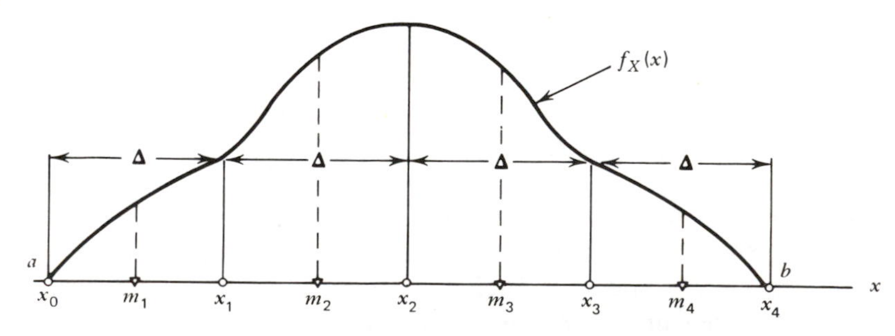

Figure 10.12 Example of uniform quantizing. Step size = Δ, $Q = 4$.

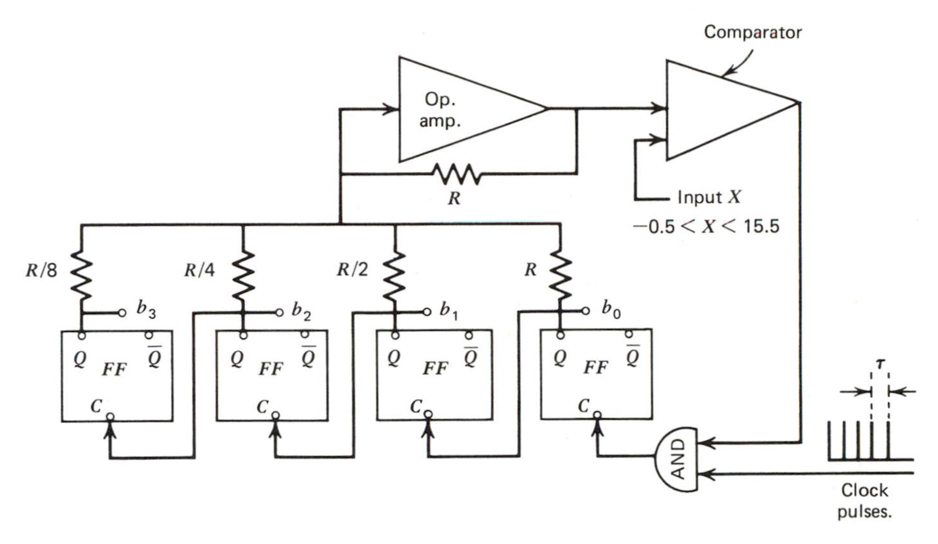

Figure 10.13 A sixteen level uniform quantizer. Step size = 1, and $b_3b_2b_1b_0$ gives the binary codeword for the level X_q. The clock rate is assumed to be much higher than the sampling rate and $Q = 1$ volt.

where

$$x_i = a + i\Delta \tag{10.26c}$$

and

$$m_i = \frac{x_{i-1} + x_i}{2}, \quad i = 1, 2, \ldots, Q \tag{10.26d}$$

A uniform quantizer (A/D converter) that generates binary codes for the output levels is shown in Figure 10.13. It consists of a binary counter, a resistor matrix and summing device, and a comparator. The quantizer input X is assumed to be in the range -0.5 volt to 15.5 volts (if the range of X is outside this interval, then scaling and level shifting are necessary). When the quantizing is started, the counter is at zero and $X_q = 0$. As the count is increased (while the AND gate is open) the value of X_q increases. As soon as X_q comes within $\frac{1}{2}$ volt of X, the comparator outputs a zero and closes the AND gate and blocks the clock pulses from incrementing the counter. The output of the operational amplifier represents the quantized value X_q of X, and the outputs of the flip-flops $b_3b_2b_1b_0$ provide a binary codeword for the output level. In this example, the numerical value of the binary codeword is equal to the value of X_q.

The quantizing noise power N_q for the uniform quantizer is given by

$$\begin{aligned} N_q &= E\{(X - X_q)^2\} \\ &= \int_a^b (x - x_q)^2 f_X(x)\, dx \\ &= \sum_{i=1}^{Q} \int_{x_{i-1}}^{x_i} (x - m_i)^2 f_X(x)\, dx \end{aligned} \tag{10.27a}$$

where $x_i = a + i\Delta$ and $m_i = a + i\Delta - \Delta/2$. The signal power S_q at the output of the quantizer can be obtained from

$$\begin{aligned} S_q &= E\{(X_q)^2\} \\ &= \sum_{i=1}^{Q} (m_i)^2 \int_{x_{i-1}}^{x_i} f_X(x)\, dx \end{aligned} \tag{10.27b}$$

The ratio S_q/N_q gives us a measure of fidelity of the uniform quantizer. This ratio can be computed if the pdf of X is known.

Example 10.1. The input to a Q-step uniform quantizer has a uniform pdf over the interval $[-a, a]$. Calculate the average signal to quantizer noise power ratio at the output.

Solution. From Equation (10.27a) we have

$$\begin{aligned} N_q &= \sum_{i=1}^{Q} \int_{x_{i-1}}^{x_i} (x - m_i)^2 \left(\frac{1}{2a}\right) dx \\ &= \sum_{i=1}^{Q} \int_{-a+(i-1)\Delta}^{-a+i\Delta} \left(x + a - i\Delta + \frac{\Delta}{2}\right)^2 \frac{1}{2a}\, dx \\ &= \sum_{i=1}^{Q} \left(\frac{1}{2a}\right)\left(\frac{\Delta^3}{12}\right) \\ &= \frac{Q\Delta^3}{(2a)12} = \frac{\Delta^2}{12}, \quad \text{since } Q\Delta = 2a. \end{aligned}$$

Now, the output signal power S_q can be obtained using Equation (10.27b) as

$$\begin{aligned} S_q &= \sum_{i=1}^{Q} (m_i)^2 \left(\frac{1}{2a}\right) \\ &= \frac{Q^2 - 1}{12} (\Delta)^2 \end{aligned}$$

and hence the average signal to quantizer noise power ratio is

$$\begin{aligned} \frac{S_q}{N_q} &= Q^2 - 1 \\ &\approx Q^2, \quad \text{when } Q \gg 1 \end{aligned} \tag{10.28a}$$

and

$$(S_q/N_q)_{\text{dB}} = 20 \log Q \tag{10.28b}$$

Equation (10.28) indicates that the fidelity of the quantizer increases with Q, the number of quantizer levels. If a large number of levels of small spacing are employed, then the output X_q can be made as near as desired to X, the input. The number of levels (Q) is determined by the desired transmission fidelity. It has been established experimentally that 8 or 16 levels are just sufficient to obtain good intelligibility of speech. But, the quantizer noise (whose power is more or less uniformly distributed throughout the signal band) can be easily heard in the background. For commercial use in standard voice telephony, a minimum of 128 levels are used to obtain a signal-to-noise ratio of 42 dB. This will require seven bits to represent each quantized sample and hence a larger transmission bandwidth than the unquantized analog voice signal.

The uniform quantizer yields the highest (optimum) average signal to quantizer noise power ratio at the output if the signal has a uniform pdf. The rms value of the quantizer noise is fixed at $\Delta/\sqrt{12}$ regardless of the value of the sample X being quantized. Hence if the signal $X(t)$ is small for extended periods of time, the apparent signal-to-noise ratio will be much lower than the design value. This effect will be particularly noticeable if the signal waveform has a large *crest factor* (the ratio of peak to rms value). For quantizing such signals it is advantageous to taper the spacing between quantizer levels with small spacings near zero and larger spacing at the extremes.

10.3.2 Nonuniform Quantizing

A nonuniform quantizer uses a *variable step size.* It has two important advantages over the uniform quantizer described in the preceding section. First, it yields a higher average signal to quantizing noise power ratio than the uniform quantizer when the signal pdf is nonuniform—which is the case in many practical situations. Secondly, the rms value of the quantizer noise power of a nonuniform quantizer is substantially proportional to the (instantaneous) sampled value X and hence the effect of quantizer noise is masked.

An example of nonuniform quantizing is shown in Figure 10.14. The input to the quantizer is a Gaussian random variable and the quantizer output is determined according to

$$X_q = m_i \quad \text{if } x_{i-1} < X \leq x_i, \quad i = 1, 2, \ldots, Q$$
$$x_0 = -\infty, \quad x_Q = \infty \tag{10.29}$$

The step size $\Delta_i = x_i - x_{i-1}$ is variable. The quantizer end points x_i and the output

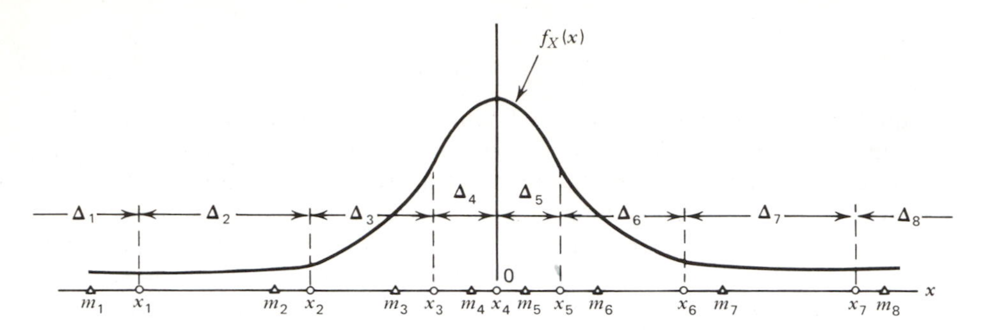

Figure 10.14 A nonuniform quantizer for a Gaussian variable. $x_0 = -\infty$, $x_Q = \infty$, $Q = 8$, and $\Delta_i = \Delta_{Q+1-i}$, $(i = 1, 2, 3, 4)$.

levels m_i are chosen to maximize the average signal to quantizing noise power ratio.

In practice, a nonuniform quantizer is realized by sample compression followed by a uniform quantizer. Compression transforms the input variable X to another variable Y using a nonlinear transformation

$$Y = g(X)$$

such that $f_Y(y)$ has a uniform pdf. Then, Y is uniformly quantized and transmitted (see Figure 10.15). At the receiver, a complementary expander with transfer characteristic g^{-1} restores the quantized values of X. The compresser and expander taken together constitute a *compander.* The most commonly used compander uses a logarithmic compression, $Y = \log X$, where the levels are crowded near the origin and spaced farther apart near the peak values of X.

Two commonly used logarithmic compression laws are the so-called μ and A compression laws defined by

$$|y| = \frac{\log(1 + \mu|x/x_{\max}|)}{\log(1 + \mu)}$$

and

$$|y| = \begin{cases} \dfrac{A|x/x_{\max}|}{1 + \log(A)}, & 0 \leqslant |x/x_{\max}| \leqslant 1/A \\[2ex] \dfrac{1 + \log(A|x/x_{\max}|)}{1 + \log(A)}, & 1/A \leqslant |x/x_{\max}| \leqslant 1 \end{cases}$$

Practical values of A and μ tend to be in the vicinity of 100. These two compression laws yield an average quantizing noise power that is largely independent of signal statistics (see Problems 10.11 and 10.12).

The design of an optimum nonuniform quantizer can be approached as

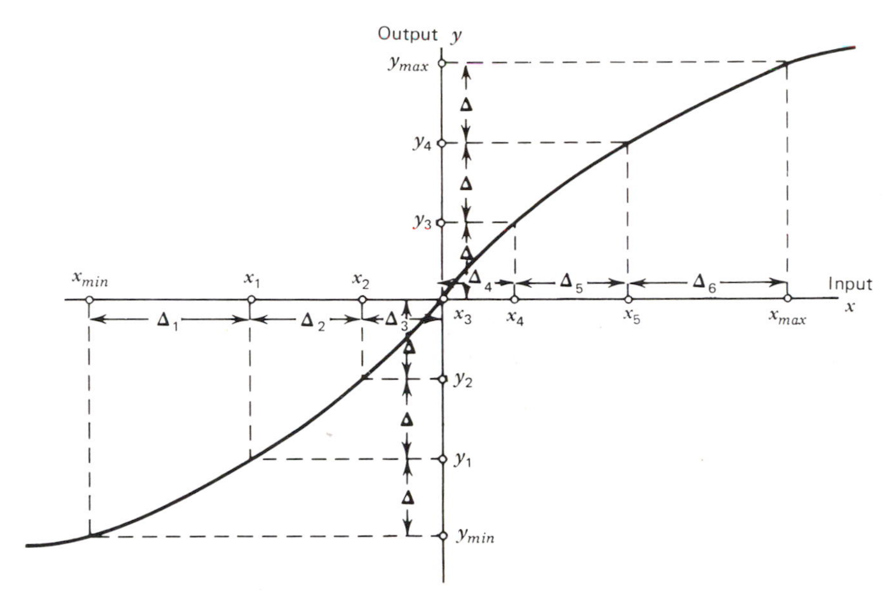

Figure 10.15 A compressor for converting a nonuniform quantizer to a uniform quantizer.

follows. We are given a continuous random variable X with a pdf $f_X(x)$. We want to approximate X by a discrete random variable X_q according to Equation (10.29). The quantizing intervals and the levels are to be chosen such that S_q/N_q defined in Equation (10.24) is maximized. If the number of levels Q is large, then $S_q \approx E\{X^2\}$ and the ratio S_q/N_q is maximized when N_q is minimized. This minimizing can be done as follows. We start with

$$N_q = \sum_{i=1}^{Q} \int_{x_{i-1}}^{x_i} (x - m_i)^2 f_X(x)\, dx, \quad x_0 = -\infty \quad \text{and} \quad x_Q = \infty$$

Since we wish to minimize N_q for a fixed Q, we get the necessary* conditions by differentiating N_q with respect to x_i's and m_i's and setting the derivatives equal to zero:

$$\frac{\partial N_q}{\partial x_j} = (x_j - m_j)^2 f_X(x_j) - (x_j - m_{j+1})^2 f_X(x_j) = 0, \quad j = 1, 2, \ldots, Q-1 \qquad (10.30a)$$

$$\frac{\partial N_q}{\partial m_j} = -2 \int_{x_{j-1}}^{x_j} (x - m_j) f_X(x)\, dx = 0, \quad j = 1, 2, \ldots, Q \qquad (10.30b)$$

*After finding all the x_i's and m_i's that satisfy the necessary conditions, we may evaluate N_q at these points to find a set of x_i's and m_i's that yield the absolute minimum value of N_q. In most practical cases we will get a unique solution for Equations (10.30a) and (10.30b).

From Equation (10.30a) we obtain

$$x_j = \tfrac{1}{2}(m_j + m_{j+1})$$

or

$$m_j = 2x_{j-1} - m_{j-1}, \quad j = 2, 3, \ldots, Q \tag{10.31a}$$

Equation (10.30b) reduces to

$$\int_{x_{j-1}}^{x_j} (x - m_j) f_X(x)\, dx = 0, \quad j = 1, 2, \ldots, Q \tag{10.31b}$$

which implies that m_j is the centroid (or statistical mean) of the jth quantizer interval. The above set of simultaneous equations cannot be solved in closed form for an arbitrary $f_X(x)$. For a specific $f_X(x)$, a method of solving (10.30a) and (10.30b) is to pick m_1 and calculate the succeeding x_i's and m_i's using Equations (10.31a) and (10.31b). If m_1 is chosen correctly, then at the end of the iteration, m_Q will be the mean of the interval $[x_{Q-1}, \infty]$. If m_Q is not the centroid or the mean of the Qth interval, then a different choice of m_1 is made and the procedure is repeated until a suitable set of x_i's and m_i's is reached. The reader can write a computer program to iteratively solve for the quantizing intervals and the means.

The end points of the quantizer intervals and the output levels for a normal random variable have been computed by J. Max [1]. Attempts have also been made to determine the functional dependence of N_q on the number of levels Q. For a normal random variable with a variance of 1, Max [1] has found that N_q is related to Q by

$$N_q \approx 2.2Q^{-1.96}, \text{ when } Q \gg 1$$

If the variance is σ_X^2, then the preceding expression becomes

$$N_q \approx (2.2)\sigma_X^2 Q^{-1.96} \tag{10.32}$$

Now, if we assume X to have zero mean, then $S_q \approx E\{X^2\} = \sigma_X^2$, and hence

$$S_q/N_q \approx (0.45)Q^{1.96} \tag{10.33}$$

Equation (10.33) can be used to determine the number of quantizer levels needed to achieve a given average signal to quantizer noise power ratio.

10.3.3 Differential Quantizing

In the preceding sections we saw that a continuous random process can be adequately represented by a sequence of its sampled values $\{X(kT_s)\}$ and that the individual samples $X(kT_s)$ can be approximated by a set of quantized levels. In the quantizing techniques we had considered thus far, each sample in the sequence $\{X(kT_s)\}$ was quantized independently of the value of the preceding sample. In many practical situations, due to the statistical nature of the message signal $X(t)$ and due to oversampling, the sequence $\{X(kT_s)\}$ will consist of samples that are correlated with each other. Differential quantizing schemes take into account the sample to sample correlation in the quantizing process. For a given number of levels per sample, differential quantizing schemes yield a lower value of quantizing noise power than direct quantizing schemes. Before we look at differential quantizing schemes, let us consider the following example that illustrates the main advantage of differential quantizing schemes.

Example 10.2. The message signal $X(t)$ in a communication system is a zero mean stationary Gaussian random process, and it is sampled at a rate of 10,000 samples per second ($T_s = 0.1$ msec). The normalized autocorrelation function $R_{XX}(\tau)/R_{XX}(0)$ has a value of 0.8 when $\tau = 0.1$ msec. Two quantizing schemes being considered are:

(a) a nonuniform quantizer with $Q = 32$, operating on each sample independently,
(b) a differential quantizer with $Q = 32$ which operates on successive differences $\{X(kT_s) - X((k-1)T_s)\}$.

Assuming that the mean squared error due to quantizing a normal random variable with variance σ^2 is $2\sigma^2Q^{-2}$, find the mean squared error of the quantizing schemes given above.

Solution.
(a) For the quantizer operating independently on each sample, the mean squared error is given by

$$N_q = 2\sigma_X^2Q^{-2} \approx (2)(10^{-3})\sigma_X^2$$

(b) In the differential quantizing scheme, the variable being quantized is

$$Y = X(kT_s) - X[(k-1)T_s]$$

and the variance of Y is given by

$$\begin{aligned}\sigma_Y^2 &= \sigma_X^2(kT_s) + \sigma_X^2[(k-1)T_s] - 2E\{X(kT_s)X[(k-1)T_s]\}\\ &= 2\sigma_X^2\left(1 - \frac{R_{XX}(T_s)}{R_{XX}(0)}\right)\\ &= 0.4\sigma_X^2\end{aligned}$$

Hence the mean squared error of the differential quantizing is given by

$$\begin{aligned}N_q &= 2(0.4)\sigma_X^2Q^{-2}\\ &= 0.8\sigma_X^2(Q^{-2}) \approx (0.8)(10^{-3})\sigma_X^2\end{aligned}$$

which is considerably less than the error associated with the direct quantizing scheme.

The preceding example illustrates that differential quantizing yields a lower mean squared error than direct quantizing if the samples are highly correlated. This error reduction is always possible as long as the sample to sample correlation is nonzero. The largest error reduction occurs when the differential quantizer operates on the difference between $X(kT_s)$ and the minimum mean squared error estimator $\hat{X}(kT_s)$ of $X(kT_s)$. Such a quantizer using a linear minimum mean squared estimator* of $X(kT_s)$ based on the quantized

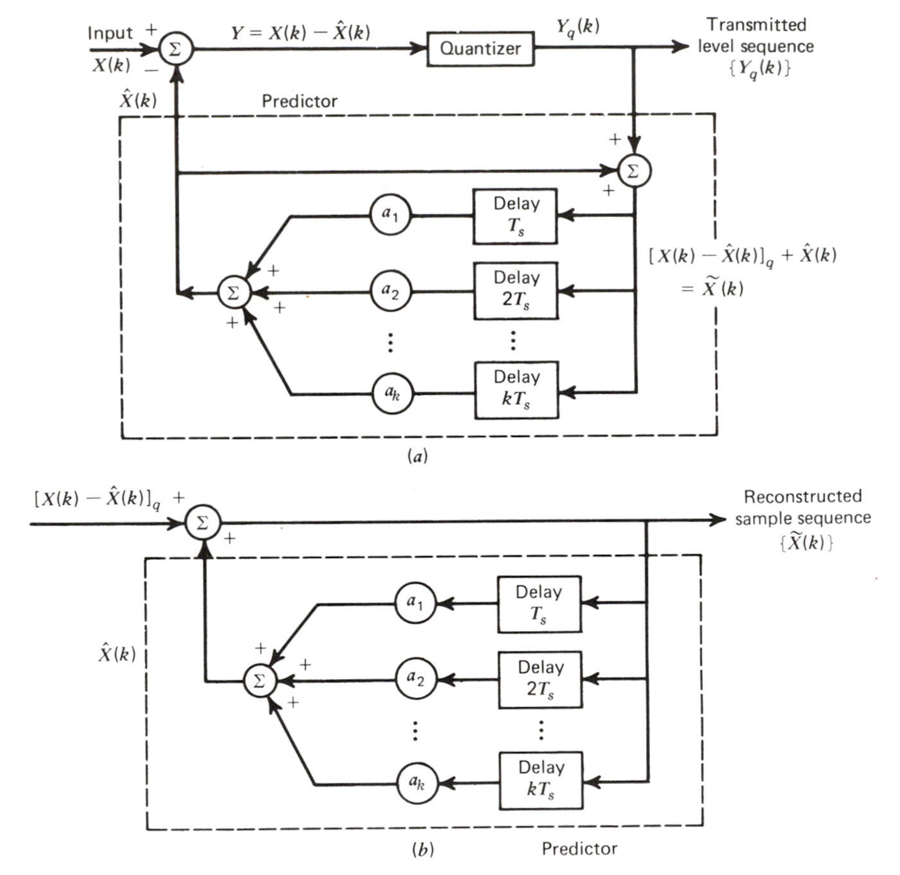

Figure 10.16 Differential quantizing scheme. (*a*) Differential quantizer at the transmitter. (*b*) Sample reconstruction scheme at the receiver.

*For a good discussion of minimum mean squared error estimation, see Papoulis [2].

values of preceding samples is shown in Figure 10.16. The difference $Y(kT_s) = X(kT_s) - \hat{X}(kT_s)$ is quantized and transmitted. Both the transmitter and the receiver use a predictor of the form

$$\hat{X}(kT_s) = a_1\tilde{X}[(k-1)T_s] + a_2\tilde{X}[(k-2)T_s] + \cdots + a_n\tilde{X}[(k-n)T_s]$$

where

$$\tilde{X}(kT_s) = \hat{X}(kT_s) + [X(kT_s) - \hat{X}(kT_s)]_q$$

In the preceding equation $\tilde{X}$ denotes the reconstructed value of X and the subscript q denotes quantized values. The coefficients $a_1, a_2, \ldots, a_n$ are chosen such that $E\{[X(kT_s) - \hat{X}(kT_s)]^2\}$ is minimized. The differential quantizer discussed in Example 10.2 uses a predictor of the form $\hat{X}(kT_s) = X[(k-1)T_s]$.

The mean squared error of a differential quantizing scheme will be proportional to $E\{[X(kT_s) - \hat{X}(kT_s)]^2\}$, whereas the mean squared error of a direct quantizing scheme will be proportional to $E\{[X(kT_s)]^2\}$. If the predictor is good, which will be the case if the samples are highly correlated, then the mean squared error of the differential quantizer will be quite small. However, it must be pointed out here that the differential quantizer requires more hardware.

10.4 CODED TRANSMISSION OF ANALOG SIGNALS

After the output of an analog information source is sampled and quantized, the sequence of output levels $\{X_q(kT_s)\}$ can be transmitted directly using a Q-ary PAM. Alternatively, we may represent each quantized level by a code number and transmit the code number rather than the sample value itself. Source coding techniques discussed in Chapter 4 could be used to arrive at an optimum way of representing levels by code words. This system of transmission in which sampled and quantized values of an analog signal are transmitted via a sequence of codewords is called *Pulse Code Modulation* (PCM).

The important features of PCM are shown in Figure 10.17 and Table 10.1. We assume that an analog signal $x(t)$ with $\max|x(t)| < 4$ volts is sampled at the rate of r_s samples per second. The sampled values are quantized using a uniform quantizing rule with 16 steps ($Q = 16$) of equal step size $\Delta = 0.5$ volt. The quantizer end points are $-4, -3.5, -3, \ldots, 3.5, 4$ volts and the output levels are $-3.75, -3.25, \ldots, 3.25$, and 3.75 volts. Table 10.1 shows a sequence of sample values and the corresponding quantized levels. The 16 output levels are arbitrarily assigned level numbers $0, 1, 2, \ldots, 15$. These level numbers are

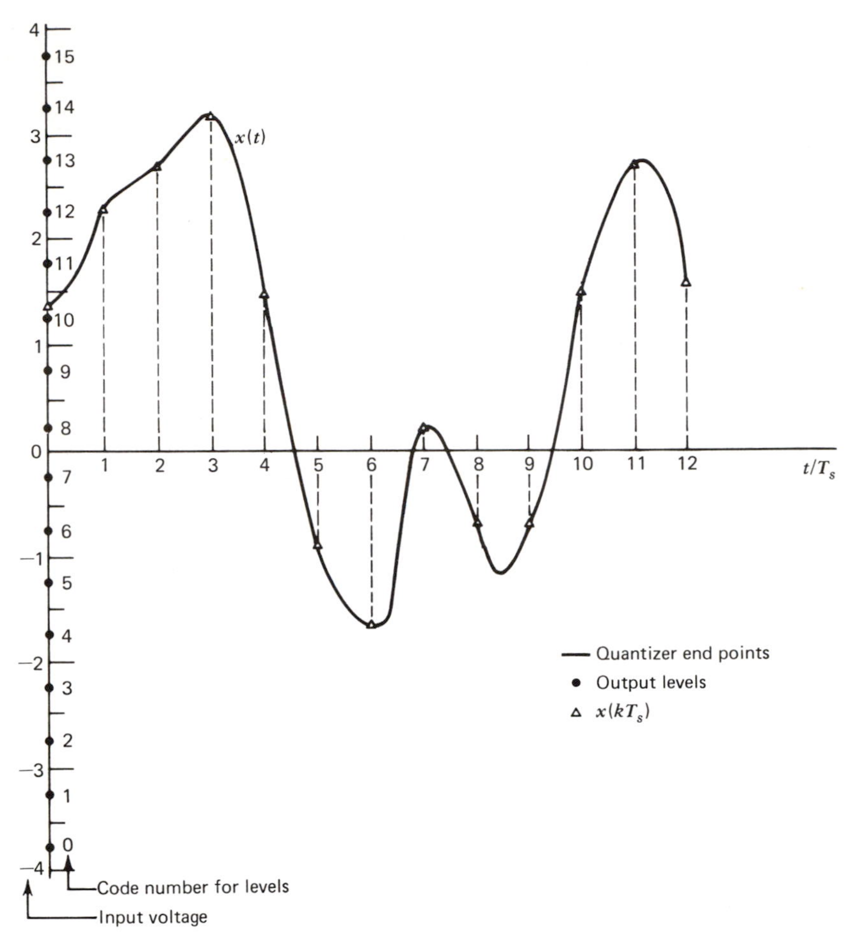

Figure 10.17 PCM example. Coded representation is given in Table 10.1.

shown encoded in binary and quarternary form in Table 10.1. The binary code is the binary representation of the level numbers. The quarternary code is easily derived from the binary code by segmenting each 4-bit binary word into two 2-bit binary words, and then converting each group of two binary digits to real integers. Now, if we are transmitting the sampled analog signal directly using analog PAM, we would transmit the sampled values 1.3, 2.3, 2.7, ... The symbol rate will be equal to the sampling rate r_s. If we are transmitting the quantized sample values using a 16-level discrete PAM, we would transmit the quantized levels 1.25, 2.25, 2.75, . . . , at a rate of r_s levels per second. In binary PCM, we would transmit the bit sequence 101011001101 . . . , at a bit

Table 10.1. Quantizing and coding of an analog signal

Sampled values of an analog signal	1.3	2.3	2.7	3.2	1.1	−1.2	−1.6	0.1	−1.2
Nearest quantizer level	1.25	2.25	2.75	3.25	1.25	−1.25	−1.75	0.25	−1.25
Level number	10	12	13	14	10	5	4	8	5
Binary code	1010	1100	1101	1110	1010	0101	0100	1000	0101
Quarternary code	22	30	31	32	22	11	10	20	11

rate of $4r_s$ bits/sec. Finally, if we use quarternary PCM, we will transmit the digit sequence 22303132 . . . , at a rate of $2r_s$ digits/sec. Each digit in this sequence can have one of four values.

Several versions of PCM schemes are currently being used; two most commonly used versions are the differential pulse code modulation (DPCM) schemes and the delta modulation (DM) schemes. DPCM systems use differential quantizers and PCM encoders. DM schemes use a differential quantizer with *two* output levels Δ or $-\Delta$; these two levels are encoded using a single binary digit before transmission. Thus, DM is a special case of DPCM.

In the following sections, we will discuss PCM, DPCM, and DM schemes in detail, and derive expressions for signal-to-noise ratios at the output of the receivers. Finally, we will compare the performance of these coded transmission schemes with the performance of analog modulation schemes such as AM and FM.

10.4.1 The PCM System

A PCM communication system is shown in Figure 10.18. The analog signal $X(t)$ is sampled and then the samples are quantized and encoded. For the purposes of analysis and discussion we will assume that the encoder output is a binary sequence. In the example shown in Figure 10.17 the binary code has a numerical significance that is the same as the order assigned to the quantized levels. However, this feature is not essential. We could have used arbitrary ordering and codeword assignment as long as the receiver knows the quantized sample value associated with each code word.

The combination of the quantizer and encoder is often called an *analog to*

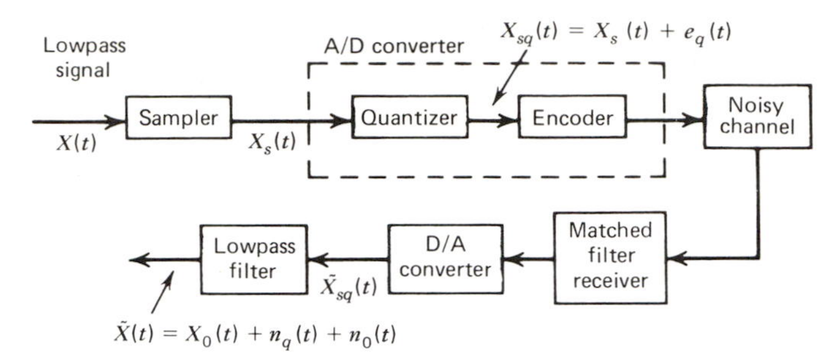

Figure 10.18 Block diagram of a PCM system. $n_q(t)$ is the noise due to quantizing and $n_0(t)$ is the noise due to bit errors caused by channel noise.

digital (A to D or A/D) *converter.* The sampler in practical systems is usually a *sample and hold* device. The combination of the sample and hold device and the A/D converter accepts analog signals and replaces it with a sequence of code symbols. A more detailed diagram of this combination, sometimes called a *digitizer,* is shown in Figure 10.19.

The digitally encoded signal is transmitted over the communication channel to the receiver (shown in Figure 10.20). When the noisy version of this signal arrives at the receiver, the first operation performed is the separation of the signal from the noise. Such separation is possible because of the quantization of the signal. A feature that eases this task of separating signal and noise is

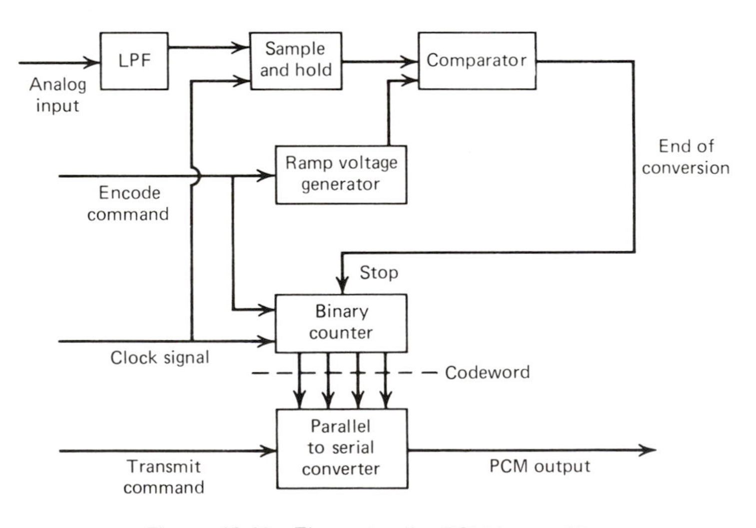

Figure 10.19 Elements of a PCM transmitter.

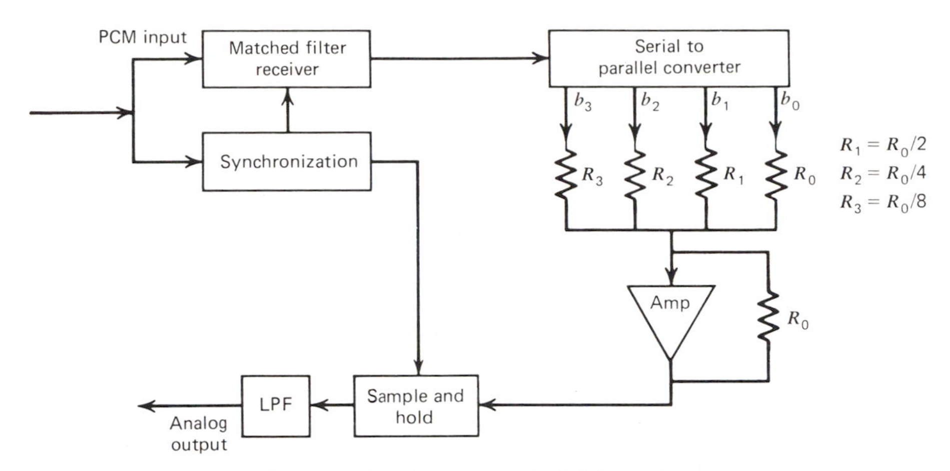

Figure 10.20 Elements of a PCM receiver.

that during each bit interval the receiver (matched filter) has only to make the simple decision of whether a 0 or a 1 has been received. The relative reliability of this decision in binary PCM over the multivalued decision required for direct Q-ary PAM is an important advantage for binary PCM.

After decoding a group of binary digits representing the codeword for the quantized value of the sample, the receiver has to assign a signal level to the codeword. The functional block that performs this task of accepting sequences of binary digits and generating appropriate sequences of levels is called a *digital to analog* (D/A) *converter.* The sequence of levels that appear at the output of the D/A converter as a Q-level PAM waveform is then filtered to reject any frequency components lying outside of the baseband. The reconstructed signal $\tilde{X}(t)$ is identical with the input $X(t)$ except for the quantization noise $n_q(t)$ and another noise component $n_0(t)$ that results from decoding errors due to the channel noise.

Figures 10.18–10.20 do not show signal companding components, and timing recovery networks.

10.4.2 Bandwidth Requirements of PCM

Since PCM requires the transmission of several digits for each message sample, it is apparent that the PCM bandwidth will be much greater than the message bandwidth. A lower bound on the bandwidth can be obtained as follows. If the message bandwidth is f_x, then the quantized samples occur at a rate $f_s(\geq 2f_x)$ samples per second. If the PCM system uses M channel symbols

(M-ary transmission) to represent the Q quantizer levels, then each codeword would consist of γ digits, where

$$\gamma = \log_M(Q), \quad M \leq Q$$

since there are M^γ different possible codewords and $M^\gamma \geq Q$ for unique coding. Thus the channel symbol rate is $r_s = \gamma f_s \geq 2\gamma f_x$. Recalling that for discrete baseband PAM signalling we need a bandwidth $\geq r_s/2$ Hz, we obtain the bandwidth of the PCM signal as

$$B_{\text{PCM}} \geq \gamma f_x \tag{10.34}$$

For binary PCM, the bandwidth required is greater than or equal to $f_x \log_2 Q$.

As an illustration of the bandwidth requirements let us consider the digital transmission of telephone-quality voice signal. While the average voice spectrum exceeds well beyond 10 kHz, most of the energy is concentrated in the range 100 to 600 Hz and a bandwidth of 3 kHz is sufficient for intelligibility. As a standard for telephone systems, the voice signal is first passed through a 3 kHz lowpass filter and then sampled at $f_s = 8000$ samples per second. Each sample is then quantized into one of 128 levels. If these samples are transmitted using binary PCM, then the bandwidth required will be larger than $(8000)(\frac{1}{2})(\log_2 128) = 28$ kHz, which is considerably greater than the 3 kHz bandwidth of the voice signal.

10.4.3 Noise in PCM Systems

It is shown in Figure 10.18 that the output $\tilde{X}(t)$ in a PCM system can be written as

$$\tilde{X}(t) = X_0(t) + n_q(t) + n_0(t) \tag{10.35}$$

where $X_0(t) = kX(t)$ is the signal component in the output; $n_q(t)$ and $n_0(t)$ are two noise components. The first noise waveform $n_q(t)$ is due to quantization and the additional noise waveform $n_0(t)$ is due to the additive channel noise. The overall signal-to-noise ratio at the baseband output, which is used as a measure of signal quality, is defined as

$$\left(\frac{S}{N}\right)_0 = \frac{E\{[X_0(t)]^2\}}{E\{[n_q(t)]^2\} + E\{[n_0(t)]^2\}} \tag{10.36}$$

The average noise power at the output, $E\{[n_q(t)]^2\}$ and $E\{[n_0(t)]^2\}$, can be calculated as follows.

Quantization Noise in PCM Systems. If we assume that ideal impulse sampling is used in the PCM system, then the output of the sampler is

$$X_s(t) = X(t) \sum_{k=-\infty}^{\infty} \delta(t - kT_s)$$

The quantized signal $X_{sq}(t)$ can then be expressed as

$$\begin{aligned} X_{sq}(t) &= X_q(t) \sum_k \delta(t - kT_s) \\ &= X(t) \sum_k \delta(t - kT_s) + [X_q(t) - X(t)] \sum_k \delta(t - kT_s) \\ &= \sum_k [X(kT_s)\delta(t - kT_s) + e_q(kT_s)\delta(t - kT_s)] \end{aligned}$$

where $e_q(t)$ is the error introduced by the quantizing operation. Using the results derived in Chapter 3, we can obtain the power spectral density of e_q as

$$G_{e_q}(f) = \frac{1}{T_s} E\{e_q^2(kT_s)\} \tag{10.37}$$

assuming that $E[e_q(kT_s)] = 0$ and $E\{e_q(kT_s)e_q[(k+j)T_s]\} = 0$. The mean squared error due to quantizing, $E\{e_q^2(kT_s)\}$, will depend on the signal statistics and the method of quantizing. For comparison purposes, let us assume a uniform quantizer operating on $X(t)$ having a uniform pdf over the interval $[-a, a]$. Then we have

$$E\{e_q^2(kT_s)\} = \Delta^2/12$$

where Δ is the step size, and

$$G_{e_q}(f) = \frac{1}{T_s}\left(\frac{\Delta^2}{12}\right)$$

If we ignore the effects of channel noise temporarily, then the noise component $n_q(t)$ has a power spectral density

$$G_{n_q}(f) = G_{e_q}(f)|H_R(f)|^2$$

where $H_R(f)$ is the transfer function of the lowpass filter used for reconstructing the signal. Assuming $f_s = 2f_x$ and $H_R(f)$ to be an ideal lowpass filter with a bandwidth f_x, we have

$$G_{n_q}(f) = \begin{cases} G_{e_q}(f) & |f| < f_x \\ 0 & \text{elsewhere} \end{cases}$$

Hence,

$$E\{n_q^2(t)\} = \int_{-f_x}^{f_x} G_{n_q}(f)\, df = \frac{1}{T_s^2}\left(\frac{\Delta^2}{12}\right)$$

The output signal component $X_0(t)$ is the response of the lowpass filter to

$X(t)\sum_k \delta(t-kT_s)$. We can calculate $E\{[X_0(t)]^2\}$ as

$$E\{[X_0(t)]^2\} \approx \frac{Q^2}{T_s^2}\left(\frac{\Delta^2}{12}\right) \tag{10.38}$$

where Q is the number of quantizer levels. Thus, the average signal to quantizer noise power ratio at the output of the PCM system is given by

$$\frac{E\{[X_0(t)]^2\}}{E\{[n_q(t)]^2\}} = Q^2 \tag{10.39}$$

This result is the same as the normalized mean square error due to quantizing (Equation (10.28*a*)), that is,

$$\frac{E\{[X_q(kT_s)]^2\}}{E\{[X(kT_s) - X_q(kT_s)]^2\}} = Q^2$$

This coincidence is due to the assumption that ideal impulse sampling is used in the system.

Channel Noise in PCM Systems. Channel noise causes the matched filter detector to make an occasional error in decoding whether a binary 0 or 1 was transmitted. The probability of error depends on the type of signaling used and the average signal-to-noise power ratio at the receiver input.

Typically, binary PCM systems operate with small word sizes and low probabilities of error. Hence, the likelihood of more than a single bit error within a codeword can be ignored. As an example, if the bit error probability is $P_e = 10^{-4}$ and a word has eight bits, we may expect on the average one word error for every 1250 words transmitted.

The probability of more than one bit error per word in this example would be of the order of $\binom{8}{2}P_e^2$ or of the order of 10^{-7}. When a bit error occurs in a PCM system, the decoder incorrectly identifies the transmitted level, and the quantized value of the signal is thus incorrectly determined. The magnitude of the error will be small if the bit error occurred in the least significant bit position, and the error will be large if the bit error occurred in the most significant bit position within the codeword.

In order to calculate the effects of bit errors induced by channel noise, let us consider a PCM system using N-bit codewords ($Q = 2^N$). Let us further assume that a codeword used to identify a quantization level is in the order of numerical significance of the word, that is, we assign $00\ldots00$ to the most negative level, $00\ldots01$ the next level, and $111\ldots11$ to the most positive level. An error that occurs in the least significant bit of the codeword corresponds to an error in the quantized value of the sampled signal by amount Δ. An error in the next significant bit causes an error of 2Δ, and an error in the ith bit position causes an error of $(2^{i-1})\Delta$. Let us call the error Q_Δ.

Then, assuming that an error may occur with equal likelihood in any one of the N bits in the codeword, the variance of the error is

$$E\{Q_\Delta^2\} = \frac{1}{N}\left(\sum_{i=1}^{N} (\Delta 2^{i-1})^2\right)$$

$$= \frac{2^{2N}-1}{3N}\Delta^2 \approx \frac{2^{2N}}{3N}\Delta^2$$

for $N \geq 2$. The bit errors due to channel noise lead to incorrect values of $X_q(KT_s)$. Since we are treating $X_q(t)$ as an impulse sequence, these errors appear as impulses of random amplitude and of random times of occurrence. An error impulse occurs when a word is in error. The mean separation between bit errors is $1/P_e$ bits. Since there are N bits per codeword, the mean separation between words that are in error is $1/(NP_e)$ words, and the mean time between word errors is

$$T = T_s/(NP_e)$$

Using the results derived in Chapter 3, we can obtain the power spectral density of the thermal noise error impulse train as

$$G_{th}(f) = \frac{1}{T} E\{Q_\Delta^2\}$$

$$= \left(\frac{NP_e}{T_s}\right)\left(\frac{2^{2N}}{3N}\right)\Delta^2$$

At the output of the ideal lowpass filter, the thermal noise error impulse train produces an average noise power N_0 given by

$$N_0 = \int_{-f_x}^{f_x} G_{th}(f)\, df = \frac{2^{2N}\Delta^2 P_e}{3T_s^2} \tag{10.40}$$

Output *S/N* Ratio in PCM Systems. The performance of the PCM system, when used for transmitting analog signals, is measured in terms of the average signal-to-noise power ratio at the receiver output. Combining Equations (10.36), (10.38), (10.39), and (10.40), we have

$$\left(\frac{S}{N}\right)_0 = \frac{2^{2N}}{1 + 4P_e 2^{2N}} \tag{10.41}$$

In Equation (10.41), P_e denotes the probability of a bit error which depends on the method of transmission. For example, if PSK signaling scheme is used, we have

$$P_e = Q\left(\sqrt{\frac{2S_{av}T_b}{\eta}}\right) = Q\left(\sqrt{\frac{2S_{av}T_s}{\eta N}}\right)$$

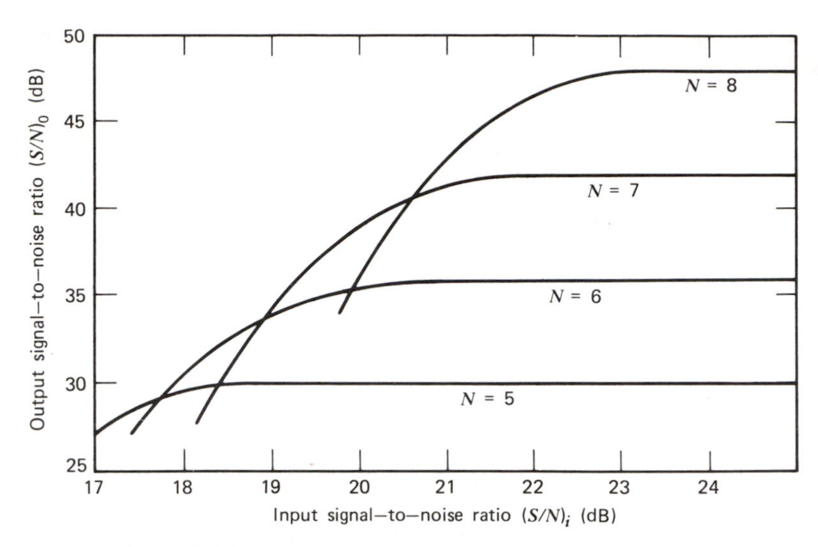

Figure 10.21 Output signal-to-noise ratio in PCM systems.

where S_{av} is the average signal power at the receiver input and T_b is the bit duration,

$$T_b = \frac{T_s}{N} = \frac{1}{2f_x N}$$

Hence,

$$P_e = Q\left(\sqrt{\frac{S_{av}}{\eta f_x N}}\right) \tag{10.42}$$

Similar expressions can be derived for other transmission schemes.

Plots of average signal-to-noise power ratio at the receiver input $(S/N)_i$ (defined as $S_{av}/\eta f_x$) versus the signal-to-noise ratio at the output $(S/N)_0$ for a PCM–PSK system are shown in Figure 10.21.

Plots shown in Figure 10.21 clearly indicate that the PCM system exhibits a threshold effect. For large values of $(S/N)_i$, P_e is small and hence

$$1 + 4P_e 2^{2N} \approx 1$$

and

$$(S/N)_0 = 2^{2N} \approx (6N)\text{ dB} \tag{10.43a}$$

When $(S/N)_i$ is small, then we have

$$\left(\frac{S}{N}\right)_0 \approx \frac{2^{2N}}{4P_e 2^{2N}} \approx \frac{1}{4P_e} \tag{10.43b}$$

The threshold point is arbitrarily defined as the $(S/N)_i$ at which $(S/N)_0$ given in Equation (10.43*b*) falls 1 dB below the value given in Equation (10.43*a*). The onset of threshold in PCM will result in a sudden increase in the output noise power. As the input signal power is increased, the output signal-to-noise power ratio $(S/N)_0$ reaches a value $(6N)$ dB that is independent of the signal power. Thus, above threshold, increasing signal power yields no further improvement in the $(S/N)_0$. The limiting value of $(S/N)_0$ depends only on the number of quantizer levels.

10.4.4 Differential PCM Systems

So far, we have discussed PCM systems using a fairly straightforward digital code for the transmission of analog signals. Several variations of PCM systems have been developed in recent years. We briefly describe two such types of systems here. Both systems use a differential quantizing scheme.

These systems are particularly more efficient when the sampled message signal has high sample to sample correlation. For example, in the transmission of picture (video) information, appreciable portions of the signal describe background information containing very little tonal variations. In such situations, if we use PCM, the codewords describe the value of the average background level; if these tonal values do not change appreciably, then we are essentially transmitting repeated sample values. One way to improve the situation is to send only the digitally encoded differences between successive samples. Thus a picture that has been quantized to 256 levels (eight bits) may be transmitted with comparable fidelity using 4-bit differential encoding. This reduces the transmission bandwidth by a factor of 2. PCM systems using differential quantizing schemes are known as differential PCM (DPCM) systems.

A differential PCM system that is particularly simple to implement results when the difference signal is quantized into two levels. The output of the quantizer is represented by a single binary digit, which indicates the sign of the sample to sample difference. This PCM system is known as delta modulation (DM). Delta modulation systems have an advantage over M-ary PCM and M-ary DPCM systems in that the hardware required for modulation at the transmitter and demodulation at the receiver are much simpler.

10.4.5 Delta Modulation Systems

The functional block diagram of a delta modulation system is shown in Figure 10.22. At the transmitter, the sampled value $X(kT_s')$ of $X(t)$ is compared with a predicted value $\hat{X}(kT_s')$ and the difference $X(kT_s') - \hat{X}(kT_s')$ is quantized into

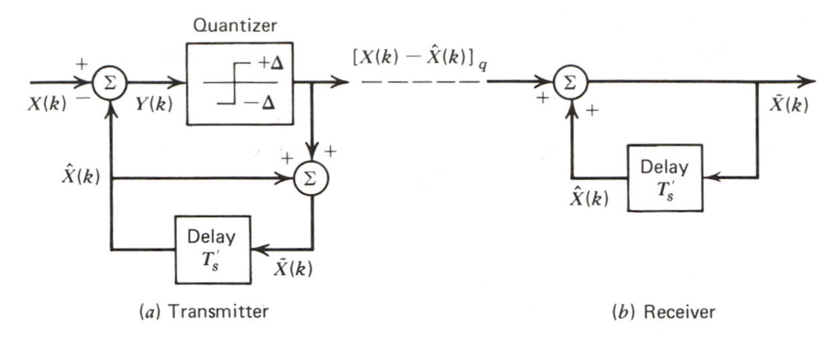

Figure 10.22 Discrete time model of a DM system. (a) Transmitter. (b) Receiver. Sampling rate $= f_s' = 1/T_s'$.

one of two values $+\Delta$ or $-\Delta$. The output of the quantizer is encoded using one binary digit per sample and sent to the receiver. At the receiver, the decoded value of the difference signal is added to the immediately preceding value of the receiver output. The operation of the delta modulation scheme shown in Figure 10.22 is described by the following equations:

$$\hat{X}(kT_s') = \tilde{X}((k-1)T_s') \tag{10.44}$$

where $\tilde{X}((k-1)T_s')$ is the receiver output at $t = (k-1)T_s'$ and

$$\tilde{X}(kT_s') = \hat{X}(kT_s') + [X(kT_s') - \hat{X}(kT_s')]_q = \tilde{X}((k-1)T_s') \pm \Delta \tag{10.45}$$

The delay element and the adder in Figures 10.22*a* and 10.22*b* can be replaced by an integrator whose input is an impulse sequence of period T_s' and strength $\pm\Delta$. This results in the system shown in Figure 10.23.

The operation of the delta modulation scheme shown in Figure 10.23 may be seen using the waveforms shown in Figure 10.24. The message signal $X(t)$ is compared with a stepwise approximation $\hat{X}(t)$ and the difference signal $Y(t) = X(t) - \hat{X}(t)$ is quantized into two levels $\pm\Delta$ depending on the sign of

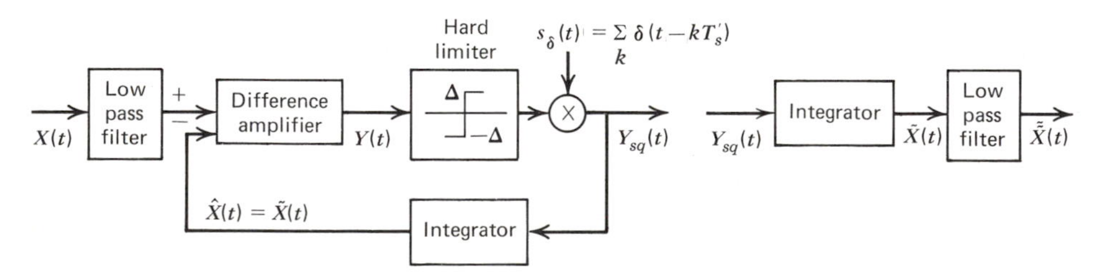

Figure 10.23 Hardware implementation of a DM system. (a) Modulator. (b) Demodulator.

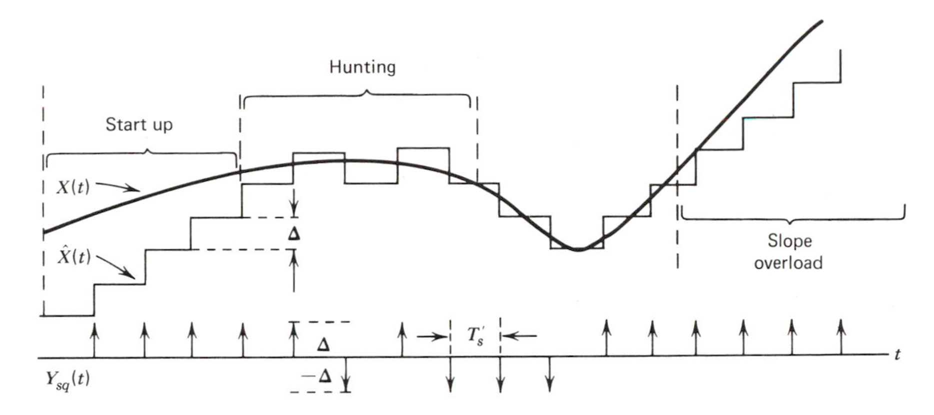

Figure 10.24 Delta modulation waveforms; $\hat{X}(t) = \tilde{X}(t)$.

the difference. The output of the quantizer is sampled to produce

$$Y_{sq}(t) = \sum_{k=-\infty}^{\infty} \Delta \operatorname{sgn}[X(kT_s') - \hat{X}(kT_s')]\delta(t - kT_s') \tag{10.46}$$

The stepwise approximation $\hat{X}(t)$ is generated by passing the impulse waveform given in Equation (10.46) through an integrator that responds to an impulse with a step rise. Since there are only two possible impulse weights in $Y_{sq}(t)$, this signal can be transmitted using a binary waveform. The demodulator consists of an integrator and a lowpass filter.

In a practical delta modulation system, the lowpass filter in the receiver will, by itself, provide an approximate measure of integration. Hence we can eliminate the receiver integrator and depend on the filter for integration. At the transmitter, the sampling waveform $s_\delta(t)$ need not be an impulse waveform but a pulse waveform with a pulse duration that is short in comparison with the interval between pulses. Furthermore, the transmitter integrator need not be an ideal integrator—a simple RC lowpass filter will be adequate. These simplifications reduce the complexity of the hardware in DM systems considerably.

Some of the problems that occur when we use delta modulation to transmit an analog signal can be seen in the waveforms shown in Figure 10.24. Initially, let us assume that $\hat{X}(t) < X(t)$ so that the first impulse has a weight of Δ. When this impulse is fed back through the integrator, it produces a step change in $\hat{X}(t)$ of height Δ. This process continues through the *start up interval* until $\hat{X}(t)$ exceeds $X(t)$. During the start up interval the receiver output will differ considerably from the message signal $X(t)$. After the start up period, $\hat{X}(t)$ exhibits a *hunting* behavior when $X(t)$ remains constant. Hunting leads to idling noise. The sampling rate in a delta modulation scheme will normally be much higher than the Nyquist rate and hence the rectangular idling noise waveform can be filtered or smoothed out by the receiver filter.

Slope Overloading. A serious problem in delta modulation schemes arises due to the rate of rise overloading. When $X(t)$ is changing, $\tilde{X}(t)$ and $\hat{X}(t)$ follow $X(t)$ in a stepwise fashion as long as successive samples of $X(t)$ do not differ by an amount greater than the step size Δ. When the difference is greater than Δ, $\hat{X}(t)$ and $\tilde{X}(t)$ can no longer follow $X(t)$. This type of overload is not determined by the amplitude of the message signal $X(t)$ but rather by its slope as illustrated in Figure 10.25; hence, the name slope overload.

To derive a condition for preventing slope overload in DM systems, let us assume that $X(t) = A\cos(2\pi f_x t)$. Then, the maximum signal slope is

$$\left[\frac{dX(t)}{dt}\right]_{\max} = A2\pi f_x$$

The maximum sample to sample change in the value of $X(t)$ then is $A2\pi f_x T_s'$. To avoid slope overload, this change has to be less than Δ, that is,

$$2\pi f_x T_s' A < \Delta$$

or, the peak signal amplitude at which slope overload occurs is given by

$$A = \frac{\Delta}{2\pi}\frac{f_s'}{f_x} \tag{10.47}$$

where $f_s' = 1/T_s'$ is the sampling rate of the DM system. For a signal $X(t)$ with a continuous spectrum $G_X(f)$, we can still use Equation (10.47) to determine the point of slope overload if f_x is taken to be the frequency beyond which $G_X(f)$ falls off at a rate greater than $1/f^2$. It has been determined experimentally that delta modulation will transmit speech signals without noticeable slope overload provided that the signal amplitude does not exceed the maximum sinusoidal amplitude given in Equation (10.47) with $f_x = 800$ Hz.

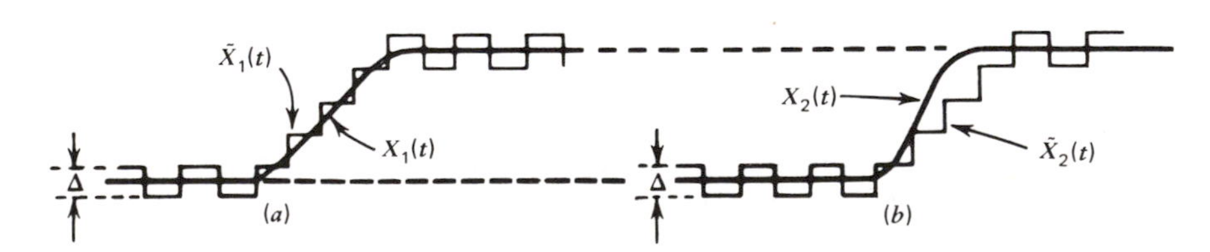

Figure 10.25 Slope overload in DM systems. Signals $X_1(t)$ and $X_2(t)$ have the same amplitude range. However, because of greater rate of rise, $X_2(t)$ causes a slope overload.

The problem of slope overloading in delta modulation systems can be alleviated by filtering the signal to limit the maximum rate of change or by increasing the step size and/or the sampling rate. Filtering the signal and increasing the step size will result in poor signal resolution, and increasing the sampling rate will lead to larger bandwidth requirements. A better way to avoid slope overload is to detect the overload condition and make the step size larger when overloading is detected. Systems using signal dependent step sizes are called adaptive delta modulation systems (ADM).

Adaptive Delta Modulation.* Hunting occurs in DM systems when the signal changes very slowly, and slope overloading occurs when the slope of the signal is very high. Both of these problems can be alleviated by adjusting the step size, in an adaptive fashion, in accordance with the signal being encountered. Ideally, the step size should be kept small when signal changes are small while increasing the step size in order to avoid slope overload when signal changes are large.

A DM system that adjusts its step size according to signal characteristics is shown in Figure 10.26. The step size is varied by controlling the gain of the integrator, which is assumed to have a low gain when the control voltage is zero and a larger gain with increasingly positive control voltage. The gain control circuit consists of an RC integrator and a square law device. When the

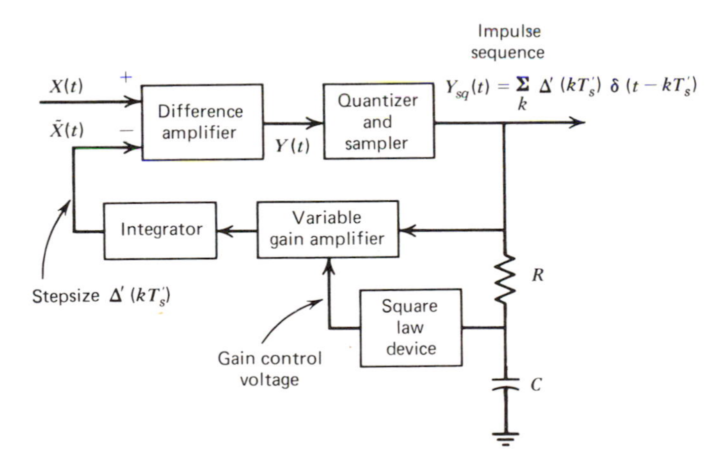

Figure 10.26 An adaptive delta modulator. The strength of the impulse $|\Delta'(kT_s')|$ depends on the slope of the signal; the sign of $\Delta'(kT_s')$ will be the same as the sign of $Y(kT_s')$.

*Adaptive delta modulation is also known by the name continuous variable-slope delta modulation (CVSDM).

input signal is constant or slowly varying, the DM will be hunting and the modulator output will be a sequence of alternate polarity pulses. These pulses when integrated by the RC filter yield an average output of almost zero. The gain control input and hence the gain and the step size are small.

In case of a slope overload, the output of the quantizer will be a train of all positive or all negative pulses (see Figure 10.24). The integrator now provides a large control voltage and the gain of the amplifier is increased. Because of the squaring circuit, the amplifier gain will be increased no matter what the polarity of the pulses are. The net result is an increase in step size and a reduction in slope overload. The demodulator in an adaptive DM system will have an adaptive gain control circuit similar to the one used in the modulator.

10.4.6 Noise in Delta Modulation Systems

The output of the demodulator $\tilde{\tilde{X}}(t)$ differs from the input to the modulator $X(t)$ because of quantizing noise $n_q(t)$ and the noise due to transmission errors $n_0(t)$, that is,

$$\tilde{\tilde{X}}(t) = X_0(t) + n_q(t) + n_0(t) \tag{10.48}$$

where $X_0(t)$ is the output signal component (assumed to be equal to $X(t)$), and $n_0(t)$, $n_q(t)$ are the noise components at the output of the baseband filter. The overall signal quality in DM systems, as in PCM systems, is measured in terms of the signal-to-noise ratio at the output of the baseband filter. This ratio is defined as

$$\left(\frac{S}{N}\right)_0 = \frac{E\{[X_0(t)]^2\}}{E\{[n_q(t)]^2\} + E\{[n_0(t)]^2\}}$$

The average power content of the noise components can be calculated as follows.

Quantization Noise in DM Systems. To arrive at an estimate of the quantization noise power we write $X(t) = \tilde{X}(t) + e_q(t)$ where $|e_q(t)| = |X(t) - \tilde{X}(t)| \leq \Delta$ in the absence of slope overloading (Figure 10.24). The quantizing noise component $n_q(t)$ in Equation (10.48) is the response of the baseband filter to $e_q(t)$. If we assume a uniform pdf for $e_q(t)$, then

$$\begin{aligned} E\{[e_q(t)]^2\} &= \int_{-\Delta}^{\Delta} \frac{1}{2\Delta} e^2 \, de \\ &= \Delta^2/3 \end{aligned}$$

It has been experimentally verified that the normalized power of the waveform $e_q(t)$ is uniformly distributed over the frequency interval $[0, f_s']$, where f_s' is the sampling rate. Thus, the power spectral density $G_{eq}(f)$ of $e_q(t)$ is given by

$$G_{eq}(f) = \begin{cases} \Delta^2/6f_s', & |f| < f_s' \\ 0, & \text{elsewhere} \end{cases}$$

Since $n_q(t)$ is the response of the baseband filter to $e_q(t)$, the normalized average power of the waveform $n_q(t)$ is given by

$$\begin{aligned} E\{[n_q(t)]^2\} &= \int_{-f_x}^{f_x} G_{eq}(f)\, df \\ &= \left(\frac{\Delta^2}{3}\right)\left(\frac{f_x}{f_s'}\right) \end{aligned} \tag{10.49}$$

In order to compute the signal to quantizing noise power ratio, we need to calculate the signal power $E\{X_0^2(t)\}$. To simplify the calculation of signal power, let us take the worst case for delta modulation where all of the signal power is concentrated at the upper end of the baseband, that is, let us take

$$X(t) = A \cos 2\pi f_x t$$

Then,

$$X_0(t) = A \cos 2\pi f_x t$$

and

$$E\{X_0^2(t)\} = A^2/2 \tag{10.50a}$$

and to avoid slope overload we have, from Equation (10.47),

$$A = \frac{\Delta}{2\pi} \frac{f_s'}{f_x} \tag{10.50b}$$

Combining Equation (10.49) with (10.50), we obtain the output signal to quantizer noise power ratio as

$$\frac{E\{X_0^2(t)\}}{E\{n_q^2(t)\}} = \left(\frac{3}{8\pi^2}\right)\left(\frac{f_s'}{f_x}\right)^3 \tag{10.51}$$

We will see later on that the performance of DM systems as measured by signal to quantizer noise power ratio falls below the performance of PCM system using comparable bandwidth.

Channel Noise in DM Systems. When channel noise is present, the polarity of the transmitted waveform will be occasionally decoded incorrectly. Since the transmitted waveform is an impulse sequence of strength $\pm\Delta$, a sign error will result in an error impulse of strength 2Δ; the factor of 2 comes from the fact that an error reverses the polarity of the pulse. This channel-error noise appears at the receiver integrator input as a sequence of impulses with random times of occurrence and strength $\pm 2\Delta$. The mean time of separation between these impulses is T_s'/P_e, where P_e is the bit error probability. The power spectral density of this impulse train can be shown to be white, with a magnitude of $4\Delta^2 P_e f_s'$. If we take the transfer function of the integrator to be $1/j\omega$, then the power spectral density of channel-error noise at the input to the baseband filter is given by

$$G_{th}(f) = \frac{4\Delta^2 P_e f_s'}{(2\pi f)^2} \tag{10.52}$$

It would appear now that to find the channel-error noise power at the output, $E\{n_0^2(t)\}$, all we need to do is to integrate $G_{th}(f)$ over the passband of the baseband filter. However, $G_{th}(f) \to \infty$ as $f \to 0$, and the integral of $G_{th}(f)$ over a range of frequencies including $f = 0$ is infinite. Fortunately, baseband filters have a low-frequency cutoff $f_1 > 0$; further, f_1 is usually very small compared to the high-frequency cutoff f_x. Hence

$$\begin{aligned} E\{n_0^2(t)\} &= 2\int_{f_1}^{f_x} G_{th}(f)\, df \\ &= \frac{2\Delta^2 P_e f_s'}{\pi^2}\left[\frac{1}{f_1} - \frac{1}{f_x}\right] \\ &\approx \frac{2\Delta^2 P_e f_s'}{\pi^2 f_1} \end{aligned} \tag{10.53}$$

since $f_1 \ll f_x$. Equation (10.53) shows that the output noise power due to bit errors depends on the low-frequency cutoff f_1 rather than the high-frequency cutoff f_x. Combining Equations (10.50), (10.51), and (10.53) we obtain the overall output signal-to-noise power ratio in a DM system as

$$\begin{aligned} \left(\frac{S}{N}\right)_0 &= \frac{E\{X_0^2(t)\}}{E\{n_q^2(t)\} + E\{n_0^2(t)\}} \\ &= \frac{(3f_s'^3/8\pi^2 f_x^3)}{1 + (6P_e f_s'^2/\pi^2 f_x f_1)} \end{aligned} \tag{10.54}$$

10.4.7 Comparison of PCM and DM Systems

We can now compare the performance of PCM and DM systems in terms of overall signal quality and equipment complexity. To ensure that the comparison is done under identical conditions, let us assume that both systems use approximately the same bandwidth for transmitting a baseband analog signal. If we use f_s and f_s' to denote the sampling rates of an N-bit PCM system and a DM system, then the bit rates for the systems are Nf_s and f_s', respectively. If the signal spectrum extends up to f_x Hz, then $f_s = 2f_x$ and identical bandwidth requirements imply that

$$f_s' = 2Nf_x$$

Signal-to-Noise Ratio. If the channel signal-to-noise ratio is high, then the performance of PCM and DM is limited by the quantization noise. The signal to quantizing noise power ratio for the PCM system is obtained from Equation (10.43),

$$(S_0/N_q)_{\text{PCM}} = Q^2 = 2^{2N}\ ; \quad N \geqslant 2$$

where $Q = 2^N$ is the number of quantizer levels. For the DM system, the corresponding ratio is given by Equation (10.51)

$$\left(\frac{S_0}{N_q}\right)_{\text{DM}} = \frac{3}{8\pi^2}\left(\frac{f'_s}{f_x}\right)^3$$
$$\approx 0.3N^3$$

The preceding equations show that for a fixed bandwidth the performance of DM is always poorer than PCM. By way of an example, if the channel bandwidth is adequate for an 8-bit PCM code, then

$$(S_0/N_q)_{\text{PCM}} \approx 48\ \text{dB}$$

and

$$(S_0/N_q)_{\text{DM}} \approx 22\ \text{dB}$$

The performance of DM can be considerably improved by using a variable step size. Indeed, for speech transmission, it has been found that there is little difference in the performances of adaptive DM and PCM systems operating at a bit rate of about 64 kbits/sec.

The overall signal-to-noise ratio of a DM system is also lower than the overall S/N ratio of a PCM system using the same bandwidth. The extent of the difference in the signal quality depends on the characteristic of the signal. An example is given below:

Example 10.3. Compare the overall output S/N ratio for 8-bit PCM and DM systems used for transmitting a baseband signal whose spectrum is confined from 300 to 3000 Hz. Assume that both systems operate at a bit rate of 64 kbits/sec and use a PSK signaling scheme with $(S_{av}/\eta f_x) = 20$ dB.

Solution.

(a) PCM system. We have, $1/T_b = 64{,}000$, $(S_{av}/\eta f_x) = 100$, and

$$P_e = Q\left(\sqrt{\frac{2S_{av}T_b}{\eta}}\right) = Q(\sqrt{9.375}) \approx 10^{-3}$$

Hence,

$$\frac{S_0}{N_0} = \frac{2^{2N}}{1 + 4P_e 2^{2N}} \approx 24\ \text{dB}$$

(b) DM system. $f_1 = 300$, $\quad f_x = 3000$, $\quad f'_s = 64{,}000$, and

$$P_e = Q(\sqrt{9.375}) \approx 10^{-3}$$

From Equation (10.54) with $f_s' = 64{,}000$, $f_1 = 300$, and $f_x = 3000$, we have

$$S_0/N_0 \approx 20 \text{ dB}$$

Bandwidth Requirements. Since PCM and DM are now considered primarily for use in speech transmission, let us compare the BW requirements of these systems for speech transmission. With the use of PCM, speech transmission is found to be of good quality when $f_s = 8000$ and $N = 8$. The corresponding bit rate is 64 kbits/sec. To obtain comparable quality using delta modulation, the sampling rate has to be about 100 kbits/sec. However, it has been recently shown that with continuous variable slope delta (CVSD) modulation it is possible to achieve good signal quality at about 32 kbits/sec. It is reasonably accurate to conclude that PCM and (CVS) DM require approximately the same bandwidth for most analog signal transmission applications.

Equipment Complexity. The hardware required to implement DM is much simpler than that required for implementing PCM. Single integrated circuit chip (continuously-variable delta modulation) coder/decoders (called CODECS) are rapidly becoming available. In comparison, PCM coder/decoders require two chips for implementation: one chip for processing the analog signal and the second one to encode the sampled analog signal. Thus the PCM hardware is more expensive than the DM hardware.

10.4.8 *Q*-Level Differential PCM Systems

We conclude our treatment of digital transmission methods for analog signals with a brief description of a technique that combines the differential aspect of DM with the multilevel quantization of PCM. This technique, known as differential PCM (DPCM) or delta-PCM, uses a Q-level quantizer to quantize the difference signal $X(t) - \tilde{X}(t)$ (Figure 10.23). Thus the output of the sampler $Y_{sq}(t)$ is an impulse train in which the strength of the impulses can have one of Q possible values. (If $Q = 2$, then DPCM reduces to DM.) The value of the quantized error sample is represented by an N bit codeword ($Q = 2^N$) and transmitted over the communication channel as a binary waveform.

The approximation $\tilde{X}(t)$ of $X(t)$ in a DPCM system has a variable step size ranging from $\pm\Delta$ to $\pm Q\Delta/2$, so $\tilde{X}(t)$ follows $X(t)$ more accurately. Thus, there will be much lower hunting noise, faster start-up, and less chance of slope overload, especially if a nonuniform quantizer is used.

The DPCM system combines the simplicity of DM and the multilevel quantizing feature of PCM. It has been found that, in many applications, the

DPCM system with $Q \geqslant 4$ yields a higher signal to quantizer noise power ratio than ordinary PCM or DM using the same bit rate. In recent years DPCM systems have been used in the encoding and transmission of video signals (for example, in the Picturephone® system developed by AT&T). For broadcast quality black and white pictures, DPCM with $Q = 8 = 2^3$ gives acceptable video-signal reproduction, whereas straight PCM must have $Q = 256 = 2^8$ levels. Thus DPCM reduces transmission bandwidth by a factor of $\frac{3}{8}$. Comparable bandwidth reduction can be obtained for speech transmission also.

10.5 TIME-DIVISION MULTIPLEXING

Time-division multiplexing (TDM) is a technique used for transmitting several analog message signals over a communication channel by dividing the time frame into slots, one slot for each message signal. In comparison, frequency division multiplexing (FDM) divides the available bandwidth into slots, one slot for each message signal. The important features of TDM are illustrated in Figure 10.27.

Four input signals, all bandlimited to f_x by the input filters, are sequentially sampled at the transmitter by a rotary switch or *commutator*. The switch makes f_s revolutions per second and extracts one sample from each input during each revolution. The output of the switch is a PAM waveform containing samples of the input signals periodically interlaced in time. The samples from adjacent input message channels are separated by T_s/M, where

®Picturephone is a registered service mark of AT & T.

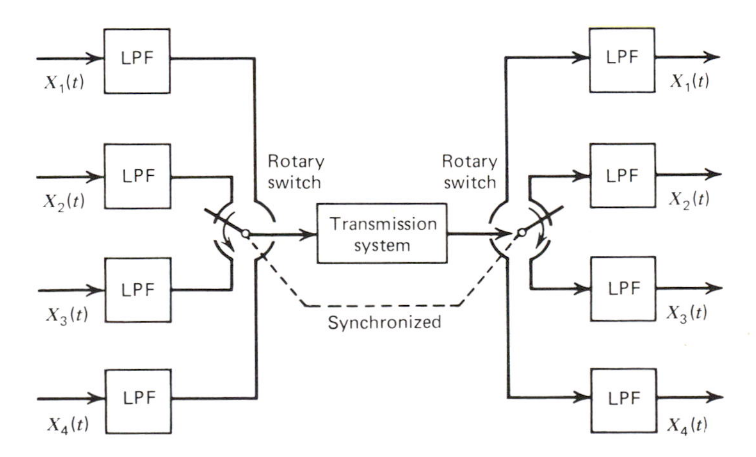

Figure 10.27 Block diagram of a four channel TDM system.

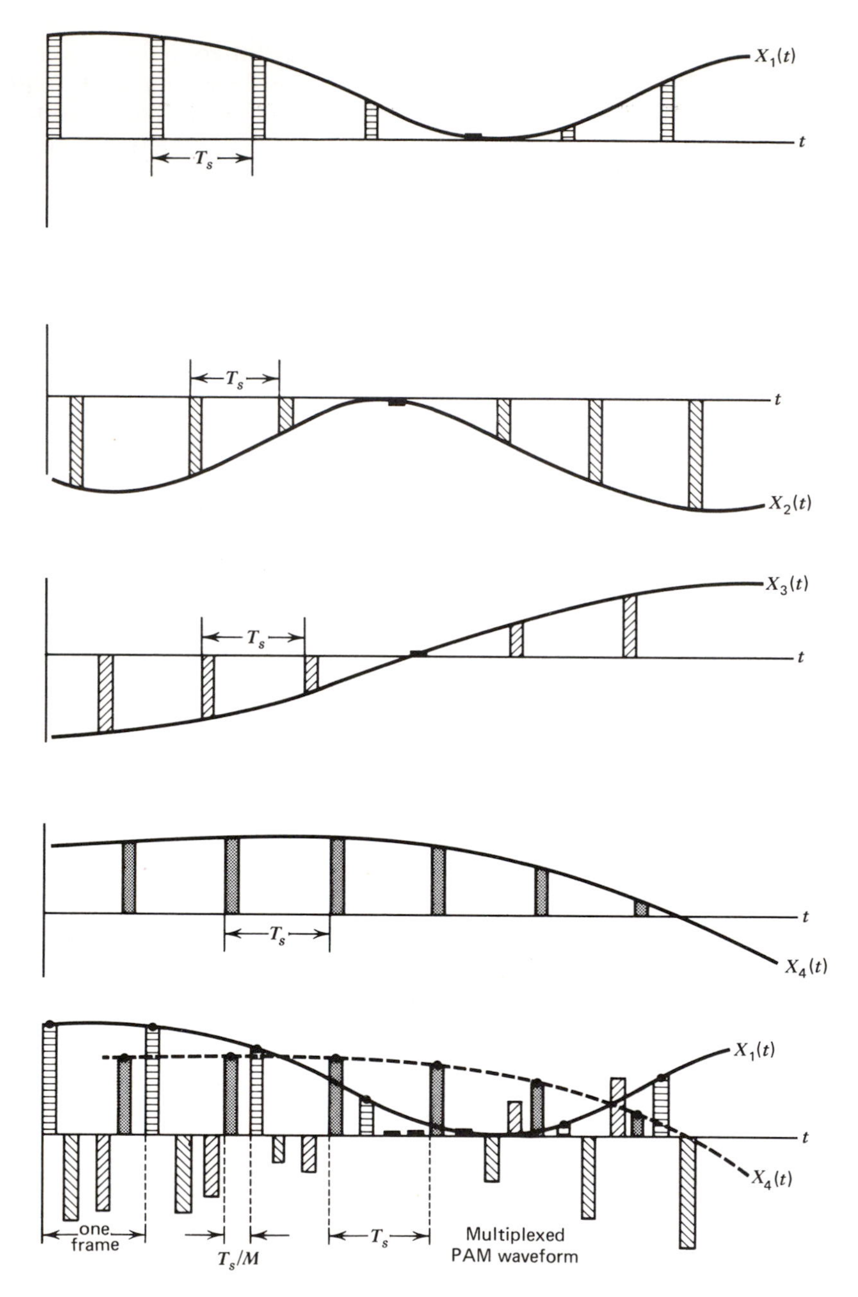

Figure 10.28 TDM waveforms; $T_s = 1/f_s$, where f_s is the number of revolutions per second of the rotary switch.

M is the number of input channels. A set of M pulses consisting of one sample from each of the M-input channels is called a *frame.* (See Figure 10.28.)

At the receiver, the samples from individual channels are separated and distributed by another rotary switch called a *distributor or decommutator.* The samples from each channel are filtered to reproduce the original message signal. The rotary switches at the transmitter and receiver are usually electronic circuits that are carefully synchronized. Synchronizing is perhaps the most critical aspect of TDM. There are two levels of synchronization in TDM: frame synchronization and sample (or word) synchronization. Frame synchronization is necessary to establish when each group of samples begin and word synchronization is necessary to properly separate the samples within each frame.

The interlaced sequence of samples may be transmitted by direct PAM or the sample values may be quantized and transmitted using PCM. Time-division multiplexed PCM is used in a variety of applications, the most important one is PCM telephone systems where voice and other signals are multiplexed and transmitted over a variety of transmission media including pairs of wires, wave guides, and optical fibers.

10.5.1 TDM–PCM Telephone System

The block diagram of a modular TDM–PCM telephone system designed by the American Telephone and Telegraph company is shown in Figure 10.29*a*. A 24-channel TDM multiplexer is used as the basic system, known as the T1 carrier system. Twenty-four voice signals are sampled at a rate of 8 kHz and the resulting samples are quantized and converted to 7-bit PCM codewords. At the end of each 7-bit codeword, an additional binary bit is added for synchronizing purposes. At the end of every group of twenty-four 8-bit codewords, another additional bit is inserted to give frame synchronization. The overall frame size in the T1-carrier is 193 bits, and the overall bit rate is 1.544 Mbits/sec. (See Figure 10.29*b*.)

The T1 system is designed primarily for short distance and heavy usage in metropolitan areas. The maximum length of the T1 system is now limited to 50 to 100 miles with a repeater spacing of 1 mile. The overall T-carrier system is made up of various combinations of lower order T-carrier subsystems designed for accommodating voice channels, Picturephone® service, TV signals, and (direct) digital data from data terminal equipment. A brief summary of the T-carrier TDM/PCM telephony system is given below in Table 10.2.

In addition to using metallic cable systems for transmission, optical fibers

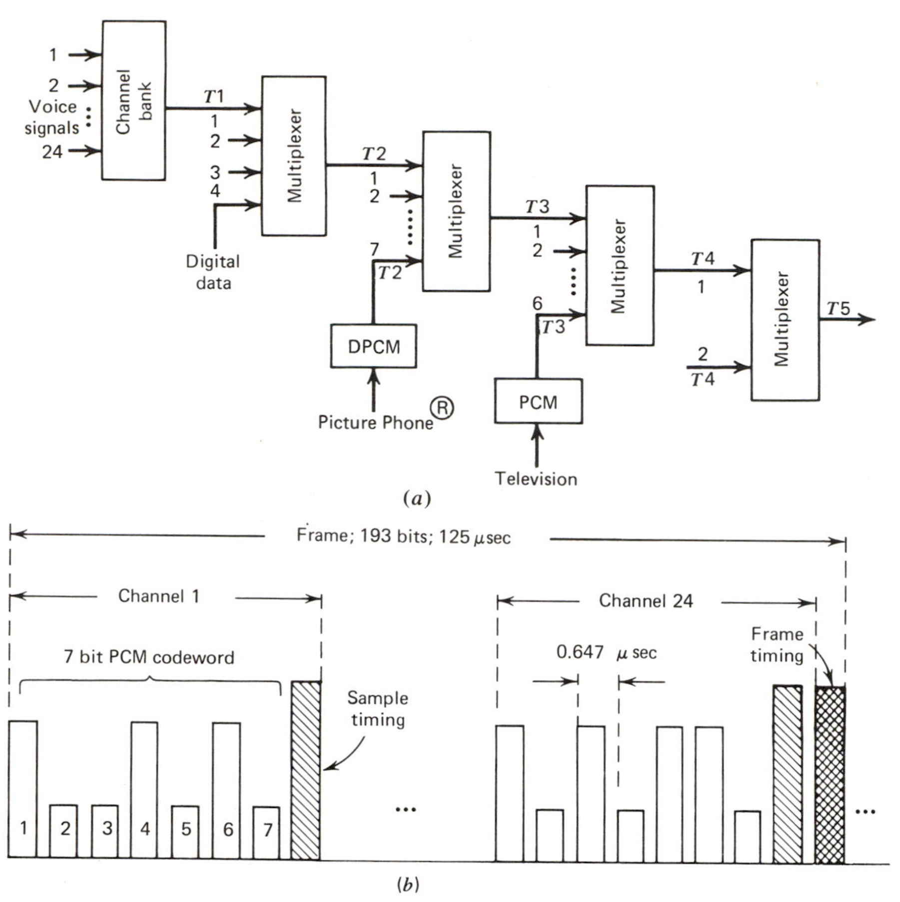

Figure 10.29 (*a*) The Bell system (AT & T) TDM/PCM telephony system. (*b*) T1-frame. Note that the timing pulses have larger amplitude than data pulses.

Table 10.2. T-Carrier Telephony System Specifications

System	Bit rate (Mbits/sec)	Medium	Repeater spacing	Maximum length	System error rate
T1	1.544	wire pair	1 (mile)	50 (miles)	10^{-6}
T2	6.312	coax	2.5	500	10^{-7}
T3	44.736	coax	multi-plexing only	—	—
T4	274.176	coax	1	500	10^{-6}
T5	560.160	coax	1	500	$(0.4)(10^{-8})$

with repeaters have been used to transmit binary data at speeds of 1.5, 3.6, 45, and 274 Mbits/sec corresponding to the speeds of the digital transmission hierarchy shown in Table 10.2 (see IEEE Spectrum, Feb., 1977).

10.5.2 Comparison of TDM and FDM

TDM and FDM techniques accomplish the same signal processing task. In TDM, the analog signals are separated in time but jumbled together in the frequency domain. In FDM the signals are separated in frequency domain but mixed together in time domain. From a theoretical point of view, the two systems may be viewed as dual techniques with neither one having any significant advantage over the other. However, from a practical viewpoint, TDM seems to be superior to FDM in at least two respects.

First, the TDM circuitry is much simpler than the FDM circuitry. FDM equipment consists of analog circuits for modulators, carrier generators, bandpass filters, and demodulators for *each* channel. In comparison, TDM circuitry is digital, consisting of a commutator and distributor. The digital circuitry is highly modular in nature and provides reliable and efficient operation.

A second advantage of TDM systems is the relatively small interchannel cross talk arising from nonlinearities in the circuits that handle the signals in the transmitter and the receiver. These nonlinearities produce intermodulation and harmonic distortion that affect both high-frequency and low-frequency channels in FDM systems. Thus the phase and amplitude linearity requirements of FDM circuits become very stringent when the number of channels being multiplexed is large. In contrast, there is no cross talk in TDM due to circuit nonlinearities if the pulses are completely isolated and nonoverlapping since signals from different channels are not handled simultaneously but are allotted different time intervals. Hence the linearity requirements of the TDM circuits are not quite as stringent as the FDM circuits. However, TDM cross talk immunity is contingent upon a wideband response and the absence of delay distortion. Disadvantages of TDM include the fact that pulse accuracy, timing jitter, and synchronization become major problems at high bit rates.

Finally, to complete our comparison of FDM and TDM, let us consider their bandwidth requirements. Let us assume that we have M input signals bandlimited to f_x Hz. With FDM using SSB modulation, the bandwidth of the multiplexed signal will be Mf_x. With TDM, if we assume a sampling rate of f_s for each channel, then the multiplexed signal consists of a series of sample points separated in time by $1/Mf_s$ sec (Figure 10.30a). By virtue of the sampling theorem, these points can be completely described by a continuous waveform $X_b(t)$ that is bandlimited to $Mf_s/2$ Hz. This waveform, even though

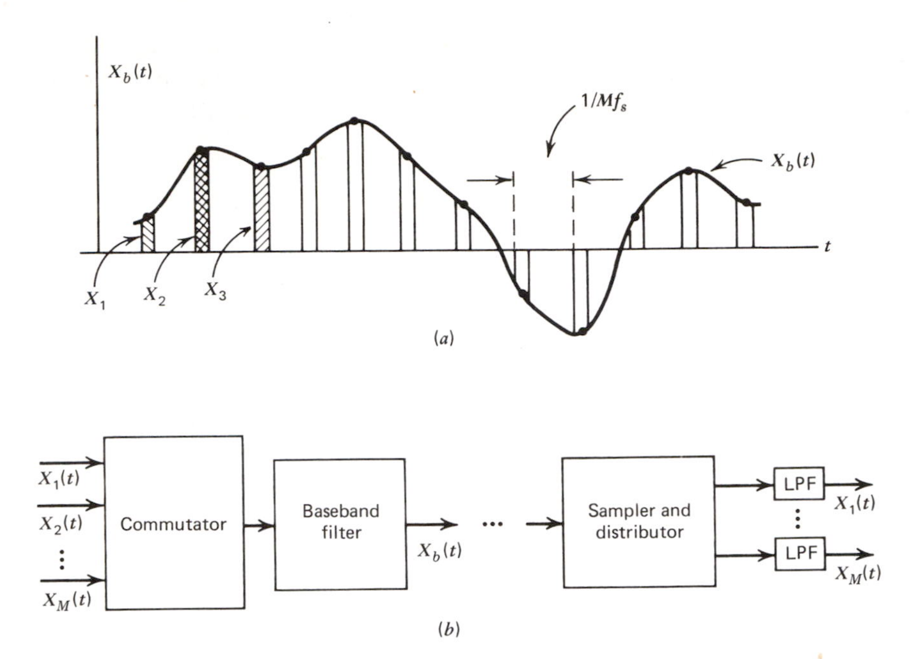

Figure 10.30 Baseband filtering of a TDM waveform.

it has no direct relation to the original messages, passes through the correct sample values at sampling times. $X_b(t)$ is obtained by lowpass filtering of the interleaved sample sequence. At the receiver, $X_b(t)$ is sampled and the sample values are distributed to appropriate channels by the distributor. If the sampling frequency is close to the Nyquist rate, that is, $f_s \approx 2f_x$, then the bandwidth of the filtered TDM waveform is Mf_x Hz, which is the same as the bandwidth of the FDM waveform.

10.5.3 Asynchronous TDM

In the preceding discussion of TDM systems we had assumed that the signals being multiplexed have comparable bandwidths and hence the sampling rate for each signal is the same. However, in many applications the signals to be time-division multiplexed have different bandwidths and hence they have to be sampled at different sampling frequencies. In these situations, we cannot multiplex these signals using the technique described previously, which employs a common clock rate for all the channels.

One method of combining a group of asynchronously sampled time-division multiplexed signals uses *elastic store* and *pulse stuffing*. An elastic storage device, which is essential for multiplexing asynchronous signals, stores a digital bit stream in such a manner that the bit stream may be read out at a rate different from the rate at which it was read in. One example of such a device is a tape recorder. Data can be recorded onto the tape and read out at a different rate by adjusting the tape speed during replay. Another example is a large (digital) buffer into which data can be read in at one rate and read out at a different rate.

To illustrate the use of elastic store and pulse stuffing in asynchronous TDM, consider the example of a satellite that records the results of a number of experiments and transmits them to the earth. Let us suppose that three experiments each lasting a duration of one second are performed simultaneously, and that their signals are sampled and stored in three separate digital storage devices. Let us assume that the three signals are sampled at rates 2000, 3000, and 5000 samples per second, respectively, and the samples are encoded using 8-bit PCM codewords. At the end of each 1-sec interval, the experiments are halted for one second during which time all of the data collected are transmitted to earth.

During transmission, each storage device can be emptied (played back) at the same rate (5000 samples per second), synchronously time-division multiplexed and a single TDM signal can be transmitted to earth. There is one major problem associated with this procedure. The first 2000 words of each signal can be multiplexed without any trouble. During the multiplexing of the next 2000 words there is no contribution from the first signal, and during the last 1000 words there is no contribution from the first and second signals. However, because of noise, the receiver will be reading words when no words from channels 1 and 2 are being transmitted. To avoid this erroneous interpretation of noise as signal, the time slots corresponding to signals that have already terminated are filled with dummy sequences of bits. These dummy sequences are carefully chosen and encoded so that the receiver recognizes them without difficulty. This technique is called pulse stuffing since it requires that digits or pulses be stuffed into spaces provided for the missing message bits. (See Figure 10.31.)

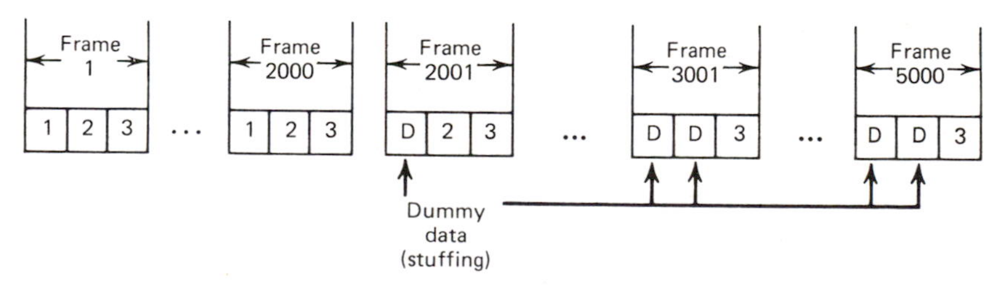

Figure 10.31 Example of pulse stuffing.

10.6 COMPARISON OF METHODS FOR ANALOG SIGNAL TRANSMISSION

In this section we provide a comparison of analog and digital methods for transmitting analog signals. First we will compare analog modulation methods with PCM methods in a qualitative manner. Later we will attempt to compare these methods in a quantitative manner in terms of signal-to-noise ratio at the receiver output and power bandwidth requirements using an information theoretic approach. In order to do this, we will define an ideal (but unrealizable) communication system using the Shannon–Hartley law. The performance of practical analog and digital modulation schemes will then be compared with the bounds set by the ideal system.

10.6.1 PCM versus Analog Modulation

PCM systems have certain inherent advantages over analog modulation schemes for transmitting analog signals. Some of these advantages are the following:

(1) In long distance communications, PCM signals can be completely regenerated at each repeater station if the repeater spacing is such that the magnitude of the noise is less than half the separation between levels (with a high probability). An example of signal regeneration at a repeater is shown in Figure 10.32. With the exception of occasional errors, a noise- and distortion-free signal is transmitted at each repeater. Further, the effect of noise does not accumulate and in designing repeaters one needs to be concerned only about the effects of channel noise between repeater stations. Repeaters for analog modulation schemes consist of amplifiers that raise the signal level at each transmitting station. While raising the signal level, the amplifier also raises the level of accompanying noise at each repeater station.

(2) At low input signal-to-noise ratios, the output signal-to-noise ratio of PCM systems is better than analog modulation schemes. A quantitative discussion of the effect of noise in various modulation schemes will be presented later on in this chapter.

(3) PCM systems can be designed to handle a variety of signals. A PCM system designed for analog message transmission can be readily adapted to handle other signals, particularly digital data.

(4) Modulation and demodulation circuitry in PCM systems are all digital thus affording high reliability and stability. Advances in integrated circuits have lowered the cost of these circuits considerably.

(5) It is easy to store and time scale PCM signals. For example, PCM data

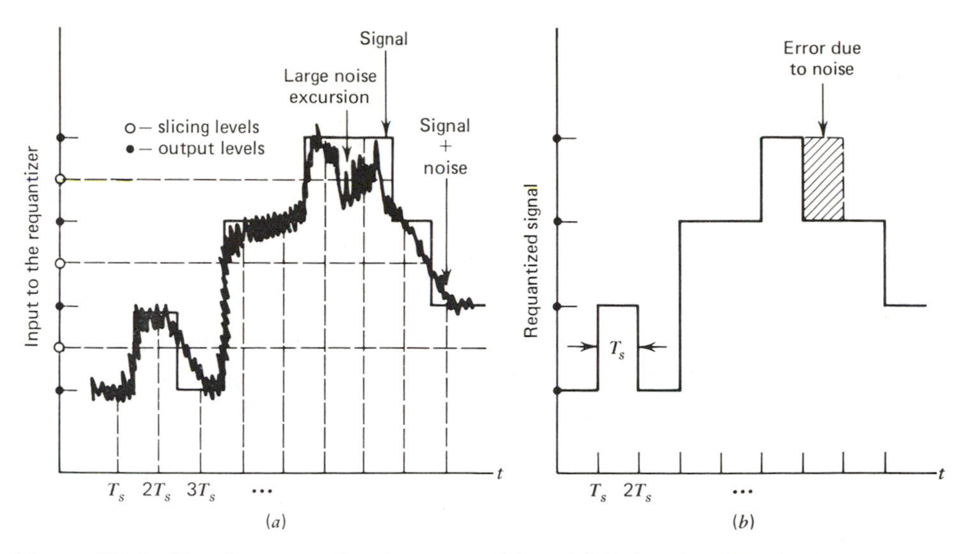

Figure 10.32 Signal regeneration by requantizing. (*a*) Noisy signal at the input of the repeater. (*b*) Signal at the repeater output; errors occur when the noise amplitude is large.

gathered intermittently from an orbiting satellite over a period of hours may be transmitted to the ground station in a matter of seconds when the satellite appears in the field of view of the receiving antenna. While this may also be accomplished using analog techniques, digital memories can perform the required storage very efficiently. Further, PCM signals can be time-division multiplexed easily.

(6) With PCM systems, source coding and channel coding techniques can be used to reduce unnecessary repetition (redundancy) in messages and to reduce the effects of noise and interference. Coding may also be used to make the digital communication channel more secure.

(7) As we will see (in the next section) the exchange of bandwidth for power is easy to accomplish in PCM systems. Since PCM systems can be easily time scaled, time can also be exchanged for signal power. Thus the communication systems designer has added flexibility in the design of a PCM system to meet a given performance criteria.

Some of the advantages of PCM systems are offset by the fact that the complexity of a PCM system is greater than that required for other types of modulation systems. However, the complexity of a PCM system varies little as the number of message channels is increased. Hence, PCM systems can compare quite favorably with other systems when the number of channels is large.

10.6.2 Comparison of Communication Systems: Power-Bandwidth Exchange

In communication systems designed to handle analog message signals, the signal-to-noise ratios at various points in the systems are used to measure the signal quality at these points. Of particular interest are the signal-to-noise ratio at the input to the receiver and the signal-to-noise ratio at the receiver output. The signal-to-noise ratio at the input depends on the transmitted power and the ambient noise appearing at the receiver antenna. The output signal-to-noise ratio depends on the input signal-to-noise ratio and the type of modulation/demodulation processes used in the system. The ratio of the signal-to-noise ratio at the output and the input signal-to-noise ratio, called the detection gain (a measure of noise immunity), is widely used as a figure of merit for communication systems. We will use this figure of merit to compare the performance of several communication systems. We will first investigate the performance of an ideal (but unrealizable) communication system implied by the Shannon–Hartley law. We will then examine various practical communication systems to see how they measure up against the ideal system—particularly in the exchange of bandwidth for signal-to-noise ratio (or transmitted power).

An Ideal Communication System. Suppose that we have a communication system (Figure 10.33) for transmitting an analog message signal $X(t)$ bandlimited to f_x Hz. Further, suppose that an ideal system is available for this purpose and that the channel bandwidth is B_T and the noise power spectral density is $\eta/2$. Also, let us assume that the average signal power at the receiver is S_r and that the desired value of the output signal-to-noise ratio is $(S/N)_d$.

Now, the channel capacity of the system is given by the Shannon–Hartley law as

$$C = B_T \log_2[1 + (S/N)_r] \tag{10.55}$$

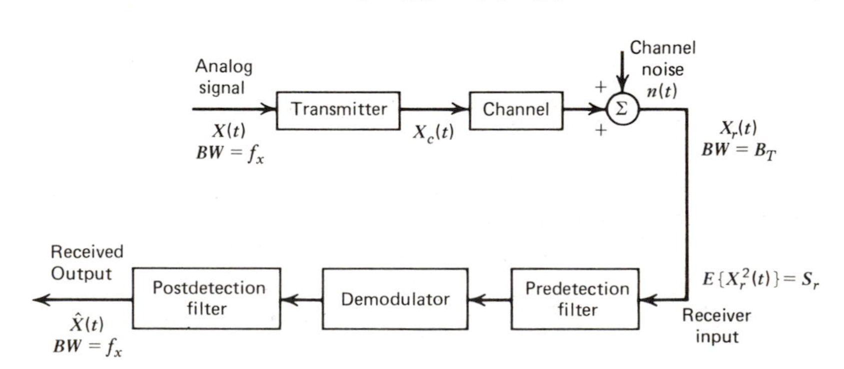

Figure 10.33 Block diagram of a communication system.

where $(S/N)_r$ is the signal-to-noise ratio at the receiver input. At the receiver output, the information rate can be no greater than

$$R_{\max} = f_x \log_2[1 + (S/N)_d] \tag{10.56}$$

An optimum or ideal system is defined as one that is operating at its capacity, with maximum output rate. That is, for the ideal system, we have

$$R_{\max} = C$$

or

$$B_T \log_2[1 + (S/N)_r] = f_x \log_2[1 + (S/N)_d]$$

We can solve for $(S/N)_d$ at the output of the ideal system as

$$\begin{aligned}(S/N)_d &= [1 + (S/N)_r]^{B_T/f_x} - 1 \\ &\approx [1 + (S/N)_r]^{B_T/f_x} \end{aligned} \tag{10.57}$$

when the signal-to-noise ratios are large. In Equation (10.57) the input signal-to-noise ratio $(S/N)_r$ is given by

$$\left(\frac{S}{N}\right)_r = \frac{S_r}{\eta B_T} \tag{10.58}$$

The ratio of transmission bandwidth B_T to message bandwidth f_x is called the *bandwidth expansion ratio* (or *bandwidth compression ratio* if the ratio is less than 1). If we let

$$\beta' = B_T/f_x$$

and

$$\alpha = \frac{S_r}{\eta f_x}$$

then we can rewrite Equation (10.57) as

$$\begin{aligned}\left(\frac{S}{N}\right)_d &= \left[1 + \frac{\alpha}{\beta'}\right]^{\beta'} - 1 \\ &\approx \left(\frac{\alpha}{\beta'}\right)^{\beta'} \end{aligned} \tag{10.59}$$

when the signal-to-noise ratios are large.

Equation (10.59) shows that, in an ideal system, the signal-to-noise ratio at the output and the bandwidth are *exponentially* related. This means that doubling the transmission bandwidth of an ideal system squares the output signal-to-noise ratio. Alternately, since $\alpha = S_r/\eta f_x$ is proportional to the transmitted power S_T, the transmitted power can be reduced to the square root of its original value without reducing $(S/N)_d$ by increasing the bandwidth by a factor of 2.

Example 10.4. Consider an ideal system, designed for transmitting an analog message signal, with the following parameters.

$$(S/N)_d = 60 \text{ dB}$$

Inband noise power $\eta f_x = 10^{-7}$ watt, and $f_x = 15$ kHz. Compare the power requirements (S_r) of the ideal system for the following transmission bandwidths: (a) $B_T = 15$ kHz. (b) $B_T = 75$ kHz. (c) $B_T = 5$ kHz.

Solution

(a) With $B_T = 15$ kHz, $\beta' = 1$; and with

$$(S/N)_d = 60 \text{ dB} = 10^6$$

we have (from Equation (10.59))

$$(\alpha/\beta')^{\beta'} \approx 10^6 \quad \text{or} \quad \alpha \approx 10^6$$

Since

$$\alpha = S_r/\eta f_x$$

we have

$$S_r = \alpha \eta f_x = (10^6)(10^{-7}) = 0.1 \text{ watt}$$
$$= 20 \text{ dBm}$$

(b) With $B_T = 75$ kHz, we have $\beta' = 5$ and

$$(\alpha/5)^5 \approx 10^6 \quad \text{or} \quad \alpha \approx 79.24$$

Hence,

$$S_r = \alpha \eta f_x = (79.24)(10^{-7})$$
$$\approx -21.02 \text{ dBm}$$

(c) With $B_T = 5$ kHz (bandwidth compression), the reader can verify that $\beta' = 0.333$ and

$$\alpha = (\tfrac{1}{3})(10)^{18}$$

or

$$S_r = (\tfrac{1}{3})(10)^{11} \text{ watts (a colossal amount of power!)}$$

The preceding example illustrates that bandwidth expansion leads to a considerable reduction in power requirements. However, bandwidth compression leads to extremely large power requirements. We may generalize this conclusion and say that optimum bandwidth to power exchange is practical in one direction only, namely, the direction of increasing bandwidth and decreasing power.

Comparison of Communication Systems. We are now ready to compare the performance of existing communication systems with the ideal system discussed in the preceding section. When comparing existing systems with the ideal system we should remember the following points. The ideal system discussed in the preceding section was arrived at via an information-theoretic approach; the primary goal of the system was reliable information transfer in the sense of information theory. In systems designed for transmitting analog message signals it is very difficult to assess the information rate. Furthermore, the primary concern in such applications might be signal-to-noise ratios, threshold power, (no threshold effect in ideal systems!), and bandwidth requirements rather than channel capacity and its utilization.

A comparison of the performance of many practical systems with that of an ideal system is shown summarized in Table 10.3 and Figure 10.34. The results for SSB, AM, and DSB modulation are taken from Chapters 6 and 7. It is assumed that signal-to-noise ratios are large, all systems are above threshold, and that the message signal is normalized with $E\{X^2(t)\} = E\{\hat{X}^2(t)\} = \frac{1}{2}$. The result for the PCM system is obtained from Equation (10.41). The performance of the PCM system operating above threshold is limited by quantizing noise, and

$$(S/N)_d = Q^2$$

where Q is the number of quantizer levels. Now if we use an M-ary PCM, and if the sampling rate is $f_s = 2f_x$, then the transmission bandwidth B_T is $r_s/2$, where r_s is the channel symbol rate given by

$$r_s = (2f_x)\log_M Q$$

Table 10.3. Performance of communication systems. $\beta' = B_T/f_x$; $\alpha = S_r/\eta f_x$.

System	Bandwidth expansion	$(S/N)_d$
Ideal	β'	$(\alpha/\beta')^{\beta'}$
SSB Baseband	$\beta' = 1$	α
DSB	$\beta' = 2$	α
AM	$\beta' = 2$	$\alpha/3$
WBFM	$\beta' > 1$	$\frac{3}{8}\alpha\beta'^2$
PCM	$\beta' = \log_M(Q)$	$M^{2\beta'}$

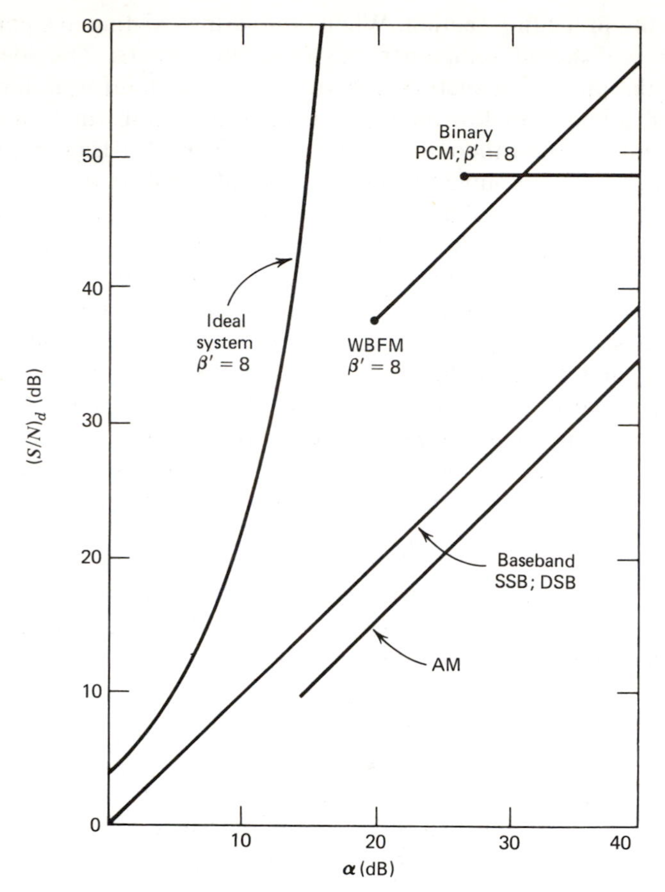

Figure 10.34 Signal-to-noise ratios in communication systems. $\alpha = (S/N)_r$; $\beta' = B_T/f_x$.

Hence

$$B_T = f_x \log_M Q$$

or

$$Q = M^{(B_T/f_x)} = M^{\beta'}$$

Thus the output signal-to-noise ratio for the PCM above threshold is

$$(S/N)_d = Q^2 = M^{2\beta'} \tag{10.60}$$

The results shown in Table 10.3 and Figure 10.34 indicate that none of the practical systems can match the signal-to-noise ratio improvement that is possible with an ideal system. This is due to the fact that practical systems, with the exception of PCM systems, do not have an exponential power-bandwidth dependence like the ideal system. While PCM does have an exponential power-bandwidth relationship, increasing transmitted power beyond threshold yields no further improvement in $(S/N)_d$ since the limiting value of $(S/N)_d$ is determined by quantization.

The SSB system offers as good a performance as an ideal system with $\beta' = 1$. However, the bandwidth ratio for SSB is fixed at $\beta' = 1$ and there is no possibility of trading bandwidth for noise reduction. AM and DSB systems with $\beta' = 2$ do not perform as well as an ideal system with $\beta' = 2$. Furthermore, like in SSB modulation, there are no possibilities of wideband noise reduction in AM and DSB systems.

Wideband FM systems offer good possibilities for noise reduction. But the performance of WBFM, like PCM and AM systems (using noncoherent demodulation procedures), falls short of the performance of ideal systems because of threshold effects, especially at low input signal-to-noise ratios.

Figure 10.35 shows the minimum values of input signal-to-noise ratios

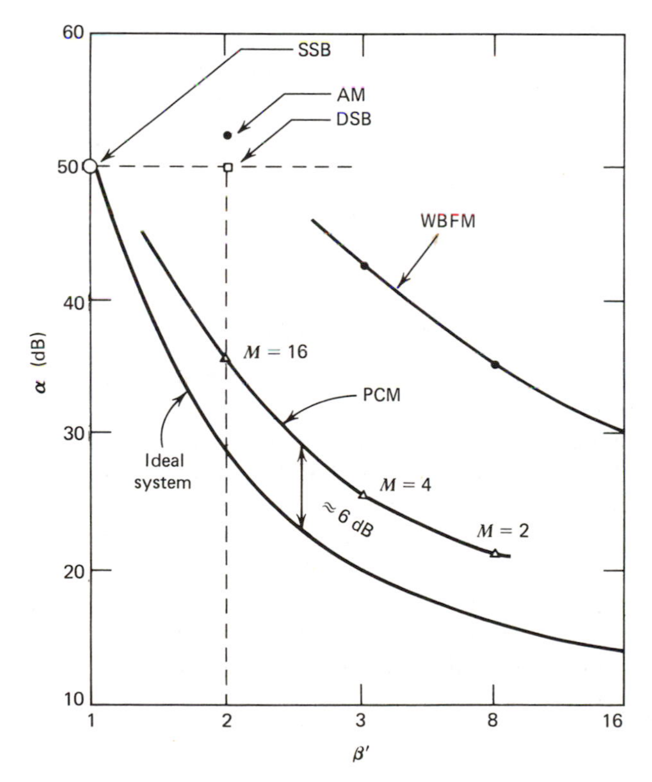

Figure 10.35 Bandwidth and power required for $(S/N)_d =$ 50 dB. $\alpha = (S/N)_r$.

required to produce $(S/N)_d = 50\,\text{dB}$ as a function of bandwidth ratio for various systems. Formulas given in Table 10.3 were used in computing $(S/N)_r$ for AM, DSB, SSB, FM, and the ideal system. For PCM systems, the minimum $(S/N)_r$ needed to produce $(S/N)_d = 50$ dB is calculated by arbitrarily defining the PCM threshold as the point at which symbol errors due to channel noise occur with a probability $P_e < 10^{-4}$. For an M-ary baseband PCM system, P_e is obtained from Equation (5.56) with $S_T = S_r$ as

$$P_e = 2\left(\frac{M-1}{M}\right)Q\left(\sqrt{\frac{6S_r}{(M^2-1)r_s\eta}}\right)$$

where S_r is the average signal power at the receiver input, $r_s = 2f_x \log_M(q)$ is the channel symbol rate, and q is the number of quantizer levels.* For $P_e < 10^{-4}$, we need

$$Q\left(\sqrt{\frac{3\alpha}{(M^2-1)\log_M q}}\right) < \frac{M}{2(M-1)}(10^{-4})$$

If z_0 satisfies $Q(z_0) = M(10^{-4})/2(M-1)$, then we have

$$\alpha \geqslant \left(\frac{(M^2-1)\log_M(q)}{3}\right)z_0^2 \tag{10.61}$$

Above threshold, $(S/N)_d = q^2$, and for $(S/N)_d = 10^5$ we need $q \approx 316$. Knowing the value of q, we can compute β' as

$$r_s = (2f_x)\log_M(q),$$
$$B_T = r_s/2 = f_x \log_M(q)$$

or

$$B_T/f_x = \beta' = \log_M(q) \tag{10.62}$$

Values of α and β' for $M = 2$, 4, and 16 are shown plotted in Figure 10.35 for the PCM system.

The plots in Figure 10.35 show that the power-bandwidth exchange in PCM is considerably better than wideband FM. The PCM system requires about 6 dB more power than the ideal system. In summary, we can say that FM and PCM offer wideband noise reduction and PCM is somewhat better than FM systems at low input signal-to-noise ratios. The performance of all practical systems from a power-bandwidth viewpoint is an order of magnitude below the performance of the ideal system. At low input signal-to-noise ratios SSB and DSB are better than other practical modulation schemes that suffer from threshold effects.

*q is used to denote the number of quantizer levels rather than Q, since Q is used here to denote the area under a normal pdf.

10.7 SUMMARY

We discussed several schemes for transmitting analog message signals using digital transmission techniques. Methods of sampling, quantizing, and encoding analog message signals were discussed. Pulse code modulation and delta modulation schemes were analyzed and the effects of quantizing noise and thermal noise in these systems were discussed. Finally, the performance of PCM, FM, SSB, DSB, and AM systems were compared with the performance of an ideal system. The results derived in this chapter clearly indicate that PCM can be used effectively for trading bandwidth for power. Also, PCM can be used for time division multiplexing a number of analog message signals.

REFERENCES

Additional discussion of the noise performance of PCM and DM systems may be found in books by Panter (1965) and Cattermole (1969). Recent techniques for digital transmission of voice and data are summarized in two articles published in the IEEE Spectrum (1977) and the IEEE Proceedings (1977).

The minimum mean squared error quantizer design discussed in Section 10.3.2 is based on Max's work (1960). His original paper contains tables of quantizer end points and output levels for several values of Q.

1. J. Max. "Quantizing for Minimum Distortion." *IRE Transactions on Information Theory*, Vol. IT-6 (1960), pp. 7–12.
2. A. Papoulis. *Probability, Random Variables and Stochastic Processes.* McGraw-Hill, New York (1965).
3. "Optical Transmission of Voice and Data," and "Digital Telephones." *IEEE Spectrum*, Vol. 14, February (1977).
4. P. F. Panter. *Modulation, Noise and Spectral Analysis.* McGraw-Hill, New York (1965).
5. K. W. Cattermole. "Principles of Pulse Code Modulation." *American Elsevier*, New York (1969).
6. *The Philosophy of PCM.* Bell System Monograph; (also in *PROC. IRE* (Nov. 1948))
7. B. Gold. "Digital Speech Networks." *IEEE Proceedings*, Vol. 65, December (1977).

PROBLEMS

Section 10.2

10.1. A lowpass signal $x(t)$ has a spectrum $X(f)$ given by

$$X(f) = \begin{cases} 1 - |f|/200, & |f| < 200 \\ 0, & \text{elsewhere} \end{cases}$$

(a) Assume that $x(t)$ is ideally sampled at $f_s = 300$ Hz. Sketch the spectrum of $x_\delta(t)$ for $|f| < 200$.
(b) Repeat part (a) with $f_s = 400$ Hz.

10.2. A signal $x(t) = 2\cos 400\pi t + 6\cos 640\pi t$ is ideally sampled at $f_s = 500$ Hz. If the sampled signal is passed through an ideal lowpass filter with a cutoff frequency of 400 Hz, what frequency components will appear in the output?

10.3. A bandpass signal with a spectrum shown in Figure 10.36 is ideally sampled. Sketch the spectrum of the sampled signal when $f_s = 20$, 30, and 40 Hz. Indicate if and how the signal can be recovered.

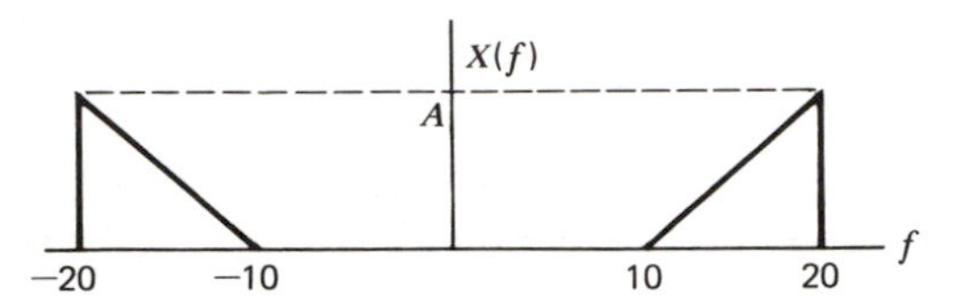

Figure 10.36 Signal spectrum for Problem 10.3.

10.4. A bandpass signal with a spectrum shown in Figure 10.37 is ideally sampled.
(a) Show that the signal can be reconstructed when $f_s = f_{xu} = 2.5B_x$.
(b) Show that the signal can be reconstructed when $f_s > 2f_{xu} = 5B_x$.
(c) Show that aliasing takes place when $f_s = 3.5B_x$.

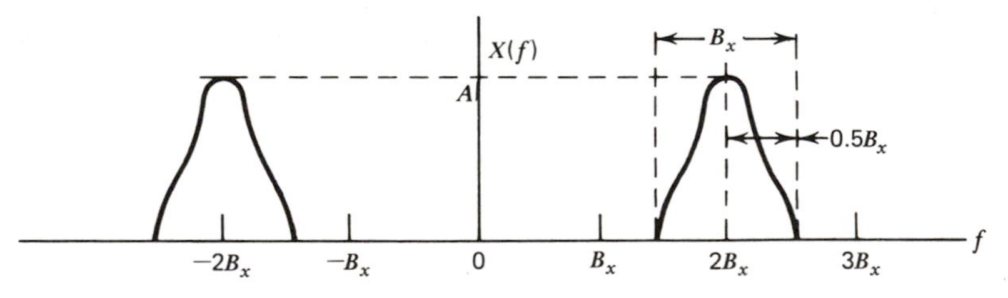

Figure 10.37 Signal spectrum for Problem 10.4.

10.5. Consider the signal $x(t) = e^{-\alpha t}u(t)$, which is not bandlimited. Determine the minimum sampling rate (in terms of the number of -3 dB bandwidths of $x(t)$) such that the magnitude of the largest aliasing frequency component introduced by sampling is at least 10 dB below the magnitude of the largest spectral component of $x(t)$.

10.6. A rectangular pulse waveform is sampled once every T_s seconds and reconstructed using an ideal LPF with a cutoff frequency of $f_s/2$ (see Figure 10.38). Sketch the reconstructed waveform for $T_s = \frac{1}{6}$ sec and $T_s = \frac{1}{12}$ sec.

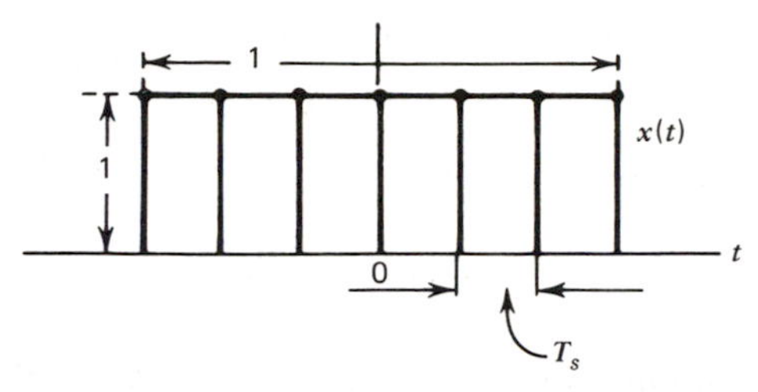

Figure 10.38 Sampling of a rectangular waveform.

10.7. The uniform sampling theorem says that a bandlimited signal $x(t)$ can be completely specified by its sampled values in the time domain. Now, consider a time limited signal $x(t)$ that is zero for $|t| \geq T$. Show that the spectrum $X(f)$ of $x(t)$ can be completely specified by the sampled values $X(kf_0)$, where $f_0 \leq 1/2T$.

10.8. Show that $\sum_{k=-\infty}^{\infty} x(kT_s) = f_s \sum_{m=-\infty}^{\infty} X(mf_s)$, where $x(t)$ is bandlimited to f_x and $f_s = 2f_x$.

Section 10.3

10.9. The probability density function of the sampled values of an analog signal is shown in Figure 10.39. Design a four-level uniform quantizer and calculate the signal to quantizing noise power ratio.

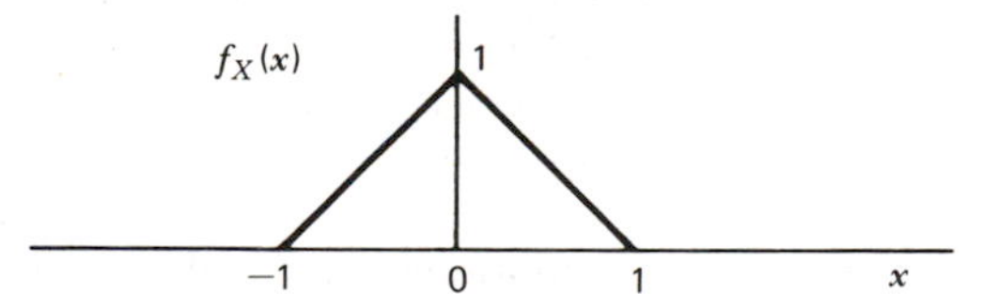

Figure 10.39 Signal pdf for Problems 10.9, 10.10, and 10.11.

10.10. (a) For the pdf shown in Figure 10.39, design a four-level minimum mean squared error nonuniform quantizer.

(b) Compute the signal to quantizing noise power ratio for the nonuniform quantizer.

(c) Design a compressor and expander so that the nonuniform quantizing can be done using a compressor and uniform quantizer.

10.11. Redraw the compressor shown in Figure 10.15, using a piecewise linear approximation. Let $y = g(x)$ be the resulting transfer characteristic and let $m_1, m_2, \ldots, m_6$ be the midpoints of the intervals on the x axis, and let the $\Delta_1, \Delta_2, \ldots, \Delta_6$ be the step sizes.

(a) Show that the piecewise linearity assumption implies

$$\Delta_i = \Delta / g'(m_i)$$

where Δ is the output step size and

$$g'(m_i) = \left.\frac{dg(x)}{dx}\right|_{x=m_i}$$

(b) If $f_X(x)$ is approximately constant throughout the step, show that the quantizing noise power is given by

$$N_q \approx \frac{1}{12}\sum_{i=1}^{6} (\Delta_i)^3 f_X(m_i)$$

(c) Using the result of (b), show that if the number of quantization levels is large, then

$$N_q \approx \frac{\Delta^2}{12}\sum_i \frac{\Delta_i f_X(m_i)}{[g'(m_i)]^2} \approx \frac{\Delta^2}{12}\int_{x_{\min}}^{x_{\max}} \frac{f_X(x)\,dx}{[g'(x)]^2}$$

(d) The *companding improvement* factor c_f is defined as the ratio of the quantizing noise power with no companding to the quantization error with companding. Obtain an expression for c_f.

10.12. The logarithmic compression used in processing speech signals has the characteristic

$$y = \begin{cases} x_{\max}\dfrac{\log_e(1 + \mu x/x_{\max})}{\log_e(1+\mu)}, & 0 \le x \le x_{\max} \\[2ex] -x_{\max}\dfrac{\log_e(1 - \mu x/x_{\max})}{\log_e(1+\mu)}, & -x_{\max} \le x \le 0 \end{cases}$$

(a) Sketch the compression characteristic with $\mu = 0, 5, 10, 100$.

(b) Plot the corresponding expander characteristics.

(c) If there are 64 quantization levels, discuss the variation of step size versus the input voltage x.

(d) Assuming X to be uniformly distributed between $-x_{\max}$ to $x_{\max}$, show that the companding improvement factor is

$$c_f = \left(\frac{\mu}{\log_e(1+\mu)}\right)^2\left(\frac{1}{1 + \mu + \mu^2/3}\right)$$

10.13. Signals are sometimes quantized using an equal probability quantizing (maximum entropy) algorithm wherein the quantizer levels are made to occur with equal probability, that is, $P(X_q = m_i) = 1/Q$ for $i = 1, 2, 3, \ldots, Q$.

(a) Design a four-level equal probability quantizer for the pdf shown in Figure 10.39.

(b) Compare the signal to quantizing noise power ratio of the equal probability quantizer with that of the minimum mean squared error quantizer (Problem 10).

(c) Compare the entropies of the output levels for the equal probability quantizer and the minimum mean squared error quantizer.

10.14. Nonbandlimited signals are usually filtered before being sampled (see Figure 10.40*a*). Filtering distorts the signal even without quantizing (see Figure 10.40*b*).

(a) Show that the distortion due to filtering, N_d, is given by

$$N_d = E\{[X(t) - X_f(t)]^2\}$$
$$= 2\int_{f_x}^{\infty} G_X(f)\, df$$

(b) Assume that $X(t)$ has a Gaussian pdf and

$$G_X(f) = e^{-|f/f_1|}$$

If $X(t)$ is sampled at a rate $f_s = 2f_x$, find the output signal-to-noise

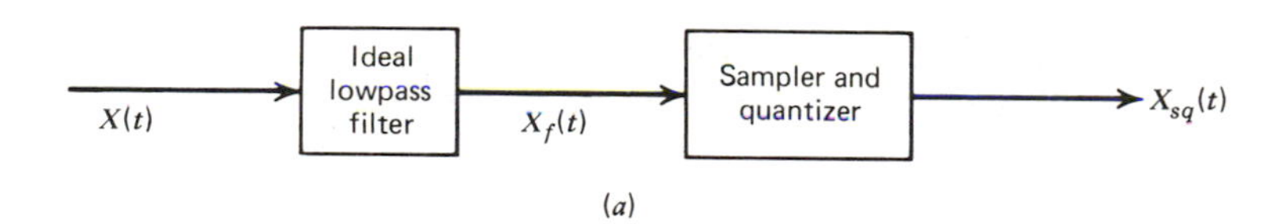

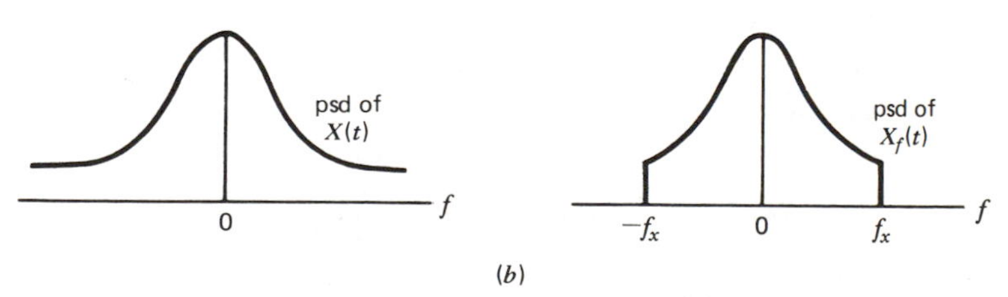

Figure 10.40 (*a*) Quantizing of nonbandlimited signals. (*b*) Effect of filtering.

ratio

$$\left(\frac{S}{N}\right)_d = \frac{S_0}{N_d + N_q}$$

where N_q is the average noise power due to quantizing and S_0 is the average signal power at the quantizer output. [*Hint*: For random signals,

$$\begin{aligned} E\{[X(t) - X_f(t)]^2\} &= R_{XX}(0) + R_{X_fX_f}(0) - 2R_{XX_f}(0) \\ &= \int_{-\infty}^{\infty} G_X(f)\,df + \int_{-\infty}^{\infty} G_X(f)|H(f)|^2\,df \\ &\quad -2\int_{-\infty}^{\infty} G_X(f)H(f)\,df \end{aligned}$$

where $H(f)$ is the filter transfer function.]

10.15. Consider the differential quantizer shown in Figure 10.41. The signal to be quantized is a zero mean Gaussian random process with an autocorrelation function

$$R_{XX}(\tau) = e^{-6000|\tau|}$$

The signal is sampled at a rate of 12,000 Hz and differentially quantized using a minimum mean squared error quantizer. The error due to quantizing can be approximated by

$$E\{[Y(kT_s) - Y_q(kT_s)]^2\} = (2.2)\sigma_Y^2 Q^{-1.96} = N_q$$

and the performance of the differential quantizer is measured by the signal to quantizing noise power ratio defined as

$$\frac{S_0}{N_q} \triangleq \frac{\sigma_X^2}{N_q}$$

where Q is the number of quantizer levels.

(a) With $\hat{X}(kT_s) = X[(k-1)T_s]$, find the minimum number of quantizer levels needed to obtain a signal to quantizing noise power ratio of 40 dB.

(b) Repeat (a) with

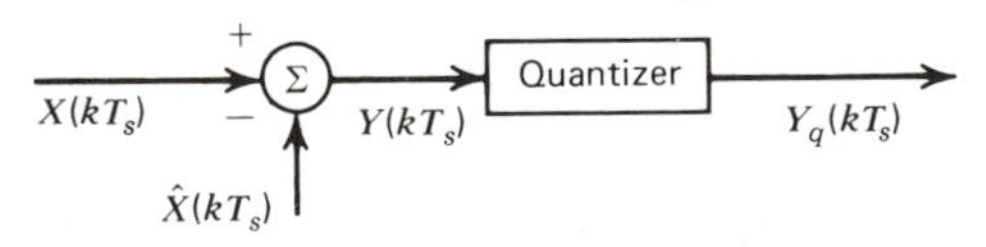

Figure 10.41 Differential quantizer.

$$\hat{X}(kT_s) = \frac{R_{XX}(T_s)}{R_{XX}(0)} X[(k-1)T_s]$$

(c) Repeat with $\hat{X}(kT_s) = 0$ (i.e., direct quantizing).

10.16. Repeat 10.15 (a) and (b) with $f_s = 24{,}000$ Hz.

Section 10.4

10.17. The threshold value of the input signal-to-noise ratio $(S/N)_i$ in PCM systems is defined as the value of $(S/N)_i$ for which the value of $(S/N)_0$ is 1 dB below its maximum.

(a) Show that the threshold occurs when

$$P_e \approx 1/[(16)2^{2N}]$$

(b) Plot P_e versus N, for $N = 2, 4, 6, 8$, and 10.

(c) Assuming that a PSK signaling scheme is used, sketch the threshold values of $(S/N)_i$ versus N for $N = 2, 4, 6, 8$, and 10.

10.18. A signal $X(t)$ bandlimited to 15 kHz is sampled at 50 kHz and the samples are transmitted using PCM/PSK. An output S/N of at least 40 dB is desired. (Assume that $X(t)$ has a uniform pdf).

(a) Find the bandwidth requirements of the system.

(b) Find $(S/N)_i$ if the system is to operate above threshold.

(c) Find $(S/N)_0$ if the system is operating with a $(S/N)_i$ that is 3 dB below the threshold value.

10.19. A nonbandlimited signal $X(t)$ has a power spectral density

$$G_X(f) = e^{-|f/3000|}$$

The signal is bandlimited to f_x Hz by ideal lowpass filtering.

(a) Find the value f_x such that the filter passes at least 90% of the signal power at its input.

(b) If the filter output is converted to 6-bit PCM and transmitted over a binary symmetric channel with $P_e = 10^{-5}$, find the overall signal-to-noise power ratio defined as

$$\left(\frac{S}{N}\right)_0 = \frac{S_0}{N_d + N_q + N_{th}}$$

N_d, N_q, and N_{th} are the distortion due to filtering and the average noise power due to quantizing, and channel bit errors, respectively.

10.20. The output of an analog message source is modeled as a zero mean random process $X(t)$ with a uniform pdf and the power spectral

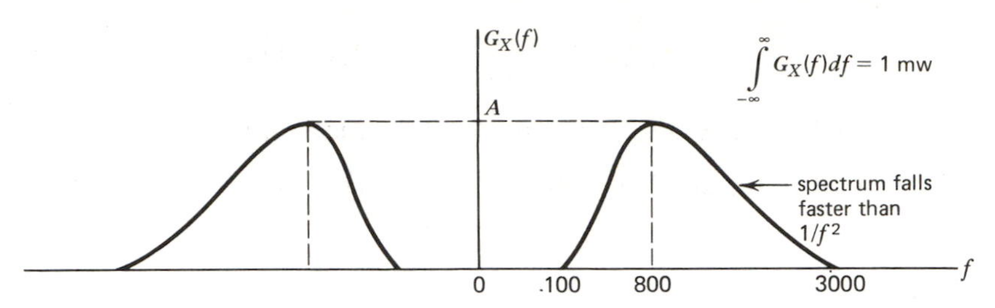

Figure 10.42 Psd of the output of an analog information source.

density shown in Figure 10.42. This signal is to be transmitted using DM.

(a) Find the sampling rate and the step size required to maintain a signal to quantizing noise power ratio of 40 dB.

(b) Compare the bandwidth of the DM with a PCM system operating with the same signal to quantizing noise power ratio. Assume that $f_s = 6000$ Hz for the PCM system.

10.21. The threshold value of P_e in DM systems is defined as the value of P_e for which the value of $(S/N)_0$ is 1 dB below its maximum. Express the value of P_e in terms of f'_s, f_x, and f_1.

10.22. Compare the threshold values of P_e for the DM and PCM systems discussed in Problem 10.20. Assuming a PSK signaling scheme, find the threshold value of the signal-to-noise ratios at the input for the DM and PCM receivers.

10.23. Plot $(S_0/N_q)_{DM}$ and $(S_0/N_q)_{PCM}$ versus B_T for equal transmission bandwidths $B_T = 4f_x$ to $32f_x$. Assume $f_x = 3000$ Hz and $f_1 = 300$ Hz.

Section 10.5

10.24. Two lowpass signals of equal bandwidth are impulse sampled and time multiplexed using PAM. The TDM signal is passed through a lowpass filter and then transmitted over a channel with a bandwidth of 10 kHz.

(a) What is the maximum sampling rate for each channel to insure that each signal can be recovered at the receiver?

(b) What is the maximum frequency content allowable for each signal?

(c) Sketch a block diagram for the transmitter and receiver.

10.25. Eight input signals are sampled and time multiplexed using PAM. The time multiplexed signal is passed through a lowpass filter before transmission. Six of the input signals have a bandwidth of 4 kHz and the other two are bandlimited to 12 kHz.

(a) What is the minimum overall sampling rate if all channels are sampled at the *same* rate?
(b) Design an asynchronous TDM for this application.
(c) Compare the transmission bandwidth requirements of (a) and (b).

10.26. Twenty-four analog signals, each having a bandwidth of 15 kHz, are to be time-division multiplexed and transmitted via PAM/AM. A guard band of 5 kHz is required for signal reconstruction from the PAM samples of each signal.
(a) Determine the sampling rate for each channel.
(b) Determine the transmission bandwidth.
(c) Draw functional block diagrams of the transmitter and receiver.

10.27. A number of 20 kHz channels are to be time-division multiplexed and transmitted using PAM. The sampling pulses are 4 μsec in width. The TDM pulse train is passed through a lowpass filter with a time constant $RC = 1$ μsec. (In previous examples we assumed that this baseband filtering was done by an ideal lowpass filter.) This filtering introduces crosstalk (or intersymbol interference) between channels, that is, a certain amount of signal energy from one pulse "spills" over into the time slot of the next pulse. Define the crosstalk factor as the ratio of signal energy from a pulse that spills over into the next time slot and the signal energy within the time slot allotted for the pulse. Using this criteria:
(a) Find the crosstalk factor for five channels.
(b) If the crosstalk factor is to be less than 0.01, find the pulse width for a five-channel system.

Section 10.6

10.28. Is it possible to design a communication system to yield $(S/N)_d = 60$ dB, with $(S/N)_r = 20$ dB, and $\beta' = 4$?

10.29. What is the minimum bandwidth expansion ratio required to obtain $(S/N)_d = 60$ dB with $(S/N)_r = 20$ dB?

10.30. A video signal having a bandwidth of 6 MHz is to be transmitted from the moon using 8-bit PCM/PSK. The thermal noise power spectral density at the receiving antenna is $\eta/2 = 10^{-12}$ watt/Hz. Assuming a power loss of 40 dB in the channel, calculate the threshold power requirements of the transmitter on the moon and compare it with the power requirements of an ideal system using the same bandwidth as the PCM system.

10.31. Suppose that black and white still pictures from the moon were digitized using $(25)(10^4)$ samples per picture (500×500 array) and transmitted to the earth using 8-bit PCM/PSK. Assume that 10 minutes are allotted for the transmission of each picture. Using a channel attenuation of 90 dB and $\eta/2 = 10^{-12}$ watt/Hz, calculate the threshold power requirements of the transmitter on the moon.

10.32. Consider an ideal system and an 8-bit PCM/PSK system operating with $(S/N)_d \approx 40$ dB. Assume that the ideal system uses the same bandwidth as the PCM system. With $\eta/2 = 10^{-12}$ watt/Hz and $f_x = 15$ kHz, find the signal power at the receiver input for the ideal system and the PCM system. [*Hint*: For the PCM system, find P_e and then $(S/N)_r$ such that $(S/N)_d = 40$ dB.]

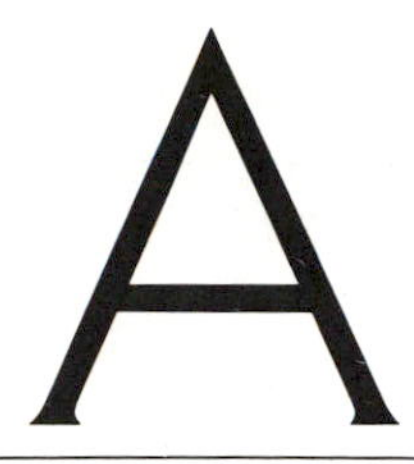

APPENDIX A HISTORY OF ELECTRICAL COMMUNICATION

Time Period	*Significant Developments*
1800–1850	Experiments on electricity by Oersted and Faraday; Ampere and Henry; Ohm's law (1826); Morse's telegraph (1838); Kirchoff's laws (1845).
1850–1900	Maxwell's equations on electromagnetic radiation (1864); Alexander Graham Bell's patent on the telephone (1876); Edison's discoveries of the microphone and phonograph; Marconi's wireless telegraph (1897); magnetic tape recorder.
1900–1920	Vacuum tubes (1904, 1906); transcontinental telephone, carrier telephony (1915); superheterodyne AM radio (Armstrong, 1918); commercial broadcasting (KDKA, Pittsburgh, 1918).
1920–1940	Landmark papers on the theory of signal transmission and noise (Carson, Nyquist, Hartley); beginning of teletype service (1931); perfection of cathode ray tubes; demonstration of TV and beginning of TV broadcasting (1938); FM radio (Armstrong, 1936); conception of PCM (Alec Reeves, 1937).

1940–1950	World War II leads to the development of radar and microwave systems; development of statistical theory of communication (Shannon, Weiner, and Kalmogoroff); invention of the transistor (Bardeen, Brattain, and Shockley); improved electronics hardware.
1950–1960	Time domain multiplexing applied to telephony (1950); transoceanic cable (1956); long distance data transmission systems for the military (1958); planning for satellite communication systems; demonstration of laser (Maiman, 1960).
1960–1970	Integrated circuits (1961); solid state microwave devices; satellite communication—Telstar I (1962); color TV; PCM for voice and TV transmission; major developments in the theory and implementation of digital transmission (Bose, Chaudhuri, Wozencraft, Lucky, and others); live transmission of voice, TV pictures and data from the moon and from Mars (Apollo and Mariner series); high speed digital computing; Picturephone®; handheld calculators.
1970–	Large scale integrated circuits, commercial satellite communication; intercontinental communication networks; operational optical communication links (laser/fiber optics); microprocessors.

B

APPENDIX B
BROADCAST FREQUENCY BANDS

Frequency	Wavelength	Designation	Transmission Media/Method of Propagation	Applications
3 Hz–30 kHz	10^8–10^4 m	Very low frequency (VLF)	Wire pairs; longwave radio	Audio; telephone; data terminals; long range navigation, undersea navigation; timing standards.
30 kHz–300 kHz	10^4–10^3 m	Low frequency (LF)	Wire pairs; longwave radio	Navigational aids; radio beacons; industrial (powerline) communication.
300 kHz–3 MHz	10^3–10^2 m	Medium frequency (MF)	Coaxial cable; longwave radio	AM broadcasting (540–1600 kHz); civil defense; amateur radio.
3 MHz–30 MHz	10^2–10 m	High frequency (HF)	Coaxial cable; shortwave radio	Amateur radio; mobile radio telephone; military communication.
30 MHz–300 MHz	10–1 m	Very high frequency (VHF)	Coaxial cable; shortwave radio	VHF television; FM radio; air traffic control; taxicab; police; navigational aids.

300 MHz–3 GHz ($GHz = 10^9$ Hz)	100–10 cm	Ultra High frequency (UHF)	Shortwave radio; waveguides; line of sight microwave radio	UHF-TV; space telemetry; radar; military; CB radio.
3 GHz–30 GHz	10–1 cm	Super high frequency (SHF)	Waveguides, microwave radio	Radar (airborne, approach, surveillance and weather); satellite and space communication; common carrier microwave relay.
30 GHz–300 GHz	10–1 mm	Extremely high frequency (EHF)	Waveguides, microwave radio	Radio astronomy; railroad service; radar landing systems, experimental systems.
10^{14}–10^{16} Hz	$(3)10^{-4}$–$(3)(10^{-6})$ cm	Ultraviolet visible infrared	Optical fibers (laser beams)	Experimental optical communication links (primarily for data transmission).

C

APPENDIX C TRIGONOMETRIC IDENTITIES AND FOURIER TRANSFORMS

1 TRIGONOMETRIC IDENTITIES

$$e^{\pm jA} = \cos A \pm j \sin A$$

$$\sin(A \pm B) = \sin A \cos B \pm \cos A \sin B$$

$$\cos(A \pm B) = \cos A \cos B \mp \sin A \sin B$$

$$\sin A \sin B = \tfrac{1}{2}[\cos(A - B) - \cos(A + B)]$$

$$\cos A \cos B = \tfrac{1}{2}[\cos(A + B) + \cos(A - B)]$$

$$\sin A \cos B = \tfrac{1}{2}[\sin(A + B) + \sin(A - B)]$$

$$\sin 2A = 2 \sin A \cos A$$

$$\cos 2A = 2 \cos^2 A - 1 = 1 - 2 \sin^2 A = \cos^2 A - \sin^2 A$$

$$\sin^2 A = \tfrac{1}{2}(1 - \cos 2A); \quad \cos^2 A = \tfrac{1}{2}(1 + \cos 2A)$$

$$\sin^3 A = \tfrac{1}{4}[3 \sin A - \sin 3A]$$

$$\cos^3 A = \tfrac{1}{4}[3 \cos A + \cos 3A]$$

$$\sin A = \frac{1}{2j}(e^{jA} - e^{-jA}); \quad \cos A = \frac{1}{2}(e^{jA} + e^{-jA})$$

2 FOURIER TRANSFORMS

$$X(f) = \int_{-\infty}^{\infty} x(t) \exp(-j2\pi ft)\, dt$$

$$x(t) = \int_{-\infty}^{\infty} X(f) \exp(j2\pi ft)\, df$$

$$\int_{-\infty}^{\infty} |x(t)|^2\, dt = \int_{-\infty}^{\infty} |X(f)|^2\, df$$

Table C.1. Transform theorems.

Name of theorem	Signal	Fourier transform
(1) Superposition	$a_1x_1(t) + a_2x_2(t)$	$a_1X_1(f) + a_2X_2(f)$
(2) Time delay	$x(t - t_0)$	$X(f) \exp(-j2\pi ft_0)$
(3) Scale change	$x(at)$	$\lvert a\rvert^{-1}X(f/a)$
(4) Frequency translation	$x(t) \exp(j2\pi f_0t)$	$X(f - f_0)$
(5) Modulation	$x(t) \cos 2\pi f_0t$	$\frac{1}{2}X(f - f_0) + \frac{1}{2}X(f + f_0)$
(6) Differentiation	$\dfrac{d^nx(t)}{dt^n}$	$(j2\pi f)^nX(f)$
(7) Integration	$\int_{-\infty}^{t} x(t')\, dt'$	$(j2\pi f)^{-1}X(f) + \frac{1}{2}X(0)\delta(f)$
(8) Convolution	$\int_{-\infty}^{\infty} x_1(t - t')x_2(t')\, dt' = \int_{-\infty}^{\infty} x_1(t')x_2(t - t')\, dt'$	$X_1(f)X_2(f)$
(9) Multiplication	$x_1(t)x_2(t)$	$\int_{-\infty}^{\infty} X_1(f - f')X_2(f')\, df' = \int_{-\infty}^{\infty} X_1(f')X_2(f - f')\, df'$

Table C.2. Fourier transform pairs.

	Signal $x(t)$	Transform $X(f)$
(1)	A; $-\tau/2$, 0, $\tau/2$; t	$A\tau \dfrac{\sin \pi f\tau}{\pi f\tau} \triangleq A\tau \operatorname{sinc} f\tau$
(2)	B; $-\tau$, 0, τ; t	$B\tau \dfrac{\sin^2 \pi f\tau}{(\pi f\tau)^2} \triangleq B\tau \operatorname{sinc}^2 f\tau$
(3)	$e^{-\alpha t}u(t)$	$\dfrac{1}{\alpha + j2\pi f}$
(4)	$\exp(-\|t\|/\tau)$	$\dfrac{2\tau}{1+(2\pi f\tau)^2}$
(5)	$\exp[-\pi(t/\tau)^2]$	$\tau \exp[-\pi(f\tau)^2]$
(6)	$\dfrac{\sin 2\pi Wt}{2\pi Wt} \triangleq \operatorname{sinc} 2Wt$	$\frac{1}{2}W$; $-W$, 0, W; f
(7)	$\exp[j(2\pi f_c t + \phi)]$	$\exp(j\phi)\delta(f - f_c)$
(8)	$\cos(2\pi f_c t + \phi)$	$\frac{1}{2}\delta(f - f_c)\exp(j\phi) + \frac{1}{2}\delta(f + f_c)\exp(-j\phi)$
(9)	$\delta(t - t_0)$	$\exp(-j2\pi f t_0)$
(10)	$\sum_{m=-\infty}^{\infty} \delta(t - mT_s)$	$\dfrac{1}{T_s}\sum_{n=-\infty}^{\infty} \delta\left(f - \dfrac{n}{T_s}\right)$
(11)	$\operatorname{sgn} t = \begin{cases} +1, & t>0 \\ -1, & t<0 \end{cases}$	$-\dfrac{j}{\pi f}$
(12)	$u(t) = \begin{cases} 1, & t>0 \\ 0, & t<0 \end{cases}$	$\frac{1}{2}\delta(f) + \dfrac{1}{j2\pi f}$

D

APPENDIX D GAUSSIAN PROBABILITIES

(1) $P(X > \mu_X + y\sigma_X) = Q(y) = \int_y^\infty \frac{1}{\sqrt{2\pi}} e^{-z^2/2}\, dz$

(2) $Q(0) = \frac{1}{2}$; $\quad Q(-y) = 1 - Q(y)$, $\quad$ when $y \geqslant 0$

(3) $Q(y) \approx \frac{1}{y\sqrt{2\pi}} e^{-y^2/2}$ when $y \gg 1$ (approximation may be used for $y > 4$)

(4) $\operatorname{erfc}(y) \triangleq \frac{2}{\sqrt{\pi}} \int_y^\infty e^{-z^2}\, dz = 2Q(\sqrt{2}y)$, $y > 0$.

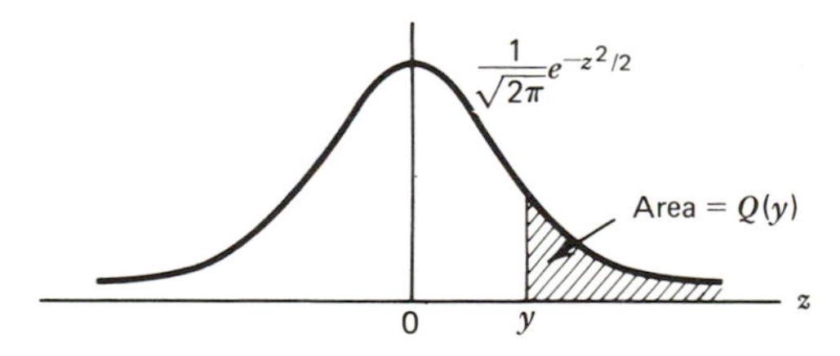

Figure D.1 Gaussian probabilities.

y	$Q(y)$	y	$Q(y)$	y	$Q(y)$
.05	.4801	1.05	.1469	2.10	.0179
.10	.4602	1.10	.1357	2.20	.0139
.15	.4405	1.15	.1251	2.30	.0107
.20	.4207	1.20	.1151	2.40	.0082
.25	.4013	1.25	.0156	2.50	.0062
.30	.3821	1.30	.0968	2.60	.0047
.35	.3632	1.35	.0885	2.70	.0035
.40	.3446	1.40	.0808	2.80	.0026
.45	.3264	1.45	.0735	2.90	.0019
.50	.3085	1.50	.0668	3.00	.0013
.55	.2912	1.55	.0606	3.10	.0010
.60	.2743	1.60	.0548	3.20	.00069
.65	.2578	1.65	.0495	3.30	.00048
.70	.2420	1.70	.0446	3.40	.00034
.75	.2266	1.75	.0401	3.50	.00023
.80	.2119	1.80	.0359	3.60	.00016
.85	.1977	1.85	.0322	3.70	.00010
.90	.1841	1.90	.0287	3.80	.00007
.95	.1711	1.95	.0256	3.90	.00005
1.00	.1587	2.00	.0228	4.00	.00003

$Q(y)$	y
10^{-3}	3.10
$\frac{10^{-3}}{2}$	3.28
10^{-4}	3.70
$\frac{10^{-4}}{2}$	3.90
10^{-5}	4.27
10^{-6}	4.78

E

APPENDIX E GLOSSARY OF SYMBOLS, NOTATIONS, AND ABBREVIATIONS

1 GENERAL NOTATIONS

Deterministic Signals

(1) Deterministic signals are denoted by lower case letters with t as their arguments; for example, $x(t)$ and $y(t)$ (Chapter 2).

(2) $C_x(nf_0)$	Fourier series coefficients of a periodic signal $x(t)$ (Section 2.1).
(3) $R_{xx}(\tau)$	Autocorrelation function of a deterministic signal $x(t)$ (Section 2.4).
(4) $G_x(f)$	Power spectral density function of $x(t)$ (Section 2.4).
(5) E_x	Normalized energy content of $x(t)$ (Section 2.1).
(6) S_x	Normalized average power content of $x(t)$ (Section 2.1).
(7) $X(f)$	Fourier transform of $x(t)$ (Section 2.3).

Random Variables

(1) Uppercase letters denote random variables and lowercase letters denote particular values of random variables; for example, X and Y are random variables; x and y denote particular values of X and Y (Chapter 3).

(2) $f_X(x)$ Probability density function (pdf) of a continuous random variable at $X = x$ (Section 3.4.1).

(3) $f_{X,Y}(x, y)$ Joint pdf of two continuous random variables X and Y at $X = x$ and $Y = y$ (Section 3.4.1).

(4) $f_{X|Y}(x|y)$ Conditional pdf of a continuous random variable X at $X = x$ given that the random variable Y has a value y (Section 3.4.1).

(5) $P(X = x)$ Probability mass function of a discrete random variable X at $X = x$ (Section 3.3.1).

(6) $P(X = x, Y = y)$ Joint probability mass function of two discrete random variables X and Y at $X = x$ and $Y = y$ (Section 3.3.1).

(7) $P(X = x|Y = y)$ Conditional probability mass function of a discrete random variable X at $X = x$, given that the discrete random variable Y has a value y (Section 3.3.1).

(8) $E\{\ \}$ Expected value or statistical average of a function of one or more random variables (Sections 3.3.2 and 3.4.1).

(9) μ_X Mean of a random variable X (Sections 3.3.2 and 3.4.1).

(10) σ_X^2 Variance of a random variable X (Sections 3.3.2 and 3.4.1).

(11) σ_X Standard deviation of a random variable X (Sections 3.3.2 and 3.4.1).

(12) ρ_{XY} Correlation between two random variables X and Y (Sections 3.3.2 and 3.4.1).

Random Processes

(1) Functions denoted by uppercase letters with t as their arguments represent random processes; for example, $X(t)$ and $Y(t)$ denote random processes whereas $x(t)$ and $y(t)$ denote particular member functions of $X(t)$ and $Y(t)$, respectively. One exception is $n(t)$, which denotes the random noise process (Chapter 3).

(2) $E\{g(X(t))\}$ Ensemble average of $g(X(t))$ (Section 3.5.1).

(3) $\langle g(X(t))\rangle$ Time average of $g(X(t))$ (Section 3.5.2).

(4) $\mu_X(t_1)$ Mean of the random process at $t = t_1$ (Section 3.5.1).

(5) $R_{XX}(t_1, t_2)$ $= E\{X(t_1)X(t_2)\}$—the autocorrelation function of $X(t)$ (Section 3.5.1).

(6) $R_{XX}(\tau)$ Autocorrelation function of a stationary random process $X(t)$ (Section 3.5.2).

(7) $R_{XY}(\tau)$ Cross correlation function of two jointly stationary random processes $X(t)$ and $Y(t)$ (Section 3.6.2).

(8) $G_X(f)$ Power spectral density function of a stationary random process $X(t)$ (Section 3.5.2).

(9) $\eta/2$ — Two-sided noise power spectral density of stationary Gaussian bandlimited white noise (Section 3.7).

Linear Systems

(1) $h(t)$ — Impulse response of a linear system (Section 2.5.1).
(2) $H(f)$ — Transfer function of a linear time invariant system (Section 2.5.2).

2 SYMBOLS

Symbol	*Meaning*
$\triangleq$	Equals to by definition
$\approx$	Approximately equals
$\longleftrightarrow$	Fourier transform pair
$[a,b]$	Closed interval from a to b
(a,b)	Open interval from a to b
H^*	*denotes complex conjugate
$x(t)*y(t)$	*denotes convolution
$\oplus$	Modulo-2 addition
$\sim$	$\tilde{x}(t)$ denotes an approximate version of $x(t)$
$\hat{}$	$\hat{b}_k$ denotes an estimated value of b_k
$F\{x(t)\}$	Fourier transform of $x(t)$
$F^{-1}\{X(f)\}$	Inverse Fourier transform of $X(f)$
$\{a_k\}$	An infinite sequence whose kth member is a_k
Σ_k	Sum over the index k; sum is over all permissible values of k
$\cup$	Union of two random events
$\cap$	Intersection of two random events
$\notin A$	element of a set A
$\in A$	not an element of set A

3 FUNCTIONS

(1) Bessel function of the first kind of order n

$$J_n(\beta) = \frac{1}{2\pi} \int_{-\pi}^{\pi} \exp(j\beta \sin \theta - jn \theta)\, d\theta$$

(2) Modified Bessel function of the first kind and zero order

$$I_0(x) = \frac{1}{2\pi} \int_{-\pi}^{\pi} \exp(x \cos \theta)\, d\theta$$

(3) Binomial coefficient (factorials)

$$\binom{n}{k} = \frac{n!}{(n-k)!k!}$$

$$k! = k(k-1)(k-2)\ldots(3)(2)(1)$$

(4) Delta function

$$\delta(t) = 0 \quad t \neq 0$$

$$\int_{-\infty}^{\infty} \delta(t)\, dt = 1, \quad \text{or}$$

$$\int_{-\infty}^{\infty} g(t)\delta(t - t_0)\, dt = g(t_0)$$

(5) Gaussian probability

$$Q(x) = \int_x^{\infty} \frac{1}{\sqrt{2\pi}} \exp\left(\frac{-z^2}{2}\right) dz$$

(6) Exponential function

$$\exp(x) = e^x$$

(7) Sinc function

$$\operatorname{sinc}(t) = \frac{\sin(\pi t)}{(\pi t)}$$

(8) Sign (signum) function

$$\operatorname{sgn}(t) = \begin{cases} 1 & t > 0 \\ -1 & t < 0 \end{cases}$$

(9) Step function

$$u(t) = \begin{cases} 1 & t > 0 \\ 0 & t < 0 \end{cases}$$

(10) Rectangle function

$$\operatorname{Rect}\left(\frac{t}{\tau}\right) = \begin{cases} 1 & |t| < \frac{\tau}{2} \\ 0 & |t| > \frac{\tau}{2} \end{cases}$$

4 ABBREVIATIONS

A/D	Analog to digital (converter)
ACK	Acknowledgment
AM	Amplitude modulation
ASK	Amplitude-shift keying
AGC	Automatic gain control
BPF	Bandpass filter
BSC	Binary symmetric channel
BCH	Bose–Chaudhuri–Hocquenghem (codes)

BW	Bandwidth
CW	Continuous wave
dB	Decibel
dBm	Power in milliwatts expressed in decibels
DM	Delta modulation
DPCM	Differential pulse code modulation
DPSK	Differential phase-shift keying
DSB	Double sideband modulation
FM	Frequency modulation
FDM	Frequency division multiplexing
FFT	Fast Fourier transform
FMFB	Frequency demodulator with feedback
FSK	Frequency-shift keying
Hz	Hertz
kHz	Kilohertz
MHz	Megahertz
HDB	High density bipolar coding
HF	High frequency
IF	Intermediate frequency
ISI	Intersymbol interference
LF	Low frequency
LPF	Lowpass filter
MF	Medium frequency
ms	Millisecond
μs	Microsecond
NBFM	Narrowband frequency modulation
NACK	Negative acknowledgment
PM	Phase modulation
PAM	Pulse amplitude modulation
PCM	Pulse code modulation
pdf	Probability density function
psd	Power spectral density
PSK	Phase shift-keying
QPSK	Quadrature phase shift-keying
RF	Radio frequency
s	Second
S/N	Signal-to-noise power ratio
$(S/N)_i$	Signal-to-noise power ratio at the receiver input
$(S/N)_0$	Signal-to-noise power ratio at the receiver output
SSB	Single sideband modulation
TV	Television
TDM	Time division multiplexing

UHF	Ultra high frequency
USB	Upper sideband modulation
VCO	Voltage controlled oscillator
VHF	Very high frequency
VLF	Very low frequency
VSB	Vestigial sideband modulation
WBFM	Wideband frequency modulation

INDEX